DYNAMICS AND MECHANISMS OF PHOTOINDUCED ELECTRON TRANSFER AND RELATED PHENOMENA

North-Holland
Delta Series

NORTH-HOLLAND
AMSTERDAM • LONDON • NEW YORK • TOKYO

Dynamics and Mechanisms of Photoinduced Electron Transfer and Related Phenomena

Proceedings of the Yamada Conference XXIX on
Dynamics and Mechanisms of Photoinduced Electron Transfer
and Related Phenomena
Senri, Osaka, Japan
May 12–16, 1991

Edited by

Noboru Mataga
Tadashi Okada
Hiroshi Masuhara
Department of Applied Physics
Osaka University
Japan

1992

NORTH-HOLLAND
AMSTERDAM • LONDON • NEW YORK • TOKYO

North-Holland
ELSEVIER SCIENCE PUBLISHERS B.V.
Sara Burgerhartstraat 25
P.O. Box 211, 1000 AE Amsterdam, The Netherlands

Library of Congress Cataloging-in-Publication Data

Yamada Conference on Dynamics and Mechanisms of Photoinduced Electron Transfer and Related Phenomena (1991 : Senri Nyū Taun, Japan)
Dynamics and mechanisms of photoinduced electron transfer and related phenomena : proceedings of the Yamada Conference XXIX on Dynamics and Mechanisms of Photoinduced Electron Transfer and Related Phenomena, Senri, Osaka, Japan, May 12-16, 1991 / edited by Noboru Mataga, Tadashi Okada, Hiroshi Masuhara.
p. cm. -- (North-Holland delta series)
Includes indexes.
ISBN 0-444-89191-9 (alk. paper)
1. Photochemistry--Congresses. 2. Charge exchange--Congresses. 3. Energy transfer--Congresses. I. Mataga, Noboru, 1947- . II. Okada, Tadashi, 1939- . III. Masuhara, Hiroshi, 1944- . IV. Title. V. Series.
QD701.Y36 1991
541.3'5--dc20 92-10487
CIP

ISBN: 0 444 89191 9

This book is printed on acid-free paper.

Printed in The Netherlands

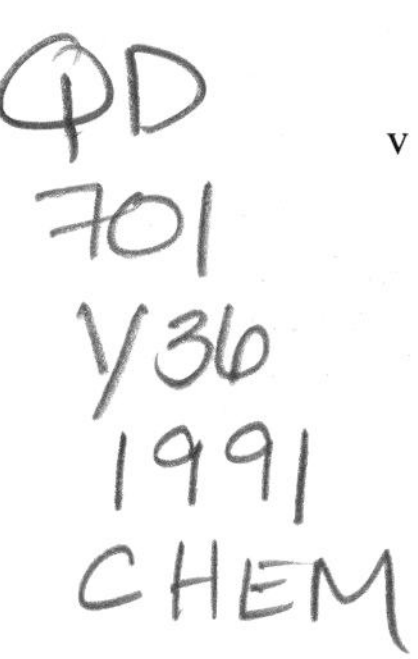

PREFACE

The 29th Yamada Conference on "Dynamics and Mechanisms of Photoinduced Electron Transfer and Related Phenomena" was held at Senri Hankyu Hotel, Osaka, Japan, from May 12 to May 16, 1991. The meeting was attended by 100 scientists from 7 countries.

In the Conference, we undertook to discuss the fundamental aspects of photoinduced electron transfer reactions, excited state proton transfers, dynamic behaviors of geminate radical ion pairs and cation–electron pairs, excitation energy transfers and related problems in condensed phase including pure liquids, solutions, various molecular assemblies and biological systems.

Of course, these are the most fundamental and important problems in the photochemical and photobiological primary processes, and recent progress in both experimental and theoretical investigations in these fields is quite remarkable. In this respect, we thought that it was very fruitful to have thorough discussions on these fundamental problems among the scientists of different disciplines and from various countries.

We organized this meeting expecting that, through the lectures and discussions following each lecture as well as discussions at posters, the participants will gain perspectives on the present status of the investigations on the fundamental aspects of electron transfer and related phenomena. In view of this, we decided to include discussions in the Proceedings of the Conference as far as possible.

Actually, many excellent and very interesting lectures as well as posters on investigations of the very fundamental problems related to some simple systems and also more complex systems were given and followed by vigorous discussions. Those presentations and discussions, I believe, contributed profoundly to make the problems clear or to get some perspectives for the further developments of these fields. I suppose also that all participants have got such strong impressions that most fundamental problems of the photoinduced electron transfer and related reaction processes are common throughout the photochemistry, radiation chemistry and photobiology.

In the arrangements of the manuscripts of the lectures and those of the posters for the

Proceedings of the Conference, we followed the Program rather closely and included the discussions in order to keep, as far as possible, the atmosphere of the Conference. This led to the result that the Proceedings contain four Parts; each Part contains the manuscripts of several lectures, followed by discussions, and the manuscripts of some posters appropriate to that Part of the Proceedings.

Roughly speaking, Parts I and II are mainly concerned with the fundamental aspects of the inter- and intra-molecular charge transfer, electron transfer and related phenomena such as the solvent effects, solvation dynamics, energy gap dependences as well as radical pair dynamics, etc. Part III is concerned with electron transfer and energy transfer phenomena mainly in polymers, films, crystals, and other confined systems. In Part IV, dynamics and mechanisms of the energy and electron transfer in biological photosynthetic systems, proteins and reaction center models are discussed. Nevertheless, all of these four Parts are very closely connected to each other and it is rather difficult to divide them.

Before closing this preface, I would like to express my sincere thanks to all participants who gave excellent lectures, those who presented interesting and important posters, and those who contributed to the vigorous discussions as well as the Yamada Science Foundation for supporting the Conference, leading to the success of this meeting.

Noboru Mataga
Chairman of the Conference

YAMADA CONFERENCE XXIX ON DYNAMICS AND MECHANISMS OF PHOTOINDUCED ELECTRON TRANSFER AND RELATED PHENOMENA - May 12-16, 1991

The Organizing Committee of the 29th Yamada Conference:
Shigeru ITOH, Toshiaki KAKITANI, Hiroshi MASUHARA,
Noboru MATAGA (Chairman), Takeshi OHNO,
Tadashi OKADA (Secretary)

PARTICIPANTS LIST

ARAI, Tatsuo (Univ. Tsukuba)
ARAKI, Shigeru (Osaka Univ.)
ASAHI, Tsuyoshi (Microphotoconversion Project)
AZUMI, Tohru (Tohoku Univ.)
BARBARA, Paul F. (Univ. Minnesota)
BAUMANN, Wolfram (Univ. Mainz)
BORVKOV, Victor (Osaka Univ.)
CALDWELL, Richard A. (Univ. Texas)
DE SCHRYVER, Frans C. (Univ. Leuvan)
FAURE, Jean (Univ. Paris-Sud)
FLEMING, Graham R. (Univ. Chicago)
FUJIHIRA, Masamichi (Tokyo Inst. Tech.)
FUKUMURA, Hiroshi (Osaka Univ.)
FUKUZUMI, Shunichi (Osaka Univ.)
FURUE, Masaoki (Osaka Univ.)
HARRIMAN, Anthony (Univ. Texas)
HASHIMOTO, Nobuhisa (Osaka Univ.)
HAYASHI, Hisaharu (Inst. Phys. & Chem. Res.)
HIRATA, Yoshinori (Osaka Univ.)
HIROTA, Noboru (Kyoto Univ.)
HONMA, Kenji (Himeji Inst. Tech.)
HYNES, James T. (Univ. Colorado)
ICHIKAWA, Musubu (Osaka Univ.)
IKEDA, Noriaki (Osaka Univ.)
ISHIGAKI, Miyuki (Osaka Univ.)
ISHIKAWA, Masashi (Osaka Women's Univ.)
ISODA, Satoru (Mitsubishi Electric Co.)
ITAYA, Akira (Kyoto Inst. Tech.)
ITOH, Michiya (Kanazawa Univ.)
ITOH, Shigeru (Nat. Inst. Basic Biol.)
IYODA, Tomokazu (Kyoto Univ.)
JORTNER, Joshua (Tel Aviv Univ.)
KAJIMOTO, Okitsugu (Kyoto Univ.)
KAKITANI, Toshiaki (Nagoya Univ.)

KATO, Shigeki (Kyoto Univ.)
KAYA, Koji (Keio Univ.)
KIKUCHI, Koichi (Tohoku Univ.)
KINOSHITA, Shuichi (Hokkaido Univ.)
KIRA, Akira (Inst. Phys. & Chem. Res.)
KITAHARA, Kazuo (Tokyo Inst. Tech.)
KITAMURA, Noboru (Microphotoconversion Project)
KLAFTER, Joseph (Tel Aviv Univ.)
KOTANI, Masahiro (Gakushuin Univ.)
MASUHARA, Hiroshi (Osaka Univ.)
MATAGA, Noboru (Osaka Univ.)
MATSUO, Taku (Kyushu Univ.)
MIMURO, Mamoru (Nat. Inst. Basic Biol.)
MIYASAKA, Hiroshi (Osaka Univ.)
MIZUNO, Kazuhiko (Univ. Osaka Prefecture)
MOBIUS, Dietmer (Max Planck Inst.)
MORISHIMA, Yotaro (Osaka Univ.)
MORITA, Akio (Univ. Tokyo)
MUKAMEL, Shaul (Univ. Rochester)
MURAI, Hisao (Osaka Univ.)
NAGAKURA, Saburo (Graduate Univ. Adv. Studies)
NAKAHARA, Masaru (Kyoto Univ.)
NAKASHIMA, Nobuaki (Osaka Univ.)
NOSAKA, Yoshio (Nagaoka Univ. Tech.)
NOZAKI, Koichi (Osaka Univ.)
NOZAWA, Tsunenori (Tohoku Univ.)
OHKOCHI, Masaya (Osaka Univ.)
OHNO, Takeshi (Osaka Univ.)
OKADA, Tadashi (Osaka Univ.)
OHMINE, Iwao (Inst. Mol. Sci.)
OSUKA, Atsuhiro (Kyoto Univ.)
OTSUJI, Yoshio (Univ. Osaka Prefecture)
PAC, Chyongjin (Kawamura Inst. Chem. Res.)
RETTIG, Wolfgang (Tech. Univ. Berlin)
SAITO, Minoru (Protein Eng. Res. Inst.)
SAKAGUCHI, Yoshio (Inst. Phys. & Chem. Res.)
SAKATA, Yoshiteru (Osaka Univ.)
SEGAWA, Koji (Kyoto Univ.)
SHIDA, Tadamasa (Kyoto Univ.)
STEINER, Urlich E. (Univ. Konstanz)
SUMI, Hitoshi (Univ. Tsukuba)
SUNDSTROM, Villy (Univ. Umea)
SUZUMOTO, Takeshi (Fuji Photo Film Co.)

TABATA, Akihiro (Osaka Univ.)
TACHIYA, Masanori (Nat. Chem. Lab. Industry)
TAGAWA, Seiichi (Univ. Tokyo)
TAKAHASHI, Yasuo (Tohoku Univ.)
TAKIGAWA, Mitsuko (Osaka Univ.)
TAMAI, Naoto (Microphotoconversion Project)
TANAKA, Fumio (Mie Nursing College)
TANAKA, Jun (Production Eng. Res. Lab., Hitachi)
TANAKA, Yoshinori (Osaka Univ.)
TANIMOTO, Yoshifumi (Hiroshima Univ.)
TERAZIMA, Masahide (Kyoto Univ.)
TOKUMARU, Katsumi (Univ. Tsukuba)
TOMINAGA, Keisuke (Univ. Mennesota)
TOYOZAWA, Yutaka (Chuo Univ.)
TRIFUNAC, Alexander (Argonne Nat. Lab.)
TSUBOI, Yasuyuki (Osaka Univ.)
TSUCHIYA, Masahiro (Univ. Tsukuba)
USHIDA, Kiminori (Inst. Phys. & Chem. Res.)
WASIELESKI, Michael R. (Argonne Nat. Lab.)
YAMAMOTO, Masahide (Kyoto Univ.)
YAMAZAKI, Iwao (Hokkaido Univ.)
YONEZAWA, Yoshiro (Kyoto Univ.)
YOSHIHARA, Keitaro (Inst. Mol. Sci.)

May 15, 1991

Welcome Address

Good Evening, Ladies and Gentlemen!
It is my great honor and pleasure to extend a welcome address to all the participants to the 29th Yamada Conference on "Dynamics and Mechanisms of Photoinduced Electron Transfer and Related Phenomena". In particular my special gratitude must go to those who travelled great distances to here from overseas.

On behalf of the Yamada Science Foundation we are very happy for being privileged to sponsor the present conference.

The Yamada Science Foundation was established in 1977 as a result of the donation of the private holdings of the late Kiro Yamada who was the former President of Rohto Pharmaceutical Company in Osaka. Mr. Yamada recognized that creative basic research in Natural Science is indispensable for the welfare and prosperity of mankind. In this respect the foundation has been supporting various innovative basic researches, mainly in physics, chemistry and biology, assisting international exchange of scientists and also holding Yamada Conferences with full financial support, three or four times every year recently.

All the conferences held so far have been rewarding, stimulating and successful.

Unfortunately I could not give a welcome address at the opening session, the day before yesterday, due to an inevitable circumstance. This evening, however, I am very happy to be here with all the participants.

Incidentally, I have been a physical chemist in the field of structural chemical thermodynamics and have been working on the study of phase transition phenomena in solid which are mainly caused by thermally induced atomic and/or molecular rearrangements and proceed with much slower time scales than that of the electron transfer treated in this conference. Although I am not the specialist in the field of subjects discussed in the present conference, I am very much convinced that the photoinduced electron transfer problems are one of the most fascinating and frontier topics which are interrelating physics, chemistry and biology and are just in accordance with the aim of Yamada Conference.

I am told by Professor Mataga that the conference has been proceeding in quite an active and marvellous fashion up to now. I do hope such a wonderful situation will continue and all the participants will enjoy a friendly contact with each other and lead to the successful final result of the conference.

Last but not least, I wish all the foreign participants a pleasant and enjoyable stay during the most beautiful season of the year in Japan and hope they may have ample opportunity to appreciate the beautiful scenery as well as the harmony between the western and oriental cultures and also the cordial hospitality of our people.

Thank you for your attention

Syûzô Seki

Syûzô Seki
On the Board of Directors
Yamada Science Foundation

YAMADA SCIENCE FOUNDATION
AND
THE SCOPE OF YAMADA CONFERENCE

Yamada Science Foundation was established in February 1977 in Osaka through the generosity of Mr. Kiro Yamada. Mr. Yamada was President of Rohto Pharmaceutical Company, Limited, a well–known manufacturer of medicines in Japan. He recognized that creative, unconstrained, basic research is indispensable for the future welfare and prosperity of mankind and he has been deeply concerned with its promotion. Therefore, funds for this Foundation were donated from his private holdings.

The principal activity of the Yamada Science Foundation is to offer financial assistance to creative research in the basic natural sciences, particularly in interdisciplinary domains that bridge established fields. Projects which promote international cooperation are also favored. By assisting the exchange of visiting scientists and encouraging international meetings, this Foundation intends to greatly further the progress of science in the global environment.

In this context, Yamada Science Foundation sponsors international Yamada Conferences once or twice a year in Japan. Subjects to be selected by the Foundation should be most timely and stimulating. These conferences are expected to be of the highest international standard so as to significantly foster advances in their respective fields.

EXECUTIVE MEMBERS OF YAMADA SCIENCE FOUNDATION

TABLE OF CONTENTS

Part 1: Fundamental Aspects of Electron Transfer and Related Processes I

Dynamics and Mechanisms of
Photoinduced Transfer and Related Phenomena
N. Mataga, T. Okada and H. Masuhara (Editors)

Photoinduced Electron Transfer and Dynamics of Transient Ion Pair States

Noboru Mataga

Department of Chemistry, Faculty of Engineering Science, Osaka University, Toyonaka, Osaka 560, Japan

Abstract

The important factors regulating the photoinduced charge separation (CS) and charge recombination (CR) of produced ion pair (IP) state such as the electronic interaction between donor (D) and acceptor (A) responsible for electron transfer (ET), the energy gap ($-\Delta G^{o}$), the reorganization energies (λ) and the solvent dynamics have been examined on the basis of the systematic femtosecond–picosecond laser photolysis studies on ET processes of various linked and unlinked D, A systems as well as ET from solute in fluorescent state to transient aggregate of solvent in polar solutions. It has been demonstrated that, depending on the strength of D, A electronic interactions and the magnitude of $-\Delta G^{o}$, the photoinduced ET mechanism changes, leading to the formation of different kinds of IP's. Those various kinds of IP's play very important roles in determining the mechanisms of successive physical and chemical processes.

1. INTRODUCTION

The mechanisms and dynamics of the photoinduced charge separation (CS) and charge recombination (CR) of the produced ion pair (IP) state are the most fundamental and important central problems in the photophysical and photochemical primary processes in condensed phases [1].

It is generally believed that the rate of the photoinduced CS and that of the CR of the produced geminate IP state are regulated by: (a) the magnitude of the electronic interaction responsible for the ET (electron transfer) between D (electron donor) and A (electron acceptor), (b) the FC (Franck–Condon) factor which is related to the free energy gap ($-\Delta G^{o}$) between the initial and final states of ET, (c) the reorganization energies (λ) of D and A as well as the surrounding solvent, and (d) solvent dynamics [2–4]. In the usual theoretical treatments of ET reactions, relatively weak D, A interactions are assumed, leading to the simple two state model, $D\cdots A \rightarrow D_S^{+}\cdots A_S^{-}$.

When the electronic interaction between D and A is very weak, the ET process is considered non–adiabatic, and it will become adiabatic when the interaction becomes fairly strong. When the electronic interaction becomes sufficiently strong in the adiabatic case and the $-\Delta G^{o}$ relation is also appropriate, the reaction becomes almost barrierless. In such a case, it is believed also that the reaction is governed mainly by the reorientation motions of polar solvent surrounding D and A, and the longitudinal dielectric relaxation time τ_L or solvation time τ_S determined by the measurement of the dynamic Stokes shift of fluorescence probe molecule will be important as a factor controlling the ET rate.

However, in the actual D, A systems in solution, we must take into account various cases of different strengths of D, A electronic interactions, different $-\Delta G^{o}$ and λ values leading in some cases to different mechanisms of photoinduced ET and to the formation of different kinds of IP's depending on those parameters [5].

For example, in some D, A systems combined by (insulating) small spacer and interacting rather strongly, the photoinduced CS takes place with time constant shorter than τ_S or even shorter than τ_L in some aprotic polar solvents [6]. On the other hand, in some more strongly interacting D–A systems combined directly by single bond, photoinduced CS process which contains component with time constant considerably longer than τ_S has been observed [6]. Analogous results have been obtained also in the case of the photoinduced CS of some CT (charge transfer) complexes in polar solvents.

In those strongly interacting D, A systems, it might be possible that the rapid photoinduced CT process will be strongly coupled with solvent motions which are not homogeneous in the neighborhood of the D, A system, leading to the nonlinear CS process containing various components from very rapid to very slow ones. It may be also possible that, not only the orientation motions of polar solvent molecules but also the intramolecular vibrations are coupled with the CT process, which will also lead to the nonlinear CS process containing very rapid component [7]. On the other hand, not only the orientation motions of polar solvent molecules but also the intramolecular vibrations are coupled with the CT process, which will also lead to the nonlinear CS process containing very rapid component [7]. On the other hand, in the case of the strongly interacting D–A system or the CT complex, some intramolecular or intracomplex structural rearrangements including the surrounding solvent are necessary to minimize the electronic delocalization interaction between D and A and to attain the complete CS. Such structural changes will make the photoinduced CS process of these systems slower than in the case of some D, A systems separated by a small insulating spacer.

Moreover, the nature of the different kinds of geminate IP's produced depending on the different parameters regulating photoinduced ET process [5] determines the mechanisms of the subsequent chemical reaction as it has been observed very clearly in the case of the IP's of benzophenone–tertiary aromatic amine systems [8,9].

In addition, there is an extreme case of slow electron transfer by very weak interaction from a solute fluorescent state to transient aggregate of polar solvent molecules formed by rapid fluctuations of solvent without definite A molecule [10,11].

In the following, above problems concerning the fundamental aspects of dynamics and mechanisms of photoinduced ET and produced transient CT or geminate IP states will be discussed on the basis of our femtosecond–picosecond laser photolysis studies on various D, A systems.

2. PHOTOINDUCED CS IN LINKED SYSTEMS

The D, A systems combined by spacer or directly by single bond are very suitable for the studies on the effect of various factors regulating the photoinduced ET processes such as the strength of electronic interaction between D and A as well as the solvent orientation dynamics. Nevertheless, results of systematic investigations on those D, A combined systems which seem to be appropriate for such purpose are very few. From such viewpoints, we are examining aromatic hydrocarbon–amine systems combined by methylene chains and also the systems where the two chromophores are combined directly by single bond, by means of femtosecond–picosecond time–resolved spectroscopy. We discuss here mainly the results of time–resolved transient absorption spectral measurements on the following systems: p–$(CH_3)_2$N–Ph–$(CH_2)_n$–(1–pyrenyl) (Pn, n=1,2,3), p–$(CH_3)_2$N–Ph–$(CH_2)_n$–(9–anthryl) (An, n=1,2,3), Ph–N(CH_3)–CH_2–(9–anthryl) (9–AnMe), 9,9'–bianthryl (BIAN) and (9–anthryl)–(N–carbazolyl) (C9A) in alkanenitrile solutions.

In the case of Pn, An and 9–AnMe, where the D, A groups are separated by methylene chain, we can clearly observe the CS process starting from the excited state localized in aromatic hydrocarbon moiety to the intramolecular IP state [12]. The time constants of photoinduced CS, τ_{CS}, obtained by the analysis of absorbance rise curves of the intramolecular IP states of Pn and An have been confirmed to be considerably longer than the longitudinal dielectric relaxation time, τ_L, of solvents, acetonitrile, butyronitrile and hexanenitrile. This comparison suggests that the photoinduced CS in these systems is not directly controlled by the solvent reorientation dynamics, but it seems possible to give a satisfactory account of these results in terms of the usual nonadiabatic ET mechanisms [12]. That is, the photoinduced CS process can be described by the simple two state model.

$$A^{*}\text{–S–D} \longrightarrow (A^{-}\text{–S–}D^{+})_S \qquad (1)$$

where S represents the spacer.

Even if we use the solvation time, τ_S [13,14], estimated from the dynamic fluorescence Stokes shift of the polar probe molecule, this conclusion is not altered except in the case of A_1 where τ_{CS} is rather close to τ_S suggesting the possibility of control by solvation dynamics. Our results of femtosecond laser photolysis measurements on 9–AnMe show clearly that τ_{CS}=0.7 ps in hexanenitrile [6]. This τ_{CS} value is much shorter than τ_S (hexanenitrile)$\sim$3.5–4.5 ps. Moreover, it is shorter than τ_L of hexanenitrile (0.98–1.1 ps). Although we have observed clearly by femtosecond absorption spectroscopy that the excitation is initially localized in anthracene moiety also in the case of 9–AnMe, the electronic interaction between the two chromophores in 9–AnMe seems to be much stronger compared with that in A_1, because the charge density on N–atom is much larger than that on carbon atom at p–position. The coupling of the intramolecular vibration with ET reaction [7] seems to be contributing to enhance the reaction rate, and the large change of the electron density around N–atom of 9–AnMe in the course of the ET might be facilitating the coupling, though its detailed mechanism is not clear at the present stage of investigation.

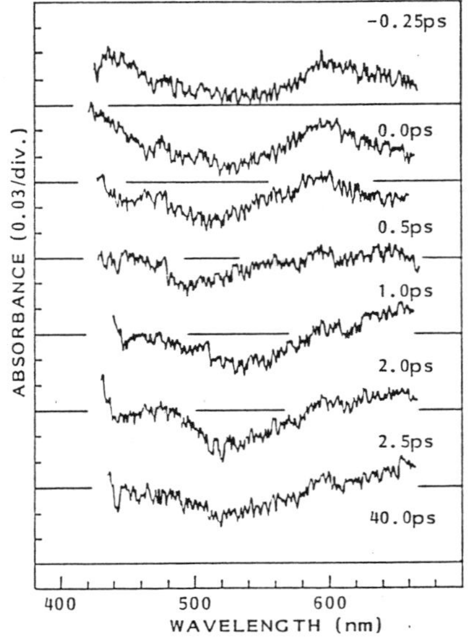

Figure 1. Femtosecond time–resolved absorption spectra of 9–AnMe in hexanenitrile solution. τ_{CS}=0.7 ps, the rise time observed at absorption band of CT state (480 nm).

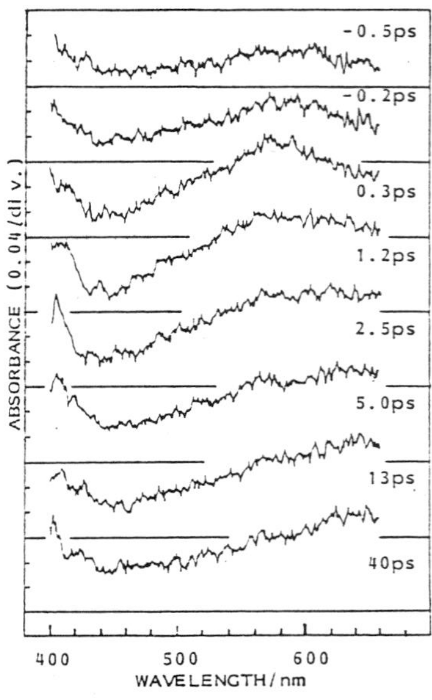

Figure 2. Femtosecond time–resolved absorption spectra of C9A in butyronitrile solution. τ_{CS}=3.8 ps, rise time observed at absorption band of CT state (600 nm).

On the other hand, we have observed previously that the rise time τ_{CS} of solvent induced broken symmetry state of BIAN in viscous alcohol (1–pentanol) is close to the solvent τ_L [15]. However, according to our detailed femtosecond absorption spectral studies in alkanenitrile solutions [6], the rise curve of the intramolecular CT state contains rather slow component with time constant τ_{CS} longer than τ_S and much longer than τ_L of the solvent. That is, τ_{CS}=1.8 ps, 3.4 ps and 7.5 ps compared with τ_S=0.4–0.9 ps, 1.5–2.1 ps and 3.5–4.5 ps, in acetonitrile, butyronitrile and hexanenitrile, respectively.

It should be noted here that, though two anthracene planes in BIAN are close to perpendicular to each other due to steric hindrance, the absorption spectrum of S_1 state is much broader and shifted compared with the superposition of spectra of anthracene anion and cation radical even in acetonitrile solution. This means that there is a considerable electronic delocalization interaction between two chromophores in the CT state. One of the possible cause for the slow component in the rise of CT state may be the structural change necessary to minimize the delocalization interaction between two moieties and to attain the more complete CS, as discussed already in **1. INTRODUCTION**. We have recognized similar slow component also in the case of C9A in alkanenitrile. Moreover, τ_{CS} of C9A is longer than that of BIAN in general: τ_{CS}=3.8 ps and 9.0 ps in butyronitrile and hexanenitrile, respectively [6]. Due to the more planer configuration of C9A compared with BIAN, the electronic delocalization interaction between two moieties may be stronger in C9A. Therefore, τ_{CS} of C9A will become longer than that of BIAN according to the above reasoning and this slow CS process may be described as a gradual change of electronic structure via multiple intermediate states accompanied with some structural change.

$$(A^{-\delta}-D^{+\delta})^*_S \longrightarrow \cdots \longrightarrow (A^{-\delta'}-D^{+\delta'})^*_{S'} \longrightarrow \cdots \longrightarrow (A^{-\delta''}-D^{+\delta''})^*_{S''} \quad (2)$$

The linked systems we have discussed above show the photoinduced CS only in rather strongly polar solvents, because $-\Delta G^o$ value becomes too small to attain the CS in less polar or nonpolar solvents. Of course, if the energetics is more favorable, the photoinduced CS can be attained even in nonpolar or only slightly polar solvents. Actually, we have observed efficient photoinduced CS of etioporphyrin (P)- and Zn-etioporphyrin (ZnP)-quinone derivative (Q) systems combined by $-Ph-CH_2-$ spacer [16] not only in tetrahydrofurane and butyronitrile but also in benzene and toluene solutions. We can examine in detail the $-\Delta G^o$ dependence and effects of the solvent reorganization energy on the photoinduced CS rate by using these P-S-Q or ZnP-S-Q systems [16]. At the present stage of the investigation, we can observe clearly the normal and top regions in the $-\Delta G^o$ dependence of k_{CS} ($=\tau_{CS}^{-1}$) as well as effects of solvent reorganization energy on k_{CS} as expected from the ET theories assuming a weak electronic interaction between D and A. Investigations on the systems with larger $-\Delta G^o$ values are now going on in our laboratory.

3. PHOTOINDUCED CS IN UN-LINKED D, A SYSTEMS

3.1. CS in Weakly Interacting D, A Systems

As discussed already in **2.**, we can examine effects of D, A electronic interactions, $-\Delta G^o$ and reorganization energies as well as solvent dynamics on the photoinduced CS process by employing the D, A combined systems. However, it is not easy to examine the energy gap dependence of k_{CS} over a wide range of $-\Delta G^o$ because of the difficulties in synthesizing appropriate linked systems especially for large $-\Delta G^o$ values. It may be possible, however, to examine the energy gap dependence of k_{CS} over wide $-\Delta G^o$ value by using various unlinked D, A systems with different redox potentials in the case of the fluorescence quenching reaction in acetonitrile solutions. On the other hand, it is well-known that the bimolecular rate constant of fluorescence quenching due to the CS at encounter shows a steep rise around zero energy gap to the diffusion limited value and no decrease within the $-\Delta G^o$ value as large as 2.5 eV examined hitherto [17]. It has been a long standing problem that we cannot observe the inverted region in this fluorescence quenching reaction.

Nevertheless, it seems possible to examine whether the "true" CS rate constant, k_{CS}, in the energy gap region where the fluorescence quenching rate constants obtained by stationary measurements are diffusion controlled, shows bell-shaped energy gap dependence or not, with k_{CS} values obtained by analyzing the transient effect in the fluorescence quenching reaction [18]. That is, the time dependent "rate constant" k(t) of fluorescence quenching reaction of eq 3, is given by eq 4 [19].

$$F^* + Q \xrightarrow{k(t)} F_S^{+}\cdots Q_S^{-} \text{ or } F_S^{-}\cdots Q_S^{+} \qquad (3)$$

$$F^* \xrightarrow{\tau_0^{-1}} F + Q \qquad [F]<<[Q]$$

$$k(t) = \left(\frac{1}{k_D^{-1}+k_{CS}^{-1}}\right)\left[1 + \frac{k_{CS}}{k_D}\cdot \exp(x^2)\mathrm{erfc}(x)\right]\cdot N' \qquad (4)$$

$$x = (Dt)^{1/2}\cdot R^{-1}\cdot[1 +(k_{CS} / k_D)]$$

where, erfc: the error function, k_D: diffusion rate constant, R: critical distance of reaction, N': Avogadro number. By using this k(t), the fluorescence decay curve is given by,

$$I(t) = I_0\exp(-t / \tau_0)\exp\{-[Q]\int_{t=0}^{t} k(t)dt\}$$

$$= I_0\exp\{\frac{N'[Q]k_{CS}^2}{2c^2(k_{CS}+k_D)}\}\exp\{-at-bt^{1/2}-\left(\frac{\pi^{1/2}\cdot b}{2c}\right)\exp(c^2t)\mathrm{erfc}(ct^{1/2})\} \qquad (5)$$

where $a = \frac{1}{\tau_0} + 4\pi\left(\frac{k_{CS}\cdot R}{k_{CS}+k_D}\right)\cdot D\cdot N'[Q]$, $b = 8\left(\frac{k_{CS}\cdot R}{k_{CS}+k_D}\right)^2(\pi\cdot D)^{1/2}\cdot N'[Q]$,

$$c = \left(\frac{D^{1/2}}{R}\right)\left[1 + \frac{k_{CS}}{k_D}\right].$$

The fluorescence decay curves obtained for various D, A systems in acetonitrile by means of picosecond dye laser for excitation and single photon counting measurements have been simulated with eq 5. From this simulation k_{CS} values have been obtained for various D, A pairs as indicated in Figure 3. These k_{CS} values are ca. $10^{11}\sim10^{12}$ M^{-1} s^{-1}.

As we discuss in the next section (4), the photoinduced CS at encounter between fluorescer and quencher in the energy gap region where we have obtained the "true" k_{CS} in acetonitrile solution, leads to the formation of the loose IP(LIP) or solvent–separated IP(SSIP). Nevertheless, the k_{CS} values obtained by the analysis of the transient effect in the fluorescence quenching process are fairly large and comparable to the rate constant of the intramolecular photoinduced CS of the D, A systems linked with a relatively short chain in acetonitrile solutions.

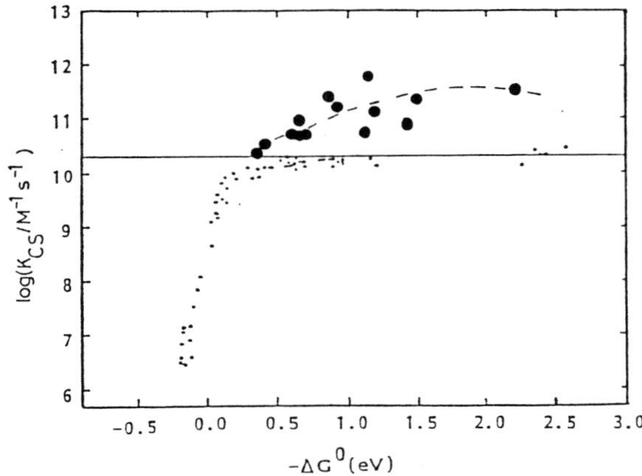

Figure 3. Plot of k_{CS} values obtained by the analysis of the transient effect in the fluorescence decay curves against the $-\Delta G^o$ values of CS reaction.

The k_{CS} values in Figure 3 show only the normal region and very flat top regin but no inverted region in the range of $-\Delta G^o$ from 0.37 eV to 2.2 eV. On the other hand, the rate constant k_{CR} of the CR decay of the LIP produced by fluorescence quenching reaction in acetonitrile solution was confirmed first by the present author to show a rather typical bell–shaped energy gap dependence [20,21]. One of the plausible interpretation for this discrepancy between the energy gap dependences of k_{CS} and k_{CR}, and the lack of the inverted region in the former, is to assume the formation of the excited state of the radical ions in the course of the CS [21]. For the system, 9,10–diphenylanthracene(DPA)–tetracyanoethylene(TCNE), with the largest $-\Delta G^o$ (2.21 eV) in Figure 3, the energy gap $-\Delta G^{o*}$ by this pathway has been estimated to be 0.5 eV using the energy of the lowest excited state of ions. However, the k_{CS} value estimated from the results given in Fig. 3 by assuming this $-\Delta G^{o*}$ value is much smaller than the k_{CS} value (3.16×10^{11} M^{-1} s^{-1}) observed for DPA–TCNE system. Therefore, the participation of the excited states of ion radicals in the CS process is not probable in the DPA–TCNE in acetonitrile solution.

The discrepancy between the energy gap dependences of k_{CS} in the fluorescence quenching process and k_{CR} of the produced LIP seems to be interpreted by taking into consideration the distribution of the distance between D and A [22–23] and a moderate extent of nonlinear polarization of the solvent around the charged ions [23–25]. This distance distribution is related to the energy gap dependence of k_{CS} through the dependence of the solvent reorganization energy λ on the D, A distance at which ET takes place [18,23,24]. That is, for the CS reaction corresponding to the larger $-\Delta G^o$, a little larger distance for ET is favorable owing to the larger solvent λ as can be seen from the following simple equations for one electron transfer [2].

$$k_{CS} = \kappa Z \exp\{-\frac{(\lambda+\Delta G^o)^2}{4\lambda kT}\}, \quad \lambda = e^2\{\frac{1}{2a_1} + \frac{1}{2a_2} - \frac{1}{R}\}\{\frac{1}{\varepsilon_\infty} - \frac{1}{\varepsilon_S}\} \quad (6)$$

where κ: the transmission coefficient, Z: the frequency factor a_1, a_2: the radii of reactants, R: the center to center distance between reactants at encounter, ε_S, ε_∞: the static and high frequency dielectric constants, respectively. This means that the inter–ionic distance in the geminate LIP will increase a little with increase of $-\Delta G^o$ for CS.

By taking into account such $-\Delta G^o$ dependence of the inter–ionic distance in the geminate LIP immediately after its formation and a little change of the distance distribution before the CR decay of the IP and also taking into account a moderate amount of the nonlinear polarization of solvent around the ions, the discrepancy between the energy gap dependence of the k_{CS} in the fluorescence quenching reaction and k_{CR} of the geminate LIP seems to be interpreted [24]. Moreover, we have found recently a more direct experimental result which seems to support such $-\Delta G^o$ dependence of the inter–ionic distance in the case of the hydrogen abstraction reaction between triplet benzophenone ($^3BP^*$) and various tertiary aromatic amine (AH) in acetonitrile [8,9]. In this case, the hydrogen abstraction occurs by the mechanism of ET followed by proton transfer in the geminate ^{3}LIP.

$$^3BP^* + AH \longrightarrow {}^3(BP_S^{-}\cdots AH_S^{+}) \xrightarrow{k_{PT}} (BPH\cdots A) \longrightarrow \quad (7)$$

By directly observing the reaction processes with femtosecond–picosecond laser spectroscopy, we have found that k_{PT} significantly decreases with increase of the $-\Delta G^o$ for the CS reaction, indicating the increase of the inter–ionic distance in LIP in this order [8,9]. Namely, the rate constants of the proton transfer (k_{PT}) within ^{3}LIP in acetonitrile were determined to be: 9.5×10^9 s^{-1}, 5.4×10^9 s^{-1}, 7.3×10^8 s^{-1}, and $<<2\times10^8$ s^{-1} when N–methyldiphenylamine, N,N–dimethylaniline, N,N–diethylaniline and N,N–diethyl–p–toluidine were used as electron donors, respectively. Their oxidation potentials decrease in this order, i.e. 0.86, 0.76, 0.72 and 0.69 V vs. SCE, respectively.

3.2. CS in Strongly Interacting D, A Systems

An extreme case of photoinduced CS in strongly interacting D, A systems is the photoexcitation of the CT complexes in polar solutions. According to Mulliken's original treatment, the excited state of the CT complex has usually much stronger CT character compared with the ground state. Therefore, the D, A configuration in the FC (Franck–Condon) excited state is not the most stable one for the excited state electronic structure, and the relaxation takes place from the excited FC state toward the excited equilibrium (eql) state accompanied with the change of the geometrical and electronic structures. The CT character is usually increased by this relaxation process. In strongly polar solvent like acetonitrile, the solvation of the complex together with the D, A structural rearrangement enhances the CT character leading to the formation of the geminate IP states and, in the case of some weak complexes, to the ionic dissociation competing with the CR deactivation [1].

It should be noted here that, although there were so may works on the absorption spectra and structures of CT complexes in the ground state only a few studies on the excited state by means of fluorescence and nanosecond transient spectral measurements were reported previously [1]. The above described relaxation processes from the excited FC to the eql states leading to the formation of the so-called CIP (contact IP), LIP, CR decay and dissociation of these IP's into free ions, have been directly observed only recently by our detailed femtosecond–picosecond laser photolysis investigations [5,26,27]. We summarize here briefly important aspects of the photoinduced CS process of CT complexes revealed by those investigations.

The photoinduced CS in the excited state of CT complexes in the case of TCNB (1,2,4,5–tetracyanobenzene) in benzene and methyl–substituted benzene (toluene, mesitylene) solutions occurs by two steps. First, a slight configurational rearrangement within 1:1 complex from an asymmetric configuration in FC state toward a more symmetric overlapped one accompanied with a slight enhancement of CT degree takes place within ca. 1 ps. However, for the complete CS, the intra-complex configuration rearrangement in this step is not sufficient but much larger structural change including the formation of the 1:2 complex (which takes ca. 30 ps) is of crucial importance [5,26].

$$(A^{-\delta}\cdot D^{+\delta}) \xrightarrow{h\nu} (A^{-\delta'}\cdot D^{+\delta'})^* \xrightarrow[\text{structural change within 1:1 complex}]{1\text{ ps}} (A^{-\delta''}\cdot D^{+\delta''})^* \xrightarrow[D]{30\text{ ps}} (A^{-}\cdot D_2^{+})^* \qquad (8)$$

In strongly polar solvent, acetonitrile, the photoinduced CS of TCNB–Tol (toluene) complex takes place to a considerable extent due to solvent reorientation within ca. 1 ps but not completely, and for a complete CS, further relaxation process with time constant τ_{CS} of ca. 20 ps is necessary. This fact indicates that further intracomplex structural change is necessary for the complete CS in this strongly interacting D, A system.

We have observed similar CS process in the case of the pyromellitic dianhydride (PMDA)–Tol complex in acetonitrile solution [5,27]. In this case, we have confirmed that it takes ca. 7 ps for the complete CS. This time constant is shorter than that of TCNB–Tol system in acetonitrile. We have examined many other CT complexes of various D and A with different ionization potential and electron affinity, respectively, and have confirmed that τ_{CS} values of many of them are longer than the solvent τ_L or τ_S values indicating some structural change necessary for the CS. We have observed also that, with decrease of the ionization potential of D and/or with increase of the electron affinity of A, τ_{CS} becomes shorter [5,26,27].

The energy of the CT absorption band becomes lower with decrease of ionization potential of D and increase of electron affinity of A, which will decrease the mixing between the LE (locally excited) and CT electronic configurations in the FC excited state and will decrease the extent of the geometrical configuration change necessary to realize the complete CS state. On the other hand, if one uses very strong A like TCNE and D with fairly low ionization potential like pyrene and perylene, mixing of the CT configuration with the ground configuration will arise because of the very small energy gap. The very small energy gap and the interaction of the CT state with the ground state will induce also the very rapid CR deactivation as discussed in the next section, which will make difficult the complete CS.

4. CR DECAY OF GEMINATE IP'S PRODUCED BY PHOTOINDUCED CS

4.1. Energy Gap Dependence of k_{CR} of LIP

We have given the first experimental confirmation of the bell–shaped energy gap dependence of k_{CR} of LIP (SSIP) produced by CS at encounter between fluorescer and quencher in acetonitrile solution including both the inverted region and the normal region by directly observing the CR decay of LIP competing with its dissociation [20,21].

As we have discussed in the previous section (**3.1**), the interionic distance in the geminate LIP immediately after CS between the fluorescer and quencher seems to have a distribution and an average distance becomes a little longer with increase of $-\Delta G^{\circ}$ for CS [23,24], for which we have given a direct experimental evidence in the case of the geminate ^{3}LIP of benzophenone–tertiary aromatic amine systems in acetonitrile [9]. The inter–ionic distance distribution will be a little modified but the averages distance will not significantly change before the CR and dissociation (several hundreds ps) [24]. These considerations together with the effect of the nonlinear polarization of solvent around ions especially in CR process seem to be of crucial importance in the interpretation of the different energy gap dependence between the CS and CR processes, as we have described already in **3.1**.

4.2. CR Decay of CIP and Energy Gap Dependence of k_{CR}

We have made a systematic investigations on the CR decay of CIP's formed by excitation of the CT complexes in acetonitrile solutions by directly observing the photoinduced CS leading to CIP formation followed by its CR decay and, in the relatively weak D, A systems, also the change to LIP as well as dissociation, with quantitative femtosecond–picosecond laser spectroscopy [5,26–28].

Moreover, we have made comparative studies on the behaviors of IP's of the same D, A systems where one is produced by CT complex excitation (CIP) and the other is produced by CS at encounter in fluorescence quenching reaction [27]. CIP shows generally much faster CR decay compared with LIP. For example, $k_{CR}^{CIP}=1.3\times10^{10}$ s^{-1}, $k_{CR}^{LIP}=2.2\times10^{9}$ s^{-1} in anthracene(D)–phthalic anhydride(PA)(A), and $k_{CR}^{CIP}=2.0\times10^{12}$ s^{-1}, $k_{CR}^{LIP}=2.6\times10^{9}$ s^{-1} in pyrene (D)–TCNE (A). By this sort of measurements on various CT complexes, we have examined the energy gap dependence of k_{CR}^{CIP} for wide ranges of systems including TCNB, acid anhydrides, TCNA (tetracyanoanthracene) TCNE, TCNQ (tetracyanoquinodimethane) as A and aromatic hydrocarbons and methyl derivatives as D, and have found $k_{CR}^{CIP} \sim -\Delta G^{o}$ relation quite different from the bell-shaped one of LIP (Figure 4), which is given approximately as,

$$k_{CR}^{CIP} = \alpha \cdot \exp[-\gamma \,|\, \Delta G^{o} \,|\,] \tag{9}$$

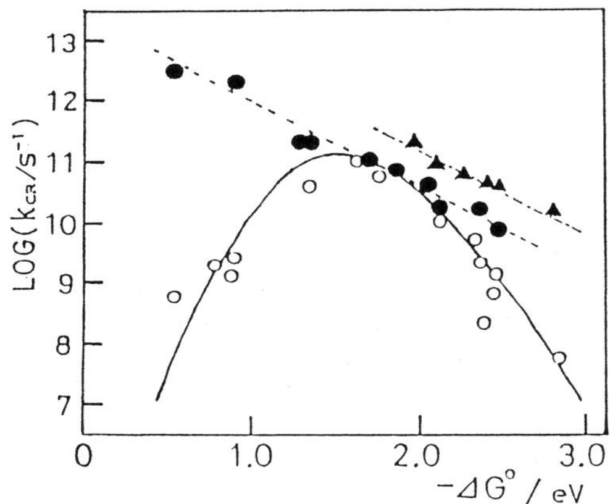

Figure 4. Energy gap dependence of k_{CR} of LIP(○) and CIP (●,▲) in acetonitrile solution.
●: polycyclic aromatic hydrocarbon (D)–several kinds of acceptor systems.
▲: benzene and methylsubstituted benzene (D)–PMDA (A) systems.

where $-\Delta G^{o}$: the free energy gap between IP and ground state, α and γ : constant independent of ΔG^{o}. We have observed this type of relation for a wide range of D, A systems. For example, the γ value for the PMDA–methylsubstituted benzene (MB) series is almost the same as those of acid anhydrides–polycyclic aromatic hydrocarbon series as well as TCNA–MB series [27,28], although the α values of each series are a little different from each other. It should be noted here that results of $k_{CR}^{CIP} \sim |\, \Delta G^{o} \,|$ relation for TCNA–MB with much larger γ value than ours was previously reported by Gould et al. [29], which may be ascribed to some complications in their method of evaluation of k_{CR}^{CIP} indirectly from the yield of dissociated ions.

The $k_{CR}^{CIP} \sim |\, \Delta G^{o} \,|$ relation in eq 9 is qualitatively similar to the energy gap dependence of the intramolecular radiationless transition probability in the weak coupling limit [30], indicating nonradiative transition in a supermolecule, $(D^{+} \cdot A^{-})$ ->(D·A). In relation to the mechanism of this peculiar energy gap dependence of k_{CR}^{CIP}, we have examined the solvent effect on $k_{CR}^{CIP} \sim |\Delta G^{o}|$ relation by changing the solvent from acetonitrile ($\varepsilon\sim37$) to ethylacetate ($\varepsilon\sim6$). The $k_{CR}^{CIP}\sim|\Delta G^o|$ relation shifted to the side of smaller ΔG^o value in less polar solvent although the effect is very small [31]. This result can be interpreted by assuming the potential energy surfaces for CIP and ground states in the weak coupling limit (similar to the case of inverted region) against the coordinate including solvation and assuming a shift of potential minimum and stabilization due to solvation of CIP state in more polar solvent.

The nature of the CR process in CIP can be examined also by measurements of the temperature dependence of k_{CR}^{CIP}. We have investigated mainly the CIP's of PMDA–aromatic hydrocarbon in ethylacetate, where CIP's do not undergo the ionic dissociation. The activation energies (E_a) for CR process were very small or zero as follows [31].

E_a=0, D: pyrene ($-\Delta G^o$=1.98 eV); E_a=0, D: durene ($-\Delta G^o$=2.37 eV); E_a=0.40 kcal/mol, D: Tol ($-\Delta G^o$=2.76 eV); E_a=0.92 kcal/mol, D: naphthalene ($-\Delta G^o$=2.32 eV).

It is, however, difficult to examine the temperature dependence of k_{CR} of LIP's in the case of the unlinked D, A system which are formed only in strongly polar solvent like acetonitrile and undergo dissociation quite easily. Therefore, we have examined it in the inverted region by using linked D, A system with relatively weak interaction between chromophores, ZnP–Ph–CH_2–Q, in toluene solution. In both cases with $-\Delta G^o$ =1.35 eV and $-\Delta G^o$=1.69 eV, we have confirmed that E_a of k_{CR} is approximately zero. All results of the above studies on temperature dependence of k_{CR} indicate that CR process in IP state is predominantly quantum mechanical tunneling owing to the intramolecular high frequency mode.

5. SPECIAL CASE OF PHOTOINDUCED CS BY VERY WEAK INTERACTION

We have discussed already examples of photoinduced CS in unlinked and linked systems with regards to the various cases of electronic interactions between D and A and also various cases of the strengths of D and A (various values of the ionization potential of D and electron affinity of A). Depending on those interactions and strengths of D and A, not only the mechanisms of photoinduced CS but also the dynamic behaviors of produced IP's are affected profoundly.

We discuss here an extreme case of photoinduced CS due to electron transfer from solute fluorescent state to transient aggregate of solvent molecules formed by their fluctuation motions, without any definite acceptor molecule [10,11,33], which we found 1982 and for which we are establishing a detailed mechanism. The most typical solute molecules with low ionization potential which undergo electron ejection from relaxed S_1 state in strongly polar solvents are such aromatic diamine as 2,7-bis(dimethylamino)-4,5,9,10-tetrahydropyren (BDATP) and N,N,N',N'-tetramethyl-p-phenylenediamine (TMPD). In acetonitrile solution, ET from these solute in fluorescent state to the transient aggregate (S_n) of near-by solvent molecules takes place with time constants (τ_{CS}) of a few ns at room temperature.

$$\mathrm{BDATP^*(S_1)} \xrightarrow[\tau_{CS}=2.3\ \mathrm{ns}]{} \mathrm{BDATP_S^{+}\cdots S_n^{-}}$$
$$\mathrm{TMPD^*(S_1)} \xrightarrow[\tau_{CS}=1.2\ \mathrm{ns}]{} \mathrm{TMPD_S^{+}\cdots S_n^{-}} \qquad (10)$$

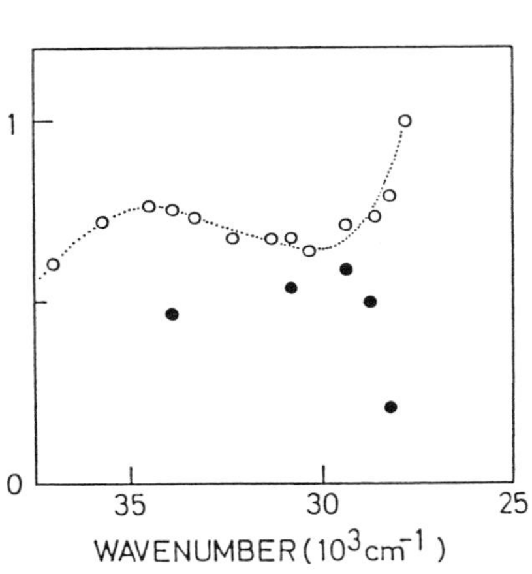

Figure 5. Excitation energy dependence of the monophotonic ionization yield (●) at 100 ps after the excitation and the relative fluorescence yield (O) of TMPD in acetonitrile solution.

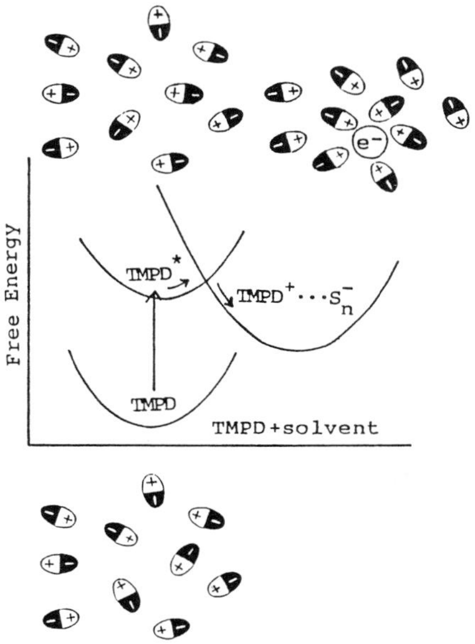

Figure 6. Schematic diagram for ET from relaxed S_1 state of TMPD to transient aggregate of solvent molecules in acetonitrile solution.

In the case of TMPD, we have examined the effect of the excess vibrational energy on this ET reaction by changing the wavelength of picosecond dye laser [10]. The excitation of TMPD at 355 nm is very near the 0–0 transition. We have confirmed that the monophotonic ionization yield increases significantly with increasing excitation energy near the origin of S_1 state and it is almost constant in the region around 3×10^4 cm^{-1}. It should be emphasized here that the yield obtained by exciting at 3.39×10^4 cm^{-1} (295 nm) is less than that obtained by exciting around 3.08×10^4 cm^{-1} (325 nm) (Figure 5). If ivr (intramolecular vibrational redistribution) is much faster than the ionization from the vibrationally unrelaxed state, the ionization yield should be determined simply by the amount of the excess vibrational energy contrary to the observed result, which indicates that the ionization can compete with ivr. The lower ionization yield when excited at 3.39 10^4 cm^{-1} (295 nm) which is the onset of the S_2 state, suggests that the ionization yield depends on the optically excited vibraitonal mode or, if the internal conversion occurs before ionization, the accepting mode of S_2->S_1 internal conversion is different from the promoting mode of ionization.

For the electron ejection from the vibrationally relaxed S_1, formation of the transient aggregate of near–by solvent molecules which can stabilize sufficiently the accepted electron is necessary. The frequency of the formation of such transient aggregate of near–by solvent may be low and the activation energy for its formation may be high leading to the slow CS. However, for the CS from the vibrationally excited state, demand to such favorable transient solvent aggregate may be less severe, where the excess vibrational energy seems to be efficiently used for CS process.

Since the transient aggregate formation seems to take place owing to the very rapid dynamic fluctuations of solvent molecules, the photoinduced CS in this system will show a rather large decrease by temperature lowering. We have actually observed such temperature effects on CS, and obtained the activation energy for the CS from the relaxed S_1 state in acetonitrile to be 7.4 kcal/mol [11]. Solvent fluctuations play both roles of the orientation motions promoting ET by very weak interaction and the transient acceptor of electron in this extreme cases as indicated schematically in Figure 6. This is a very interesting and important example in relation to the most fundamental aspects of the ET reaction.

ACKNOWLEDGMENT

The investigations, results of which are discussed in this paper were partially supported by a Grant–in–Aid for Specially Promoted Research (No. 6265006) from the Ministry of Education, Science, and Culture of Japan to the author.

REFERENCES

1. N.Mataga, in: Molecular Interactions, eds. H. Ratajczak and W. J. Orville–Thomas (John Wiley & Sons, Chichester, 1981) pp. 509–570; idem, in: Photochemical Energy Conversion, eds. J. R. Norris and D. Meisel (Elsevier, New York, 1989) pp. 32–46.
2. R. A. Marcus and N. Sutin, Biochim. Biophys. Acta 811 (1985) 265.
3. M. Sparpaglione and S. Mukamel, J. Phys. Chem. 91 (1987) 3938.
4. M. Moroncelli, J. McInnis and G. R. Fleming, Science 243 (1989) 1674.
5. N. Mataga, in: Electron Transfer in Inorganic, Organic, and Biological Systems, eds. J. R. Bolton, N. Mataga, and G. McLendon, Advances in

Chemistry Series, 228, ACS, 1991, Chapt. 6.
6. S. Nishikawa, T. Okada, N. Mataga and W. Rettig, to be published.
7. H. Sumi and R. A. Marcus, J. Chem. Phys. 84 (1986) 4894.
8. H. Miyasaka, K. Morita, K. Kamada and N. Mataga, Bull. Chem. Soc. Jpn. 63 (1990) 3385.
9. H. Miyasaka, K. Kamada, K. Morita, T. Nagata, M. Kiri and N. Mataga, Bull. Chem. Soc. Jpn., submitted.
10. Y. Hirata, M. Ichikawa and N. Mataga, J. Phys. Chem. 94 (1990) 3872.
11. Y. Hirata, Y. Tanaka and N. Mataga, in preparation.
12. N. Mataga, S. Nishikawa, T. Asahi and T. Okada, J. Phys. Chem. 94 (1990) 1443.
13. M. A. Kahlow, T. J. Kang and P. F. Barbara, ibid. 91 (1987) 6452.
14. I. Rips, J. Klafter and J. Jortner, ibid. 94 (1990) 8557.
15. N. Mataga, H. Yao, T. Okada and W. Rettig, ibid. 93 (1989) 3383.
16. T. Asahi, M. Ohkohchi, N. Mataga, A. Osuka and K. Maruyama, to be published.
17. D Rehm and A. Weller, Israel J. Chem. 8 (1979) 259.
18. S Nishikawa, T. Asahi, T. Okada, N. Mataga and T. Kakitani, Chem. Phys. Lett., submitted.
19. F. C. Collins and G. E. Kimball, J. Colloid Sci. 4 (1949) 425.
20. N. Mataga, T. Asahi, Y. Kanda, T. Okada and T. Kakitani, Chem. Phys. 127 (1988) 249, and papers cited there-in.
21. N. Mataga, Y. Kanda, T. Asahi, H. Miyasaka, T. Okada and T. Kakitani, ibids. 127 (1988) 239.
22. R. A. Marcus and P. Siders, J. Phys. Chem. 86 (1982) 622.
23. T. Kakitani, A. Yoshimori and N. Mataga, in: Electron Transfer in Inorganic, Organic and Biological Systems, eds. J. R. Bolton, N. Mataga, and G. McLendon, Advances in Chemistry Series 228, ACS, 1991, Chapt. 4.
24. T. Kakitani, A. Yoshimori and N. Mataga, in preparation.
25. A. Yoshimori, T. Kakitani, Y. Enomoto and N. Mataga, J. Phys. Chem. 93 (1989) 8316, and papers cited there-in.
26. H. Miyasaka, S. Ojima and N. Mataga, J. Phys. Chem. 93(1989) 3380; S. Ojima, H. Miyasaka and N. Mataga, ibid. 94 (1990) 4147, 5834, 7534.
27. T. Asahi and N. Mataga, J. Phys. Chem. 93 (1989) 6575; idem., ibid.. 95 (1991) 1956.
28. T. Asahi, N. Mataga, Y. Takahashi and T. Miyashi, Chem. Phys. Lett. 171 (1990) 309.
29. I. R. Gould, M. Rogesr and S. Farid, J. Am. Chem. Soc. 110 (1988) 7242.
30. R. Englman and J. Jortner, Mol. Phys. 18 (1970) 145.
31. T. Asahi, M. Ohkohchi and N. Mataga, in preparataion.
32. N. Mataga and Y.Hirata, in: Advances in Multiphoton Processes and Spectroscopy, Vol. 5, ed. S. H. Lin (World Scientific, Singapore, 1989) pop. 175-279.

DISCUSSION

Tachiya

You say that in the inverted region electron transfer occurs at longer distances. From the analysis of the transient effect of the fluorescence decay curve which you used to determine the intrinsic electron transfer rate, you can also estimate the encounter distance. Is it really longer in the inverted region than in the normal region?

Mataga

I did not say that the inverted region electron transfer occurs at longer distances but I said that the electron transfer distance in the fluorescence quenching reaction due to the charge separation a little increases with increase of the energy gap for the charge separation. The encounter distance obtained by the analysis of the transient effect in the fluorescence quenching reaction seems to increase a little as an average but the data are rather scattered.

Jortner

The linear free-energy relationship discovered by Professor Mataga and his coworkers is significant and interesting. The general description of electron transfer in terms of a nonadiabatic, multiphonon radiationless transition involving the donor(D)-acceptor(A) pair and the entire medium, implies the existence of two limits.
(I) The solvent-dominated limit, which is characterized by the conventional Gaussian free energy relationship, manifesting strong coupling to solvent vibrational modes. Of course, also in this limit, the role of intramolecular high-frequency vibrational modes of the D and A may be significant, resulting in connections to the parabolic free energy relationship.
(II) The molecular limit where "intramolecular" radiationless transition dominant. Now the major nuclear coupling involves the high-frequency intramolecular vibrational modes of the D and A centres (which may be intramolecular or possibly also due to ion-solvent motion). The free-energy relationship in this limit can be (exponentially) linear, as observed by Mataga.
On the basis of general arguments we would expect that the molecular limit (II) would be realized when the medium reorganization energy is small. This can occur in a large D-A system in a weakly polar, or even nonpolar solvent. The interesting issue is what is unique about the particular systems studied by Mataga to reveal the free-energy relationship of the molecular limit?

Mataga

The system which shows the Gaussian or parabolic free energy relationship is the loose ion pair (LIP) or the so-called solvent-separated ion pair (SSIP) which is formed by charge separation (CS) at encounter between fluorescer and quencher in strongly polar solution. In the LIP or SSIP, each ion in the pair is rather strongly solvated and the IP is strongly coupled to solvent mode. Contrary to this, the IP which shows the linear (exponential) free energy relationship is made by excitation of the ground state charge transfer complex and believed to have no intervening polar solvent molecules between ions in the pair (contact or compact IP, CIP). We have confirmed that the linear free energy relationship of CIP is rather insensitive to solvent polarity which means that the solvent reorganization energy is very small due to the absence of the intervening solvent molecules between the pair in the "supermolecule" of CIP.

Harriman

We have observed a shallow linear dependence between rate of charge recombination and thermodynamic driving force for quaternary polycyclic dyes intercalated in polynucleotides. Rapid CS is followed by fast (∿100 ps) CR but the rate of CR is hardly affected by ΔG^o between −2.5 and −1.2 eV.

Mataga

The free energy relationship of your system seems to be somewhat similar to those we observed but its $-\Delta G^o$ dependence is more shallow. It might be related to the large size of molecules composing the system undergoing charge recombination decay.

Yoshihara

In relation to your observation with electron transfer slightly faster than τ_L (9-AnMe in hexanenitrile), we have recently found an order of magnitude faster electron transfer (∿100 fs) of a dye in electron donating solvent (τ_L∿a few picosecond), and explained by the vibrational contribution to the electron transfer. Do you also consider a vibrational promotion for electron transfer effect in your system?

Mataga

We are considering that possibility also.

De Schryver

In the case of CR of CIP a linear relation between ln k_{CR} and $-\Delta G^o$ is observed. How does coupling with the ground state affects this behavior?

Mataga

In some of the CIP's we examined the energy gap between IP and ground states is small (∿0.5 eV, for example, in the pyrene–TCNE, perylene–TCNE systems). In such case, certainly there is some electronic interaction between IP and ground configuration which may facilitate in some way the rapid CR decay, but its effects on the energy gap itself may be rather small because donor molecules are large, which will make the electronic interaction matrix element small.

Hynes

My question concerns your considerations of the changing separation of an ion pair in the inverted regime. Your discussion focused on the role of changing reorganization energy with changing separation. However, the electronic character of the reaction could change as well. In particular, the electronic coupling should decrease with increasing separation. Thus the electron transfer could change from adiabatic to nonadiabatic, for example. Have you considered this possibility and its consequences?

Mataga

We have considered both possibility in our analysis of the energy gap dependence of electron transfer distance in the fluorescence quenching reaction. As a consequence of such analysis, we have concluded that a slight increase of the electron transfer distance which results in the increase of the solvent reorganization energy is advantageous for the electron transfer at larger energy gap region.

Baumann

In systems like BIAN, is there any shift of the transient absorption bands, and if so can this be used to interprete the slower rising component?

Mataga

In the course of the relaxation from Franck–Condon excited state of the CT complex to the contact ion–pair, we have observed in some systems the small shift of the ion band in addition to the change of the ion band shape from broad one to the sharper one, but in other cases, only the sharpening of the band shape without the shift of the ion band. The behavior of the transient absorption band seems to depend on the nature of the relevant electronic state concerned with light absorption of the ionic state of each donor and acceptor. In the case of the BIAN and similar molecules with directly combined donor and acceptor which we have examined, detection of the shift of the ion like band is rather difficult partly because of their very broad bands, and also because the shift seems to be very small.

Dynamics and Mechanisms of
Photoinduced Transfer and Related Phenomena
N. Mataga, T. Okada and H. Masuhara (Editors)

Ultrafast Studies on Electron Transfer

P.F. Barbara, K. Tominaga, and G.C. Walker

Department of Chemistry, University of Minnesota
Minneapolis, Minnesota 55455

Abstract

Excited state intramolecular charge separation in organic electron donor/acceptor (D-A) compounds has become an important model reaction for studying the mechanism of small barrier ($\Delta F < RT$) electron transfer (et) reactions. In this paper we present new ultrafast fluorescence measurements on the excited states dynamics of 4-(9-anthryl)-N,N'-dimethylaniline (ADMA), especially on the excited state et process ($D\text{-}A^* \rightarrow D^+ - A^-$, conventionally denoted by LE $\rightarrow$ CT). The variation on the static absorption and emission spectra of ADMA, as a function of solvent, is primary due to simple "solvent coordinate" effects, rather than large intramolecular structural changes. In polar solvents the excited state et (LE $\rightarrow$ CT) of ADMA is in the Marcus inverted regime. There are two distinct et kinetic components, a faster (< 150 fs) component which is not solvent controlled and probably involves intramolecular bath states: and a slower solvent controlled component with a limiting rate constant of $\sim 1/\langle\tau_s\rangle$, where $\langle\tau_s\rangle$ is the average solvation time.

1. INTRODUCTION

One major focus in et research in the last few years has been the role of solvation dynamics in et kinetics.[1-10] Theory and experiment show that for approximate models for et reactions in the adiabatic limit, the rate constant k_{et}, can be inversely proportional to the solvation relaxation time, τ_s. For simple continuum treatments of the solute/solvent, τ_s is given by the longitudinal relaxation, τ_L,

$$\tau_L \simeq \tau_D(\epsilon_\infty/\epsilon_0) \tag{1}$$

where τ_D is the dielectric relaxation time and ε_{00}, and ε_0 are the infinite frequency and static dielectric constants, respectively. For barrierless reactions the et process is controlled by a diffusional process along the "solvent coordinate", and k_{et} is on the order of $1/\tau_s$. The et reaction kinetics are controlled by solvation dynamics in this limit.

Recent theories[1-10] on dynamic solvent effect show that a number of important effects must be considered to attain an accurate level of description. A distribution of solvation times should be considered for real solvents.[2,3] The coupling of vibrational modes and the solvent coordinate can significantly alter the dependence of k_{et} on τ_s.[7,8] Solvent effects on k_{et} are diminished as a reaction becomes more nonadiabatic. For strongly adiabatic reactions, the reaction occurs over a broad region along the reaction coordinate, and conventional treatments of k_{et}, and et mechanisms are inappropriate. Instead, new approaches must be derived, see Kang *et al.* [9]

$$\text{D–A} \xrightarrow{h\nu} \text{D–A}^* \tag{2}$$

$$\text{D–A}^* \xrightarrow{k_{et}} \text{D}^+\text{–A}^- \tag{3}$$

The molecule we study in this paper is 4-(9-anthryl)-N,N'-dimethylaniline (ADMA):

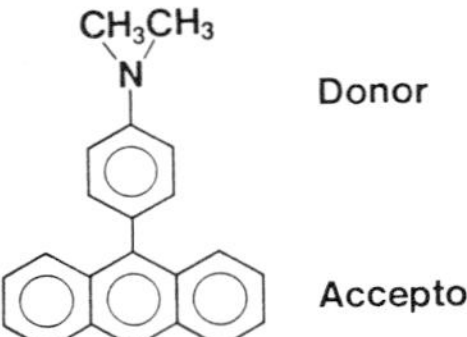

For this molecule, the aniline and anthracene moieties correspond to the electron donor and acceptor, respectively. D-A* represents the lowest $\pi\pi^*$ excited singlet state of the anthracene ring of D-A, and D$^+$ -A$^-$ signifies an intramolecular ion pair in the singlet state. Historically, D-A, D-A*, and D$^+$ -A$^-$ are denoted by S_0, LE and CT, respectively.[11] Figure 1 portrays how the energy of the three states varies for ADMA as a function of the "solvent coordinate", z, in a polar solvent, according to Marcus theory[12]. Analogous energy profiles have been shown to be accurate for bianthryl (BA), a closely related et reaction.[9]

There are a number of theoretical issues that are relevant to all ultrafast et reactions. For example, a very important question is "What et regime applies, i.e., normal, barrierless or inverted regime, for the excited state et of ADMA in polar solvents?" Another important issue is concerned with the nature of the state that is prepared by laser excitation. Since there exist two optical transitions, $S_0 \rightarrow$ LE and $S_0 \rightarrow$ CT, one might suspect that both states could be excited simultaneously by an excitation laser pulse. If so, "How should this effect be included in a description of the kinetics and spectroscopy?" It is also important to establish whether the excited state et kinetics are affected by solvation dynamics, as has been the case for the other excited et reactions examples, such as bianthryl.

Finally, it is interesting to explore what role, if any, vibrational modes play in altering the et rate and the manifestation of the dynamic solvent effect on k_{et} for ADMA.

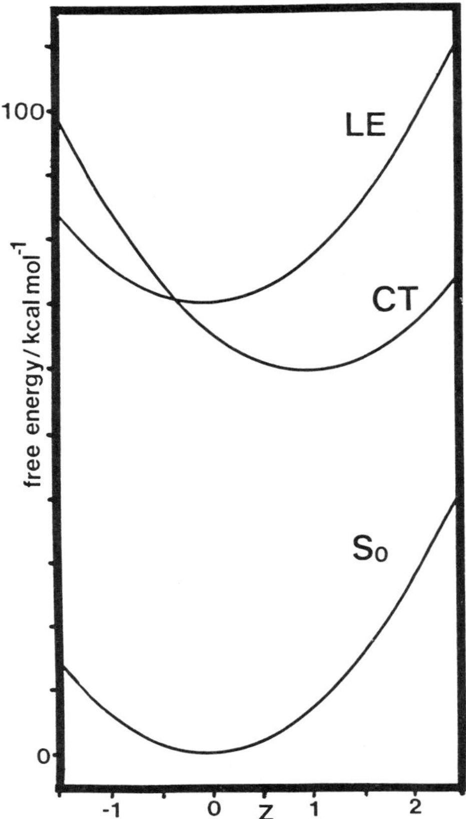

FIGURE 1.

Theoretical estimates for the three diabatic free energies of ADMA in N,N-dimethylformamide as a function of the solvent coordinate: the ground state (S_0), locally excited state (D-A* ≡ LE) which results from τ-τ* excitation of the anthracene ring, and charge transfer sate (D^+ -A^- ≡ CT) which is an intramolecular ion-pair state. The free energy parameters are as follows: k_s = 13.0 kcalmol-1 and ΔF^0 = 10.5 kcalmol-1.

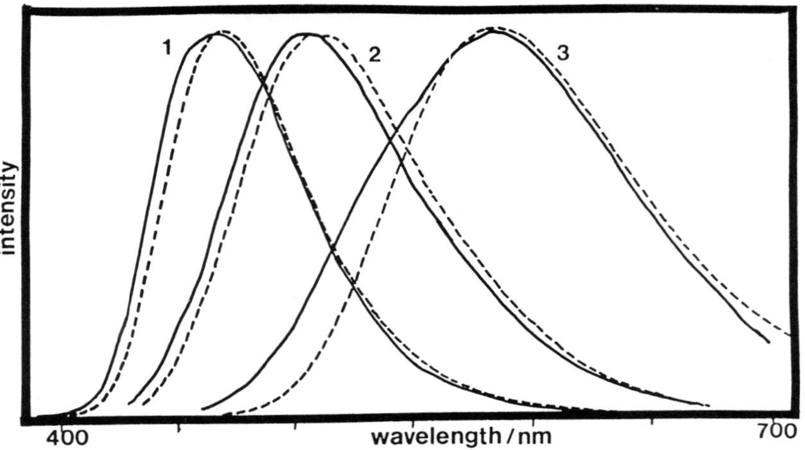

FIGURE 2.

Observed (solid lines) and simulated (broken lines) static fluorescence spectra for ADMA in a less polar (diethyl ether, 1), moderately polar (ethyl acetate, 2), and very polar (N,N-dimethylformamide) solvents. The peak intensities are normalized.

ADMA is an interesting excited state et example because the change in charge distribution associated with the LE $\rightarrow$ CT conversion is extraordinarily large (i.e. $\mu_{LE} \sim \mu S_0 \sim 1.3$ D and $\mu_{CT} \sim 5.5$ D).[11i-j] Consequently, the solvent sensitivity of the fluorescence spectrum of this compound is great, as discussed in the pioneering work on this molecule by Mataga and co-workers and other groups.[11]

2. A SOLVENT COORDINATE FOR ADMA

According to a linear response theory for the solvent coordinate, the non-equilibrium free energy dependence of the S_0, LE, and CT diabatic states are as follows.

$$F_{S_0}(z) = (1/2)k_s z^2. \tag{4}$$

$$F_{LE}(z) = (1/2)k_s z^2 + F_0. \tag{5}$$

$$F_{CT}(z) = (1/2)k_s(z-1)^2 + F_0 - \Delta F^0, \tag{6}$$

where k_s is the solvent force constant, E_0 is the energy of the spectroscopic transition between S_0 and LE states, and ΔF_0 is the equilibrium free energy change for the LE $\rightarrow$ CT charge separation. z is the solvent coordinate. Figure 1 shows plots of these functions employing empirically determined k_s, F_0, and ΔF_0.

Based on the diabatic model there are two types of optical transitions, namely

$$S_0 \rightarrow \text{LE } (\pi\pi^*), \text{ and} \tag{7}$$

$$S_0 \rightarrow \text{CT (a charge transfer absorption)}. \tag{8}$$

As stated, the former corresponds to a strongly allowed transition, e.g. $^1A_{1g} \rightarrow {}^1B_{2u}$ of anthracene with typical $\varepsilon_{max} \sim 10^4$.[11a] The latter band should be significantly weaker since it involves a transfer of charge between the two rings.

3. EXPERIMENTAL RESULTS

The static fluorescence spectra of ADMA are shown in Figure 2 by solid lines. These spectra reflect the relaxed excited state of ADMA since the excited state photodynamics are over on the order of picoseconds[11c,m], while the lifetime is many nanoseconds.[11a] The fluorescence band width and fluorescence Stokes shift increase dramatically as the solvent polarity is increased. These are well known manifestations of charge transfer fluorescence bands. It is reasonable to assign the relaxed form of the excited state of ADMA to CT in polar solvents.

The absorption spectra in various solvents are considerably more complex as shown in Figure 3. The $\lambda < 380$ nm region is almost similar to anthracene derivatives for which charge separation is not energetically feasible, except for a slight broadening in the vibronic

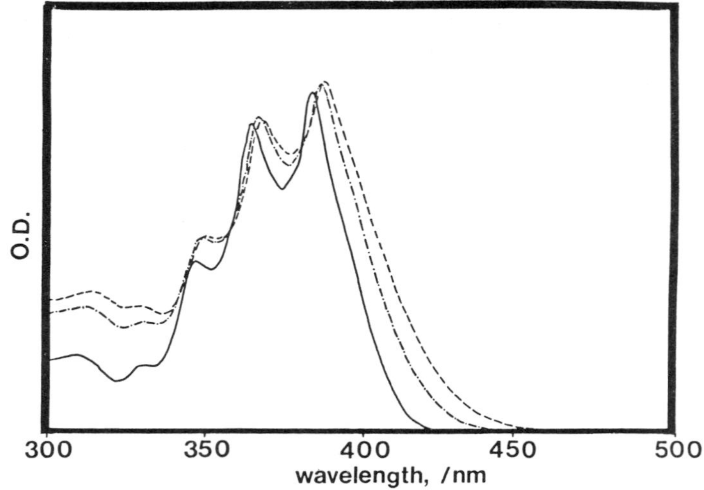

FIGURE 3.

Steady-state absorption spectra of ADMA in nonpolar (3-methylpentane, solid line), moderately polar (1.4-dioxane, broken-dotted line), and very polar (N,N-dimethylformamide, broken line) solvents.

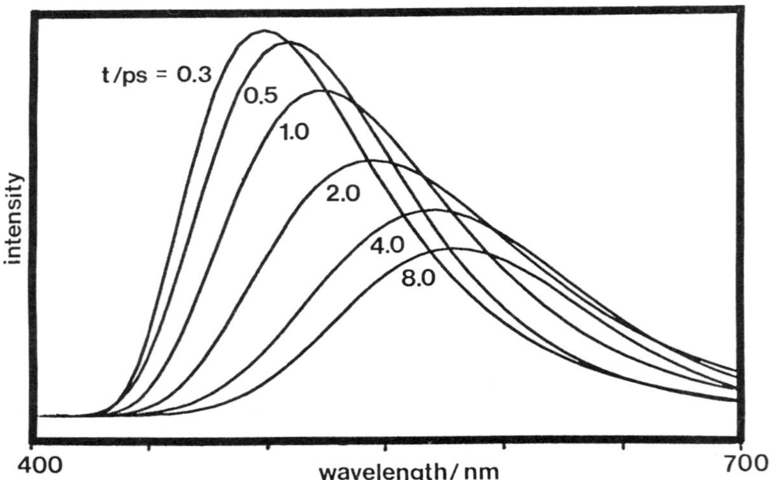

FIGURE 4.

Time-resolved emission spectra of ADMA in N,N-dimethylformamide. The curves are log-normal function fits to the data. The times in the figure represent the delay times after the laser excitation.

structure which is more apparent in polar solvent. In contrast, the absorption spectrum in the 380 - 450 nm region has a very prominent and peculiar "red tail", which has been noted previously.[11a,i] We will show below that the red tail is a consequence of $S_0 \rightarrow$ LE and $S_0 \rightarrow$ CT.

Figure 4 portrays time-dependent fluorescence spectra of ADMA in the polar solvent (N,N'-dimethylformamide, DMF) in the time range 0.3 - 8 picoseconds (ps). The dynamics at early times (< 0.3 ps) will be discussed below. At times longer than 8 ps the spectra are indistinguishable from the relaxed spectrum, except for a slow excited state population decay.

The special dynamics in Figure 6 exhibit the features of dipolar solvation, as observed in the coumarins. The spectra are broad, and the emission maximum evolves continuously, i.e. a transient Stokes shift. We have analyzed the data by the transient Stokes shift method, using the solvation energy relaxation function C(t).

The experimental average C(t) relaxation time of ADMA is very similar to the average solvation time obtained by transient Stokes shift measurements on coumarins in this solvent, $\langle\tau_s\rangle = 1.0 - 1.5$ ps.[1.b]

In summary, the transient fluorescence data at $t > 0.3$ ps have the appearance and dynamic behavior of a charge transfer band undergoing the transient Stokes shift phenomenon. On Figure 1 this would correspond to an initially excited distribution of molecules in the CT state near $z = 0$. The transient Stokes shift we observe corresponds to the relaxation of the initially excited CT molecules toward the equilibrium position in CT, namely $z = 1$.

The static and dynamic spectroscopy of ADMA gives clear evidence of strong mixing between the nonadiabatic LE and CT states. The integral of each spectrum in Figure 5 can be related to the oscillator strength f in each solvent. The oscillator strength decreases regularly as the solvent polarity is increased. This is due to an increase in the fraction of CT character in the excited state. It is important to note, however, that the increase in CT character is not simply due to a solvent dependent equilibrium between two weakly coupled states, i.e. LE and CT. If a weakly coupled model were correct, then a separate LE emission band would be apparent in the various solvents. In fact, a LE band is only apparent in the least polar solvent, 3-methylpentane.

4. AN ADIABATIC LE/CT MODEL

The electronic coupling between LE and CT can be treated approximately in terms of two diagonalized states, S_1 and S_2,

$$|S_1(z)\rangle = C_{LE}^{(1)}(z)|LE\rangle + C_{CT}^{(1)}(z)|CT\rangle \tag{9}$$

$$|S_2(z)\rangle = C_{LE}^{(2)}(z)|LE\rangle + C_{CT}^{(2)}(z)|CT\rangle \tag{10}$$

where $C_{LE}^{(i)}$ and $C_{CT}^{(i)}$ are coefficients of the LE and CT diabatic states, respectively, and i signifies the adiabatic state, i = 1 or 2.

Figure 5 portrays energy profiles for ADMA S_0, S_1, and S_2 in DMF. The parameters have been adjusted to optimize the agreement between simulated and observed spectra, see Section 5.2.

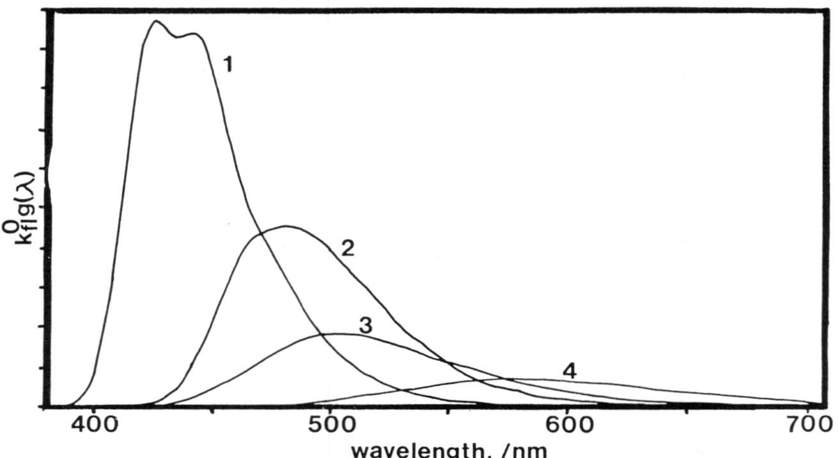

FIGURE 5.

The observed static fluorescence spectrum of ADMA in nonpolar (3-methylpentane, 1), less polar (ethyl ether, 2), moderately polar (ethyl acetate, 3), and very polar (N,N-dimethylformamide, 4) solvent. The vertical axis is $k_{fl}g(\lambda)$, where k_{fl} is the radiative rate constant of the fluorescence, and $g(\lambda)$ is a normalized fluorescence shape function.

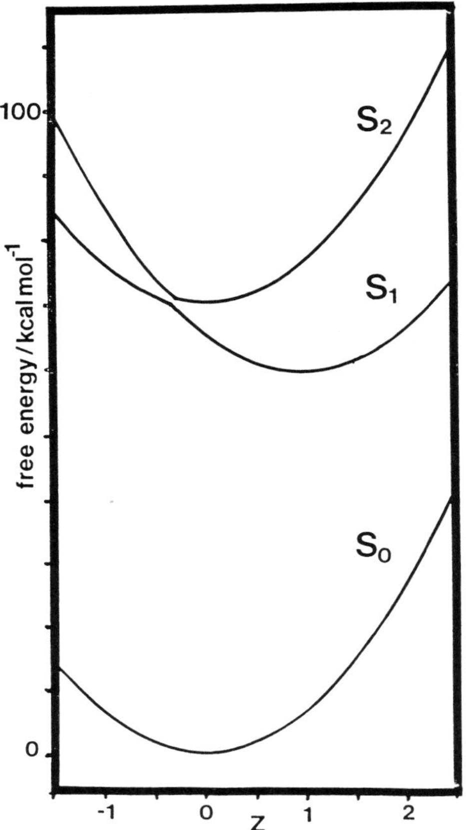

FIGURE 6.

Theoretical estimates for the adiabatic free energies of the ground state (S0) and the two excited states (S_1 and S_2) of ADMA in N,N-dimethylformamide as a function of the solvent coordinate. The free energy parameters were optimized to simulate the static absorption and fluorescence spectra simultaneously. The obtained values are k_s = 13.0

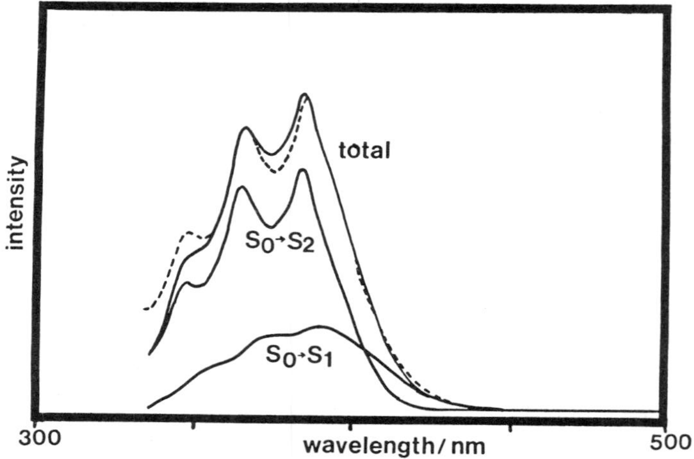

FIGURE 7.

Simulated static absorption spectrum of ADMA in ethyl acetate. The broken line is the observed spectrum. "S_0-S_1" and "S_0-S_2" correspond to the transitions from S_0 to S_1 and S_0 to S_2, respectively. "total" signifies the sum of the two absorption bands.

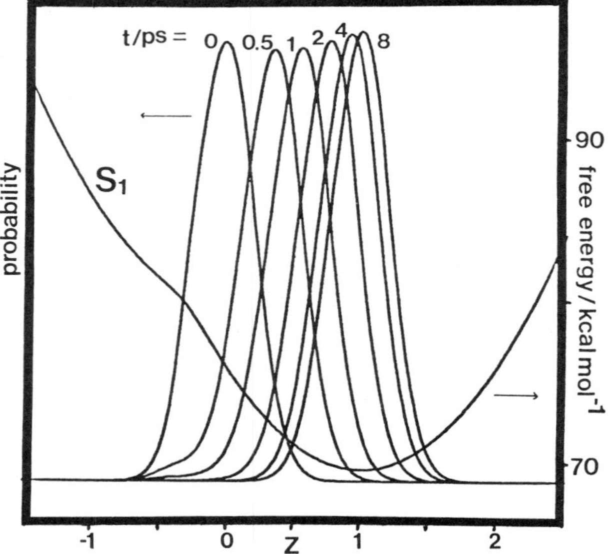

FIGURE 8.

Time-dependent probability distribution function, p(zt), for the excited state electron transfer on the S_1 surface of ADMA in N,N-dimethylformamide obtained from a generalized Smoluchowski equation. A two exponential function of Δ (t) (eqn 6-7) with $A_1 = 0.55$, $A_2 = 0.45$, $\tau_1 = 0.75$ ps, and $\tau_2 = 2.5$ ps was used for the solvation dynamics. The S_1 potential surface is also shown in the figure.

The potential of Figure 5 can be used in the procedure of Kang *et al.* [9] to predict the absorption and emission spectra I(ν) of charge transfer compounds like ADMA in polar solvents. The approach is based on a classical model for the solvent coordinate.

Figure 2 compares observed emission (solid lines) spectra of ADMA in various solvents to simulated spectra employing a set of best-fit parameters. The electronic coupling $H_{LE,CT}$ is set at 0.8 $kcalmol^{-1}$ for all the solvents as a parameter used to simulate the absorption and emission spectra in all the solvents. Thus, in the simulation for each solvent ΔF^0 and k_s are the parameters adjusted to simultaneously fit the absorption and emission spectra of ADMA, as shown in Figure 7. The simulated spectrum ("total" solid line) is compared to the observed spectrum (dashed line). Note that the simulated absorption spectrum is the sum of two components, $S_0 \rightarrow S_1$ and $S_0 \rightarrow S_2$ transitions, which are largely $S_0 \rightarrow$ CT like and $S_0 \rightarrow$ LE like, respectively.

The simulated spectra agree reasonably well with the observed spectra. This seems to indicate that the basic concepts of the Kang *et al.* model[9], a classical solvent coordinate and an aidabatic model for the LE/CT interactions, are valid.

Interestingly, that the simulation in Figure 7 agrees with the experiment in both the structured and the red tail regions of the spectrum. The different components of the spectrum ($S_0 \rightarrow S_1$ and $S_0 \rightarrow S_2$) reveal that optical excitation near the maximum of the spectrum (in fact, the laser induced absorption region, ~ 396 nm) excites both adiabatic excited states.

5. DYNAMICAL SIMULATIONS AND THE PHOTODYNAMICS

The results of the previous two sections have important implications for understanding of photodynamics of ADMA and related molecules. Foremost, the analysis of the absorption spectrum reveals that optical excitation at the wavelength of our femtosecond laser, produces both S_1 and S_2 population, in a roughly 1/2 ratio (see Figure 7). This provides novel access to the mechanistic understanding of excited state intramolecular et reactions.

It follows from these arguments that there are two distinct et processes that should be considered, namely

$$S_2(z\approx 0) \xrightarrow{k_{et}(S_1\rightarrow S_2)} S_1(z\approx 0), \tag{11}$$

which is a diabatic with respect to the S_1/S_2 representation, and

$$S_1(z\approx 0) \xrightarrow{k_{et}(S_1\ \text{adiabatic})} S_1(z\approx 1) \tag{12}$$

which is an adiabatic et process on the S_1 surface. The latter process resembles the Kang *et al.* model for the et in BA.[9] The rapid rate of the $S_2 \rightarrow S_1$ et process is extraordinary! It exceeds the average solvation time of DMF ($\langle\tau_s\rangle$ = 1.0 - 1.5 ps) by over an order of magnitude. It is less surprising when one takes the point of view that this et process resembles an internal conversion on a large polyatomic molecule with a small S_2/S_1 gap. In the presence of a stationary configuration of the polar solvent molecules around ADMA (which is appropriate at very short times) the S_2/S_1 energy gap is $\approx$1400 cm^{-1}, for z = 0, the most probable value of z immediately after excitation by the laser pulse, and for an internal conversion process with such a gap the rapid rate is reasonable.

We now consider a quantitative model for predicting the transient emission of ADMA. The model is based on the adiabatic potentials of Figure 6. Employing the spectroscopic model, we can predict easily the initial probability distribution of $S_1(\rho_{S1}(z,t=0))$ and $S_2(\rho_{S2}(z,t=0))$. The next stage in the simulation is to add the S_2 population to the S_1, which is based on the assumption that $k_{et}(S_2 \rightarrow S_1)$ is unresolvably rapid on the time scale of our measurements.

The next stage in the simulation is to calculate the evolution of $\rho_{S1}(z,t)$ due to "polarization diffusion" along the reaction coordinate. The results of this simulation are shown in Figure 8 along with an expanded region of the S_1 potential.

The predicted evolution of the probability distribution $\rho_{S1}(z,t)$ is represented in Figure 8 superimposed on a plot of the S_1 potential, $F_{S1}(z)$. The initial coefficients for the simulation which corresponds to the instantaneous distribution after the unresolvably rapid $k_{et}(S_2 \rightarrow S_1)$ process. The evolution in S_1 of ADMA is simultaneously an et process and solvation process. The et character of the evolution of the probability distribution is emphasized by considering the dependence of $C_{LE}^{(i)}$ on the solvent coordinate z, as shown in Figure 9. Note that the LE "character" of S_1 varies as z varies because of the strong mixing ($H_{LE,CT}$) and the avoided crossing of the CT and LE states.

We now turn to the simulation of the fluorescence dynamics. Figure 10 shows a simulation of fluorescence using the probability distribution functions of Figure 8. For times t > 0.3 ps the simulation is qualitatively very similar to the experimental spectra in Figure 4, although the band maximum, ν_{max}^{fl}, at t = 0.3 ps is at 469 nm, a shorter wavelength than that in the observed spectrum (498 nm).

The similarity of the evolution in Figure 8 and a simple solvation process has been noted above. For a simple solvation process, the potential would be exactly harmonic according to linear response theory. Furthermore, the average solvation time would be exactly $\int_0^\infty dtC(t)$, i.e. $<\tau>$. In the present case, $<\tau_s>$ is an input to our simulation from solvation dynamics measurements, and $<\tau_s> = 1.5$ ps.

The agreement between experiment and simulation is strong supporting evidence for the two assumptions: First, $k_{et}(S_2 \rightarrow S_1)$ must be more rapid than ~1/(150 fs) because 433 nm is in the region where we should have observed any long lived S_2 emission. Second, after the rapid internal conversion, the et reaction is solvent controlled.

6. SUMMARY

The photodynamics of ADMA has been investigated by ultrafast fluorescence measurements in order to explore the role of the solvation dynamics in charge separation processes of an excited states of the organic electron donor and acceptor molecules with a strong driving force (ΔF_0). All manifestations of the et process, including static absorption, static emission, and time-resolved emission spectra, can be rationalized in terms of an extension of the strongly adiabatic model of Kang *et al.* [9] The solvent reorganizational effects, rather than the intramolecular vibrational motion, predominantly

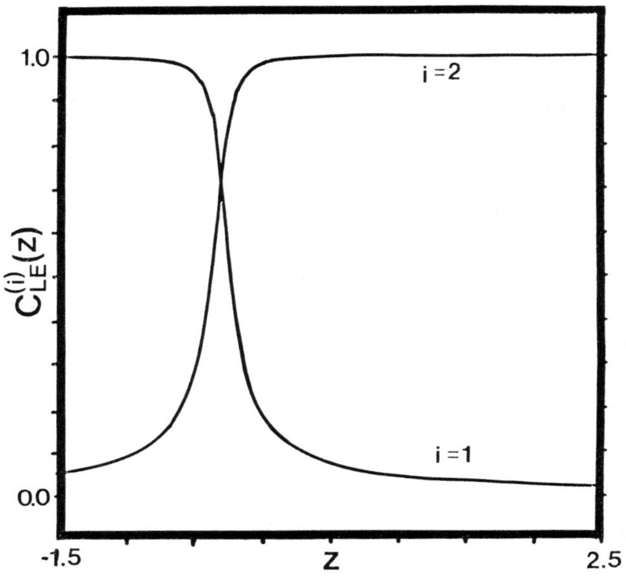

FIGURE 9.

Coefficients of the LE state character in the adiabatic (i = 1 or 2) states as a function of the solvent coordinate.

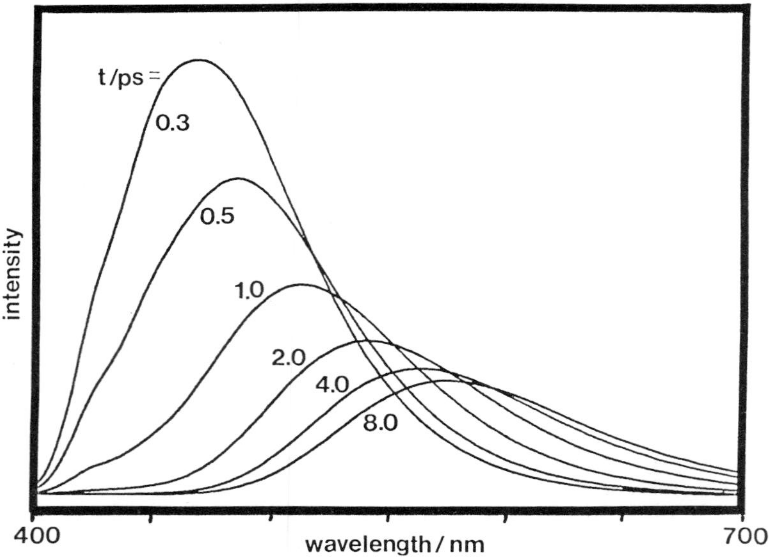

FIGURE 10.

Simulated time-resolved fluorescence spectra for ADMA in N,N-dimethylformamide obtained from the probability distribution function shown in Figure 14. The free energy parameters are as follows: k_s = 13.0 kcalmol^{-1}, ΔF^0 = 10.5 kcalmol^{-1}, and $H_{LE,CT}$=0.8 kcalmol^{-1}. A two exponential function of $\Delta(t)$ (eqn 6-7) with A_1 = 0.55, A_2 = 0.45, τ_2 = 2.5 ps was used to represent the solvation dynamics.

cause the variation of static absorption and emission spectra as a function of solvent polarity. The simulation of the static spectra shows that the excited state et in polar solvents is in the Marcus inverted regime.

ACKNOWLEDGEMENT

This work was supported by the Office of Naval Research. KT and PFB would like to thank Prof. N. Mataga for helpful discussions.

REFERENCES

1 (a) J.D. Simon, Acc. Chem. Res. 21 (1988) 128. (b) P.F. Barbara, W. Jarzeba, Adv. Photochem. 15 (1990) 1. (c) M. Maroncelli, J. MacInnis, G.R. Fleming, Science 243 (1988) 1674.
2 J.T. Hynes, J. Phys. Chem. 90 (1986) 3701.
3 L.D. Zusman, Chem. Phys. 49 (1980) 295. ibid. 80 (1983) 29. ibid. 119 (1988) 51.
4 B. Bagchi, G.R. Fleming, J. Phys. Chem. 94 (1990) 9.
5 M.J. Weaver, G.E. McManis III, Acc. Chem. Res. 23 (1990) 294.
6 I. Rips, J. Klafter, J. Jortner, J. Phys. Chem. 94 (1990) 8557.
7 H. Sumi, R.A. Marcus, J. Chem. Phys. 84 (1986) 4894.
8 J. Jortner, M. Bixon, J. Chem. Phys. 88 (1988) 167.
9 T.J. Kang, W. Jarzeba, P.F. Barbara, T. Fonseca, Chem. Phys. 149 (1990) 81.
10 T.J. Fonseca, Chem. Phys. 91 (1989) 2869.
11 (a) T. Okada, T. Fujita, N. Mataga, Z. Phys. Chem. N.F. 101 (1976) 57. (b) T. Okada, N. Mataga, W. Baumann, A. Siemiarczuk, J. Phys. Chem, 91 (1987) 4490. (c) N. Mataga, S. Nishikawa, T. Ashai, T. Okada, J. Phys. Chem. 94 (1990) 1443. (d) M. Gordon and W.R. Ware (eds.), The Exciplex, Academic, New York, 1975. (e) A. Siemiarczuk, Z.R. Grabowski, A. Krowczynski, M. Asher, M. Ottolenghi, Chem. Phys. Lett 51 (1977) 315. (f) Z.R. Grabowski, K. Rotkiewicz, A. Siemiarczuk, J. Luminescence 18/19 (1979) 420. (g) Z.R. Grabowski, K. Rotkiewicz, A. Siemiarczuk, D.J. Cowley, W. Baumann, Nouv. J. Chim. 3 (1979) 443. (h) A. Siemiarczuk, J. Koput, A. Pohorille, Z. Naturforsch. 37a (1982) 598. (i) W. Baumann, F. Petzke, K.-D. Loosen, Z. Naturforsch. 34a (1979) 1070. (j) N. Detzer, W. Baumann, B. Schwager, J.-C. Frohling, C. Brittinger, Z. Naturforsch. 42a (1987) 395. (k) A. M. Rollinson, H.G. Drickamer, J. Chem. Phys. 73 (1980), 5981. (l) V. Nagarajan, A.M. Brearley, T.J. Kang, P.F. Barbara, J. Chem. Phys. 86, (1987) 3183. (m) G.C. Walker, W. Jarzeba, T.J. Kang, A.E. Johnson, P.F. Barbara, J. Opt. Soc. Am. B. 7 (1990) 1521. (n) D. Huppert, P.M. Rentzepis, J. Phys. Chem. 92 (1988) 5466.
12 R.A. Marcus, N. Sutin, Biochem. Phys. Acta 811 (1985) 265. M.D. Newton, N. Sutin, Ann. Rev. Phys. Chem. 35 (1984) 437.

DISCUSSION

Faure

Did you apply Fokker–Planck formalism to describe the drift of the population along the potential surface in BA and ADMA?

Barbara

We used two alternative approaches. The generalized diffusion equation and the generalized Langevin equation (using a large mass, which leads to an overdamped behavior). These two approaches are closely related to the Fokker–Planck description.

Sumi

You have shown experimentally that the rate of electron transfer k_{et} can be larger than the relaxation rate of solvents $1/\tau_s$. It can be expected theoretically, too. Theoretically, however, whether $k_{et}>1/\tau_s$, or not is a boundary of whether time decay of reactants is nonexponential or single exponential, respectively. Have you checked this theoretical expectation?

Barbara

Yes, we do see nonexponential kinetics in some environments. We are now pursuing this in more detail.

Honma

I wish to ask about the electronic state of 9,9'–bianthryl (BA). I am working on the dynamics of BA in gas phase clusters with some polar molecules and found out that the electron transfer occurs even in low temperature clusters. Those results imply me that the coupling between locally excited state and the electron transferred state is very strong and, further, S_1 state could have already electron transferred character. You mentioned that in case of ADMA, S_1 and S_2 states both have ionic character and that character is enhanced with the solvent coordinate. Do you think that situation is same in BA?

Barbara

In polar liquids the most probable structure of the solvent around BA is a symmetric environment. In this situation the CT state is at higher energy than LE. Thus, if the solvent is frozen only LE emission is observed. Also, at early times in ordinary polar solvents, the emission has a large amount of LE character. On the other hand for BA at surfaces or in small clusters, the local environment is unsymmetrical and the CT state can be selectively stabilized relative to LE. This may be an explanation for your results.

Jortner

Until a short time ago we believed that an upper limit for electron transfer (ET) rates is determined by solvent-dynamics controlled ET, with the rate for an activationless process being given by $k_{ET} \sim \tau_L^{-1}(E_m/16\pi k_B T)^{1/2}$, where E_m is the medium reorganization energy. This conclusion rests on the assumption of weak variation of the nuclear Franck-Condon factors (intramolecular + solvent) in different systems. The observation of Barbara et al. that $k_{ET} >> 1/\tau_L$ may be related to the "molecular limit" for ET, which is dominated by coupling with high-frequency intramolecular donor and acceptor modes, while the role of the coupling to the solvent is negligible. This molecular limit for ET was also considered by Mataga.

Theories of intramolecular radiationless transitions in large molecules, which considered two classes of vibrational modes, expressed the rates in terms of a convolution of line shapes corresponding to the different nuclear modes. It will be interesting to extend this approach for the "molecular limit" of ET.

Barbara

I agree. The "molecular limit" of electron transfer may be relevant to a broad range of inverted regime electron transfer reactions. We look forward to theoretical progress on this problem.

Baumann

Could you fit our experimental dipole moments of ADMA (and some derivatives) and their solvent dependence to your data on varying CT contributions to the ADMA fluorescence spectra in differently polar solutions?

Barbara

We have not done this yet, but the theoretical simulations of my coworker Dr. Tominaga, could be modified in a simple way to calculate the "adiabatic" variation of the dipole moment of equilibrated S_1 ADMA as a function of different solvents. We plan to do this in the near future.

Mobius

Depending on solvation of the betain you have a different dipole moment. How does the variation of the dipole moment influence the electron transfer rate constant?

Barbara

Well, there is not a lot of information on the solvation dependence of the dipole moment. If the dipole moment is strongly dependent on the solvent (which it might be), it does not seem to greatly affect the kinetics. Roughly speaking, most of the solvent dependence of k_{et} can be ascribed to variation in the solvation time.

Kakitani

You have investigated the photoinduced ET rate of betaine by changing the solvent. When the solvent is changed, reorganization energy and free energy gap change in addition to the change of solvent relaxation time. How did you treat them in your calculations based on the Sumi–Marcus model and Jortner–Bixon model?

Barbara

We measured the absorption maximum $\tilde{\nu}_{max}$ of the betaine in the various solvents. Using $\tilde{\nu}_{max}$ we were able to extract the appropriate potential parameters from the experimental study of Kjaer and Ulstrup. These authors analyzed the CT absorption spectrum of betaine–26 and extracted a characteristic solvent force constant, an effective high frequency mode, and ΔG^{o} for various $\tilde{\nu}_{max}$.

Dynamics and Mechanisms of
Photoinduced Transfer and Related Phenomena
N. Mataga, T. Okada and H. Masuhara (Editors)

Solvent Dynamics and Charge Transfer Reactions

Barton B. Smith,[a] Hyung J. Kim,[a] Daniel Borgis[b] and James T. Hynes[a]

[a]Department of Chemistry and Biochemistry, University of Colorado, Boulder, Colorado 80309-0215, USA

[b]Laboratoire de Physique Théorique des Liquides,Université Pierre et Marie Curie,75252 Paris Cedex 05, FRANCE

Abstract

A brief account is given of some highlights of recent studies of the influence of solvation dynamics for assorted charge transfer reactions. The necessity of a generalized Langevin equation description for solvation dynamics is illustrated, for rapidly relaxing dipolar aprotic solvents. For such solvents, the solvent dynamically induced activation barrier recrossings leading to departures from a Marcus Transition State Theory description are quite different in character from those described in many recent theories. Instead a Grote-Hynes theoretical approach is appropriate. The related but quite distinct areas of S_N1 ionizations and proton transfers are also briefly described. Attention is focussed on both the aspects shared by, and the contrasts with, ET reactions.

1. INTRODUCTION

Outer sphere electron transfer (ET) reactions occupy a special place in the panoply of solution reaction classes. The reaction coordinate is a solvent coordinate, and the activation barrier owes its origin to the solvent. These key aspects have been described some time ago by Marcus [1] and Hush [2]. In particular, the formulation by Marcus for the ET reaction rate constant has seen widespread use and gained considerable acceptance [3]. In recent years, new aspects of the influence of the solvent on ET reaction rates have come under scrutiny. These include the role of the solvent dynamics in affecting the ET rate (in addition to the solvent free energetic barrier effects included in Marcus Theory). Here we give a brief account of some of our recent efforts[4-8] to elucidate these questions. We then turn to a very brief reprise of the related but actually quite distinct areas of S_N1 ionization [9,10] and proton transfer [11,12]. Here we focus on both the common and distinct features of these reactions compared to ET.

2. SOLVATION DYNAMICS

Before addressing the ET reaction problem, it is important to address the issue of solvation dynamics in the absence of a real reaction. To this end, we specialize for convenience to a model finite dipolar pair $A^{-1/2}A^{1/2}$, with fractional charges. Even though we do not address ET reactions just yet, it is useful to label that pair as "reactant" R and the symmetrically related pair $A^{1/2}A^{-1/2}$ as the "product" P. The solute is immersed in a solvent of finite dipolar molecules, which is [4] a caricature of methyl chloride, and is representative of several dipolar aprotic solvents currently under experimental investigation [13,14].

The relevant microscopic solvent coordinate is $\Delta E = H_R - H_P$ which is the difference in Coulomb interaction potential energy, for a given configuration of all the solvent molecules, between the solvent and the solute in the P and R charge states. (In a dielectric continuum solvent, this same energy difference would be expressed in terms of the orientational polarization of the solvent; see Sec. 3.). A convenient measure of the dynamics of ΔE is the normalized equilibrium time correlation function (tcf)

$$\Delta(t) = \langle(\delta\Delta E)^2\rangle^{-1} \langle\delta\Delta E \delta\Delta E(t)\rangle_R \ , \tag{1}$$

where $\delta\Delta E$ is the fluctuation $\Delta E - \langle\Delta E\rangle_R$ about the average when the solute is in the reactant state (cf. Fig. 1). There is a direct and appealing, although approximate, connection of $\Delta(t)$ to the solvation dynamics. If one imagines that in an electronic transition the dipole shift $A^{-1/2}A^{1/2} \rightarrow A^{1/2}A^{-1/2}$ is induced by laser radiation at t=0, and the "product" state $A^{1/2}A^{-1/2}$ radiates, then there will be a resulting time dependent fluorescence (TDF) shift in emission frequency as the solvent relaxes to equilibrium with the new solute charge distribution. This shift, normalized to unity at t=0, is proportional to $\Delta(t)$. This approximate relation is a linear response theory result, and has been employed in numerous studies [14]. A central point is that the MD simulated time dependence of $\Delta(t)$ shown in Fig. 1 is rather far from the simple exponential dependence which is predicted by dielectric continuum theories [15]. A convenient measure of the relaxation speed is the solvation time τ defined by the time area of $\Delta(t)$, tabulated in Table 1.

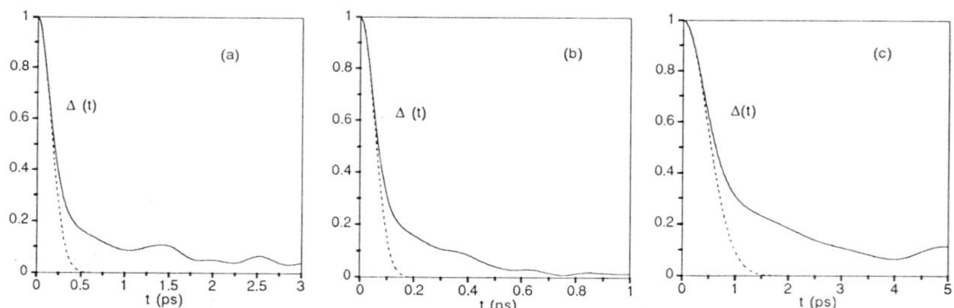

Figure 1. MD results[5] for $\Delta(t)$ for three different mass values. Case (a), all atomic masses = 40 amu; (b) 40 amu/9; (c) 9 x 40 amu. Dashed lines indicate the Gaussian approximation.

The next point of importance is that a significant fraction of the initial decay of $\Delta(t)$ is well described by the simple Gaussian behavior $\Delta_G(t) = \exp(-\omega^2 t^2/2)$ where the solvent frequency ω is given by [4,5]

$$\omega^2 = \langle(\delta\Delta E)^2\rangle^{-1}\langle(\delta\dot{\Delta E})^2\rangle \tag{2}$$

This significant feature, missing in almost all of the current dynamical theories of solvation dynamics, was first pointed out by Carter and Hynes [6]. The Gaussian decay arises physically from the initial simple inertial motion of the solvent molecules: At time t=0, those molecules will move with their translational and rotational velocities, governed by Maxwellian distributions, and change thereby the value of $\delta\Delta E$. Note that no intermolecular forces or torques govern that initial motion (although $\delta\Delta E$ itself depends on the

Table 1.
Dynamical Solvent Properties

System	τ(ps)	ω(ps^{-1})	ζ(t=0)(ps^{-2})	ζ(ps^{-1})
Light[a]	0.12	19	1680	45
Standard[b]	0.49	6.6	180	21
Heavy[c]	1.2	2.2	20	5.8

[a]m = (40/9) amu [5]; [b]m = 40 amu [4]; [c]m = (9 x 40) amu [5].

intermolecular interactions). From (2) and the equipartition theorem, the solvent frequency should scale as $m^{-1/2}$, which is corroborated in Table 1. This accounts, for example, for the much faster Gaussian decay for the lighter solvent in Fig. 1.

The influence of the intermolecular forces and torques is responsible for the longer time departures from the Gaussian behavior evident in Fig. 1 They and the shorter time dynamics can be described via a Generalized Langevin equation (GLE):

$$\delta\ddot{\Delta E}(t) = -\omega^2\delta\Delta E(t) - \int_0^t d\tau\zeta(t-\tau)\,\delta\dot{\Delta E}(\tau) \quad . \qquad (3)$$

(We suppress the "random force" term). This depicts a generalized Brownian motion of the solvent coordinate in, e.g., the reactant well whose frequency is ω, and $\zeta(t)$ is the time dependent friction on the solvent coordinate. In the simplest picture, each solvent molecule can be imagined to rotate and reorient its dipole moment and thus change $\delta\Delta E$; since such reorientational motion will be impeded by e.g., other solvent molecules, there will result a net damping on $\delta\Delta E$. The GLE was introduced to describe solvation dynamics in Ref. [16], in a dielectric continuum context (and has found a number of interesting applications and extensions since [17]). However, (3) is in fact exact for the $\delta\Delta E$ dynamics in $\Delta(t)$ [18].

Two more restricted approximations to (3) are of interest as simpler descriptions of the solvation dynamics. The first is a simple Langevin equation(LE), $\delta\ddot{\Delta E}(t) = -\omega^2\delta\Delta E(t) - \zeta\delta\dot{\Delta E}(t)$. Here the friction constant ζ , the full time integral of $\zeta(t)$, measures the overall, long-time influence of the damping of the $\delta\Delta E$ motion. The LE completely ignores the time dependent aspects of $\zeta(t)$ in the GLE, assuming that they are rapid compared to the time scale of the velocity $\delta\dot{\Delta E}$, so that the whole friction acts instantly. A further and more draconian approximation would be to assume that $\delta\dot{\Delta E}$ changes rapidly compared to $\delta\Delta E$ itself in such a way that the "potential" and frictional forces cancel, so that the net acceleration is negligible and can be ignored. This is a diffusive limit in which $\Delta(t)$ decays exponentially $\Delta(t) \approx \exp(-\omega^2 t/\zeta)$. This overdamped description is the most commonly employed characterization of solvent dynamics; in it, the solvation time is $\tau = \zeta/\omega^2$. In a dielectric continuum treatment [15], τ is proportional to the longitudinal dielectric relaxation time of the solvent. Fig. 2 shows that neither the LE nor the diffusive approximations--which differ little from each other--well portray the solvation dynamics. A full GLE description, including the time dependent friction, is required. Unfortunately there currently exists no analytic theory which tells us what the time dependent friction $\zeta(t)$ looks like. We therefore have to fall back on MD results to discover this.

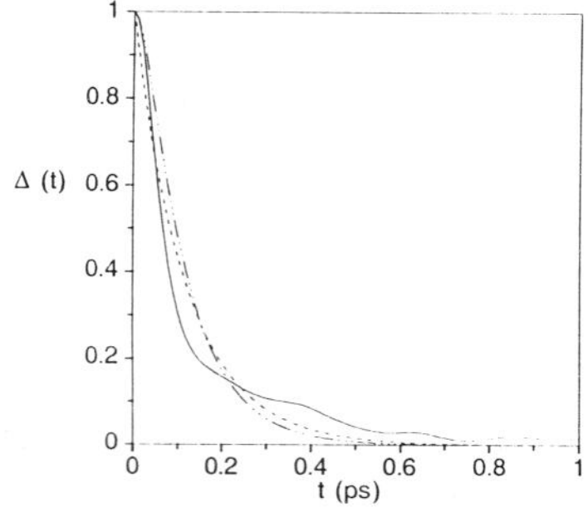

Figure 2. The Langevin equation (—•••—) and diffusive (- - -)approximations to $\Delta(t)$ (——) for the light mass case.

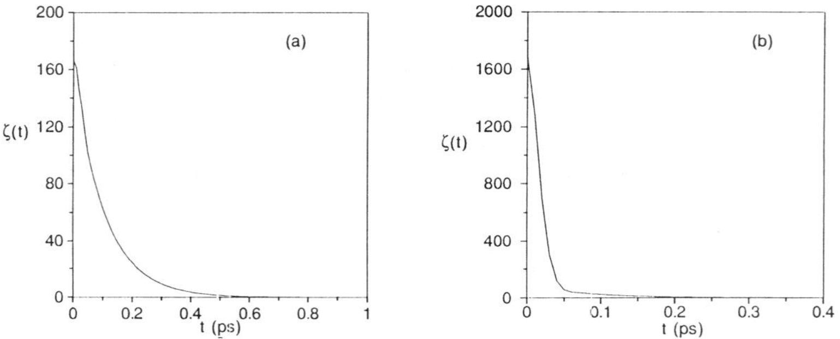

Figure 3. The time dependent friction coefficient $\zeta(t)$ for (a) the 40 amu mass case [4] and (b) the 40 amu/9 mass case [5].

The friction $\zeta(t)$ can be extracted [4-6,18] from $\Delta(t)$ and the GLE via Fourier transform techniques and is displayed in Fig. 3 for several mass cases. For the light solvent, the initial amplitude $\zeta(t=0)$ of the friction is very large, while it is very small for the massive solvent: $\zeta(t=0)$ scales [5] as $\zeta(t=0) \propto m^{-1}$. On the other hand, the "lifetimes" of $\zeta(t)$ vary in the opposite direction, e.g., dynamics governing the fluctuating "force" on the solvent coordinate are the most rapid for the lighter solvent. In consequence, the friction constant changes less rapidly than m^{-1} (cf. Table 1); since ω^2 scales as m^{-1}, this indicates that the solvation time $\tau=\zeta/\omega^2$ lengthens as the solvent becomes more massive. [We stress here that (3) holds [6,16] even with a GLE, even though $\Delta(t)$ is nonexponential.] At a more detailed level, it turns out[5,6,18] that the longer time "tail" in the tcf $\Delta(t)$ apparent in Fig. 1 originates from the tail in $\zeta(t)$, which is therefore essential in accounting for the bimodal time structure of $\Delta(t)$.

The conclusion to be drawn from these (and other [19]) results is that, for rapidly relaxing polar aprotic solvents, the relaxation dynamics of the solvent differs significantly from the exponential behavior expected from a dielectric continuum perspective, and a full GLE treatment is required. While further work is obviously needed to comprehend the molecular origins of the key features of the time dependent friction, even at the current stage this description can be usefully exploited to examine solvent dynamical effects for ET rates, which is our next topic.

3. ELECTRON TRANSFER REACTIONS

We now turn to the ET reaction problem. There has been a plethora of theoretical papers [16,20,21] dealing with the possible role of solvent dynamics in causing departures from the standard Marcus TST rate theory [1,2]. The first of these is due to Zusman [20] and for the most part, further works are variants on this. Zusman focussed on the solvent dynamics within each of the R and P wells, in the diffusive approximation so that the exponential decay Sec. 2 is assumed to hold in each well. The simplest form of Zusman Theory is for a cusped, i.e., sharply peaked, barrier reaction, for which the transmission coefficient $\kappa = k/k^{TST}$ measuring the departure of the rate constant k from its TST approximation is given by

$$\kappa_Z = (\omega\tau)^{-1}(\pi\Delta G^{\pm}_{na}/k_BT)^{1/2} \ . \qquad (4)$$

Here $\Delta G^{\pm}_{na}$ is the nonadiabatic barrier height, i.e., the free energy at the intersection of nonadiabatic R and P curves minus the free energy at the minimum of the R well, which has frequency ω. The picture here is that the rate limiting steps for the reaction are the diffusive climbing of the R well wall and the diffusive slide down the P well wall. The passage over the barrier top itself is supposed to be rapid and direct, i.e., there is no solvent-induced recrossing associated with the barrier top *per se*; instead, recrossings of the barrier supposedly arise from "reapproach" from the region of the wells before the bottom of either of the R or P wells is reached. Fig. 4 illustrates the concept.

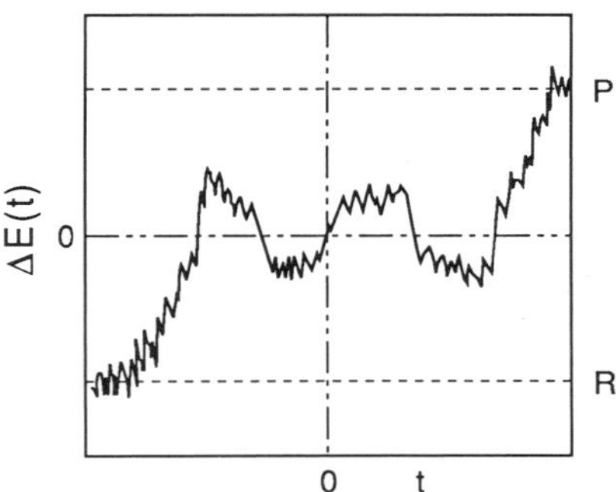

Figure 4. Schematic trajectories envisioned in the Zusman and allied theories.

The ET reaction considered is a simplified model, $A^{-1/2}A^{1/2} \rightarrow A^{1/2}A^{-1/2}$, in the solvent described above. The technical and computational rationales for this somewhat artificial fractional charge model are given in Ref. [4]; however, the model is sufficiently realistic to explicitly address the key dynamical issues.

The MD simulation of the reaction can be effected via the electronically adiabatic Hamiltonian [4] $H_{ad} = \frac{H_R + H_P}{2} - \frac{1}{2}\sqrt{(\Delta E)^2 + 4\beta^2}$,where $H_{R(P)}$ is the system Hamiltonian when the solute has the R(P) charge distribution and β is the electronic coupling. This is appropriate as representative for many adiabatic ET reactions; in addition, solvent dynamical effects are expected to be most pronounced in the electronically adiabatic limit [16]. With this Hamiltonian, the solute is always in its ground electronic state whatever the solvent configuration. The barrier is traversed as ΔE progresses from values appropriate to the neighborhood of equilibrium with R to those similarly appropriate to P. The electronic charge character evolves smoothly from that of R, through the transition state distribution A^0A^0 (which is a neutral pair), on to the P charge distribution. Reaction free energy profiles are illustrated in Fig. 5.

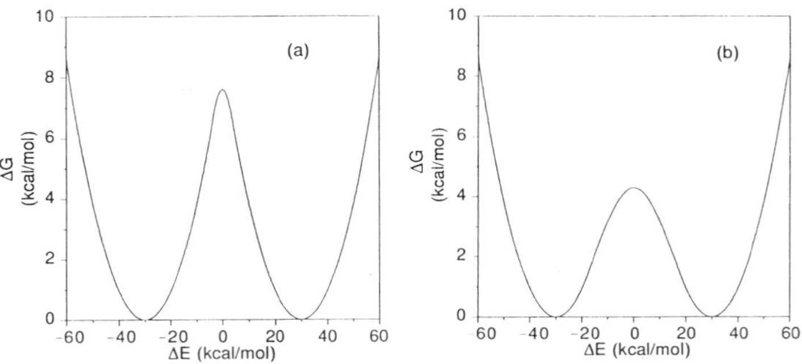

Figure 5. Free energy profiles [4]. (a) $\beta = 1$ kcal/mol; (b) $\beta = 5$ kcal/mol.

The transmission coefficient κ can be directly computed in an MD simulation for the ET reaction by procedures described elsewhere involving special constraint dynamics techniques,[4] as can the various quantities in κ_Z. The results are displayed in Table 2 for two choices of the electronic coupling β. The first important point is that for $\beta = 1$ kcal/mol, κ is quite close to unity; there are few recrossings of the barrier and the Marcus TST Theory is thus an excellent approximation.

Table 2 also shows that κ_Z gives very poor results for the $\beta = 1$ kcal/mol case. This can be improved by realizing that the rate of unidirectional crossing the barrier top itself is not infinitely fast, but rather can be estimated from the Marcus TST formula. This gives [4,16] $\kappa_Z' = (1 + \kappa_Z)^{-1}\,\kappa_Z$ and, as Table 2 shows, predicts transmission coefficients much closer to the MD values. But, despite this apparent success, there is still a serious problem. For, with either equation, the barrier recrossings responsible for the transmission coefficient are supposed to arise from trajectories as in Fig. 4, i.e., arising from individual single recrossings subsequent to erratic motion in the R and P wells before the trajectory settles down into the bottom of either well. Fig. 6 shows that the MD computed trajectories are not at all like this. Instead, the recrossings that occur are completely associated with the barrier top region. Any trajectory that makes it into the P (or R) well proper in fact continues on to the bottom of that well without returning to the transition state and recrossing it.

Table 2.
ET Transmission Coefficients

	β = 1 kcal/mol			β = 5 kcal/mol
System[a]	κ_{MD}	κ_Z	κ'_Z	κ_{MD}
Light[5]	0.87 ± 0.09	3.0	0.75	0.64 ± 0.15
Standard[4]	0.95 ± 0.04	2.2	0.68	0.59 ± 0.11
Heavy[5]	0.86 ± 0.10	2.6	0.72	0.65 ± 0.15

[a]see Table 1 for explanation.

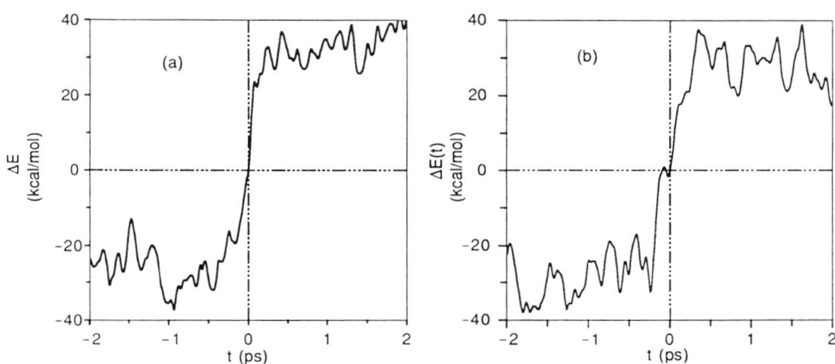

Figure 6. MD computed trajectories [5] for the light mass case for β = 1 kcal/mol.

When the coupling is increased to β = 5 kcal/mol and the barrier becomes more rounded, Table 2 indicates that the transmission coefficient is smaller and there are noticeable, though not gigantic, departures from the Marcus TST theory. The character of the barrier recrossings is illustrated in Fig. 7. Again, these recrossings occur in the immediate neighborhood of the transition state, and not in the wells. It is not appropriate to apply the Zusman approach or its variants (which assume cusped barriers) to this smooth barrier situation.

For each of the β coupling value ET reactions above, the barrier recrossings occur exclusively in the barrier top region. A theory for just such dynamical rate effects--which has proved highly successful for a wide variety of MD-simulated chemical reaction types [4,10,22] and in the interpretation of experimental rates [23]--is Grote-Hynes Theory [24]. In this theory, the transmission coefficient is given by

$$\kappa_{GH} = \left[\kappa_{GH} + \omega_b^{-1}\int_0^{\infty} dt e^{-\omega_b \kappa_{GH} t}\, \zeta^{\ddagger}(t)\right]^{-1} , \qquad (5)$$

where ω_b is the barrier frequency and $\zeta^{\ddagger}(t)$ is the time dependent friction coefficient at the barrier top. In the present context, GH Theory is based on the assumption that the GLE

$$\ddot{\Delta E}(t) = \omega_b^2 \Delta E(t) - \int_0^t d\tau \zeta^{\ddagger}(t-\tau)\dot{\Delta E}(\tau) \tag{6}$$

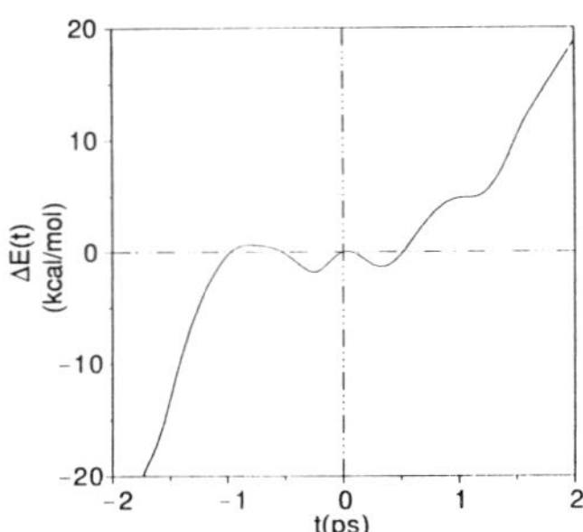

Figure 7. Typical barrier recrossing for the β = 5 kcal/mol, heavy mass case [5].

holds in the vicinity of the barrier top $\Delta E = 0$.[4,12] The barrier frequency can be estimated[4] from the formula $\omega_b = \omega[(2\Delta G^{\ddagger}/\beta) - 1]^{1/2}$. The friction $\zeta^{\ddagger}(t)$ is more difficult to obtain than the well friction, but Zichi et al. [4]. have argued that it can be approximated by the time dependent friction for the reference situation of a neutral pair. This approximate identification derives from the observation, noted above, that in the ET reaction, the transition state charge distribution is that of a neutral pair. The neutral pair friction $\zeta_{NP}(t)$ can be extracted [4,5,18] from the TDF studies of Carter and Hynes [6], and with the approximation $\zeta^{\ddagger}(t) = \zeta_{NP}(t)$, Zichi et al. [4]. showed that κ_{GH} could be estimated for the ET reaction. For example, the results [4] for the standard mass case are $\kappa_{GH} = 0.94$ and $\kappa_{GH} = 0.74$ for β = 1 kcal/mol and β = 5 kcal/mol respectively. These agree within the error bars with the MD simulation values in Table 2.

All these results suggest that the conventional dynamic approaches to ET in dipolar aprotic solvents based on well relaxation miss the correct picture, even for fairly cusped barrier reactions. Instead, it is the solvent dynamics occurring near the barrier top, and the associated time dependent friction, that are the critical features. It is however possible that, for cusped barrier ET reactions in much more slowly relaxing solvents, the well dynamics could begin to play an important role.

4. NONADIABATIC PROTON TRANSFER

We now turn to the case of nonadiabatic proton transfer reactions in solution. We consider a model symmetric proton reaction in an inter- or intramolecular H-bonded complex AH - - A $\rightleftharpoons$ A - - HA. The reactive system is described by two coordinates q_H and Q. q_H is a proton coordinate, e.g., the displacement of H from the center of the A—A bond, while Q represents the A—A distance. The reactive molecule is supposed to be immersed in the model polar solvent of Secs. 2 and 3.

Our description relies on a Born-Oppenheimer approximation [11,12] for which the time scale for the protonic motion is short compared to those of the solvent molecules' motion and of the intramolecular vibrations. For a fixed Q-coordinate and any fixed positions and orientations of the solute and of the solvent molecules (characterized by the general coordinate S), the motion of the proton is described by an electronically adiabatic double-well potential $V(q_H;Q,S)$. The shape of this potential, including the barrier height and the relative asymmetry of the wells, evolves as the heavy particles move. The proton ground-state wave functions in the reactant (R) and the product (P) wells and the associated vibrational levels, $\langle \psi_{R,P} | H | \psi_{R,P} \rangle$, thus depend parametrically on the intramolecular vibration and on the solvent.

For weak or intermediate strength H-bonded systems, the potential barrier in q_H separating the R and P wells will be always high compared to k_BT, and the proton transfer can be described as quantum-tunneling. In this case, the overall reaction process including the solvent is in the "nonadiabatic" limit for the proton (in the standard terminology of charge-transfer reactions). In this perspective, the reaction is a quantum proton tunneling modulated by the Q vibration and the solvent fluctuations, with an activation occurring in the latter degrees of freedom (but not, as more conventionally viewed, in the proton coordinate) [11,12]. The Q vibration accelerates the reaction in a compression by lowering the barrier (Fig. 8) and concomitantly increasing the tunneling probability. (This feature is unimportant for short range ET [11,12].) One influence of the solvent is to attain via fluctuations the symmetric double well of the intrinsic proton transfer system, at the cost of free energy (Fig. 8); this is closely related to the corresponding solvent effect for ET reactions [1,2].

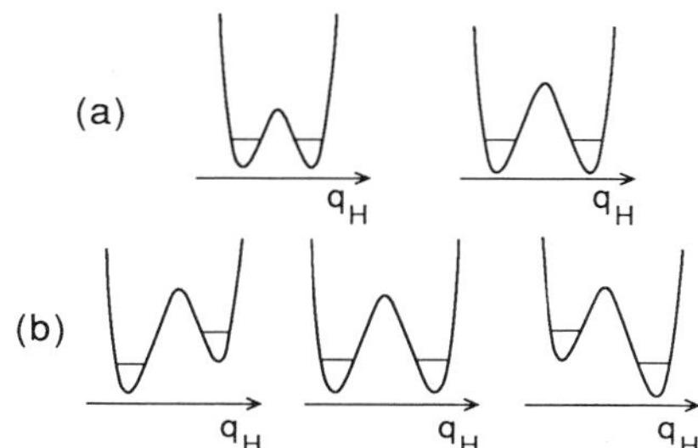

Figure 8. (a) Schematic proton transfer potentials vs. proton coordinate q_H, for different values of the vibrational coordinate δQ. The ground diabatic proton vibrational levels are indicated. (b) Proton transfer potentials for different solvent configurations, illustrating solvent-induced asymmetry. A symmetric double well occurs for certain activated solvent configurations.

Starting from the Born-Oppenheimer approximation and the Fermi golden rule, it may be shown [11,12] that the rate constant can be written as the integral of a time-dependent flux correlation function [x,x],

$$k = \int_0^\infty dt \, \langle jj(t) \rangle = (2/\hbar^2) \, \mathrm{Re} \int_0^\infty dt \left\langle C(0) \exp_{(-)} \left[i/\hbar \int_0^t d\tau \, \Delta H(\tau) \right] C(t) \right\rangle . \tag{7}$$

Here j(t) denotes a time-dependent H^+ probability flux, C(t) and ΔH(t) are the time-dependent coupling and splitting, and $\exp_{(-)}$ designates the negative time-ordered exponential. The solvent is treated classically. Finally, the equilibrium average is taken in the reactant state for the proton, i.e., with a potential energy for the heavy atoms defined by $\langle \psi_R | H | \psi_R \rangle$. The Q vibration has the primary effect of parametrically modulating the barrier height of this potential and therefore the coupling between the proton ground state in the R and P wells, i.e., $C(Q,S) = \langle \psi_P | H | \psi_R \rangle$. The solvent mainly parametrically modulates the asymmetry of the double well and thus the splitting of energy between the R and P ground-state levels, i.e., $\Delta H(Q,S) = \langle \psi_P | H | \psi_P \rangle - \langle \psi_R | H | \psi_R \rangle$.

This rate constant formula can be treated via cumulant expansion, and the rate is then reduced to the time integral of the exponential of certain integrals of the tcf's of the solvent variable ΔE and the vibrational coordinate δΔQ. Details may be found in Ref. [11,12].

In the model for simulation, the two A atoms of the reactive molecule, each of mass 40 amu, are represented by two Lennard-Jones spheres maintained at a fixed distance of 3 Å and bearing negative partial charges at their centers (q = 0.4 e). The proton is represented by a positive point charge (q = 0.8 e) which can be either in the reactant or product site at a fixed bond distance of 1 Å from one or the other A atom. The reactive molecule and the solvent molecules interact via Lennard-Jones and Coulomb potentials [11].

Fig. 9 shows the calculated MD free energy curves for ΔE for the reactant and product. The harmonic character apparent here has also been discussed for other examples of simulated charge transfer processes [6,25]. It is worth noting that the solvent force constant $k_S = k_B T / \langle \delta \Delta E^2 \rangle$ is more than an order of magnitude smaller than for the comparably sized ET solute pair in the same solvent (Sec. 3). This reflects the more delocalized, three-center charge distribution character in the proton transfer case, resulting in a weaker solute-solvent interaction.

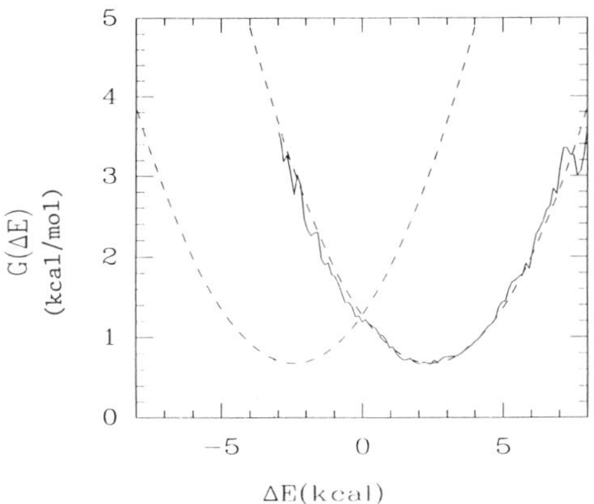

Figure 9. Free energy curves for proton transfer. Solid line is computed; dashed line is inferred by symmetry [11]. ΔE = ΔH evaluated at the equilibrium vibrational coordinate value

The rate constant formula can be approximately evaluated analytically [12] and compared to the MD results. The agreement is very good [11]. These results are discussed at length in Refs. [11] and [12], but a simple illustration can be given here. The basic structure of the analytic results, in a common regime, is $k \propto \langle C^2 \rangle \exp(-\Delta G^{\ddagger}/k_BT)$. Here $\langle C^2 \rangle$ is the thermal average of the square coupling, which contains the H^+ tunneling factors and kinetic isotope effects, while a significant contribution to the activation free energy is due to the reorganization of the solvent to allow the symmetric proton double well configuration. It is also interesting to observe that the solvent dynamics are irrelevant for this k result; instead, the distribution of ΔE values due to differing solvent configurations is important [11].

5. S_N1 IONIC DISSOCIATION

We finally consider an S_N1 ionic dissociation $RX \rightarrow R^+X^-$ in a polar solvent. In the gas phase, due to the typical magnitudes of the ionization potential of R for organic species versus the magnitude of the electron affinities of, for example, the atomic halogens, the covalent curve is more stable than the ionic curve. In a polar solution, however, RX dissociates to ions; the basic reason for this, as qualitatively explained long ago [26], is the stabilization of the ionic state by the polar solvent.

To analyze the solution reaction [9], we write the electronic wave function as a linear combination of the pure covalent and pure ionic (charge transfer) states $\Psi = c_C\psi_C(RX) + c_I\psi_I(R^+X^-)$, where the wave functions ψ_i and state coefficients c_i depend parametrically on the RX separation r. In this basis, the vacuum Hamiltonian matrix is

$$H^0 = \begin{pmatrix} H^0_C(r) & -\beta(r) \\ -\beta(r) & H^0_I(r) \end{pmatrix} , \tag{8}$$

where $\beta(r)$ is the vacuum electronic coupling. These quantities are displayed for a model of t-butyl chloride in Fig. 10. This system is analyzed by the nonlinear Schroedinger equation approach developed by Kim and Hynes for ET processes [7,8]. We do not enter the details of the analysis [9] here, but simply note that in our formulation adapted to the S_N1 case, it is convenient to introduce a solvent coordinate s via the solvent orientational polarization representation

$$\vec{P}_0(\vec{r}) = s\frac{1}{4\pi}\left(\frac{1}{\varepsilon_\infty} - \frac{1}{\varepsilon_0}\right)\varepsilon_\infty \vec{E}[\psi_I] , \tag{9}$$

such that $s = 0$ corresponds to equilibrium of $\vec{P}_0$ with the pure covalent state (assumed to have no dipole moment), and $s = 1$ to equilibrium with $\psi_I(R^+X^-)$. This is a dielectric continuum description.

The results of our theory for the t-BuCl reaction in (dielectric continuum) acetonitrile are given in the two-dimensional free energy contour map Fig. 11. The reaction system proves to be in the electronic strong coupling limit with $\beta \sim 15$-18 kcal/mol in the transition state neighborhood; there is a stable well in the solvent coordinate s in the ground electronic state. Here is an important difference from activated ET. Instead of two wells separated by a barrier in the solvent coordinate, as in Fig. 5, the large coupling completely suppresses the solvent barrier.

Two paths for the reaction are indicated in Fig. 11. The first is the equilibrium solvation reaction path (ESP), along which $\vec{P}_0$ (as well as the electronic polarization $\vec{P}_e$) is always equilibrated to the electronic charge distribution at any given RX separation r; the condition for this is $\partial G(r,s)/\partial s = 0$, which is equivalent to $\delta G/\delta\vec{P}_0 = 0$. The solution reaction path (SRP), on the other hand, is the generalization to solution [27] of the Fukui intrinsic reaction path [28]; it is the steepest descent path from the transition state (TS) in the (r, s) space. Fig. 5 shows that the SRP involves much more solvent motion than does the ESP in the initial exit from the stable reactant RX. But in the neighborhood of the TS the reverse is true and the reaction coordinate motion has a very large nuclear displacement component. The latter feature, previously emphasized in

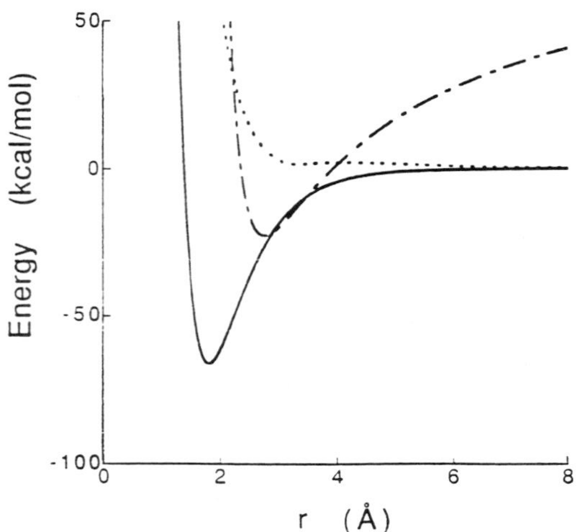

Figure 10. Model calculated gas phase potential energy curves and electronic coupling for the t-butyl chloride system. — $H_c^0(r)$; —•— $H_I^0(r)$; - - - $\beta(r)$ [9].

a model S_N1 study [29], corresponds to nonadiabatic, nonequilibrium solvation [30], in which there is a "frozen solvent" [22a], unable to provide equilibrium solvation for the dissociating solute. Note also the contrast with ET in which the S_N1 reaction coordinate near the TS is mainly a solute nuclear coordinate. As stressed elsewhere [22a,24,29,30] the nonequilibrium solvation has important consequences for the validity of conventional TST, which assumes that equilibrium solvation prevails. For example, we estimate via the nonadiabatic solvation limit of Grote-Hynes Theory [24] (based on a GLE for the solute coordinate r) that the transmission coefficient is about 0.5 for t-BuCl in CH_3CN and more polar solvents. A κ value near this has been found in an MD simulation [10] of a model of t-BuCl ionization in water; it is noteworthy that Kramers Theory [31] incorrectly predicts a very much smaller κ value of ~ 0.02, a feature anticipated in Ref. [29].

Although we do not discuss it here, the present theory predicts [9] a quite different source of the activation free energy-solvent polarity trend than does the standard Hughes-Ingold explanation .

This work was supported by grants NSF CHE88-07852, NIH R01 GM 41332, DOE DE-FG02-84 ER 13247, and the Pittsburgh Supercomputer Center. We thank E. Carter, D. Zichi, G. Ciccotti and M. Ferrario for collaboration on earlier aspects of some of the topics discussed.

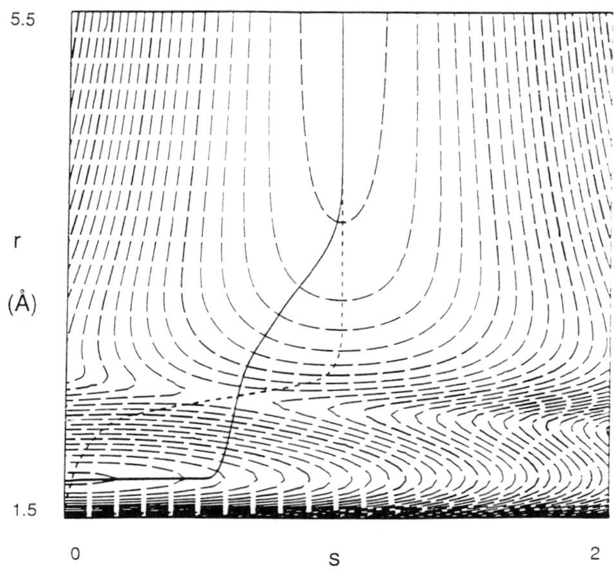

Figure 11. Calculated free energy contour map in s and r for the t-BuCl system in CH_3CN [9]. The free energy difference between two nearby contour lines is 0.1 eV. The free energy value for the contour in the upper center is -3 eV. — SRP; ······ ESP. The two reaction paths meet at the reactant state, at the TS, and merge as they approach the product state.

REFERENCES

1. R. A. Marcus, *J. Chem. Phys.* **24** (1956) 966, 979.
2. N. S. Hush, Trans. Faraday Soc. **57** (1961) 557.
3. See, e.g., the reviews by M. D. Newton and M. D. Sutin, *Annu. Rev. Phys. Chem.* **35** (1984) 437 and R. D. Cannon and J. F. Endicott, in *Mechanisms of Inorganic and Organometallic Reactions,* ed. M. V. Twigg, (Plenum, New York, 1989), Vol. 6, p. 3.
4. D. A. Zichi, G. Ciccotti, J. T. Hynes and M. Ferrario, *J. Phys. Chem.* **93** (1989) 6261. Reference may be found here to other simulation work focussed on different issues in electron transfer.
5. B. B. Smith and J. T. Hynes, unpublished.
6. E. A. Carter and J. T. Hynes, *J. Chem. Phys.*, (1991) in press; J. T. Hynes, E. A. Carter, G. Ciccotti, H. J. Kim, D. A. Zichi, M. Ferrario and R. Kapral, in *Perspectives in Photosynthesis*, eds. J. Jortner and B. Pullman, (Kluwer, Dordrecht, 1990), p. 133.
7. H. J. Kim and J. T. Hynes, *J. Phys. Chem.* **94** (1990) 2736.
8. H. J. Kim and J. T. Hynes, *J. Chem. Phys.* **93** (1990) 5194, 5211.

9. H. J. Kim and J. T. Hynes, "A Theoretical Model for S_N1 Ionic Dissociation Reactions in Solution", submitted to *J. Am. Chem. Soc.*.
10. W. Keirstead, K. R. Wilson and J. T. Hynes, submitted to *J. Chem. Phys.*
11. D. Borgis and J. T. Hynes, *J. Chem. Phys.* **94**, (1991) 3619.
12. D. Borgis and J. T. Hynes, in *The Enzyme Catalysis Process*, ed. by A. Cooper, J. Houben and L. Chien (Plenum, New York, 1989). p. 293; D. Borgis, S. Lee and J. T. Hynes, *Chem. Phys. Lett.* **162**, (1989) 19.
13. E. W. Castner, Jr., M. Maroncelli and G. R. Fleming, *J. Chem. Phys.* **86** (1987) 1090; M. A. Kahlow, W. Jarzeba, T. J. Kang and P. F. Barbara, ibid. **90** (1989) 151 and earlier papers of the Barbara group.
14. For recent reviews, see, e.g., J. D. Simon, Acc. Chem. Res. **21** (1988) 128; P. F. Barbara and W. Jarzeba, *Adv. Photochem.* **15** (1990) 1; M. Maroncelli, J. MacInnes and G. R. Fleming, *Science*, **243** (1989) 1674. These reviews summarize some of the major experimental contributions of the Simon, Barbara, Fleming and other groups.
15. See, e.g., G. van der Zwan and J. T. Hynes, *J. Phys. Chem.* **89** (1985) 4181 and references therein. For the new features that appear in electrolyte solutions, see G. van der Zwan and J. T. Hynes, *Chem. Phys.* **152** (1991) 169.
16. J. T. Hynes, *J. Phys. Chem.* **90** (1986) 3701.
17. See, e.g., P. F. Barbara, T. J. Kang, W. Jarzeba and T. Fonseca in *Perspectives in Photosynthesis*, eds. J. Jortner and B. Pullman, (Kluwer, Dordrecht, 1990), p. 273; T. J. Kang, W. Jarzeba, P. F. Barbara and T. Fonseca, *Chem. Phys.* **149** (1990) 81.
18. D. A. Zichi, H. J. Kim, E. A. Carter and J. T. Hynes, unpublished.
19. See, e.g., M. Maroncelli, *J. Chem. Phys.* **94** (1991) 2084 and T. Fonseca and B. M. Ladanyi, *J. Phys. Chem.* **95** (1991) 2116.
20. L. D. Zusman, *Chem. Phys.* **49** (1980) 295.
21. The literature is very extensive here. Some references are: I. V. Alexandrov and R. G. Gabrielyan, *Mol. Phys.* **37** (1979) 1963; H. L. Friedman and M. D. Newton, *Faraday Discuss. Chem. Soc.* **74** (1982) 73; D. F. Calef and P. G. Wolynes, *J. Phys. Chem.* **87** (1983) 3387; H. Sumi and R. A. Marcus, *J. Chem. Phys.* **84** (1986) 4272; W. Nadler and R. A. Marcus, *J. Chem. Phys.* **86** (1987) 3906;M. Sparpaglione and S. Mukamel, *J. Chem. Phys.* **88** (1988) 3263; I. Rips and J. Jortner, *J. Chem. Phys.* **87** (1987) 2090; Yu. I. Dakhnovskii and A. A. Ovchinikov, *Mol. Phys.* **58** (1986) 237; L. D. Zusman, *Chem. Phys.* **51** (1988) 119; M. Murillo and R. I. Cukier, *J. Chem. Phys.* **89**(1988) 6736 ; T. Fonseca, *J. Chem. Phys.* **91** (1989) 2869.
22. (a) B. J. Gertner, K. R. Wilson and J. T. Hynes, *J. Chem. Phys.* **90** (1989) 3537; J. P. Bergsma, B. J. Gertner, K. R. Wilson and J. T. Hynes, *J. Chem. Phys.* **86** (1987) 1356; B. J. Gertner, J. P. Bergsma, K. R. Wilson, S. Lee and J. T. Hynes, *J. Chem. Phys.* **86** (1987) 1377; (b) J. P. Bergsma, J. R.Reimers, K. R. Wilson and J. T. Hynes, *J. Chem. Phys.* **85** (1986) 5625; G. Ciccotti, M. Ferrario, J. T. Hynes and R. Kapral, *J. Chem. Phys.* **93** (1990) 7137; S. B. Zhu, J. Lee and G. W.Robinson, *J. Phys. Chem.* **92** (1988) 2401; B. J. Berne, M. Borkovec and J. E. Straub, *J. Phys. Chem.* **92** (1988) 3711; B. Roux and M. Karplus, *J. Phys. Chem.* in press.
23. B. Bagchi and D. W. Oxtoby, *J. Chem. Phys.* **78** (1983) 2735; J. Ashcroft, M. Besnard, V. Aquada and J. Jonas, *Chem. Phys. Lett.* **110** (1984) 430; D. M. Zeglinski and D. H. Waldeck, *J. Phys. Chem.* **92** (1988) 692; N. Sivakumar, E. A. Hoburg and D. H. Waldeck, *J. Chem. Phys.*, **90** (1989) 2305; N. S. Park and D. H. Waldeck, *J. Phys. Chem.*, in press and G. E. McManis and M. J.

Weaver, *J. Chem. Phys.* **90** (1989) 1720. (This last reference can be consulted for extensive references to experimental work on ET reaction rates.)
24. R. F. Grote and J. T. Hynes, *J. Chem. Phys.* **73** (1980) 2715; J. T. Hynes, in *The Theory of Chemical Reaction Dynamics*, Vol. IV, ed. M. Baer, (CRC Press, Boca Raton, Fl, 1985), p. 171.
25. J. K. Hwang and A. Warshel, *J. Am. Chem. Soc.*, **109** (1987) 715; J.. W. Halley and J. Hautman, *Phys. Rev. B* , **38** (1988) 11704; R. A. Kuharski, J. S. Bader, D. Chandler, M. Sprik and M. L. Klein, *J. Chem. Phys.* **89** (1988), 4248; E. A. Carter and J. T. Hynes, *J. Phys. Chem.* **93** (1989) 2184.
26. E. C. Baughn, M. G. Evans, and M. Polyani, *Trans. Faraday Soc.* **37** (1941) 377.
27. S. Lee and J. T. Hynes, *J. Chem. Phys.* **88** (1988) 685.
28. K. Fukui, *J. Phys. Chem.*, **74** (1970) 4161.
29. D. A. Zichi and J. T. Hynes, *J. Chem. Phys.* **88** (1988) 2513.
30. G. van der Zwan and J. T. Hynes, J. Chem. Phys. **76** (1982) 2993; *J. Chem. Phys.* **78**, 4174 (1983); *Chem. Phys.* **90** (1984) 21; *J. Chem. Phys.* **89** (1985) 4181.
31. H. A. Kramers, *Physica* , **7** (1940) 284.

DISCUSSION

Jortner

The understanding of the short–time behavior of the correlation function C(t) for solvation is of importance for the elucidation of ultrafast processes. The initial Gaussian time evolution of C(t) obtained by Hynes et al. from molecular dynamics simulation is very informative regarding the inertial dephasing, which occurs before diffusional rotation gets in. An analytic theory of C(t) which incorporates the short–time inertial Gaussian dephasing was presented some time ago [I. Rips and J. Jortner, J. Chem. Phys. 87, 2090 (1987)]. We have utilized Kubo's stochastic modulation theory, which results in

$$C(t) = \exp(-\Omega^2 t^2/2) \; ; \; t \ll \tau^1$$

$$C(t) = \exp(-t/\tau_L) \; ; \; t \gg \tau^1$$

where the effective rotational frequency

$$\Omega^2 = [(2\varepsilon_s+\varepsilon_\infty)/3\varepsilon_s g](k_B T/I)$$

is expressed in terms of the static and optical dielectric constants, ε_s and ε_∞, respectively, the Kirkwood factor g and the moment of inertia, I, of the solvent molecule. The time scale is

$$\tau^1 = (\Omega^2 \tau_L)^{-1}$$

The range of the modulation is determined by $(\Omega^2\tau_L)^{-1}$. It will be instructive to compare the theory with the molecular dynamics results.

Hynes

Thank you for pointing out the Rips–Jortner discussion, which I had missed. Actually the expression you give for the solvent frequency appeared in van der Zwan and Hynes, J. Phys. Chem., **89**, *4181* (1985) in a Langevin equation description for dynamic fluorescence. Even such an equation allows for initial Gaussian inertial dephasing. However, I consider that a key point is the following: one knows that a classical time correlation function *must* be Gaussian at short times; the important aspect of the Carter–Hynes results (and the subsequent results of Maroncelli) is the surprising *dominance* of the Gaussian in the relaxation dynamics. Note also that this is not always so, as recently shown for methanol (Fonseca and Ladanyi, J.Phys.Chem,1991). Finally, concerning the possible applicability of analytic results, I have two comments. First, the TDF calculated in J.P.C. **89**, *4181* (1985) for acetonitrile did not look strongly bimodal (whereas the Maroncelli simulations do). Second, one can see in my paper that the long time behavior is *not* exponential with the solvation time. I think a new analytical theoretical approach is required for these solvation problems.

Tominaga

Can the fast inertial part of the solvation be described by the diffusion equation with the time-dependent diffusion coefficient?

Hynes

If we limit consideration to solvation dynamics as in TDF, one could include the Gaussian inertial dephasing effect in a generalized diffusion equation with a time-dependent "diffusion constant". But I have three further comments. First, the solvent free energy curve would have to be harmonic. Second, to the extent that well relaxation is a relevant picture for cusp-like barrier ET, one cannot just insert this into the theoretical equations including non-Markovian solvation dynamics[J.Phys.Chem, 90,3701,(1986)]; The reason is that inertial effects were not even included in the formulation of the rate, and this must be addressed. Third, Fonseca [Chem.Phys.Lett.,(1989)] has shown that there are unexpected pathologies associated with a time-dependent "diffusion coefficient".

Mukamel

Zusman's equations of motion describe the evolution near the barrier top as well as in the well. His idea is to look at the off diagonal density matrix element. These equations can be solved in various degrees of approximation. Some approximate solutions (including Zusman's) do not include the barrier curvature. But if Zusman's equations are solved differently, the barrier dynamics will show up. Exact solution of his equations in terms of a continued friction for of matrices (M. Sparpaglione, Ph. D. Thesis, University of Rochester, 1988) should show the effects of the barriers curvature. Therefore the statement is that Zusman's formulation wins the barrier dynamics.

Hynes

My statement referred to the expression for the rate constant that Zusman presented, and this is the formula discussed and used in the many theoretical and experimental papers that I quoted. For that rate formula, my statement is correct. I would need to see the alternate solution you mention to comment on it. But in any event, it is certainly true that the solvent is overdamped in the Zusman formulation, and this must necessarily miss the important inertial and "memory" effects for fast solvents.

Mataga

Recently we have made femtosecond-picosecond laser spectroscopic studies on excited state proton transfer in aromatic alcohol (pyrenol)-triethylamine hydrogen bonded complexes in several solvents of different polarities. Our results show rather slow proton transfer with time constant of ca. 1 ps and no effect of deuteration has been recognized. Do you have any comment on these results.

Hynes

Of course I would have to study your results to comment in any detail, but I can offer a few comments. Your lack of an isotope effect is puzzling from either tunneling or classical activation points of view (both would lead to an isotope effect). Perhaps the proton motion is not rate limiting in your system, but rather some other motion is (internal molecular or solvent).

Barbara

Prof. Hynes I would like to comment that Strandjord et al. have reported picosecond measurements on S_1 proton transfer of 3-hydroxyflavone/methanol complexes which show a temperature independent k_H/k_D isotope effect. The effects are very similar to the general predictions of your model. On the other hand, Smith et al., have recently reported subpicosecond measurements on S_1 proton transfer reaction in the inverted regime, namely the acetylaminoanthraquinone derivatives. For these molecules $k_{PT} >> \langle\tau_s\rangle^{-1}$ and intramolecular modes apparently are the dominant accepting mode for the reaction.

Hynes

I am well aware of your very interesting results for 3-hydroxyflavone, and I also believe that they tend to support the picture that I described in my lecture. Please note, however, that I was only able to describe one limit in the talk. There are in fact several important limits of our theory. In particular, one limit discussed in the original papers is the case where the solvation activation free energy is small compared to the reaction exothermicity. In this case, the rate of proton transfer becomes independent of solvent factors, is instead governed by vibrational Franck-Condon factors, and can be quite fast. This seems at first sight similar to the second case you describe, and merits a detailed examination.

Dynamics and Mechanisms of
Photoinduced Transfer and Related Phenomena
N. Mataga, T. Okada and H. Masuhara (Editors)

Photoinduced Electron Transfer in Twisted π-Systems

Wolfgang Rettig

Iwan-N.-Stranski Institute, Techn. Univ. Berlin, Straße des 17. Juni 112, D-1000 Berlin 12

Abstract

It is shown that electron transfer kinetics in TICT systems is governed by a variety of factors including the initial conditions and the topology/dimensionality of the reactive hypersurface. The solvent relaxational properties become important only when solute relaxation is of the same order of magnitude or slower than solvent relaxation. In TICT systems possessing several flexible donor groups, competing channels are expected and exemplified for the case of benzopyrylium dyes.

1. INTRODUCTION

The dual fluorescence of dimethylaminobenzonitrile (DMABN) and its derivatives has been the subject of numerous kinetic studies because the adiabatic photoreaction populating the A* (or "Twisted Intramolecular Charge Transfer" (TICT) state /1-3/) from the primarily excited B* state can be viewed as a prototype example of an electron transfer (ET) reaction which is strongly coupled to both intramolecular motion (twisting process, controlled by solvent viscosity) and the dynamics of dipolar solvent reorganization. In particular, the question arises, which of these factors is the most important one, whether their relative influence remains approximately constant and whether there are still other factors important for the description of TICT formation dynamics. An answer to these questions would also help to elucidate the apparent inconsistency of recent ET kinetic models /4/ with experiments on TICT formation. These models predict that the ET rate constant is controlled by the solvent relaxational properties. Recent kinetic measurements of TICT formation in closely related TICT derivatives, however, reveal that the rates can differ by more than an order of magnitude for identical solvent conditions /5/.

2. SOLVATOKINETICS AND VARIATION OF INITIAL CONDITIONS IN DIALKYLAMINOBENZONITRILES

The latter study /5/ shows that besides solvent relaxation, other factors are important for the observed kinetics like the initial conditions and the presence of a conical intersection along the reaction path for DMABN. The influence of these factors can be exemplified by comparing the fluorescence kinetics of two dialkylaminobenzonitriles with different hindrance to planarity (Scheme I).

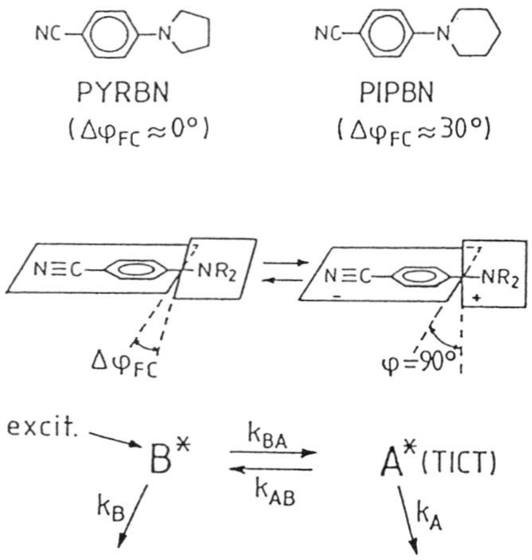

PYRBN ($\Delta\varphi_{FC} \approx 0°$) PIPBN ($\Delta\varphi_{FC} \approx 30°$)

Scheme I

For this purpose, the solvent polarity influence on the relative rates of TICT formation (differential solvatokinetics /5/) for the two nitriles of Scheme I is measured. Fig. 1 shows decay curves of the precursor fluorescence F_B of the B^* state. The pretwisted compound PIPBN exhibits faster initial decays than the non-pretwisted PYRBN, irrespective of the solvent. From this difference, it can be concluded that the TICT formation rate k_{BA} is always faster for PIPBN. The long decay components which can also be seen are indicative in this case that excited state equilibration occurs, i.e. that the back reaction k_{AB} is nonnegligeable. The TICT formation rate k_{BA} increases with increasing solvent polarity, as was shown by isoviscosity experiments by Hicks and Eisenthal et al. /6/. For the comparison of solvents used here, isoviscosity conditions could not be realized, but the differential approach, i.e. the comparison of rates for PIPBN and PYRBN, proves to be similarly or even more powerful. As can be seen from Fig. 1, the rate preference for PIPBN is reduced with increasing solvent polarity. A collection of the rate ratios r measured for different solvent polarities is given in Table 1.

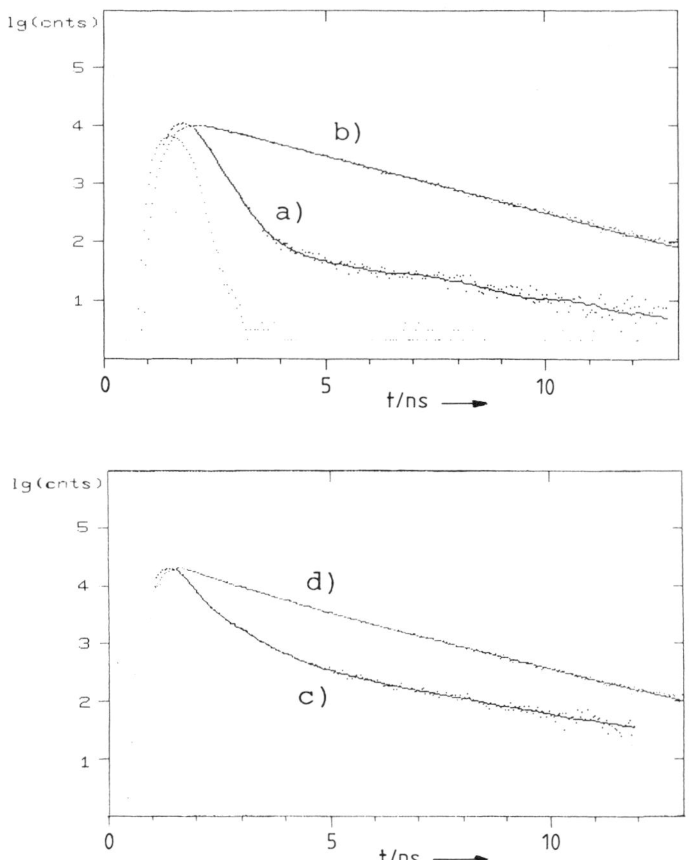

Fig.1: Decay curves of the F_B fluorescence in diethyl- ether at -105° C (a) PIPBN, b) PYRBN) and in n- butyronitrile/iso-butyronitrile (9:1) at -120°C (c) PIPBN d) PYRBN). The figure also contains the fitted curves (unbroken line) and the experimental prompt response function. The longest decay components for a) and c) (with very small weight) are attributed to excited state equilibration.

Table 1: Ratio r of the TICT formation rate constant k_{BA} for PYRBN and pretwisted PIPBN ($r = k_{BA}(PIPBN)/k_{BA}(PYRBN)$) as a function of solvent polarity ϵ /5/.

solvent	temperature	ϵ a)	r
diethylether b)	-120° C	4.34	9.5
n-butyl chloride	-120° C	7.4	19
n-butyronitrile c)	-120° C	20.3	8.3
n-propanol	-105° C	20.8	3.3

a) dielectric constant at room temperature b) diethylether/iso-pentane (9:1) c) n-butyronitrile/isobutyronitrile (9:1)

From weakly polar diethylether to medium polar n-butyl chloride, r increases, but for more strongly polar solvents, r decreases again. These results can be used to extract information on the polar-solvent induced shape changes of the excited-state potential surface for TICT formation. The conical intersection between B* and A* states (Fig. 2) is shown to move to smaller twist angles as solvent polarity increases.

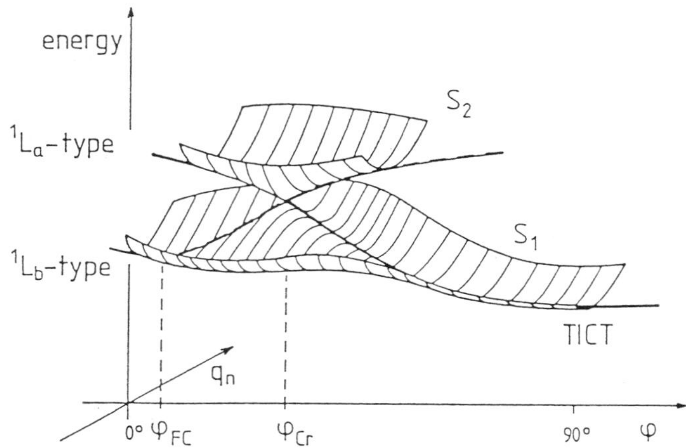

Fig. 2: Schematic S1 and S2 potential energy surfaces for DMABN and related nitriles exhibiting a conical intersection. In addition to the twist angle, a further coordinate q is important which couples the otherwise noninteracting S1 and S2 states for $q \neq 0$. This leads to an energetic stabilization of the lower energy surface and produces the pointed cone. A good candidate for q is pyramidalization at the amino nitrogen.

The molecule can "travel" round the top of the cone (Fig. 2 and 3a) and therefore initially encounters a relatively flat region of the potential energy surface, the length of which is solvent polarity dependent. This model of polarity-induced changes of the hypersurface topology is an alternative and supplement to the previously proposed model of solvent-polarity dependent activation barriers /6/.

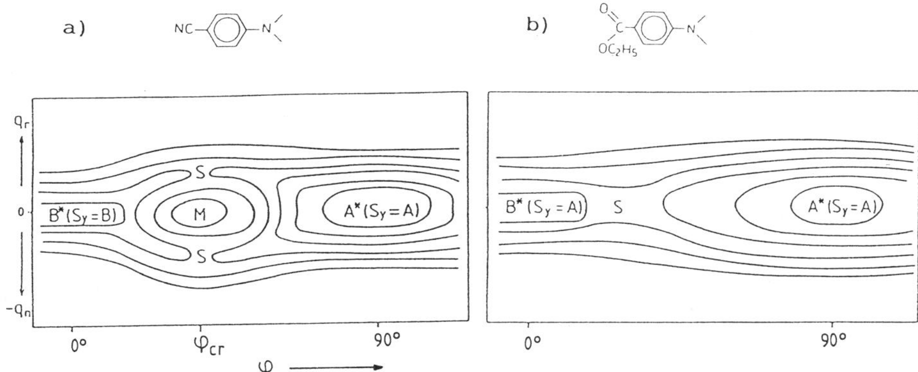

Fig. 3: Schematic S1 hypersurface for the TICT formation reaction in DMABN (a) and DMABEE (b). Because the symmetry of the primary excited state B* is different for the two compounds but similar for the A* (TICT) state /2/, an "obstacle" M (lower cone of the conical intersection) arises for the nitrile (case a) which can be circumvented, however, by activating, in addition to the twist angle, a further coordinate q. S indicates a saddle point.

The maximum value for r measured for n-butyl chloride can be interpreted within this model to signify that in this solvent of medium polarity, ϕ_{cr} (the conical intersection) is situated at approximately 30° twist, near the ground state twist angle of PIPBN /5/.

3. SOLVATOKINETICS AND THE INFLUENCE OF HYPERSURFACE TOPOLOGY

Even more important for differential solvatokinetics and dual fluorescence properties than the initial conditions is the presence or absence of this conical intersection along the reaction path of the adiabatic photoreaction. This can be shown by comparing the kinetics of DMABN with that of the corresponding ester DMABEE, where the conical intersection is absent. Although both dyes show a similar dual fluorescence in polar solvents the importance of the long wavelength TICT band is always stronger for the ester than for the nitrile /2/. Some esters (diethylamino- and dimethylpyrrolidino-derivative) exhibit even strong TICT fluorescence in nonpolar alkane solvents /2/. This difference is also reflected in TICT formation kinetics. Esters are known to react about an order of magnitude faster than the nitriles, in medium polar solvents /7/. This rate ratio also depends on solvent polarity, and again can be traced back to the presence of the conical intersection for DMABN, and its absence in the case of DMABEE, i.e. to a different reaction dimensionality (Fig. 3).

Table 2 shows the rate ratio r(DMA) for DMABN and DMABEE as a function of solvent polarity, as well as for the corresponding pyrrolidino compounds r(PYR) which react slower and can therefore be measured more accurately with our present equipment (time-correlated single photon counting using BESSY synchrotron raditation).

Table 2: Ratio r of the TICT formation rate constant k_{BA} for esters and nitriles ($r = k_{BA}(ester)/k_{BA}(nitrile)$) as a function of solvent polarity ϵ /8/. The table contains values for the dimethylamino (DMA) and pyrrolidino (PYR) pair of compounds.

solvent	temperature	$\epsilon^{a)}$	r(DMA)	r(PYR)
diethylether[b)]	-120° C	4.34	> 8	> 60
n-butyl chloride	-120° C	7.4	8	24
n-butyronitrile[c)]	-120° C	20.3	2.5	6
n-propanol	-105° C	20.8	1	5

a) dielectric constant at room temperature b) diethylether/iso-pentane (9:1) c) n-butyronitrile/isobutyronitrile (9:1)

r is especially large for weakly polar solvents (diethylether), because there the conical intersection for the nitrile (M and ϕ_{cr} in Fig. 3a) has moved to larger twist angles, and consequently the nitriles are slowed down by the initial flat portion of the potential (region between B^* and the saddle pont S) with respect to the esters. As solvent polarity increases, M and S move to the left (towards smaller twist angles), and the flat potential region is reduced. If the conical intersection is absent (esters, Fig. 3b), solvent polarity exerts a much weaker influence on the saddle point S.

4. INTERNAL AND SOLVATION COORDINATE: A MULTIDIMENSIONAL PICTURE

The flat, nearly barrierless nature of the reactive hypersurface can lead to pronounced nonexponentialities of the fluorescence decay in certain solvents (especially alcohols). An example for this is shown in Fig. 4 which contains the fluorescence decays measured for PYRBN and PIPBN in n-propanol at low temperature. As evident from this Figure (and Table 1), the strong kinetic difference between these two compounds is much weaker in this solvent of polarity similar to butyronitrile. Moreover, the decays can only be fitted using a model more complicated than biexponential kinetics as previously observed /3,9,10/. This is taken as evidence that in reality, the reaction kinetics is nonexponential in alcohols and can, for example, be interpreted with time-dependent reaction kinetics /3,9/.

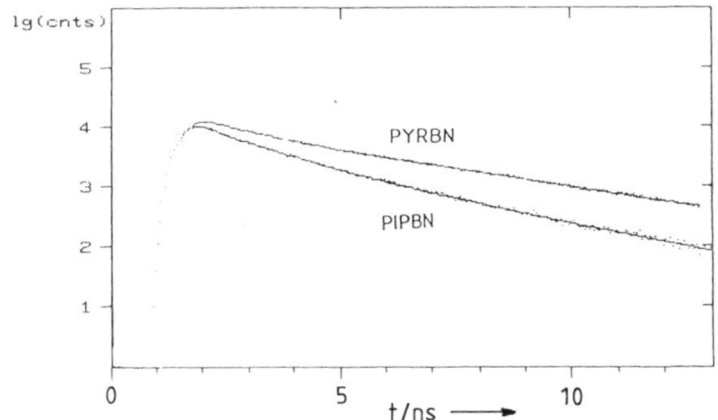

Fig. 4: Decay curves of the F_B fluorescence in n-propanol at -105° C for a) PIPBN, b) PYRBN. The decays are curved in the semilogarithmic plot, even for PYRBN although reversibility (importance of back reaction k_{AB}) can be excluded. Note the similarity for the two compounds.

The nonexponentialities as well as the solvent polarity dependence of the rate ratios of esters and nitriles can be explained /5/ within a stochastic kinetic model simulating the relevant part of the hypersurface by a barrierless region (staircase model /11/.

An additional important factor to be taken into account is the comparison of time-scales for the twisting motion and for the solvent relaxation. A different behaviour results whether solvent relaxation is fast as compared to the twisting motion (case of fast relaxing solvents, e.g. diethylether or n-butyl chloride: strong influence of pretwisting, less of solvation time) or whether it is comparable (alcohols: less influence of pretwisting, stronger correlation with solvation time). An indication which of these limits actually applies can be gained by observing the time-dependent red-shift of the TICT band and comparing it with TICT formation kinetics. For n-butyl chloride and diethylether solvents, the relaxation time of this shift (a good indication for the solvent longitudinal relaxation time /12/) is fast as compared to reaction kinetics, for n-propanol and, also but less, for butyronitrile, both are of the same order of magnitude.

Recently, this multidimensional timescale problem has also gained attention from a theoretical viewpoint: Stochastic kinetic calculations can mimic the different influences (dimensions) and model the observed kinetics using time-dependent reaction profiles /13/.

Taking together the above experiments and the calculations mentioned, it is possible to conclude that solvent controlled electron transfer as reviewed in /4/ mainly occurs under conditions where solvent relaxation is slow as compared to the twisting motion and that other factors can control electron transfer if this condition is not met.

5. MULTIDIMENSIONAL TICT FORMATION: PARALLEL PATHWAYS

Compounds with several flexible groups possess, in principle, several different TICT pathways. Whether they will be active, of course, depends on their energetics. With the example of benzopyrylium dyes, some aspects of this other kind of multidimensionality will be exemplified in the following.

Fig. 5 schematically shows an acceptor system A connected with a flexible single bond to two different donor systems D_1 and D_2. If these are linked at opposite ends of A, the dipole moment of the respective product (TICT) state will also point in opposite direction (see Fig. 5). This is especially important for coumarine acceptors /14/. Here, we will consider the systems with A = benzopyrylium cation and show how the rates of charge transfer (charge localization in this case, without properly defined dipole moment of the "TICT-like" state) are affected for variations of donor strength and flexibility /15/. In this case, only fluorescence quenching of the precursor state is observable. The product seems to be nonemitting. Its charge transfer nature, however, can be inferred from the dependence of the quenching rates on donor and acceptor properties and on bridging. Fig. 6 summarizes the main results derived from nonradiative decay/photoreaction kinetics.

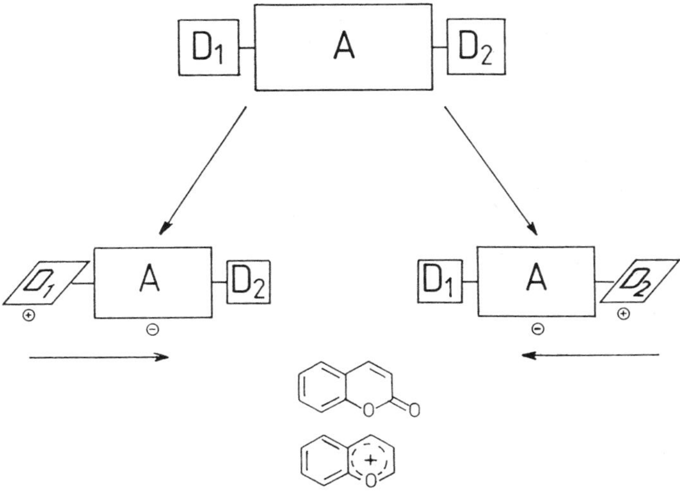

Fig. 5: Schematic representation of the two TICT reaction channels open for an acceptor system A with two different donor substituents. The resulting charge transfer directions are indicated. Examples for A = coumarine and benzopyrylium are known.

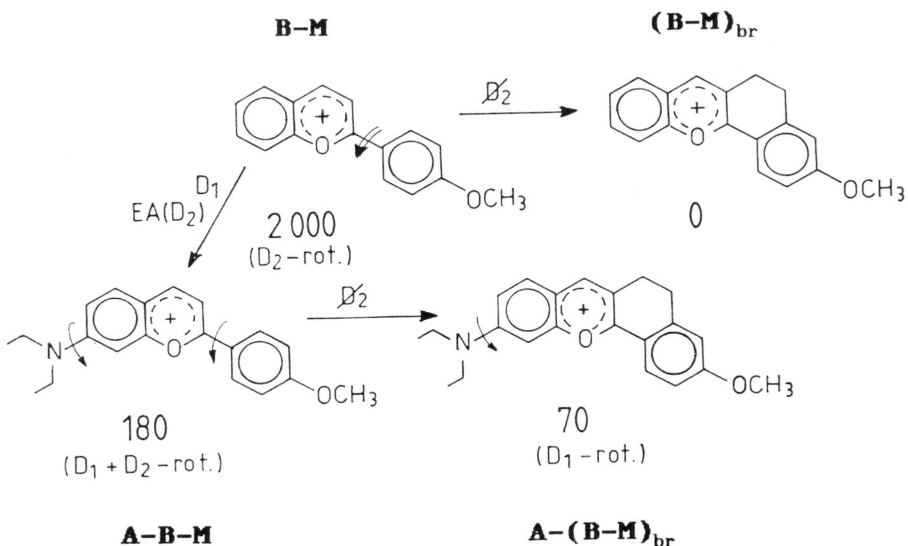

Fig. 6: Nonradiative decay rates (units of $10^7\ s^{-1}$ as measured by fluorescence kinetics for various benzopyrylium derivatives in n-butyronitrile at room temperature /15/.

The flexible parent compound B-M shows only a very low fluorescence quantum yield (ϕ_f = 0.004 in acetonitrile /16/) due to fast nonradiative losses. The bridged model compound $(B\text{-}M)_{br}$ (ϕ_f = 1 /16/) proves that this loss is connected to intramolecular rotation of the methoxyphenyl group M (= D_2). Introducing a second flexible donor group D_1 (Dialkylamino) as in A-B-M reduces the nonradiative decay rates. This is quite contrary to what one would expect from the so-called loose-bolt effect which was previously made responsible for fluorescence quenching in related dyes /17/. The "anti-loose-bolt-effect" observed here, however, can readily be explained using the TICT model (or better its extension including charged and uncharged systems, rotation around single and double bonds: The model of Biradicaloid Charge Transfer (BCT) states /18-20/).

Twisting M (= D_2) in the doubly-substituted molecule A-B-M leads to a charge-localized system with a weaker acceptor (A-B) than in the case of the comparable reaction within the singly substituted B-M (acceptor B). The D_2 reaction rate is therefore found to be significantly reduced. On the other hand, an additional channel is opened, that of A (= D_1) rotation. Here again, the acceptor (B-M) has weakened character with respect to a system lacking the donor D_2, and therefore the combined reaction rate responsible for nonradiative decay is only 180 • 10^7 s^{-1} in butyronitrile at room temperature. If D_2-rotation is blocked, as in the system $A\text{-}(B\text{-}M)_{br}$, then a reaction rate remains which is about half that of the unbridged system A-B-M. We can therefore conclude that in the unbridged system, both BCT channels contribute about equally to the observed fluorescence quenching.

6. REFERENCES

/1/ Z.R. Grabowski, K. Rotkiewicz, A. Siemiarczuk, D.J. Cowley, and W. Baumann, Nouv. J. Chim. 3 (1979) 443.

/2/ W. Rettig, Angew. Chem. Intern. Ed. Engl. 25 (1986) 971.

/3/ E. Lippert, W. Rettig, V. Bonačić-Koutecký, F. Heisel und J.A. Miehé, Adv. Chem. Phys. 68 (1987)1

/4/ E.M. Kosower and D. Huppert, Annu. Rev. Phys. Chem. 37 (1986) 127

/5/ W. Rettig, Ber. Bunsenges. Phys. Chem. 95 (1991) 259

/6/ J.M. Hicks, M. Vandersall, Z. Babarogic and K.B. Eisenthal, Chem. Phys. Lett. 116 (1985) 18; J.M. Hicks, M.T. Vandersall, E.V. Sitzmann and K.B. Eisenthal, Chem. Phys. Lett. 135 (1987) 413.

/7/ W. Rettig, M. Vogel, E. Lippert and H. Otto, Chem. Phys. 103 (1986) 381.

/8/ W. Rettig, to be published

/9/ F. Heisel and J.A. Miehé, Chem. Phys. Lett. 100, 183 (1983) and Chem. Phys. 98, 233 (1985)

/10/ S.R. Meech and D. Phillips, Chem. Phys. Lett. 116, 262 (1985); S.R. Meech and D. Phillips, J. Chem. Soc., Faraday Trans.2, 83, 1941 (1987).

/11/ B. Bagchi, Chem. Phys. Lett. 135 (1987) 558; B. Bagchi and

G.R. Fleming, J. Phys. Chem. 94 (1990) 9.
/12/ M.A.Kahlow, T.J. Kang, and P.F. Barbara, J. Phys. Chem. 91 (1987) 6452; T.J. Kang, M.A. Kahlow, D. Giser, S. Swallen, V. Nagarajan, W. Jarzeba, and P.F. Barbara, J. Phys. Chem. 92 (1988) 6800; P.F. Barbara and W. Jarzeba, Adv. Photochem. 15 (1990); M. Maroncelli and G.R. Fleming, J. Chem. Phys. 89 (1987) 6221, J. Chem. Phys. 89 (1988) 4288 and J. Chem. Phys. 92 (1990) 3251
/13/ G.K. Schenter and C.B. Duke, Chem. Phys. Lett. 176 (1991) 563
/14/ T. Runke and W. Rettig, to be published
/15/ G. Haucke, P. Czerney, D. Steen, W. Rettig and H. Hartmann, to be published
/16/ G. Haucke, P. Czerney, C. Igney, and H. Hartmann, Ber. Bunsenges. Phys. Chem. 93 (1989) 805
/17/ G.N. Lewis and M. Calvin, Chem. Rev. 25 (1939) 273; L.J.E. Hofer, R.J. Grabenstetter adn E.O. Wiig, J. Am. Chem. Soc. 72 (1950) 203
/18/ V. Bonačić-Koutecký and J. Michl, J. Am. Chem. Soc. 107 (1985) 1765
/19/ V. Bonačić-Koutecký, J. Koutecký, and J. Michl, Angew. Chem. 99 (1987) 216; Angew. Chem. Int. Edit. Engl. 26 (1987) 170
/20/ W. Rettig, in "Modern Models of Bonding and Delocalization" (Molecular Structure and Energetics Vol.6), eds. J. Liebman and A. Greenberg, VCH-Publishers, New York, 1988, S. 229

The support by BMFT within project 05414 FAB1 is gratefully acknowledged.

DISCUSSION

Mobius

Do you see the influence of double bond character of single bonds in molecules with ground state CT character? The torsion should be influenced by this effect.

Rettig

Ground state torsion may be described by this picture of mixing valence bond structures. In the excited state, the character often reverses, and a formal single bond can get extensive double bond character and vice versa. In the cases we considered, it was always sufficient to look at the energetic changes of precursor and product state to explain the change of kinetics.

De Schryver

If you have in some systems already extensive CT in the ground state of a D–A system, would you not consider in the analysis of fluorescence decay the direct excitation of ground state CT?

Rettig

If there is a double minimum potential in the excited and in the ground state, then there is indeed the possibility of non–negligible population of the ground–state TICT conformation and direct TICT excitation. Whether this occurs can readily be seen from the risetimes of TICT fluorescence: Instantaneous for direct TICT excitation, slower for indirect TICT population via the LE state, with a 50% weight of the preexponential factor in the simplest case for which I showed an example.

Nakashima

The radiative lifetime depends on the angle of the TICT state. Have you observed this effect in time resolved measurements? Have you observed any relation between this effect and hypersurface for the TICT formation processes in time resolved spectra?

Rettig

Most of the TICT fluorescence is from upper excited vibrational TICT levels ("hot fluorescence") which leads to a temperature dependence (as well as a solvent polarity dependence) of the radiative TICT rate, interpretable in terms of an increasing angular deviation from 90° twist. This is directly observable by time resolved measurements (M. Van der Auweraer et al. J. Phys. Chem. **95** (1991) *2083*). TICT risetimes faster than LE decay times, observable in some solvents, may also be explained as transient effects due to this source.

Steiner

You gave examples demonstrating the effect of ground state pretwisted configurations on the ease or efficiency of TICT state formation. Is there also experimental evidence of the possibility to select pretwisted conformations within the same type of compound by observing the excitation wavelength dependence of TICT state fluorescence in viscous media?

Rettig

Yes, experiments of TICT compounds in polymers at room temperature show an extensive red edge excitation effect which is due to this source. You can also observe excitation wavelength effects in low–temperature viscous solvents which are presumably due to this source.

Kajimoto

The dispersed fluorescence spectrum of ADMA gave no red–shifted emission like charge–transfer band. Therefore, we think that the complication is simply due to the presence of many internal rotors within the molecule. However, when we added a few acetonitrile or water molecules to DMA, slightly red–shifted emission appears. In this case, the LIF excitation spectrum may contain broad structureless background.

Rettig

Concerning the complicated features observed in the LIF excitation spectrum of ADMA, underlying broad continuum or very broad peaks may contribute to it in addition to the congested sharp transitions.

Dynamics and Mechanisms of
Photoinduced Transfer and Related Phenomena
N. Mataga, T. Okada and H. Masuhara (Editors)

On the Energy Gap Law in Electron Transfer Reaction

Toshiaki Kakitani

Department of Physics, Faculty of Science, Nagoya University, Furo-cho, Chikusa-ku, Nagoya 464-01, Japan

Abstract

Adopting a uniform distribution model of the donor-acceptor distance in the neutral state of reactants, we could theoretically reproduce the experimental data of the very broad energy gap law of the photoinduced charge separation (CS) reaction. Comparing the theoretical energy gap law of the charge recombination (CR) reaction with the experimental data, we found that the non-linear response effect could be considerable for electron transfer reactions in polar solvents such as acetonitrile. The CR reaction of the geminate radical ion-pair was found to proceed without appreciable expansion of the ion-pair distance distribution when the energy gap is moderate ($-\Delta G=1.3\sim1.8$eV) and so the CR rate is large. But, it proceeded only after considerable expansion of the ion-pair distribution for the species with very small or large energy gap and so the CR rate being small. Treating the translational motion of solvent molecules as a dynamical variable in those non-linear response systems, we obtained a theoretical result that the energy gap law of the CS reaction is little affected by this solvent dynamical effect but that the energy gap low of the CR reaction is greatly modified.

1. INTRODUCTION

Electron Transfer (ET) reactions are one of the most important elementary processes in physicochemistry and biology. Efficient energy conversions in biological systems are based on the well-designed ET processes. Therefore, it is now becoming a very important theme to elucidate the molecular mechanism which regulates the ET rate.

From a theoretical point of view, the ET rate k in the case that the thermal equilibrium is maintained in the initial state is written as

$$k=A\cdot F \tag{1}$$

where A is a factor related to a frequency of the ET reaction. In the case of the non-adiabatic mechanism, this factor has an exponential decay with a distance R between donor and acceptor molecules

$$A = A_0 e^{-\alpha R} \quad (2)$$

where α is a constant. When the adiabatic mechanism applies, A is a proper frequency factor independent of R. The second factor F might be termed the thermally averaged Franck-Condon factor of the intramolecular vibration and solvent motions. This Franck-Condon factor depends on the free energy gap between the initial and final states, and its relation is called the energy gap law.

In the present paper, we discuss mainly on the energy gap law from two points of view. How is the energy gap law modified when the non-linear solvent polarization and the donor-acceptor distance distribution work? What is the solvent dynamical effect in such non-linear systems? However, before going to those topics, we first discuss pertinent problems in the R-dependence of the factor A in the non-adiabatic case.

2. R-DEPENDENCE OF A

Based on the experimental data of the ET rate in polar solvents and protein environments, the coefficient α in the exponent of A was found to be about $1.0 Å^{-1}$ [1-5]. It should be noted that this α-value is very small and the long-range ET is made possible. Indeed, if we assume $k=10^{12} s^{-1}$ for R=7Å, we obtain $k=3\times10^8$, 2×10^6 and $10^2 s^{-1}$ for R=15, 20 and 30Å, respectively. In order to elucidate the physical reasoning of this rather universal α-value, we first estimate the R-dependence of A by means of the square of the overlap of two $2p_z$ Slater orbitals ($\mu=1.625a_0^{-1}$, a_0 being the Bohr radius) locating at the donor and acceptor molecules. We obtain a result that $\alpha(=2\mu)$ is as large as $6.14Å^{-1}$, which is much larger than the experimental value. Then, we notice the fact that the Slater's μ-value is determined for the electron cloud not so much far from the carbon nucleus. However, when R is large, the overlap of the electron cloud between donor and acceptor takes place almost in the middle region which is far from the carbon nuclei. The electron in such region will feel effectively a unit charge from the carbon atom because the other electrons belonging to the same carbon atom will almost completely screen. If such a screening takes place, the wavefunction extends very much. Putting $\mu\simeq0.50a_0^{-1}$, we obtain $\alpha=1.89Å^{-1}$. In addition to this, since the microscopic environment around this carbon atom is, more or less, polarizable, the charge of the carbon atom will be screened furthermore. The reorientation of polar groups in the environment will little takes place because the space which we are concerned is so limited. Only the electronic polarizability will work effectively. If we tentatively choose the square of the refractive index as 2, α is reduced to be $0.95Å^{-1}$, in good agreement with the experimental value. Since the refractive index may be changed somewhat among the environments, α may vary from 0.5 to 1.5. This range of the α-value appears to cover most of the experimental data. So, we qualitatively conclude that the origin of the small value of the exponent α might be two kinds of screenings: One is the screening by the other electrons belonging to the same carbon atom and the other is the screening by the environmental electronic polarizability. The mechanism of the super-exchange[6,7] where the electronic wavefunctions of solvent (or environmental molecule) are mixed with the electronic wavefunctions of donor and acceptor molecules will basically correspond to our polarizable medium model in the above treatment.

3. ENERGY GAP LAWS IN THERMAL EQUILIBRIUM

According to the Marcus theory [8] using the linear response approximation of solvent polarization in the thermal equilibrium state, the energy gap law is written as a Gauss function

$$k = A \exp\left\{-\frac{(-\Delta G-\lambda)^2}{4k_BT\lambda}\right\} \quad (3)$$

where $-\Delta G$ and λ are the free energy gap and the reorganization energy, respectively, k_B and T are the Boltzmann factor and temperature. This form does not change, depending on the kind of ET reactions. However, available experimental data of the energy gap law differ considerably, depending on the kind of reactions. In the photoinduced charge separation (CS) reaction in the steady state as studied by Rehm and Weller [9], the fluorescence quenching rate increases rapidly in the normal region which exists at the very small energy gap, and it remains constant in the moderate and large energy gaps due to the diffusion control. Recently, the intrinsic photoinduced CS rate was obtained in a wide energy gap region by an analysis of the transient effect of the fluorescence decay by Nishikawa et al. [10]. Its result was that the intrinsic CS rate is certainly larger than the diffusion limit level but the curve is rather flat around the maximum ranging over a wide region of the energy gap. On the contrary, the energy gap law of the charge recombination (CR) reaction was found to be nearly bell-shaped but its normal region is much shifted to the larger energy gap side [11], as compared to the CS reaction.

In order to explain the different energy gap laws between the CS and CR reactions, we consider two effects: One is the non-linear response effect due to the dielectric saturation of solvent around a charged solute molecule. The other is the effect of the donor-acceptor distance distribution. Factors which have distance dependences are the electron-tunneling matrix element in the non-adiabatic mechanism, local energy gap and reorganization energy. We assume a uniform distance distribution for the neutral pair of reactants which is the initial state of CS reaction and we assume a localized distance distribution for the geminate ion-pair of reactants which is the initial state of the CR reaction.

We have calculated the energy gap law by assuming the thermal equilibrium in the initial state. The details of the analytical method is given in Ref. 12.

The results are shown in Figures 1 and 2 for the CS and CR reactions, respectively. The experimental data of the CS in

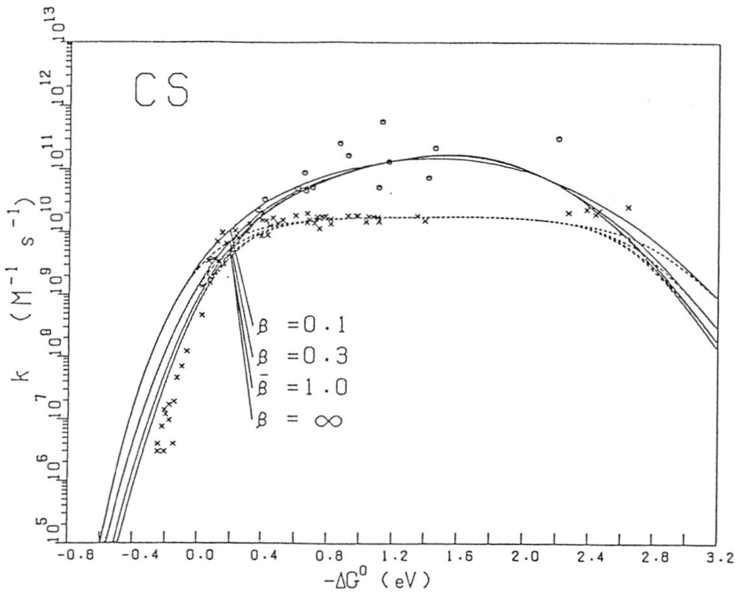

Figure 1. Energy gap laws of the CS reaction. Solid and broken curves are calculated ones. The crosses and open circles are the experimental data of Rehm and Weller [9] and Nishikawa et al. [10], respectively.

the steady state are represented by crosses and those in the non-stationary state by open circles. The solid and broken curves are the theoretical data corresponding to the stationary and non-stationary states, respectively. The triangles are the experimental data of the CR reaction. The degree of non-linearity is represented by β which is defined as

$$\beta = \frac{C_{np}}{C_{ip}-C_{np}} \tag{4}$$

where C_{ip} and C_{np} are the curvatures of free energy curves of the system in the ion-pair and neutral pair states, respectively.

It is seen from Figure 1 that the theoretical curves fit well the experimental data for both of the stationary and non-stationary states. It is also found that the non-linearity does not affect so much the energy gap law of the CS reaction except for the normal region. The broad, flat energy gap law in the intrinsic CS reaction is mainly due to the distance distribution which was assumed to be uniform. The normal region was mainly contributed from the donor-acceptor pairs in close distances and the inverted region was mainly contributed from the donor-acceptor pairs in large distances.

It is found from Figure 2 that the non-linearity affects considerably the energy gap law of the CR reaction, especially in the inverted region. The theoretical curve with β=1.0 fits best the experimental data. This fact indicates that the non-linear effect on the energy gap law is considerable even if it is not large. The narrow energy gap law is mainly due to a limited distance distribution of the ion-pair (virtually Gauss function was assumed with a maximum at the solvent separated ion-pair (SSIP) distance.

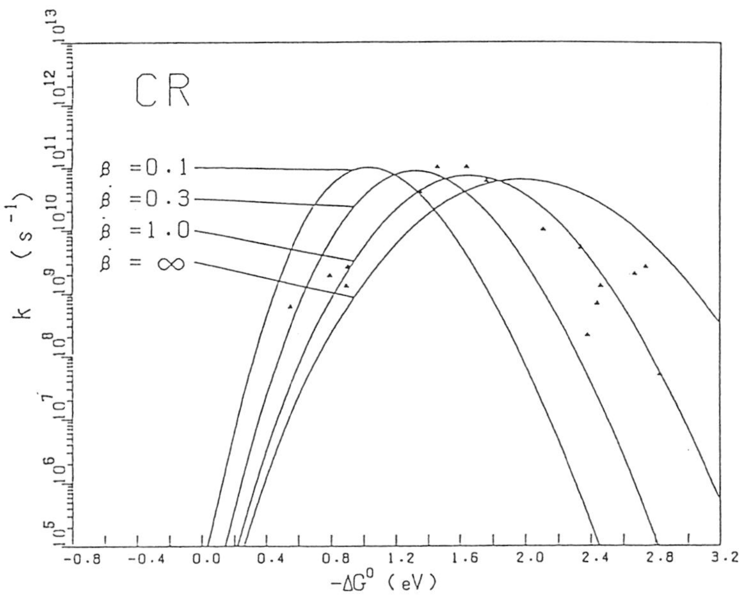

Figure 2. Energy gap laws of the CR reaction. Solid curves are calculated ones. The triangles are the experimental data of Mataga et al. [11].

4. ENERGY GAP LAW OF THE ET REACTION IN NON-THERMAL EQUILIBRIUM

So long as we assume that the thermal equilibrium is maintained, the best fit to the experimental data of the CR reaction was obtained for the distance distribution of the Gauss function with a maximum at the SSIP distance. However, according to the Monte Carlo simulations where free energy curve is calculated as a function of the ion-pair distance [13], we do not observe a deep free energy minimum at the SSIP distance. So, the assumption of the thermal equilibrium for the initial state of the CR reaction would not be appropriate.

We draw the calculated value of $k_{CS}R^2$ as a function of R in the CS reaction in Figure 3. It represents the probability

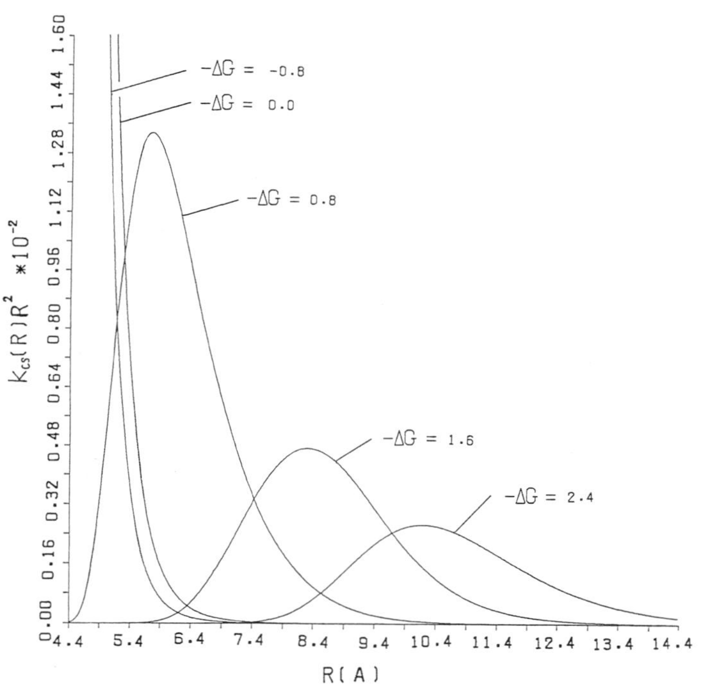

Figure 3. Calculated values of $k_{CS}R^2$ as a function of R for some values of the energy gap -ΔG in eV.

that the ion-pair was produced by the CS reaction for each value of R. We observe that the distribution has a maximum at a contact distance when the energy gap is very small, and it becomes broad and has a maximum at a larger distance when the energy gap increases. Those distributions would expanded when considerable time elapses. For this case, we assume that the position R_0 of the distance corresponding to the maximum distribution remains the same. We express the distribution function by a Gauss function as

$$g(R)=\frac{e^{-(R-R_0)^2/(d\sigma)^2}}{\int_{R_c}^{\infty} e^{-(R-R_0)^2/(d\sigma)^2}\,dR} \tag{5}$$

where d, σ and R_c are the width of the original distribution which is produced by the CS reaction, scale of the width expansion and the distance of the contact ion-pair.

We calculated the CR rate using the above distribution function with $\sigma=1$, 2 and 3 for each species of ion-pair. The calculated results are shown in Table 1 and are compared with the experimental data. We observe that ion-pairs with a small value of $-\Delta G_{CR}$ (NO.1-4) have large values of R_0 and d. Its k_{CR} for $\sigma=1$ is relatively small and increases with an increase of σ. Then, k_{CR} for a large value of σ (e.g. $\sigma=3$) agrees better with experimental data. Those ion-pairs which have a medium value of $-\Delta G_{CR}$ (NO.5-8) have medium values of R_0 and d. Its k_{CR} is very large and decreases with an increase of σ. The calculated value of k_{CR} for $\sigma=1$ agrees best with the experimental data. On the other hand, the ion-pairs which have a large value of $-\Delta G_{CR}$ (NO.9-16) have rather small values of R_0 and d. Its k_{CR} for $\sigma=1$ is small and increases with an increase of σ. Then, k_{CR} for a larger value of σ becomes to fit the experimental value.

From the above results, we conclude that for those ion-pairs which have large values of k_{CR} (e.g. $k_{CR}>5\times10^{10}\,s^{-1}$) using the distribution prepared by the CS reaction($\sigma=1$), little expansion of the distance distribution would takes place before the ET occurs, but that for those ion-pairs which have small values of k_{CR} (e.g. $k_{CR}<10^{10}\,s^{-1}$) using the distribution prepared by the CS reaction($\sigma=1$), considerable expansion (e.g. $\sigma=3$) of the distance distribution would takes place before the ET occurs and the expansion of the distribution works to enhance the ET rate as a whole. From these results and the experimental observation that the dissociation rate of the ion-pair is about $10^9\,s^{-1}$ [11], we find that the distance distribution begins to expand in about 20ps after the geminate ion-pair formation and the distribution becomes uniform (i.e., the ion-pair dissociates) in about 1ns.

5. SOLVENT DYNAMICAL EFFECTS IN THE NON-LINEAR RESPONSE SYSTEMS

From the theoretical analysis of the energy gap laws of CS and CR reactions, we found that the non-linear response effect is considerable ($\beta\sim1.0$) even if it is not large. Recently, we

Table 1
Calculated values of k_{CR} for some values of the distance expansion parameter σ, in comparison with the experimental values in Ref. 11. The donor-acceptor pair number is the same as Ref. 11.

NO	Donor	Acceptor	$-\Delta G_{CS}$(eV)	R_0(eV)	d(Å)	$-\Delta G_{CR}$(eV)	$k_{CR}(s^{-1})$			$k_{CR}(s^{-1})$
							$\sigma=1$	$\sigma=2$	$\sigma=3$	exptl.
1	Per*	TCNE	2.30	9.5	1.8	0.55	1×10^{7}	8×10^{7}	1×10^{8}	6.1×10^{8}
2	BPer*	TCNE	2.46	9.8	1.8	0.79	3×10^{8}	1×10^{9}	2×10^{9}	1.9×10^{9}
3	DPA*	TCNE	2.21	9.3	1.7	0.89	7×10^{8}	2×10^{9}	3×10^{9}	1.3×10^{9}
4	Py*	TCNE	2.44	9.8	1.8	0.90	8×10^{8}	3×10^{9}	3×10^{9}	2.6×10^{9}
5	Per*	PMDA	1.50	8.0	1.6	1.35	7×10^{10}	5×10^{10}	4×10^{10}	$>4\times10^{10}$
7	Per*	MA	1.21	7.1	1.5	1.64	1×10^{11}	1×10^{11}	7×10^{10}	$>1\times10^{11}$
8	TMPD	Per*	1.09	6.7	1.4	1.76	1×10^{11}	8×10^{10}	6×10^{10}	$>6\times10^{10}$
9	Per*	PA	0.74	5.6	1.0	2.11	9×10^{9}	2×10^{10}	2×10^{10}	1×10^{10}
10	BPer*	PA	0.91	6.1	1.2	2.34	1×10^{9}	3×10^{9}	4×10^{9}	5.2×10^{9}
11	DMA	Per*	0.47	4.9	0.66	2.38	2×10^{7}	2×10^{8}	7×10^{8}	2.1×10^{8}
12	o-DMT	Per*	0.41	4.8	0.60	2.44	1×10^{7}	1×10^{8}	4×10^{8}	6.8×10^{8}
13	Py*	PA	0.88	6.0	1.2	2.46	4×10^{8}	1×10^{9}	1×10^{9}	1.3×10^{9}
16	DMA	Py*	0.51	5.0	0.69	2.83	3×10^{4}	6×10^{5}	4×10^{6}	$<5\times10^{7}$

R_0 and d are obtained from the distribution of $k_{CS}R^2$ as a function of R for each value of the energy gap $-\Delta G_{CS}$ corresponding to the donor-acceptor pair. k_{CR} is calculated using the distribution function of Eq. (5) for each value of the energy gap $-\Delta G_{CR}$ corresponding to the ion-pair.

found by the Monte Carlo simulation study [14] that the free energy curvature is greatly increased when the translational motion of solvent molecules is frozen as compared when the translational motion is not frozen. This increase of the free energy curvature for solvent surrounding a charged solute molecule is much larger than that for solvent surrounding a neutral solute molecule. This fact indicates that the translational motion of solvent molecules has a strong non-linear coupling with the rotational motion. Generally speaking, the relaxation time of the rotational motion is much smaller than that of the translational motion. Therefore, it will often happen that the rotational relaxation is faster than the tunneling frequency A, but the translational relaxation is solwer than A. We investigate in the following what is the solvent dynamical effect in this case.

We consider a polar solution where solvent and solute molecules have a point dipole moment and a point charge, respectively. First of all, taking into account the non-linear coupling between translational and rotational motions, we define the non-linear free energy curve of the system as functions of the reaction coordinate x [15] and the translational motion n of solvent as follows [16]

$$u_n(x,n)=(n-sx^2)^2+u_{n,eq}(x) \tag{6}$$

$$u_c(x,n)=(n-sx^2)^2+u_{c,eq}(x) \tag{7}$$

with

$$u_{n,eq}(x)=tx^4+\frac{1}{2}kx^2 \tag{8}$$

$$u_{c,eq}(x)=tx^4+\frac{1}{2}kx^2-x/k_BT \tag{9}$$

where $u_n(x,n)$ and $u_c(x,n)$ are dimensionless free energy functions corresponding to the neutral pair and ion-pair, respectively. The parameters s, t and k are evaluated by comparing with the Monte Carlo simulation data [14].

Assuming the thermal equilibrium for the rotational motion at each fixed position of n, we obtain the potential energy for the translational motion as

$$e^{-\bar{u}_J(n)} = \frac{1}{k_BT}\int_{-\infty}^{\infty} e^{-u_J(x,n)}dx, \quad (J=n \text{ or } c) \tag{10}$$

and the ET rate constant as

$$k_J(n) = A\frac{e^{-u_J(x^*,n)}}{\int_{-\infty}^{\infty} e^{-u_J(x,n)}dx} \tag{11}$$

with J = n or c corresponding to the CS or CR reaction. The reaction coordinate x^* at the transition state is related to the energy gap as follows

$$x^* = \Delta G - \Delta G_s \tag{12}$$

where ΔG_s is the solvation energy of the ion-pair.

Our next problem is to solve the reaction-diffusion equation

$$\frac{\partial P(n,t)}{\partial t} = \frac{1}{\tau_n}\frac{\partial}{\partial n}\left(\frac{\partial}{\partial n} + \frac{\partial u_J(n)}{\partial n}\right)P(n,t) - k_J(n)P(n,t) \tag{13}$$

where τ_n is the translational relaxation time.

We calculate the mean passage time defined by

$$\tau_a = \int_0^\infty dt \int_{-\infty}^\infty P(n,t)dn \tag{14}$$

in the same way as Nadler and Marcus [17]. The details are given elsewhere [16].

The calculated results are shown in Figures 4 and 5. The time scale is normalized by an inverse of $k_e^J(0)$ which is the maximum rate in the energy gap law in the thermal equilibrium condition. We observe that the energy gap law little changes

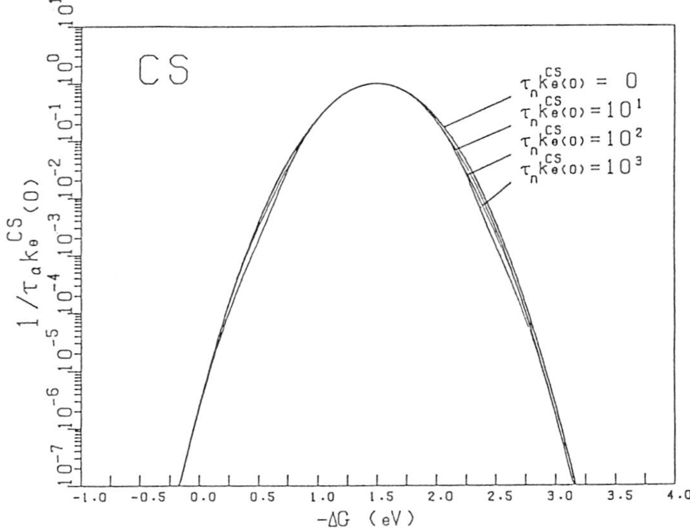

Figure 4. Dynamical solvent effect on the energy gap law of the CS reaction. τ_a and τ_n are the mean passage time and the relaxation time of the translational mode, respectively.

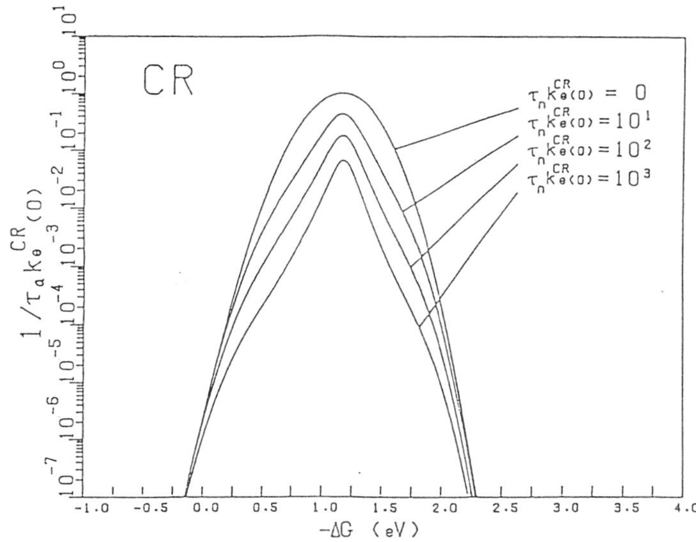

Figure. 5. Dynamical solvent effect on the energy gap law of the CR reaction.

for the CS reaction when τ_n becomes large except at the shoulder. In contrast to this, the energy gap law of the CR reaction is greatly modified around the maximum and at the shoulder when τ_n is large. The latter effect is similar to that obtained by the Sumi-Marcus model [18] where the intramolecular vibration is in thermal equilibrium and the solvent motion is the dynamical variable. The quite different solvent dynamical effect between the CS and CR reactions which we obtained is due to the non-linear coupling between the translational and rotational motions of solvent.

6. DISCUSSIONS

Comparing theoretically calculated energy gap laws with the experimental data, we could evaluate the magnitude of the non-linear response effect in the equilibrium state. The result was that the non-linear effect is appreciable ($\beta \sim 1.0$), even if it is not very large. This non-linear effect was found to be more prominently seen in the transient state as a solvent dynamical effect on the energy gap law; i.e., the energy gap law of the CS reaction is little affected but that of the CR reaction is greatly modified by the non-linear solvent dynamics. The similar kind of notions was also obtained in the molecular dynamics simulation [19]: the solvation dynamics due to a charge up of a solute from z=0 to z=1 is quite different from that due to a charge elimination from z=1 to z=0. The theoretical analysis of the solvation dynamics within a framework of non-linear response was also made elsewhere [20]. It is strongly expected that experiments examining such non-linear response effects will be done in future.

7. REFERENCES

1 J.R. Miller, J.V. Beitz, and R.K. Huddleston, J. Am. Chem. Soc. 106(1984)5057.

2 S.L. Mayo, W.R.Jr. Ellis, R.J. Crutchley, and H.B. Gray, Science 233(1986)948.

3 M.D. Johnson, J.R. Miller, N.S. Green, and G.L. Closs, J. Phys. Chem. 93(1989)1173.

4 K.S. Schanze and L.A. Cabana, J. Phys. Chem. 94(1990)2740.

5 J.M. Vanderkooi, S.W. Englander, S. Papp, W.W. Wright, and C.S. Owen, Proc. Natl. Acad. Sci. 87(1990)5099.

6 D.N. Beratan, J.N. Onuchic, and J.J. Hopfield, J. Chem. Phys. 86(1987)4488.

7 D.N. Beratan and J.N. Onuchic, Photosynthesis Research 22(1989)173.

8 R.A. Marcus, Ann. Rev. Phys. Chem. 15(1964)155.

9 D. Rehm and A. Weller, Isr. J. Chem. Phys. 63(1975)4358.

10 S. Nishikawa, T. Asahi, T. Okada, N. Mataga, and T. Kakitani, submitted to Chem. Phys. Letters.

11 N. Mataga, T. Asahi, Y. Kanda, T. Okada, and T. Kakitani, Chem. Phys. 127(1988)249.

12 T. Kakitani, A. Yoshimori, and N. Mataga, in Electron Transfer in Inorganic, Organic, and Biological Systems, Advances in Chemistry Series, J. Bolton et al.(eds.), Am. Chem. Soc., Chapt. 4, 1991.

13 G.N. Patey and J.P. Valleau, J. Chem. Phys. 63(1975)2334.

14 Y. Hatano, T. Kakitani, Y. Enomoto, and A. Yoshimori, Mol. Sim., in press.

15 A. Yoshimori, T. Kakitani, Y. Enomoto, and N. Mataga, J. Phys. Chem. 93(1989)8316.

16 A. Yoshimori and T. Kakitani, submitted to J. Chem. Phys.

17 W. Nadler and R.A. Marcus, J. Chem. Phys. 86(1987)3906.

18 H. Sumi and R.A. Marcus, J. Chem. Phys. 84(1986)4894.

19 M. Maroncelli, J. Chem. Phys. 94(1991)2084.

20 A. Yoshimori, submitted to Chem. Phys. Letters.

DISCUSSION

Wasielewski

How can you be sure of the "uniqueness" of your fit of experimental data for both CS and CR on multiple parameters such as the distance distribution of donors and acceptors and values of β? Are you fitting the CS data separately or simultaneously with the CR data?

Kakitani

We used the same parameter values for the CS and CR reactions. The distance distribution functions of the reorganization energy, mean ion–pair potential and electron–tunneling frequency factor are almost uniquely determined by the conditions that the normal region of the CS reaction is very sharp and is located at a very small energy gap (the edge of the normal region going into the diffusion limit rate is $-\Delta G \sim 0.1$ eV), the CS rate does not go below the diffusion limit rate until around $-\Delta G \sim 2.2$ eV, the normal region of the CS reaction shifts to the larger energy gap by about 0.8 eV, the inverted region of the CR reaction clearly appears for $-\Delta G > 2.0$ eV, the maximum intrinsic CS rate is about 10^{11} M^{-1} s^{-1} and the maximum rate of the CR reaction is about 10^{11} s^{-1}. We also combined with the Monte Carlo simulation data for the solvent reorganization energy of the ion–pair using spherical hard core model. The detailed procedure for fixing the parameter values step by step is shown in Ref. 12 of this paper. After fixing those distance dependencies, we theoretically calculated energy gap laws of the CS and CR reactions for $\beta = \infty$, 1.0, 0.3 and 0.1. Finally we found that β=1.0 is the best. This β–value is also consistent with our recent Monte Carlo simulation data considering the electronic polarizability of solvent molecules explicitly.

Jortner

The exponential distance dependence of the intermolecular electron exchange integral was attributed by you to direct exchange. This electronic interaction depends on the asymptotic behavior of the electronic wavefunctions and you were correct in using a lower orbital exponent for the atomic orbitals. I am, however, concerned about your argument that the direct exchange prevails at large donor–acceptor separations and that the interaction is screened (by the high–frequency dielectric constant). Evidence is currently accumulating that long–range electronic interactions in solutions and in biophysical systems are mediated by superexchange interaction via the empty orbitals of the medium molecules. Superexchange also results in an exponential distance dependence of the electronic interaction, and a careful analysis is required to distinguish between direct exchange and superexchange.

Kakitani

In my feeling, our treatment is not a direct exchange model in a usual sense, but it would conceptually correspond to the superexchange model because the screening by way of the refractive index is due to the electronic polarization of environment using the empty orbitals of the medium molecules. As a result, the atomic orbital to which tunneling electron belongs extends considerably.

Fleming

I found your discussion of nonlinearity in the Monte Carlo simulations to be rather surprising. In molecular dynamics simulations, even in cases where the energetic and dynamics are dominated by the first shell molecules, linear response theory has been found to be reasonably accurate. However it is true to say that the full extended charge distribution must be used to calculate the energetic of the first shell, rather than a continuum (point dipole) approximation. For the outer shells the point dipole approximation is quite good. Was this effect taken into account in your simulations?

Kakitani

I think that the definition of the first shell might be different between ours and yours. In our case, we divided the solvent sphere into some thin spherical shells around a charged molecule, and defined the saturation region as the 1st shell. According to our Monte Carlo simulation data, its width is 1.2∿1.5 Å, depending on the molecular parameters. This definition will be different from the definition by the others where the 1st shell is determined from the density distribution function. In the latter case, the width is about the diameter of solvent molecule and it is 4.4 A in our parameter. I do not think that the molecular model of a point dipole moment causes qualitatively different result from that of the charge distributed model. But, I will try to recalculate using the latter model.

Hynes

The relatively small nonlinear effects found by Carter and Hynes [J. Phys. Chem. 93, 2184 (1989)], for neutral and ion pair solutes in solvents of comparable size, suggests that nonlinear effects will be even smaller for large organic solutes (since the solute electric fields will typically be smaller). Can you comment?

My second question concerns your model for reactions with fixed solvent translational positions. You suggest that this is relevant for fast solvent reorientation and "slow" solvent translation. Could you indicate how slow the translation would have to be in realistic cases for this picture to be appropriate?

Kakitani

So far as only the permanent dipole moment is considered for solvent molecules, the nonlinear effect is not so large. However, if one takes into account the asymmetric, electronic polarizability, the nonlinearity increases. By the Monte Carlo simulations, we confirmed that the nonlinearity is considerable (e.g., 1.2). The physical reason of the enhancement of nonlinearity is the many body interaction among solvent molecules by the electronic polarizability. We think that the variable n will reflect the reorganizing motion of solvents around neutral and charged reactants due to ET. This motion may involve certain amount of diffusion process because the solvation number of solvent molecules around the reactant would change. In this respect, its time scale may be 1 ps∿100 ps. But, detailed examination of the experimental data relating to the present model will be necessary.

Tachiya

I do not agree with your conclusion that the dielectric saturation effect causes the broadening of the curve of the rate constant against the energy gap. I have shown that the rate constant curve is obtained by turning the free energy curve for the neutral pair upside down. (N. Tachiya, J. Phys. Chem. **93**, *7050* (1989)). You have also mentioned this in your talk, although in a different form. In the presence of dielectric saturation the free energy curve deviates upward from the parabola in the region far from the bottom. So the dielectric saturation should cause the narrowing of the rate constant curves.

Kakitani

There is no unique, definite way of going from a linear response system to the non-linear response system. Indeed, it is difficult to find the physical parameter which increases only the parameter a in the free energy function ax^4+bx^2. We changed the parameter values a and b simultaneously by keeping the reorganization energy of the CS reaction constant. However, more important thing which we obtained as a conclusion is that the energy gap law of the CS reaction is broad and that of the CR reaction is narrow. We proved it quite generally, independent of the values of a and b.

Tachiya

As I said, the rate constant curve is obtained by turning the free energy curve upside-down. If you assume a broad free energy curve, you obtain a broad rate constant curve as shown in your Figure. However, the important thing is what is the physical mechanism for having such a broad free energy curve.

Kakitani

There would be no physical sense to compare the width of the energy gap law between the linear response system and the nonlinear response system because we cannot find any good physical quantity which transfers between these two systems. (None of dipole moment, radii of solvent and solute, temperature, density of solvent alone can be such a physical quantity which plays a role of changing only the parameter a in the free energy curve.) In contrast to this, comparison of the energy gap between the CS and CR reaction in the same solvent is always possible and it has physical significance.

Kitahara

In order to clarify my understanding, please tell me what is "n", the typical translational motion in the latter part of your talk. You showed in your Figure the dependence of the free energy G on n. The function G should be dependent on the choice of "the typical motion". How does the local collisional aspect of the solvent molecules enter into your theory?

Kakitani

Mathematically, the free energy as a function of x and n is uniquely determined so that the increase of the free energy curvatures due to fixing the translational motion (i.e. n) at the minima of free energy curves which was observed in the Monte Carlo simulation may be reproduced by our model free energy function. The physical picture of the coordinate n may be the reorganizing movement of solvent molecules from positions surrounding the neutral reactants to those surrounding the charged products. Local collisions due to translational motions are reflected in the relaxation time τ_n.

Dynamics and Mechanisms of
Photoinduced Transfer and Related Phenomena
N. Mataga, T. Okada and H. Masuhara (Editors)

SOLVENT EFFECTS ON THE RATE *vs* FREE ENERGY DEPENDENCE OF PHOTOINDUCED CHARGE SEPARATION IN FIXED-DISTANCE DONOR-ACCEPTOR MOLECULES.

Michael R. Wasielewski, George L. Gaines, III, Michael P. O'Neil, Walter A. Svec, Mark P. Niemczyk, Luca Prodi, and David Gosztola

Chemistry Division, Argonne National Laboratory, Argonne, Illinois, 60439, USA

Abstract

A series of 32 fixed-distance donor-acceptor molecules using porphyrin donors, triptycene spacers, and 8 different quinone or quinonoid acceptors has been prepared. Subsets of these molecules are used to probe the dependence of photoinduced charge separation on free energy of reaction as a function of solvent both in liquid and solid solution. Data is presented on rates of charge separation in butyronitrile and toluene at 295 K and in 2-methyltetrahydrofuran at 77 K. The range of free energies explored spans three-quarters of the total energy available from excitation of the porphyrin to its lowest excited singlet state. The rate constant *vs* free energy relationships obtained show that the energy level of the ion-pair state is sensitive to solvent polarity and that semi-classical electron transfer theories can be used to model the results obtained.

1. INTRODUCTION

Recent interest in the role of the solvent in electron transfer reactions has focused on ultrafast photoinduced electron transfers and theoretical modeling of solvation.[1-10] Experimental attempts to explore rate *vs* free energy relationships involving photochemical charge separation reactions trace their heritage to the pioneering work of Mataga [11] and Weller [12]. The vast majority of work in this area involves donor-acceptor systems in which the donor and acceptor either freely diffuse in solution to form a complex following excitation, or are already complexed in their ground state prior to excitation. Many difficulties in the analysis of electron transfer rates as a function of reaction free energy arise due to translational degrees of freedom within such complexes.

A particularly controversial aspect of this work for many years was the theoretical prediction derived from the work of Marcus[13] and amplified by the work of many others[14] that a decrease in rate constant should occur when the free energy of reaction became very large. This is the so-called "inverted region" in the rate *vs* free energy relationship. Recently, Kakitani and Mataga[15] have proposed an alternative view in which selective solvation of the donor-acceptor pair influences the rate *vs* free energy dependence as a function of whether ion-pairs are created or destroyed in the reaction, or whether charge is merely shifted from the donor to the acceptor in the reaction. Experimental studies designed to test these theoretical models have to date involved mostly non-covalent donor-acceptor complexes.[16] The success with which non-photochemical charge shift reactions have been elucidated with the aid of fixed-distance, covalent donor-acceptor complexes [17] suggests that this approach is a useful one to employ for the study of photochemical charge separation reactions. We have employed this approach successfully in a number of previous studies of photoinduced charge separation and dark charge recombination reactions. We have studied the dependence of these reactions on free energy, distance, and solvation.[18-20] Our recent work focuses on the influence of solvent polarity both in liquid and solid solution on the free energy dependence of these reactions. This paper is an account of work in progress on an expanded series of porphyrin-based donor-acceptor molecules that possess fixed donor-acceptor distances and close structural relationships. The latter are important because we wish to maintain as closely as possible the same degree of electronic coupling between the donor and the acceptor within the series of molecules.

2. EXPERIMENTAL

The compounds in Figure 1 were synthesized by methods described earlier[18,19] and the syntheses generally follow methodology developed earlier[21]. The details of their preparation will be presented in a future publication.

Solvents for all spectroscopic experiments were dried and stored over 3 Å molecular sieves. HPLC grade toluene was distilled from $LiAlH_4$ (LAH). Butyronitrile was refluxed over $KMnO_4$ and Na_2CO_3, then twice distilled retaining the middle portion each time. 2-Methyltetrahydrofuran (MTHF) was freshly distilled from LAH before each experiment.

UV-visible absorption spectra were taken on a Shimadzu UV-160. Fluorescence spectra were obtained using a Perkin-Elmer MPF-2A fluorimeter interfaced to an IBM personal computer. All samples for fluorescence were purified by preparative TLC on Merck silica gel plates. Samples for fluorescence measurements were 10^{-7} M in 1 cm cuvettes. The emission was measured 90° to the excitation beam. Fluorescence quantum yields were determined by integrating the digitized emission spectra from 600 to 800 nm and referencing the integral to that for either free base or zinc porphyrin in toluene.[22]

Redox potentials for each donor acceptor molecule were determined in butyronitrile containing 0.1M tetra-n-butylammonium perchlorate using a Pt disc electrode at 21°. These potentials were measured relative to a saturated calomel electrode using ac voltammetry.[23] Both the one electron oxidations and reductions of these molecules exhibited good reversibility.

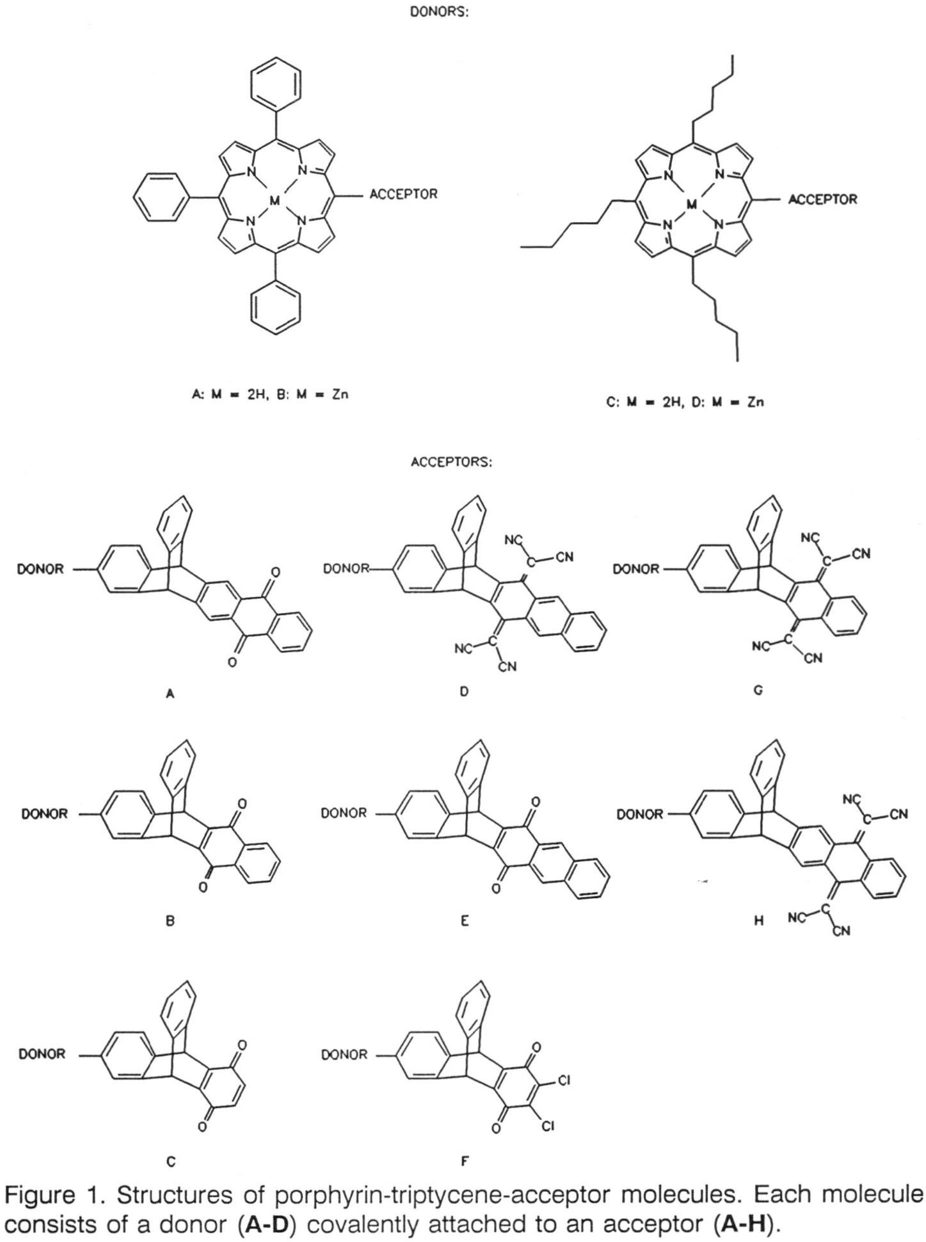

DONORS:

A: M = 2H, B: M = Zn

C: M = 2H, D: M = Zn

ACCEPTORS:

A D G

B E H

C F

Figure 1. Structures of porphyrin-triptycene-acceptor molecules. Each molecule consists of a donor (**A-D**) covalently attached to an acceptor (**A-H**).

The transient absorption spectra were obtained using a Rh-6G dye laser synchronously pumped by a frequency-doubled, mode-locked CW Nd-YAG laser. The 1.0 psec pulses of 610 nm light were amplified by a 4-stage dye amplifier (Rh-640) pumped by the frequency-doubled output (532nm) of a Nd:YAG laser. Malachite green saturable absorber dye jets between stages 2 and 3, and between stages 3 and 4 of the amplifier chain minimized the amplified stimulated emission generated in the amplifier. The amplifier typically produced a 1 mJ/pulse at a 10 Hz repetition rate. This beam was sent through a 60/40 beam splitter. The smaller portion was focused down to a 2 mm diameter beam and used as the excitation pulse. The larger portion was tightly focused into a 2 cm path length cell containing either 2/1 $CCl_4/CHCl_3$ or 1/1 H_2O/D_2O. This generated a continuum which was used as the probe light. The arrival at the sample of the probe beam was delayed relative to the excitation beam by an optical delay. The probe beam was divided into reference and measuring beams by a 50/50 beam splitter. Both probe beams passed through the sample. The reference beam passed through an area that was not illuminated by the excitation beam, while the measuring beam passed through the same portion of the sample through which the excitation beam passed. Both beams were then focused onto the slit of a monochromator. The monochromator dispersed the beams onto the face of an intensified SIT detector which was part of an optical multichannel analyzer (PAR OMA II). Solutions with an absorbance of about .3 at 610 nm (2 mm pathlength cells) were used.

Fluorescence lifetime measurements used 1.0 psec, 2 mm diameter, 200 μJ pulses from the same source as described for the transient absorbance experiments. The samples were placed in 0.5 cm cells (optical density ca. 0.03 at 610 nm) and emission 90° to the excitation was collected and focused onto the slit of a Hamamatsu C979 streak camera. The temporally dispersed image was recorded by the intensified SIT vidicon of the PAR OMA II. The geometry of the experimental set up determined the 15 ps instrument response. Decay times were obtained by iterative reconvolution of the data with least squares fitting using the Levenberg-Marquardt algorithm.

3. RESULTS

3.1 Free Energy Measurements

In polar liquids the free energies of charge separation in a donor-acceptor molecule, $\Delta G'_{cs}$, can be estimated with reasonable accuracy using the one-electron oxidation, E_{ox}, and reduction, E_{red}, potentials of the donor and acceptor, respectively, and the coulomb stabilization of the ion-pair:

$$\Delta G'_{cs} = E_{ox} - E_{red} - e_0^2/\epsilon_s r_{12} - E_s \tag{1}$$

where e_0 is the charge of the electron, ϵ_s is the static dielectric constant of the high polarity medium, r_{12} is the center-to-center distance between the ions, and E_s is the energy of the lowest excited singlet state of the porphyrin donor determined from the frequency of the (0,0) band of its fluorescence spectrum. We determined $\Delta G'_{cs}$ for the 32 compounds presented in Figure 1 in butyronitrile at 295 K containing 0.1M tetra-n-butylammonium perchlorate. Since r_{12} = 11 ± 1 Å for these compounds[24], the coulombic term is only 0.065 eV and is neglected. The values of $\Delta G_{cs}'$ are given in Table 1.

Table 1
Free energies of charge separation in butyronitrile for the compounds in Figure 1

	Donor			
	A	B	C	D
Acceptor				
A	0.05	0.43	0.18	0.62
B	0.26	0.64	0.39	0.83
C	0.40	0.78	0.53	0.97
D	0.67	1.05	0.80	1.24
E	0.19	0.57	0.32	0.76
F	0.71	1.09	0.84	1.28
G	0.92	1.30	1.05	1.49
H	0.53	0.91	0.66	1.10

3.2 Charge Separation Rate Constants in Liquid Solution.

Rate constants for electron transfer in butyronitrile were determined using picosecond transient absorption and emission spectroscopy along with fluorescence quenching measurements. The ion pair transient absorption spectra closely match those reported for the various porphyrin cation radicals.[25] For the most part the spectra of the acceptor anion radicals are of sufficiently low intensity that they are buried under the stronger porphyrin cation radical spectra. A typical transient absorption spectrum for molecule **BG** is shown in Figure 2.

The rate constants for electron transfer from the lowest excited singlet state of the porphyrin to the acceptor are plotted as a function of free energy in Figure 3. Note that eqn 1 is adequate for estimating the free energy of the ion-pair in polar liquids like butyronitrile. In this case the electrochemical determination of $\Delta G_{cs}'$ and the rate constant measurements are both performed in the same solvent. The data points in Figure 3 are not labeled with the identity of the compounds to avoid congestion. The identity of the compounds may be obtained by reference to the reaction free energies in Table 1.

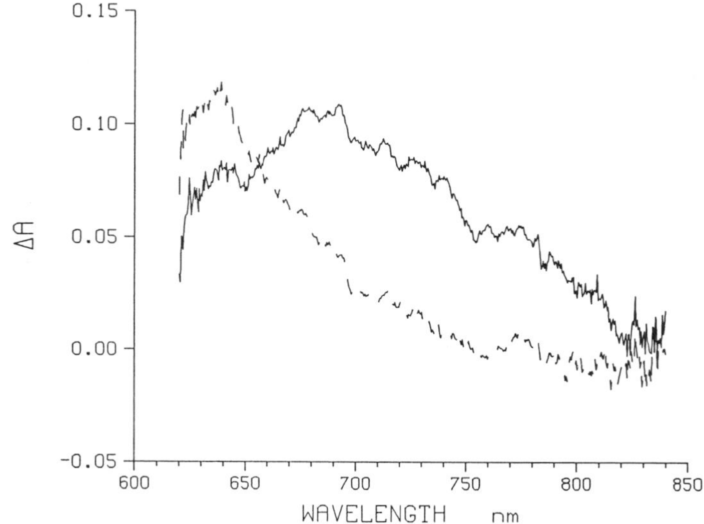

Figure 2. Transient absorption spectra of S_1 (----) and D^+-A^- (——) of **BG** in butyronitrile following a 1 ps laser flash at 610 nm.

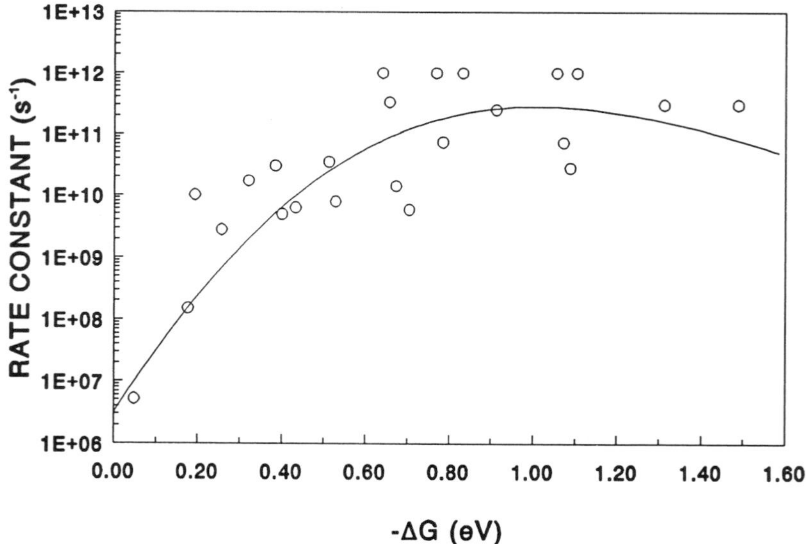

Figure 3. Rate *vs* free energy relationship for ^{1*}D-A -> D^+-A^- in butyronitrile at 295 K. The solid line is the theoretical fit to the data (see discussion).

3.3 Charge Separation Rate Constants in Glassy Solid Solution.

Most porphyrin-based donor-acceptor molecules exhibit significantly reduced efficiencies of light-initiated, singlet state electron transfer whenever they are dissolved in rigid glass media.[26] This occurs because solvent dipoles reorient around an ion-pair in a polar liquid, decreasing the energy of the ion-pair, while solvent dipoles cannot reorient around an ion-pair produced within a frozen solvent, and thus, provide little stabilization of the ion-pair.[27-30] As a result, the energy level of the ion-pair is much higher in the rigid glass than in the liquid. In fact, the ion-pair state energy may be so high that it lies above the energy of the excited state, in which case photoinduced electron transfer cannot occur. We recently reported that the ion-pair states in **BG** and **DG** are destabilized by as much as 0.9 eV in going from a polar liquid to a rigid glass.[31] To obtain a quantitative picture of the dependence of charge separation rate on free energy of reaction in a rigid glass, we measured charge separation rate constants for 14 molecules in Figure 1, which possess sufficiently large, negative free energies for charge separation to allow electron transfer to compete with excited singlet state decay in glassy MTHF at 77 K.

The free energy of charge separation, ΔG_{cs}, in the rigid glass is given by eqn 2:

$$\Delta G_{cs} \;=\; \Delta G_{ip} - E_s \qquad (2)$$

where ΔG_{ip} is the free energy of the ion-pair in the rigid glass and E_s is defined above. Since ion-pair recombination in porphyrin-acceptor molecules is non-radiative, the value of ΔG_{cs} in the rigid glass is difficult to obtain. On the other hand, in polar liquids the free energies of charge separation in these molecules can be determined from eqn 1. Rate constants for charge separation, k_{cs}, were determined from picosecond transient absorption and emission measurements at 77 K. In Figure 4 ln k_{cs} is plotted *vs* $-\Delta G'_{cs}$. Since we do not know *a priori* what the relationship between $\Delta G_{cs}'$ in polar liquids *vs* solids is, we plot ln k_{cs} *vs* the free energy determined electrochemically in butyronitrile. Strictly speaking, the total nuclear reorganization energy of the reaction obtained from the maximum of the ln k_{cs} *vs* $-\Delta G_{cs}'$ plot, 1.4 eV, is not correct. Nevertheless, the plot reveals that a smooth monotonic increase in rate constant occurs when the free energy becomes more negative. Fortunately, we are able to determine directly the relationship between the free energy measured electrochemically in a polar liquid and the true free energy in the solid state. When $\Delta G_{cs} \approx 0$, thermal repopulation of the lowest excited singlet state from the ion-pair state may yield biphasic fluorescence decays.[32] Assuming rapid equilibrium between the excited and ion-pair states, these biphasic decays can be fit analytically to a model in which the two equilibrating states decay to ground state with their own respective rate constants. Thus, the data can be used to obtain the rate constants for both the charge separation reaction, k_{cs}, and the thermal repopulation of the singlet state, k_{rep}. The value of ΔG_{cs} is obtained directly from the relationship, $\Delta G_{cs} = -RT \ln(k_{cs}/k_{rep})$. We have observed thermal repopulation of the lowest excited singlet state of the porphyrin in two of these molecules, **BC** and **DB**. These data show that $\Delta G_{cs} < 0.01$ eV for **BC** and **DB**. Assuming a linear free energy relationship the $\Delta G'_{cs}$ energy scale can be calibrated to yield ΔG_{cs} in the rigid glass using the data for **BC** and **DB**, Figure 4. Thus, the energies of the ion-pair states of these molecules in glassy MTHF are destabilized by $\Delta G_d = 0.80 \pm 0.05$ eV relative to their energies in highly polar butyronitrile ($\epsilon_s = 20$).

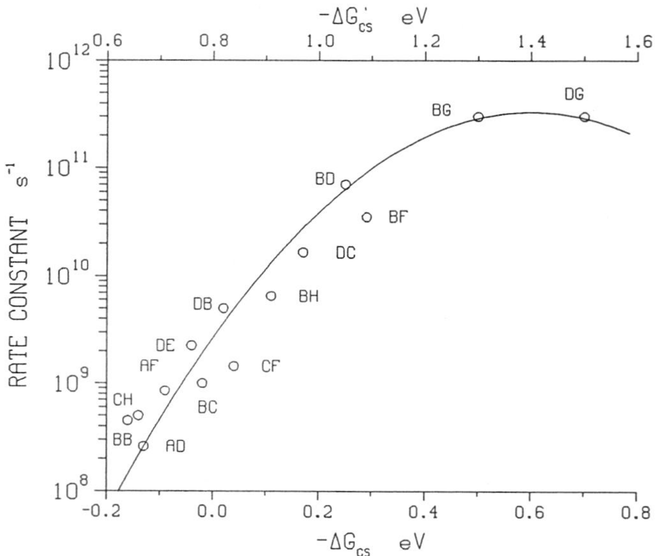

Figure 4. Rate *vs* free energy relationship for selected molecules in MTHF at 77 K. The labels correspond to the compounds depicted in Figure 1.

4. DISCUSSION

4.1 Free Energy Changes as a Function of Solvent Polarity.

Our results show that the free energy of a charge separation reaction determined from electrochemical redox potentials of the donor and acceptor in a high static dielectric constant medium are not accurate in the solid state or even in other solvents for that matter. We need to relate the change in free energy measured in polar media to that in other media such as non-polar solvents and solids. In classical Marcus electron transfer theory the activation energy for the charge separation reaction is given by eqn 3.

$$E_a = (\Delta G_{cs} + \lambda)^2/4\lambda \qquad (3)$$

where λ is the total nuclear reorganization energy of the reaction.
The true value of ΔG_{cs} in a particular medium can be given as

$$\Delta G_{cs} = E_{ox} - E_{red} - E_s + \Delta G_d \qquad (4)$$

where ΔG_d the energy by which the ion-pair is destabilized when it is taken from a medium with high ϵ_s to one in which ϵ_s is low.

Using the Born dielectric continuum model of the solvent, Weller[33] derived eqn 5 to calculate the ion-pair destabilization energy, ΔG_d, in a solvent with an arbitrary value of ϵ_s and high frequency dielectric constant ϵ_o, if the redox potentials of the donor and acceptor are measured in a medium with a high dielectric constant, ϵ_s':

$$\Delta G_d = e_o^2 [(1/2r_1)+(1/2r_2)-(1/r_{12})]/\epsilon_s - e_o^2[(1/2r_1)+(1/2r_2)]/\epsilon_s' \quad (5)$$

where r_1 and r_2 are the radii of the two ions, and the remaining parameters are defined above.

Marcus dissected the total nuclear reorganization energy for electron transfer into two terms: λ_s, the change in energy due to nuclear motion of the solvent, and λ_i, the change in energy due to nuclear motion within the donor and acceptor.[13] Thus,

$$\lambda = \lambda_s + \lambda_i \quad (6)$$

Later, Marcus used the same Born model of the solvent to derive an expression for λ_s [34]:

$$\lambda_s = e_o^2 [(1/2r_1)+(1/2r_2)-(1/r_{12})][(1/\epsilon_o) - (1/\epsilon_s)] \quad (7)$$

If we define $\Delta G_E = E_{ox} - E_{red} - E_s$ and $C = e_o^2[(1/2r_1)+(1/2r_2)-(1/r_{12})]$, eqns 4,5,6,and 7 can be combined to yield eqn 8:

$$E_a = [\Delta G_E + (C/\epsilon_o) + \lambda_i]^2 / 4[C[1/\epsilon_o]-[1/\epsilon_s] + \lambda_i] \quad (8)$$

Each term that comprises ΔG_E can be measured experimentally, while C can be estimated reasonably well from the molecular structure of the donor-acceptor system. Equation 8 predicts that plots of ln k_{cs} *vs* $-\Delta G_E$ in a series of structurally related compounds will possess maxima at the *same value of* $-\Delta G_E$ *independent of the static dielectric constant of the medium in which* k_{cs} *is measured.* This occurs because ϵ_o is always about 2 for typical organic media, and C and λ_i are values that are approximately constant for a given structural type. Moreover, since the denominator depends both on ϵ_o and ϵ_s, changing ϵ_s will generate a series of nested rate *vs* free energy curves with widths that depend on the static dielectric constant of the solvent.

4.2 Rate vs Free Energy Relationships in Liquid Solution.

Figure 3 shows that the rates of charge separation within the porphyrin-triptycene-acceptor series increase dramatically with increasing free energy of reaction and remain very fast throughout the high driving force regime. The theoretical curve through the points is obtained from semi-classical electron transfer theory [35,36] with an electronic coupling = 40 cm^{-1}, λ_i = 0.3 eV, λ_s = 0.7 eV, and a single vibrational frequency of 1500 cm^{-1}. With these parameters the semi-classical theory does not predict a steep decrease in rate constant in the inverted region, i.e. the free energy domain where $-\Delta G_{cs} < 1.0$ eV. The experimental rate constants remain above 10^{11} s^{-1} all the way out to $-\Delta G_{cs}$ = 1.5 eV, which is 3/4 of the total energy available from excitation of the porphyrin to its lowest excited singlet state.

Equation 8 predicts that changing the static dielectric constant of the solvent should not result in a change in the maximum of the rate *vs* free energy profile if the data is plotted against $-\Delta G_E$. In addition, eqn 8 predicts that a series of nested curves will result if the rate constants are measured in solvents wherein ϵ_s is varied. Figure 5 shows a series of plots using eqn 8 in the classical Marcus treatment of the activation free energy wherein ϵ_s is varied from 2 - 20. The nesting of the curves is quite apparent as is the common maximum.

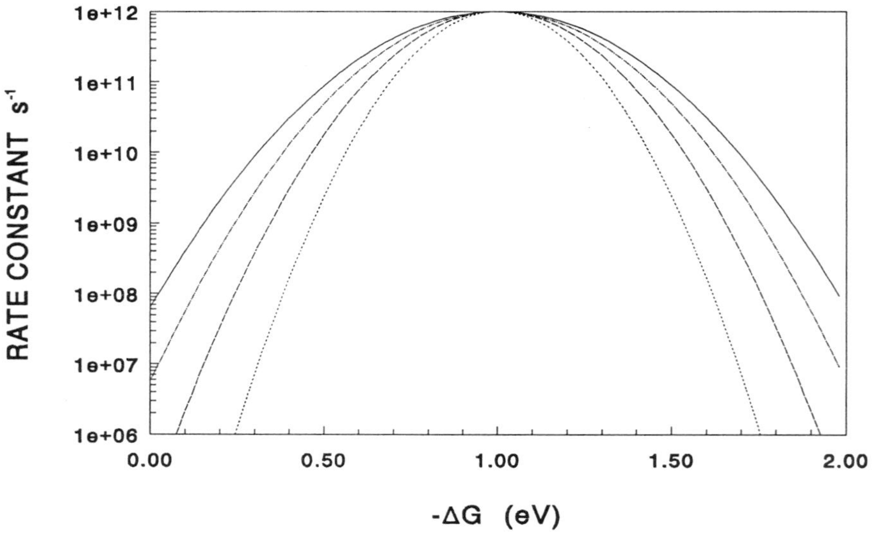

Figure 5. Rate *vs* free energy calculation using classical Marcus theory with ϵ_s = 20 ——; 7 — -; 3.5 ---; 2 ....

Figure 6 shows the same series of curves using semi-classical electron transfer theory with a single high frequency vibration of 1500 cm^{-1}. The electronic coupling is 40 cm^{-1} and λ_i = 0.3 eV. Once again the nesting of the curve with a common maximum is apparent, although in the inverted region the difference in rate as a function of ϵ_s is small.

Figure 7 shows a plot of ln k_{cs} *vs* $-\Delta G_E$ for a subset of compounds shown in Figure 1 obtained in toluene and in butyronitrile. The solid line in Figure 5 fits the butyronitrile data with the following parameters: ω = 1500 cm^{-1}, electronic coupling = 40 cm^{-1}, λ_s = 0.7 eV, λ_i = 0.3 eV, ΔG_d = 0, and T = 300 K, while the dotted line fits the toluene data with ω = 1500 cm^{-1}, electronic coupling = 40 cm^{-1}, λ_s = 0.1 eV, λ_i = 0.3 eV, ΔG_d = 0.6, and T = 300 K. Although the data set that we have obtained thus far is incomplete, the data show that the maxima of ln k_{cs} *vs* $-\Delta G_E$ curves for toluene and butyronitrile are approximately the same as predicted by eqn 8. Moreover, the plot

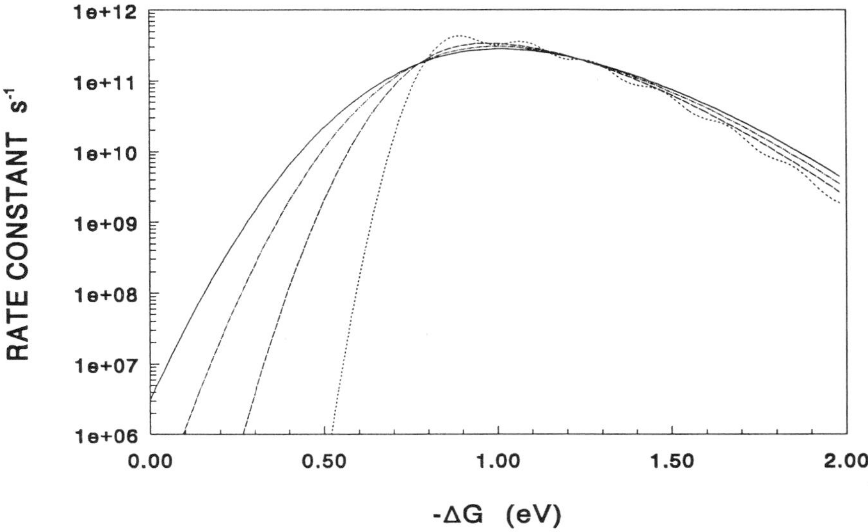

Figure 6. Rate *vs* free energy calculation using semi-classical electron transfer theory with ϵ_s = 20 ——; 7 — -; 3.5 ---; 2 ....

clearly reveals the nesting of the rate *vs* free energy curves that is predicted by the treatment derived above, and shown in Figure 6.

4.3 Rate vs Free Energy Relationships in Glassy Solid Solution.

Using the dielectric continuum model of the solvent, eqn 3 can be used to predict ΔG_d. Since ϵ_s = 2.6 for glassy MTHF at 77 K,[37] ϵ_s' = 20 for butyronitrile at 295 K, r_{12} = 11 ± 1 Å[24], r_1 for the porphyrin = 5 Å,[24] and r_2 for the quinone = 3 Å,[24] eqn 3 predicts that ΔG_d = 0.78 eV. This calculated value of ΔG_d agrees remarkably well with the value of ΔG_d = 0.8 eV that we measure directly. Electron transfer theories beginning with that of Marcus show that the electron transfer rate constant will be greatest when $\Delta G_{cs} + \lambda = 0$, where λ is the total nuclear reorganization energy for the electron transfer reaction.[13] This reorganization energy is comprised of terms describing both the solvent reorganization energy, λ_s, and the internal nuclear reorganization of the donor-acceptor molecule, λ_i, where $\lambda = \lambda_s + \lambda_i$. Semi-classical electron transfer theory[35,36] is used to fit the data in Figure **2** to a monotonic function with rate constants increasing as free energy of reaction becomes increasingly negative. The fitting parameters are ω = 500 cm^{-1}, electronic coupling = 40 cm^{-1}, λ_s = 0.3 eV, λ_i = 0.3 eV, and T = 77 K, The plot of ln k_{cs} *vs* -ΔG_{cs} in Figure **2** reaches a maximum at ΔG_{cs} = -0.6 eV. Since the dielectric continuum model of the solvent is in reasonable agreement with our experimental results, we can use eqn 7 to determine λ_s. Using the parameters given above with ϵ_{op} = 2, eqn 7 yields λ_s = 0.3

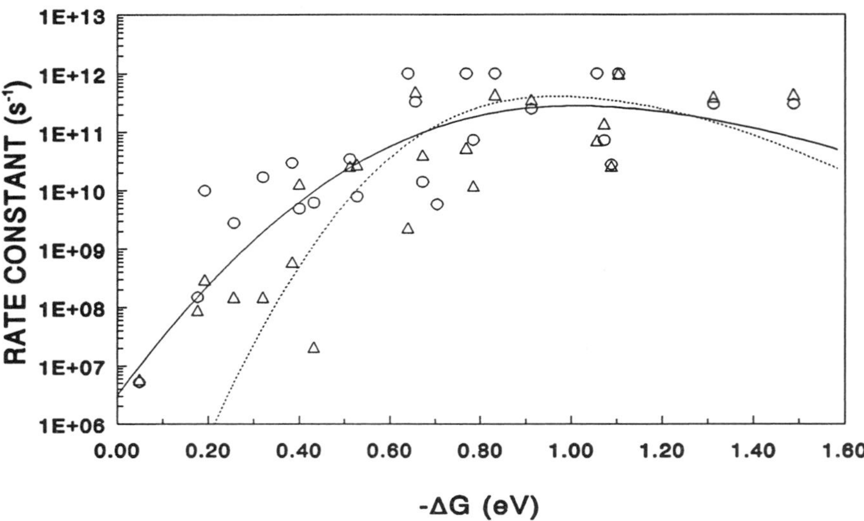

Figure 7. Rate *vs* free energy dependence of 24 compounds from Figure 1 in butyronitrile (O and ——) and toluene (△ and ---).

eV. Finally, since λ = 0.6 eV, the internal nuclear reorganization energy of the porphyrin-triptycene-acceptor system is 0.3 eV. This number is similar to values of λ_i determined for other organic π donor-acceptor molecules.[17,38]

5. CONCLUSIONS

Our studies on photoinduced charge separation within fixed-distance donor-acceptor molecules as a function of solvent polarity show that the rate constants for charge separation remain large well into the inverted region of the rate *vs* free energy profile. While the rate $_{vs}$ free energy dependence can be modeled by semi-classical electron transfer theory, a more thorough look into the theoretical models that can be used to describe this phenomenon will await completion of measurements on all 32 molecules in a wide variety of solvents. Since the energy level of the ion pair is strongly modulated by the solvent polarity, this modulation can be used to further examine the rate *vs* free energy dependence of these reactions.

Our results show that porphyrin-triptycene-acceptor molecules possess ion-pair states that are destabilized by 0.8 eV in rigid glasses relative to their energies as determined from electrochemical measurements in polar liquids. This information can be used to design multi-step electron transfer molecules to separate charge efficiently in the solid state.

6. ACKNOWLEDGEMENT

This work was supported by the Division of Chemical Sciences, Office of Basic Energy Sciences, U. S. Department of Energy under contract W-31-109-Eng-38.

7. REFERENCES

1 Brunschwig, B.; Ehrenson, S.; Sutin, N. *J. Phys. Chem.*, **1987**, *91*, 4714.

2 Castner, Jr., E. W.; Bagchi, B.; Maroncelli, M.; Webb, S. P.; Ruggiero, A. J.; Fleming, G. R. *Ber. Bunsenges. Phys. Chem.* **1988**, *92* 363.

3 Maroncelli, M.; Fleming, G. R. *J. Chem. Phys.* **1988**, *89*, 875.

4 Hynes, J. T. *J. Phys. Chem.* **1986**, *90*, 3701.

5 Rips, I.; Klafter, J.; Jortner, J. *J. Chem. Phys.*, **1988**, *89*, 4288.

6 Wolynes, P. *J. Chem. Phys.* **1987**, *86*, 5133.

7 Sparpaglione, M.; Mukamel, S. *J. Phys. Chem.* **1987**, *91*, 3938.

8 Kahlow, M. A.; Jarzeba, W.; Kang, T. J.; Barbara, P. F. *J. Chem. Phys.* **1988**, *90*, 151.;

9 Bashkin, J. S.; McLendon, G.; Mukamel, S.; Marohn, J. *J. Phys. Chem.*, **1990**, *94*, 4757.

10 Heitele, H.; Pollinger, F.; Weeren, S.; Michel-Beyerle, M. E. *Chem. Phys.* **1990**, *143*, 325.

11 Mataga, N. in "The Exciplex", M. Gordon and W. R. Ware, Eds. Academic, New York, 1975, p. 113.

12 Rehm, D.; Weller, A. *Israel J. Chem.*, **1970**, *8*, 259.

13 Marcus, R. A. *J. Chem. Phys.* **1956**, *24*, 966.

14 For a review see DeVault, D. "Quantum Mechanical Tunnelling in Biological Systems", Cambridge Univ. Press, Cambridge, 1984.

15 Kakitani, T.; Mataga, N. *J. Phys. Chem.* **1986**, *90*, 993.

16 For a review see: Wasielewski, M. R. in "Photoinduced Electron Transfer, Vol I., Fox M. A. and Chanon, M., Eds., Elsevier, Amsterdam, 1988, p. 161.

17 Miller, J. R.; Calcaterra, L. T.; Closs, G. L. *J. Am. Chem. Soc.* **1984**, *106*, 3047.

18 Wasielewski, M. R.; Niemczyk, M. P. *J. Am. Chem. Soc.* **1984**, *106*, 5043.

19 Wasielewski, M. R.; Niemczyk, M. P.; Svec, W. A.; Pewitt, E. B. *J. Am. Chem. Soc.* **1985**, *107*, 5583.

20 Wasielewski, M. R.; Niemczyk, M. P. in "Porphyrins, Excited States and Dynamics", Gouterman, M.; Rentzepis, P.,Straub, K. D., Eds. ACS Symposium Series, 1986, p. 154.

21 Lindsey, J. S.; Wagner, R. W. *J. Org. Chem.*, **1989**, *54*, 828.

22 Seybold, P. G.; Gouterman, M. *J. Molec. Spectroscopy* **1969**, *31*, 1.

23 Wasielewski, M. R.; Smith, R. L.; Kostka, A. G. *J. Am. Chem. Soc.* **1980**, *102*, 6923.

24 Determined from CPK molecular models.

25 Fajer, J.; Borg, D. C.; Forman, A.; Dolphin, D.; Felton, R. H. *J. Am. Chem. Soc.* **1970**, *92*, 3451.

26 Harrison, R. J.; Pearce, B.; Beddard, G. S.; Cowan, J. A.; Sanders, J. K. M.

Chem. Phys. **1987**, *116*, 429.
27 Miller, J. R.; Peeples, J. A.; Schmitt, M. J.; Closs, G. L. *J. Am. Chem. Soc.*, **1982**, *104*, 6488.
28 Chen, P.; Danielson, E.; Meyer, T. J. *J. Phys. Chem.* **1988**, *92*, 3708.
29 Kakitani, T.; Mataga, N. *J. Phys. Chem.* **1988**, *92*, 5059.
30 Marcus, R. A. *J. Phys. Chem.* **1990**, *94*, 4963.
31 Wasielewski M. R.; Johnson, D. G.; Svec, W. A.; Kersey, K. M.; Minsek, D. W. *J. Am. Chem. Soc.* **1988**, *110*, 7219.
32 Heitele, H.; Finckh, S.; Weeren, S.; Pollinger, F.; Michel-Beyerle, M. E. *J. Phys. Chem.* **1989**, *93*, 5173.
33 Weller, A. *Z. Phys. Chem. N.F.* **1982**, *133*, 93.
34 Marcus, R. *J. Chem. Phys.* **1965**, *43*, 679.
35 Hopfield, J. J. *Proc. Natl. Acad. Sci. USA*, **1974**, *71*, 3640.
36 Jortner, J. *J. Chem. Phys.* **1976**, *64*, 4860.
37 Furutsuka, T.; Imura, T.; Kojima, T.; Kawabe, K. *Technol. Rep. Osaka Univ.* **1974**, *24*, 367.
38 Oevering, H.; Paddon-Row, M. N.; Heppener, M.; Oliver, A. M.; Cotsaris, E.; Verhoeven, J. W.; Hush, N. S. *J. Am. Chem. Soc.* **1987**, *109*, 3258.

DISCUSSION

Faure

The spacer used in the series of chemically bond donor–acceptor systems is polarizable. Do you expect that energetic terms such as ion–induced dipole are negligible (in the calculation of the driving forces)?

Wasielewski

Yes, we have compared a series of ZnTPP–spacer–naphthoquinone molecules not shown here that do not use polarizable spacers. We find no indication of a strong dependence of rate on the presence of a polarizable spacer.

Steiner

Concerning the population of ZnTPP triplet state from the $ZnTPP^{+}$ Q^{-} CT state for which you showed an ESR spectrum with a polarization pattern different from that observed if the ZnTPP triplet is formed by intramolecular ISC, what is the polarization mechanisms and how does it account for the specific polarization?

Wasielewski

The triplet state of ZnTPP within ZnTPP–spacer–TCNQ occurs from the following mechanism: ^{1*}ZnTPP–spacer–TCNQ –> ^{1}\{Zn^{+}TPP–spacer–$TCNQ^{-}$\} –> ^{3}\{Zn^{+}TPP–spacer–$TCNQ^{-}$\} –> ^{3}ZnTPP–spacer–TCNQ. Spin inversion does not occur by the radical pair mechanism. Triplet character results from the large change in angular momentum that accompanies the electron transfer between the systems that are nearly perpendicular. This results in over–population of the spin–sublevel of ZnTPP that has its principal axis in the plane of ZnTPP.

Steiner

Concerning the nature of the spin sublevel selective ISC process, do you think that it occurs in combination with electron transfer of the electron from the quinone to the porphyrin moiety or that ISC is accomplished in the CT state before back transfer of the electron takes place?

Wasielewski

We favor the latter mechanism.

Mobius

You have observed two exponential fluorescence decay in solid solution of the porphyrin–naphthoquinone system. Such behavior could be due to heterogeneity of the solid system. Did you check the decay behavior of the porphyrin without linked quinone in solid solution?

Wasielewski

We observed fluorescence decays from most of the molecules examined in the solid state. In addition, we examined the appropriate reference porphyrins containing no electron acceptors. In glassy MTHF we did not see significant deviations from exponential behavior except for two molecules (DB and BC) noted. As the temperature is changed, the relative amplitudes of the two components change dramatically. All the other molecules show no such temperature dependence. Molecules with lower driving force for electron transfer than DB and BC show only long fluorescence decays, while those with higher driving force than DB and BC show only short components. These latter cases are consistent with dominant quenching of the excited state by the competitive electron transfer process.

Kakitani

You have obtained the energy gap law of the photoinduced charge separation (CS) rate in MTHF glass at 77 K. You corrected the energy gap for all the species by estimating the destabilization energy in rigid glass medium for the geminate ion pair as 0.8 eV. With respect to this, I would like to point out the following fact. The solvent molecules surrounding the neutral pair of reactants are randomly oriented and its manner differs from pair to pair. If solvent configurations are frozen in those forms, the stabilization energy by the solvent for the ionpair which might be formed by the photoinduced CS is not enough and its value will greatly distributes from ionpair to ionpair. That is, the energy gaps for the photoinduced CS reaction distribute very much. The distribution function of the energy gap can be uniquely determined if one knows the solvent reorganization energy in non-frozen state at room temperature. Since the magnitude of the average correction 0.8 eV you adopted is large, it is expected that the energy gap distribution is large. The energy gap distribution would affect significantly the energy gap law because the CS would proceed using the preferable energy gap part at first and less preferable energy gap part in the latter time. How do you think about this fact?

Wasielewski

The energy gap in frozen MTHF at 77 K for compounds DB and BC is accurately determined to be 0 by a direct experimental measurement. We assume, based on the fact that we have a set of molecules with a homogeneous structural type, that a linear free energy relationship exists within this set, when data at 77 K and 295 K are compared. If each molecule were experiencing significantly different selective solvation in either the excited or ion-pair states, then considerable dispersion in the ln k_{CS} vs $-\Delta G_{CS}$ curve would occur. Experimentally, this is not observed.

Rettig

Comparing the low temperature and room temperature, if you are on the top of the Marcus bell-shaped curve at 77 K, you should expect a reduction of charge separation rate upon temperature increase, due to the shift of ΔG. Do you observe that?

Wasielewski

The maximum of the rate vs free energy curve occurs at $-\Delta G = 0.6$ eV at 77 K. The rate constant is about 3×10^{11} s^{-1}. If at room temperature we measure a compound with $-\Delta G = 0.6$ eV, the rate falls to 6×10^{10} s^{-1}. Since $-\Delta G = 0.6$ eV is a relatively large driving force in either case, the rate constants are both large.

Azumi

I would like to know the magnitude of exchange integral (J) you obtained from the simulation of your spin–correlated EPR spectrum. Your abstract says that J is close to zero. However, it does not seem to be zero from your EPR spectrum.

Wasielewski

The spectra can only be simulated when $J < 1G$. The spectra are dominated by the anisotropic dipolar interaction, D, and g–factor anisotropy, especially that of the quinone. If we try to use larger values of J, the simulated spectra split to show additional lines not seen in the experimental spectra.

Mobius

Do you have the kinetic data for electron transfer from the porphyrin to the little naphthoquinone, and how do they fit in the series of data you showed in the first part of your talk?

Wasielewski

Yes, we have kinetic data for the molecule in which the O–O axis of the naphthoquinone acceptor is parallel to the plane of the porphyrin. The rate constant for the initial charge separation is 3×10^{11} s^{-1} and that of the back reaction is 8×10^{9} s^{-1}. This back reaction is 4x faster than that of the isomeric naphthoquinone. Since, the $-\Delta G$ of reaction is the same in both cases, the increase in rate is due strictly to increased electronic coupling.

Itoh

Could you explain the temperature dependence of the reaction rate?

Wasielewski

We have measured the temperature dependence of a few of the molecules that I described. Generally speaking, the electron transfer rate constants decrease gradually until the glass point for the solvent is attained. There, the rates change more dramatically, and then remain constant down to 5K.

Dynamics and Mechanisms of
Photoinduced Transfer and Related Phenomena
N. Mataga, T. Okada and H. Masuhara (Editors)

Primary Charge Separation in Photosynthetic Bacterial Reaction Centers and Femtosecond Solvation Dynamics

Chi-Kin Chan, Sandra J. Rosenthal, Theodore J. DiMagno, Lin X.-Q. Chen, Xiaoliang Xie, James R. Norris and Graham R. Fleming

Department of Chemistry and the James Franck Institute, The University of Chicago, 5735 South Ellis Avenue, Chicago, IL 60637, U.S.A.

1. INTRODUCTION.

The capability of generating reliable, tunable femtosecond pulses has given new impetus to the study of condensed phase dynamics. In this paper we describe work in two areas: the dynamics of polar solvation and the primary charge separation step in the reaction centers of photosynthetic bacteria.

2. EXPERIMENTAL.

Two different systems were used to collect the data described in this paper. The experiments on reaction centers used a 20 Hz amplified colliding pulse ring laser system. Two continua were generated, from one of which was selected (via a 10 nm bandwidth filter) the 870 nm pump pulse. This was further amplified in a single pass Bethune type cell using LDS-867 to an energy of about 4 μJ. The second continuum was used as the probe pulse. The polarization of the pump and probe pulses were generally set at the magic angle, although experiments were also performed with parallel and perpendicular polarizations. The instrument response function, determined by monitoring the bleaching of the dye IR-143, was typically 300 fs. Typically 10-15% of the reaction centers were excited by the pump pulse.

The solvation studies utilized a fluorescence upconversion spectrometer based on reflective optics. The design is similar to that of Chesnoy *et al.* [1], but an elliptical mirror is used to collect and focus the fluorescence. A diagram of the optical arrangement is shown in Figure 1. The excitation pulses were typically 75 fs duration, 1.4 nJ in energy, with center wavelength in the range of 605-608 nm. The sum frequency was generated in a 1 mm LiO_3 crystal and yielded a typical instrument response function of 125 fs (FWHM). At long fluorescence wavelengths (750-775 nm) the instrument function is lengthened by about 45 fs by group velocity mismatch in the crystal [2]. The upconversion bandwidth is about 8 nm.

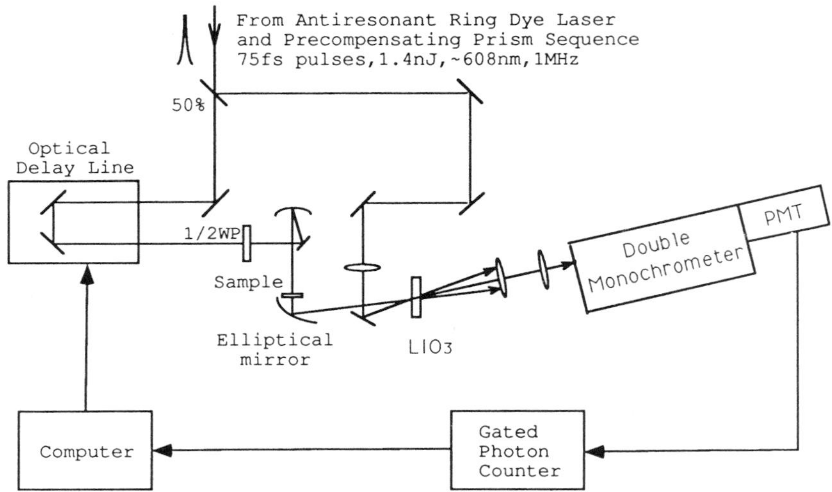

Figure 1. Fluorescence upconversion spectrometer

3. PRIMARY ELECTRON TRANSFER IN BACTERIAL REACTION CENTERS.

The primary charge separation process in bacterial photosynthetic reaction centers occurs with approximately 100% efficiency on the few picosecond time scale. The x-ray structural data [3] raise a number of intriguing issues regarding the mechanism of the primary electron transfer step. The structure shows two apparently similar branches, yet electron transfer proceeds entirely down one branch. The structure also shows an "accessory" bacteriochlorophyll molecule (B) between the primary donor (the special pair, P) and the "primary" acceptor bacteriopheophytin (H_A). The initial step of photosynthesis has been proposed to occur either by a one-step superexchange mechanism ($P^*BH \rightarrow P^+BH^-$) or by a two-step mechanism corresponding to the scheme [4-11]

$$P^*BH \rightarrow P^+B^-H \rightarrow P^+BH^-$$

Bixon *et al.* [11] have presented a theoretical model for the superposition of the two schemes presented above. According to their model the electron transfer path is dominated by the two-step sequential mechanism at room temperature. As the temperature is lowered, their model predicts that the fraction of the sequential mechanism becomes smaller and the contribution from the superexchange process becomes important. Evidence for a real B^- intermediate from femtosecond spectroscopy has been controversial. Initial studies did not reveal evidence for a B^- state, but the recent work of Zinth and coworkers does suggest a sequential process [12] whereas the parallel work of Kirmaier and Holten was interpreted as supporting the single step mechanism [13].

In interpreting the kinetic data a number of complicating factors must be considered. First, a model for the spectrum and its evolution is required. Most studies to date have (usually implicitly) assumed that while the spectra of the various species may overlap the intensity of the absorption bands can be considered as directly proportional to the concentration of the various molecular and ionic states present at a particular time. It is not *a priori* clear that such a weak coupling picture is appropriate for the reaction center [14-16], and therefore caution must be exercised in interpreting experimental data via simple kinetic schemes. Molecular orbital theory has been applied to the reaction center [17-20] but to date no calculations of the difference spectrum immediately following excitation have been presented. In the absence of further information we will interpret the data presented here in terms of a kinetic scheme, but consider this to be of uncertain validity at the shortest times. Even in the context of a kinetic scheme the possibility of shifts of the absorption bands as a result, for example, of electrochromic effects during charge separation must be considered. This bandshift effect is of particular concern in the B absorption region where a significant electrochromic shift is known to occur.

Given a kinetic picture it is then necessary to include all the species absorbing at a particular probe wavelength, along with the relative orientations of their transition moments. To illustrate this complexity Figure 2 shows data at 665 nm with the probe polarized both parallel and perpendicular to the pump pulse. This wavelength corresponds to the maximum of the bacteriopheophytin anion absorption band. The transition moment of H_A^- is oriented at 65° to that of P [13] resulting in a scaling of parallel and perpendicular curves according to $(1+2\cos^2\theta)$ / $(2-\cos^2\theta)$ for a randomly oriented sample. However, the kinetic information contained in the two curves is the same and thus when properly scaled the curves should superimpose. Figure 2 shows that this is clearly not the case and it is tempting to conclude that these data directly exclude the one-step model. However, when the contribution from the excited state of P (P^*) is included the situation changes. Figure 3 shows simulated kinetics for several values of the P^* extinction coefficient. We estimate $\varepsilon(P^*) = 1.8 \times 10^4\ M^{-1}\ cm^{-1}$ from a variety of information. Comparison of Figures 2 and 3 leads us to conclude that dichroic measurements at 665 nm in the absence of extremely precise extinction coefficients for all species involved cannot distinguish between the one-step and two-step mechanisms. Our conclusions (and data) at this wavelength are in accord with those of Kirmaier and Holten [13].

Similar experiments were also performed at 680 nm which is close to the peak of the B^- absorption band. Our parallel data look similar to those (at 665 nm) of Zinth and coworkers and lend support to the presence of detectable B^- concentration. Again, however, uncertainty over the contribution of P^* makes unambiguous interpretation difficult and for this reason we turn to a different spectral region.

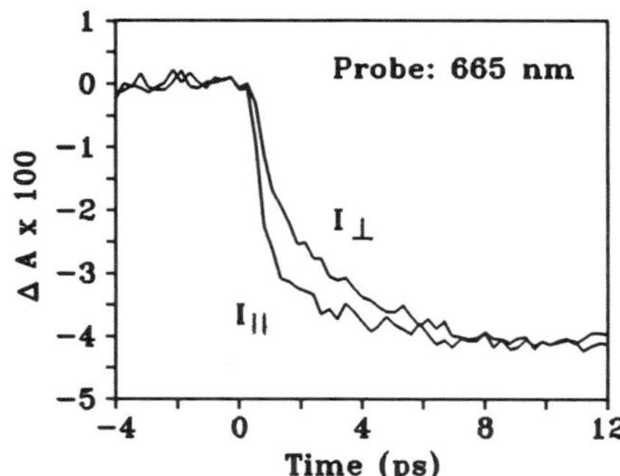

Figure 2. Transient absorption data at 665 nm with the probe pulse parallel (||) and perpendicular (⊥) to the 870 nm excitation pulse. The || and ⊥ curves are scaled to the same height for comparison.

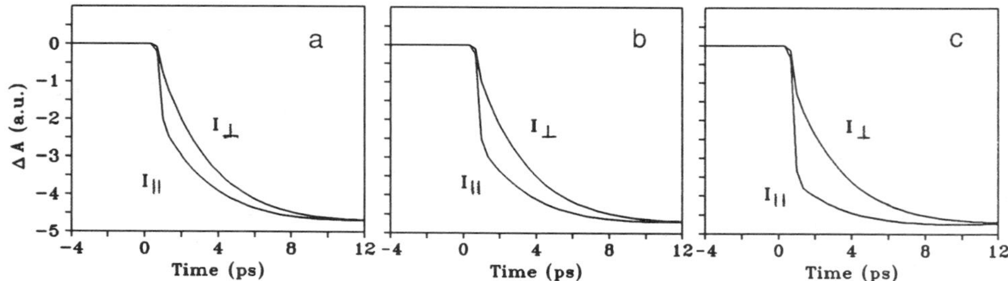

Figure 3. Simulations of the dichroic kinetics with different values of P^* extinction coefficients. (a) $\varepsilon(p^*)/\varepsilon(H_A^-)=0.24$; (b) $\varepsilon(p^*)/\varepsilon(H_A^-)=0.3$; (c) $\varepsilon(p^*)/\varepsilon(H_A^-)=0.4$. A single step kinetic model is used for the simulations and the sample is assumed to be randomly oriented. The angles of the transition moments used here for P^* and H_A^- absorption are 30° and 65° with respect to P absorption, respectively.

We consider the bacteriopheophytin Q_x absorption region (H_A at 545 nm, H_B at 525 nm) to be capable of providing much less ambiguous information on the mechanism of primary electron transfer for two major reasons. First, the bleaching of H_A (as a result of H_A^- formation) and the absorption of P^* give rise to signals with opposite signs and are much more easily disentangled from each other. Second, the fact that the inactive bacteriopheophytin, H_B, shows no bleaching allows for a much more direct measure of the P^* contribution and a small extrapolation from 526 nm to 545 nm rather than a large one from 545 nm to 665 nm in the determination of the P^* contribution. Thus simultaneous fits at 545 nm and 526 nm should prove incisive.

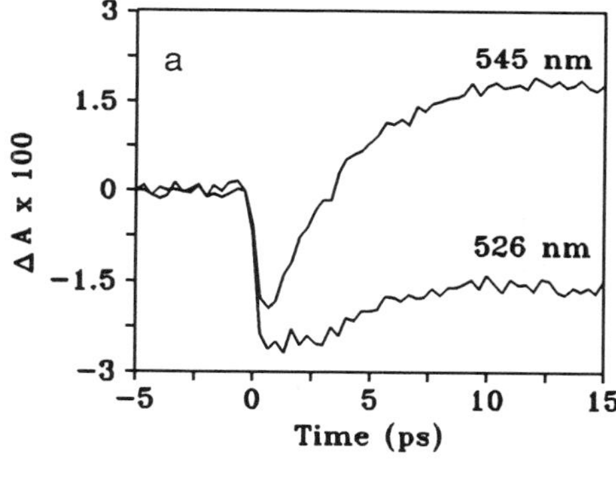

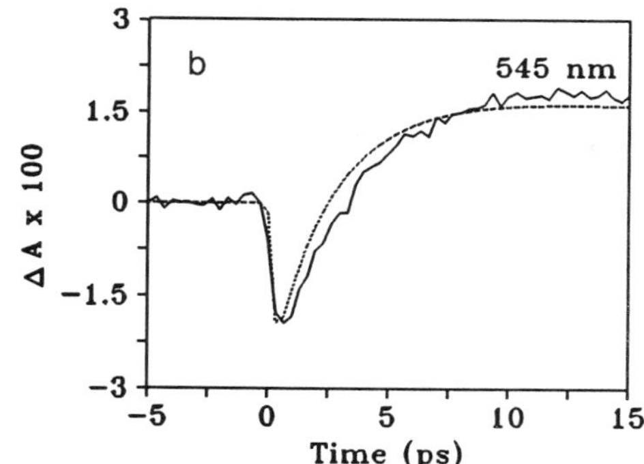

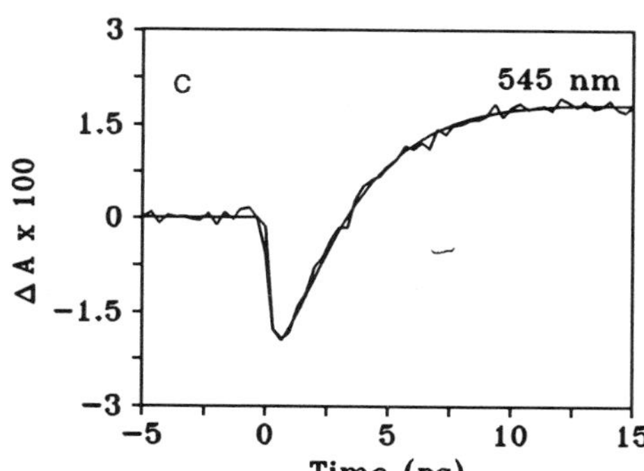

Figure 4. (a) Time-resolved absorption changes measured at 545 nm and 526 nm for *Rb. sphaeroides* R26 at 283 K with excitation at 870 nm. The probe polarization was set at ~55° with respect to the pump pulse. (b) A kinetic simulation using a one-step time constant of 2.6 ps results in a poor description of the 545 nm data. (c) Simulation of the 545 nm bleaching kinetics using a two-step model with k_1^{-1}=2.6 ps and k_2^{-1}=1.25 ps.

Figure 4 shows data at 545 nm and 526 nm for *Rb. sphaeroides* R26 at 283 K. Stimulated emission at 926 nm gives a time constant of 2.6 ± 0.2 ps for the decay of P^*. Fitting the data in Figure 4 to the one-step superexchange model gives a time constant of 4.0 ± 0.3 ps. Attempts to fit with a single 2.6 ps time constant result in very poor fits and examination of the fits suggests that a model with slower bleaching is required. As Figure 4 shows a two-step simulation with time constants of 2.6 ps for the first step and 1.25 ps for the second step fit the data well at 926 nm, 545 nm and 526 nm. (Of course these parameters will also fit the data at 665 nm and 680 nm but vide supra).

We have also examined the possibility that both sequential and super-exchange mechanisms operate at room temperature. Simulations show that the contribution of the single step process can be as much as 25% but beyond that the simulations start to deviate significantly from the data.

Given the prediction of Bixon *et al.* it is of great interest to perform the same analysis at low temperatures. The decay of P^* at 22 K gives a time constant of 1.6 ± 0.2 ps consistent with earlier studies [21]. Figure 5 shows data at 545 nm and

526 nm. Attempts to fit to a single step process give a 3.1± 0.3 ps time constant. Now, however, attempts to fit with the two-step mechanism with the first step fixed at 1.6 ps do not satisfactorily describe the data. The curve of 1.6 ps for the first step and 0.9 ps for the second step describes the early time portion of the data fairly well but all the curves show poor agreement at longer times. This finding led us to try the parallel pathway model with the superexchange rate being represented by k and the two-step process being described by k_1 and k_2. Figure 6 shows the fits for k^{-1} = 3.3 ps, k_1^{-1} = 3.1 ps and k_2^{-1} = 3.2 ps. The data are well described by this model and a series of simulations implies that the fractional contribution of the superexchange pathway is 50 ± 10%.

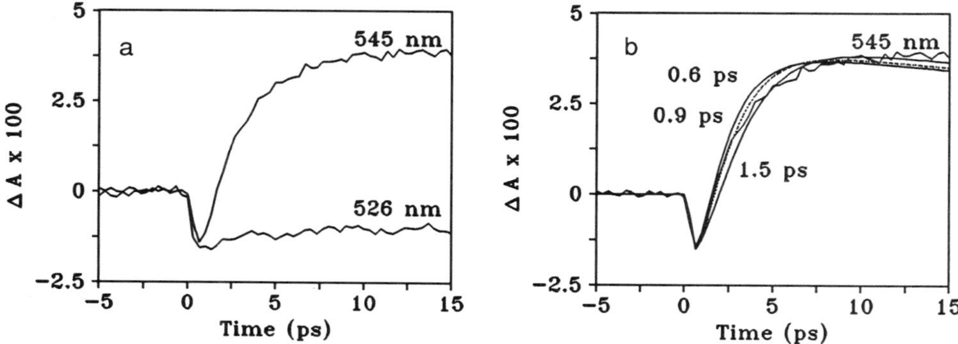

Figure 5. (a) Bleaching kinetics measured at 545 nm and 526 nm for *Rb. sphaeroides* R26 at 22 K. Excitation is at 870 nm. (b) Two-step model simulations of the 545 nm bleaching kinetics with a fixed first step time constant of 1.6 ps and a variable second step time constants of 0.6 ps, 0.9 ps and 1.5 ps.

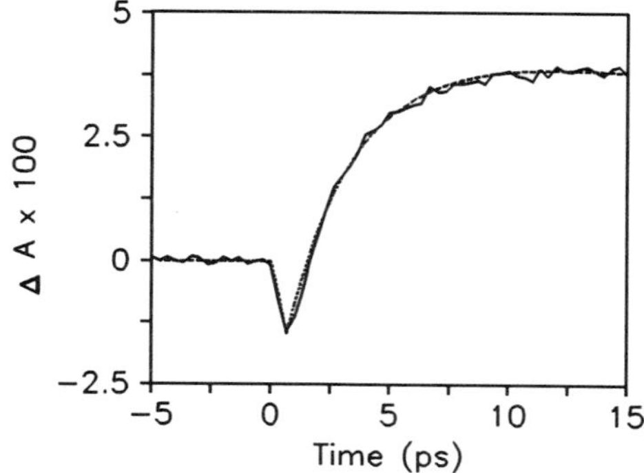

Figure 6. Fit of the 545 nm data shown in Figure 5a with the parallel electron transfer pathway model. The time constants are k^{-1}=3.3 ps, k_1^{-1}=3.1 ps and k_2^{-1}=3.2 ps.

Very similar data were obtained in the *Rb. capsulatus* mutant $Phe^{L181} \rightarrow Tyr$ with again roughly equal contributions from the two mechanisms around 20 K. Our results provide a rather different picture of the temperature dependence of the decay of P^* than the conventional one. In this picture the increased decay rate of P^* at low temperature comes about from the onset of a second electron transfer channel rather than the increase in rate in a single process. In fact, both the initial and second steps slow down slightly in the sequential process. It is of some interest to discuss the temperature dependence of the rate constants, k, k_1 and k_2. Figure 7 shows relevant potential surfaces.

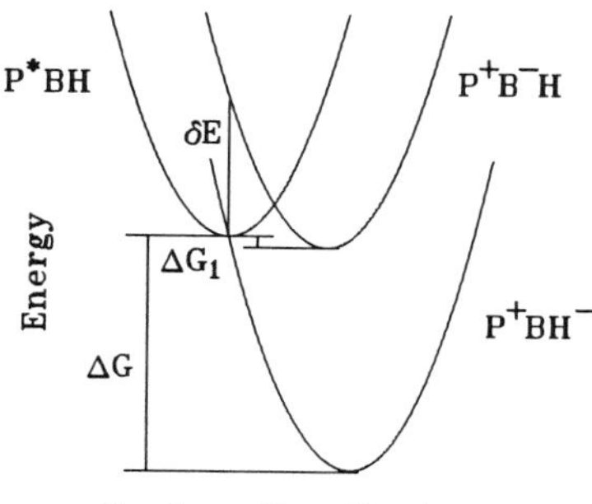

Figure 7. Schematic free energy surfaces for the three relevant states. Our data imply that ΔG_1 is negative. Note that the $P^+B^-H \rightarrow P^+BH^-$ process is in the inverted region.

According to Bixon *et al.* [11] the superexchange rate constant k is

$$k = \frac{2\pi}{\hbar}\left[\frac{1}{4\pi\lambda_1 k_B T}\right]^{\frac{1}{2}}\left(\frac{V_{PB}V_{BH}}{\delta E}\right)^2 \exp\{-E_a/k_b T\}$$

and the sequential rate constant k_1 is

$$k_1 = \frac{2\pi}{\hbar}\left[\frac{1}{4\pi\lambda_1 k_B T}\right]^{\frac{1}{2}} \frac{V_{PB}^2}{\left(1+4\pi V_{PB}^2\tau_s/\hbar\lambda_1\right)} \exp\{-E_{a1}/k_b T\}$$

with an analogous expression for k_2. The $T^{-1/2}$ terms in these expressions arise from the thermally averaged Franck-Condon factors at high temperatures ($k_B T > \hbar\omega_o$). Use of these expressions at 22 K is very likely beyond the range of their validity, however our purpose is simply to make qualitative comparison of the k and k_1 temperature dependence. From Figure 7 the simplest arrangement of the three surfaces that is consistent with our data is one in which E_a is zero and E_{a1} has a small positive value. The superexchange rate should increase by roughly 3-4 fold from the $T^{-1/2}$ term. A very small barrier for k_1 (≤ 50 cm^{-1}) is sufficient to counteract the $T^{-1/2}$ term over such a large temperature range. The second step of the consecutive process lies in the inverted region and the temperature dependence is more complex to predict. Experimental studies on model systems by Miller *et al.* [22] reveal a very weak temperature dependence in the inverted region consistent with our results. This over simplified analysis neglects changes in the electronic coupling and energy levels as the temperature is lowered and a full analysis will have considered these factors.

The fitting procedure described thus far did not include a reverse rate k_{-1} for the first step in the sequential process. This is equivalent to assuming that ΔG_1 (Fig. 7) is negative so that the $P^+B^-H \rightarrow P^*BH$ process is uphill. If limits on k_{-1} can be obtained from the data then a rather precise value for the free energy differences of P^*BH and P^+B^-H can be given. Preliminary analysis suggests that at room temperature k_{-1} cannot be larger than 0.3 k_1, otherwise the mechanism reverts to a single step process and the data are poorly fit.

Finally, a comment on why photosynthetic systems require two separate ("peacefully coexisting" [11]) electron transfer mechanisms seems in order. One answer may be that it is not possible to have one without the other and still have efficient electron transfer. Secondly, the two processes, as noted by Bixon *et al.*, allow stability of the electron transfer over a substantial range of energy gaps.

4. SOLVATION DYNAMICS ON FEMTOSECOND TIME SCALES.

The role of polar solvents in the dynamics of electron transfer reactions has been a topic of much interest over the past ten years [23-27]. Such studies have given impetus to direct experimental measures of solvation dynamics using the techniques of ultrafast spectroscopy. Accompanying the experimental work (see [25,27,28] for recent reviews) was a huge burst of theoretical activity aimed at developing molecular descriptions of solvation dynamics [29]. With the notable exception of van der Zwan and Hynes [30] the majority of the treatments did not consider any inertial contribution to the relaxation. In parallel with the experimental and theoretical studies a number of computer simulations of solvation dynamics have been carried out. These molecular dynamics simulations reveal the short-time portion of the solvent response which has been, until recently, inaccessible experimentally. Simulations in water [31], acetronitrile [32], methanol [33] and a methyl chloride like solvent [34] all reveal a 25-150 fs inertial component which contributes 60-80% of the solvent relaxation. It has been suggested that simulations, perhaps because of the neglect of polarizability or internal motions of the solvent, exaggerate the importance of the inertial component in the solvent response [35] . Very recently theoretical models incorporating inertial and viscoelastic effects have been developed by Chandra and Bagchi [36] .

Van der Zwan and Hynes [30] showed that the time-dependent fluorescence shift correlation function, C(t) [37] is directly proportional to the time-dependent dielectric friction:

$$C(t) = \zeta_D(t) / \zeta_D(0)$$

Simulations of $\zeta_D(t)$ for ions, ion pairs, electron and proton transfer and S_N2 reactions [38-42] reveal the time dependent friction to be dominated by a rapid inertial (Gaussian) component followed by a slow diffusive tail. Hynes and coworkers have demonstrated that chemical dynamics (e.g. transmission coefficients) are most strongly influenced by the short time behavior of the time dependent friction. Accordingly experimental verification of the importance of inertial contributions to solvent relaxation is highly desirable.

Experimentally the solvation correlation function C(t) is accessible from the time resolved fluorescence spectra via

$$C(t) = \frac{\nu(t) - \nu(\infty)}{\nu(0) - \nu(\infty)}$$

where ν is some characteristic frequency such as the peak or the first moment of the spectrum [43]. For our initial set of studies we used the probe molecule LDS-750 dissolved in acetonitrile. Previous work has shown that LDS-750 in polar aprotic solvents gives structureless well behaved time resolved spectra [44]. Evidence for two emitting species has recently been presented by Blanchard [45] in butanol solution so that care in solvent choice is clearly necessary. Fluorescence decays were recorded for 11 wavelengths (approximately one decay for every 10-15 nm of the steady state emission spectrum). Each decay was collected with a 6.67 fs step size and Figure 8 shows a typical data set along with the instrument response function. The time evolving spectra are shown in Figure 9.

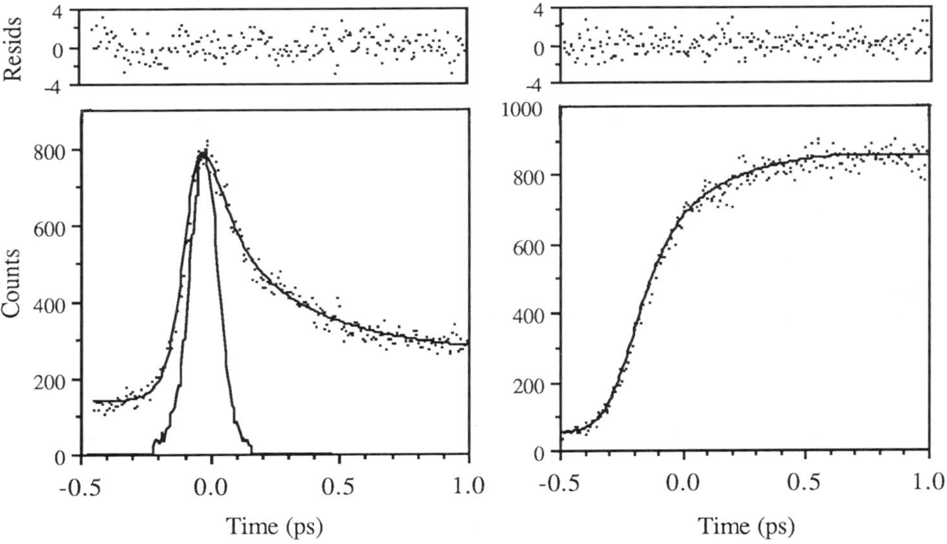

Figure 8. Upconversion data, fitted curves and residuals for fluorescence from LDS-750 in acetonitrile at 19°C. Left panel: 654 nm emission. Right panel: 779 nm emission. The instrument response function is also shown in the left panel and has a FWHM of 100 fs.

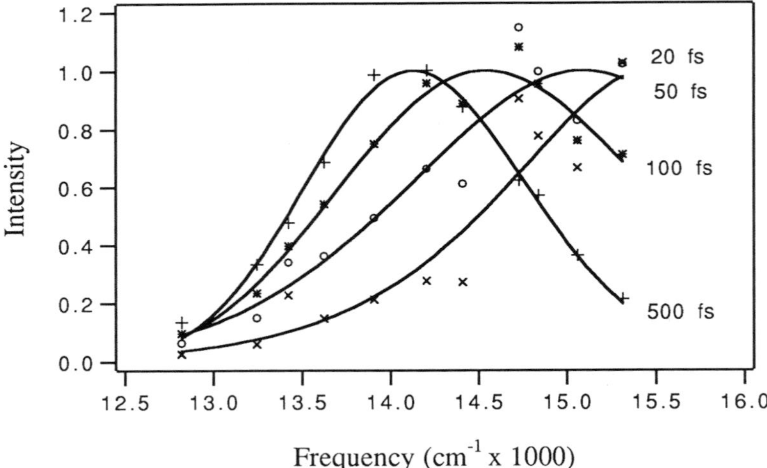

Figure 9. Raw data and log normal fits for time resolved spectra at t =20 fs, 50 fs, 100 fs, 500 fs.

The solvation time correlation function was constructed from the individual fluorescence decays as described by Maroncelli and Fleming [43]. The result is shown in Figure 10. The solvation response clearly occurs on two time scales. The initial relaxation accounts for ~80% of the amplitude and is well fit by a Gaussian of 120 fs FWHM giving a 1/e decay time of 70 fs. The slower tail appears exponential and has a decay time of ~200 fs. This latter value is probably not well determined in these experiments.

The time resolution of our instrument allows us to be confident about the presence of the ultrafast component in the spectral evolution. In fact, multi-exponential fits to the individual data sets routinely reveal a component of about 60 fs. However, since our measurements are made with significantly higher time resolution than previous solvation studies it is important to consider other possible contributions to the spectral evolution. The most likely complication is that of vibrational relaxation. Excitation at 608 nm prepares S_1 of LDS-750 with about 1000 cm^{-1} of excess vibrational energy. For the dye molecules Nile Blue and Oxazine, Chesnoy et al. [1] have shown that excess energy in the range 500-1000 cm^{-1} leads to shifts of ~5 nm with relaxation times of 500 fs and 800 fs, respectively. These molecules do not have a change in dipole moment upon excitation and the spectral shift arises primarily from vibrational relaxation. The total shift in LDS-750 is more than 70 nm, 50 nm of which occurs in the first 100 fs. It seems difficult to suggest a mechanism other than that resulting from the Coulomb interaction that could induce such a large shift. We conclude, therefore, that the data presented in Figure 10 are dominated by polar solvation dynamics.

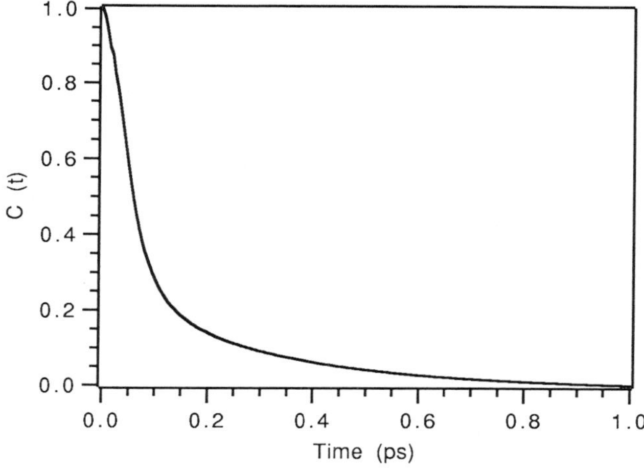

Fig. 10. The solvation time correlation function, C(t), for acetonitrile. C(t) was evaluated using the peak of the time resolved fluorescence spectra, determined from the log normal fits.

The physical origin of the two relaxation components has been discussed by Maroncelli [32], from the perspective of his molecular dynamics simulations. The simulation result for C(t) is strikingly similar to that in Figure 10 with an initial Gaussian contribution accounting for 80% of the relaxation with a decay time of ~100 fs and a slower (mildly oscillatory) decay on a time scale of 0.5 to 1 ps. Maroncelli (private communication) has also carried out a simulation with coumarin 153 in acetonitrile. Here the oscillatory behavior observed (with a frequency of ~30 ps^{-1}) is far less prominent than in the small spherical ion simulations [32]. Maroncelli assigns the rapid part of the response to small amplitude inertial rotational motion of the molecules in the first solvation shell. By an elegant series of "rigid cage" simulations in which all except one solvent molecule are fixed in orientation he is able to show that this rapid relaxation results from independent single particle motion. The equilibrium dynamics of the single free molecule are followed and the appropriate correlation function constructed. The results are then averaged over all possible choices of the free molecule. The correlation function constructed in this way is indistinguishable from the full dynamics for the first 100 fs. The uncorrelated angular displacements are in the range of 10°-20° and it is simply the small amplitude of motion that accounts for the rapid relaxation [32]. Interaction between molecules only becomes important for times longer than 0.2 ps and by this time, in acetonitrile, most of the solvation energy is relaxed. The comparatively small amount of further relaxation results from diffusive restructuring of the first shell.

The large amplitude of the Gaussian component in the solvent response has substantial implications for theoretical descriptions of chemical dynamics in solution. Solvent friction manifests itself in two ways in such dynamics. Firstly, solvent friction enters directly into electron transfer reaction rates, the connection between this friction and the solvation dynamics being made by Hynes [30]. For reactions with very small barriers C(t) is the directly relevant quantity as emphasized by Barbara, Fonseca and coworkers [25,46,47]. When a significant barrier occurs the appropriate time dependent friction must be evaluated in the barrier region [39]. This is a difficult task in general but estimates may be made from C(t) determinations on systems with the same electronic structure as the transition state [39]. Again, the inertial portion of C(t) plays a major role.

Secondly, solvent friction may also impede nuclear motion, such as reorientation, via dielectric effects and again a connection exists between C(t) and the time dependent dielectric friction [30]. This explanation has been explored experimentally [48] and in computer simulation of the reorientation of ion pairs [49]. The latter studies show that the normalized dielectric friction shows very similar time dependence to C(t) and again has a major Gaussian component.

Thus, theories of chemical dynamics in polar liquids must include a full description of the time dependent solvent response - continuum and overdamped approximations are not likely to provide realistic descriptions.

ACKNOWLEDGEMENTS.

We thank Mark Maroncelli and Casey Hynes for preprints and insightful discussions and the NSF (to GRF) and DOE (to JRN) for support. We also thank Norbert Scherer for development of the laser system and for much advice as to its use and Mei Du for experimental advice.

REFERENCES.

1 A. Mokhtari, A. Chebira, and J. Chesnoy, J. Opt. Soc. Am. B 7, (1990) 1551.
2 J. Shah, IEEE J. Quantum Electronics 24 (1988) 276.
3 J. Deisenhofer and H. Michel, EMBO J. 8 (1989) 2149.
4 N.W. Woodbury, M. Becker, D. Middendorf and W.W. Parson, Biochemistry 24 (1985) 7516.
5 S.F. Fischer and P.O.J. Scherer, Chem. Phys. 115 (1987) 151.
6 M.E. Michel-Beyerle, M. Plato, J. Deisenhofer, H. Michell, M. Bixon and J. Jortner, Biochim. Biophys. Acta 932 (1988) 52.
7 R.A. Friesner and Y. Won, Biochim. Biophys. Acta. 977 (1989) 99.
8 R.A. Marcus, Chem. Phys. Lett. 133 (1987) 471.
9 R.A. Marcus, Chem. Phys. Lett. 146 (1988) 13.
10 S. Creighton, J-K Hwang, A. Warshel, W.W. Parson and J. Norris (1988) Biochemistry 27 (1988) 774.
11 M. Bixon, J. Jortner and M.E. Michel-Beyerle, Biochim. Biophys. Acta 1056 (1991) 301.

12 W. Holzaptel, U. Finkele, W. Kaiser, D. Oesterhelt, H. Scheer, H.U. Stilz and W. Zinth, Chem. Phys. Lett. 160 (1989) 1.
13 C. Kirmaier and D. Holten, Biochemistry 30 (1991) 609.
14 R.A. Friesner and R. Wertheimer, Proc. Natl. Acad. Sci. USA 79 (1982) 2138.
15 J. Jean, R.A. Friesner and G.R. Fleming, Ber. Bunsenges. Phys. Chem. 95 (1991) 253.
16 M. Sugawara, Y. Fujinura, C.Y. Yeh and S.H. Lin, J. Photochem. Photobiol. A 54 (1990) 321.
17 A. Warshel and W.W. Parson, J. Am. Chem. Soc. 109 (1987) 6l43; W.W. Parson and A. Warshel, J. Am. Chem. Soc. 109 (1987) 6152.
18 P.O.J. Scherer and S.F. Fisher, Biochim. Biophys. Acta. 891 (1987) 157.
19 M.A. Thompson and M.C. Zerner, J. Am. Chem. Soc. 112 (1990) 7828.
20 M. Plato, K. Mobius, M.E. Michel-Beyerle, M. Bixon and J. Jortner, J. Amer. Chem. Soc. 110 (1988) 7279.
21 G.R. Fleming, J.-L. Martin and J. Breton, Nature 333 (1988) 190.
22 N. Laing, J.R. Miller and G.L. Closs, J. Amer. Chem. Soc. 112 (1990) 5353.
23 J.T. Hynes, J. Phys. Chem. 90 (1986) 3701.
24 I. Rips and J. Jortner, J Chem. Phys. 87 (1987), 2090, 6513 ibid 88 (1988) 818.
25 P.F. Barbara and W. Jarzeba, Adv. Photochem. 15 (1990) 1.
26 M.J. Weaver and G.E. McManis, Acc. Chem. Res. 23 (1990) 294.
27 M. Maroncelli, J. MacInnis and G.R. Fleming, Science 243 (1989) 1674.
28 J.D. Simon, Acc. Chem. Res. 21 (1986) 128.
29 B. Bagchi, Ann. Rev. Phys. Chem. 40 (1989) 115.
30 T. van der Zwan and J.T. Hynes, J. Phys. Chem. 89 (1985) 4181.
31 M. Maroncelli and G.R. Fleming, J. Chem. Phys. 89 (1988) 5044.
32 M. Maroncelli, J. Chem. Phys. 94 (1991) 2084.
33 T. Fonseca and B.M. Ladanyi, J. Phys. Chem. 95 (1991) 2116.
34 A.E. Carter and J.T. Hynes, J. Chem. Phys. 94 (1991) 5961.
35 Faraday Discussion 85, Solvation 85 (1988) 231.
36 A. Chandra and B. Bagchi, J Chem. Phys., in press.
37 B. Bagchi, D.W. Oxtoby and G.R. Fleming, Chem. Phys. 86, 257 (1984).
38 G. Ciccotti, M. Ferrario, J.T. Hynes and R. Kapral, J. Chem. Phys. 93 (1990) 7137.
39 D.A. Zichi, G. Ciccotti, J.T. Hynes and M. Ferrario, J. Phys. Chem. 93 (1989) 6261.
40 J.T. Hynes, E.A. Carter, G. Ciccotti, H.J. Kim, D.A. Zichi, M. Ferrario and R. Karpral in Perspectives in Photosynthesis, edited by J. and R. Karpral in Perspectives in Photosynthesis, edited by J. Jortner and B. Pullman (Kluwer Academic, Dordrecht, the Netherlands, 1990).
41 D. Borgis and J.T. Hynes, J. Chem. Phys., in press.
42 B.J. Gertner, K.R. Wilson and J.T. Hynes, J. Chem. Phys. 90 (1989) 3537.
43 M. Maroncelli and G.R. Fleming, J. Chem. Phys. 86 (1987) 6221.
44 E.W. Castner, Jr., M. Maroncelli and G.R. Fleming, J. Chem. Phys. 86 (1987) 1090.
45 G.J. Blanchard, J. Chem. Phys., submitted.

46 P. F. Barabara, T. J. Kang, W. Jarzeba and T. Fonseca in Perspectives Phototsynthesis , edited by J. Jortner and B. Pullman (Kluwer Academic, Dordrecht, the Netherlands, 1990).
47 T. J. Kang, W. Jarzeba, P. F. Barbara and T. Fonseca, Chem Phys. 81 (1990) 149.
48 J.D. Simon and P.A. Thompson, J. Chem. Phys. 92 (1990) 2891.
49 J.T. Hynes, Private communication.

DISCUSSION

Tominaga

From the static absorption and emission spectra, you can estimate the t=0s spectrum peak position very roughly, I think. My question is whether the predicted t=0 peak position is consistent with that observed experimentally.

Fleming

Yes, consistent. By estimating the t=0 peak position from the emission spectrum of LDS-750 in frozen acetonitrile we find the reconstructed t=0 peak position is consistent with the steady state estimate, within experimental error.

Tominaga

So it means that there is no more faster component than that observed.

Fleming

Yes.

Wasielewski

1. Holten's central argument for the one step superexchange model in reaction centers involves a physical inhomogeneity of reaction center samples as reflected in the 665 nm band. Do you see any wavelength dependent changes in kinetics across the Q_x bands of the pheophytins?

2. Have you examined as yet any of the mutants with somewhat slower electron transfer kinetics? The ratio of 2 step to 1 step processes may change in these cases.

Fleming

1. We do not see evidence for heterogeneity in the fast step kinetics.

2. Not yet. This will be very interesting to look at.

Barbara

It is very interesting that you have observed a large amplitude (80%) fast component of solvation in acetonitrile for the dye LDS-750. We examined C(t) for a coumarine probe in the polar aprotic solvent propylene carbonate with an instrument response function of 200 fs at FWHM (which is about three times as long as your IRF). Our data did not exhibit a very short compound in C(t) although a component as much as 40% might have been missed. Assuming our resolution was sufficient, the message here might be that in bigger more massive solvents, the inertial components are much less important than acetonitrile. Furthermore, if you consider the impact of the small amplitude fast components of solvation on solvent controlled small barrier ET in BA, you come to the conclusion from stochastic theories that the slower times still dominate picosecond time scale rate process, since the system must still traverse over a broad range of the solvent coordinate.

Fleming

It is clear from the optical Kerr data (OKE) that inertia plays a significant role for light polar liquids. Note the correspondence of the spectral density obtained by deconvolution of the OKE data of Lotshaw et al. and the spectral density fitted to our Stokes shift data. In addition we saw a very similar time constant (60 fs) for the librational modes in the OKE data for $CHCl_3$. In considering other solvents from the perspective of Mr. Cho's Langevin oscillator mode one needs to consider both the spectral density and the value of the friction constant γ. Different values of even for the same spectral density lead to different amplitudes for the inertial and diffusion contributions.

Tachiya

I would like to make a comment on Onsager's note you referred to. He made that comment in connection with the solvation of an electron, not an ion. In the case of electron solvation, the electron cloud is initially spreaded. It gradually contracts as the solvation proceeds. The longitudinal relaxation time is the relaxation time relevant to the step function change of the electric displacement. In the vicinity of the center the electric displacement gradually increases since the electron cloud flows in, and the solvent there relaxes in response to this gradual increase of the electric displacement. On the other hand, in the region far away from the center the time change of the electric displacement is the step function, and the solvent there relaxes in response to this change. So the solvation is established faster outside than inside. This effect was already pointed out by me (J. Chem. Phys. 66, 3056 (1977)) before Onsager. However, this effect does not apply to the solvation of an ion.

Fleming

You are correct that Onsager was referring to localization of an electron, not solvation of an ion. However, in reading his contemporaneous work with Hubbard on molecular solvation it seems that he believed that similar concepts apply to molecular phenomena. Further information on this topic may be found in an exchange of letters to Physics Today between Gordon Freeman (Edmonton) and Peter Wolynes and myself. (ref. Physics Today 1990 Nov.)

Jortner

A very interesting problem in the area of time-resolved solvation phenomena involves the solvation dynamics of an excess electron in polar liquids. While solvation dynamics of an ion or a dipole occurs on a single adiabatic potential surface, the electron solvation dynamics may involves nonadiabatic transitions between different electronic states. This state of affairs is similar to electron–hole recombination in semiconductors, which was explored since the pioneering studies of Kubo and Toyozawa. This process seems to involve an initial electron trapping in highly excited electronic states followed by radiationless decay to lower lying electronic configurations. A similar state of affairs pertains to the solvation dynamics of an excess electron in polar solvents, where cascading between electronically excited states occurs, resulting in the interplay between nonadiabatic electronic relaxation and adiabatic solvent relaxation in several electronic states. The multistate description of electron solvation in alcohols by Hirata and Mataga and in water by Migus et al. and by Eisenthal et al. are consistent with this cascading picture. Quantum molecular dynamics simulations of electron localization in water by Rossky et al. also address this issue.

Fleming

Thank you.

Hynes

Concerning Professor Barbara's comment on inertial Gaussian behavior being less important for more massive solvents: The Gaussian decay rate scales inversely with the square root of the solvent mass [see my contribution to the Proceedings]. The slower decay for more massive solvents allows other, dissipative processes to take over sooner in time, there by indeed reducing the importance of the Gaussian inertial dephasing.

Fleming

I agree. I would, however, like to comment that the Gaussian decay comes from the destructive interference between many frequencies. The net result can be described by a single ("average") frequency and assigned a mass.

Mukamel

1. Your discussion of the applicability of the Brownian oscillator model to the optical Kerr effect is a beautiful demonstration of the nonlinear response function for the multimode Brownian oscillator model which can be calculated exactly [Y. J. Yan and S. Mukamel, J. Chem. Phys., *89*, 5160 (1988)] and applies to any four wave mixing process. The stochastic model of line broadening is a special case of the Brownian oscillator model. That model neglects, however, the effects of solvent reorganization (since the stochastic motion is independent on the electronic state of the system) and therefore completely misses the time dependent Stokes shift. In addition it is restricted to the overdamped limit.

2. A careful analysis of the superexchange model using the density matrix shows that in general it is not possible to separate the electron transfer mechanism into a

sequential and a superexchange contributions. Interference terms of mixed origin show as well. (Y. Hu and S. Mukamel, J. Chem. Phys. *91*, 6973 (1989)) Also, that work shows a complete analogy of the sequential/superexchange branching and the Fluorescence/Raman branching in electronic spectroscopy. The two calculations are mathematically identical. As the electronic dephasing rate is decreased, the Raman component becomes dominant. The dominance of the superexchange mechanism at lower temperatures may therefore be viewed as a consequence of the reduced dephasing rate with temperature.

Fleming

1. Thank you. The short time behavior in the optical Kerr effect appears to be well described by an inhomogeneous distribution of molecular oscillators. The long time behavior (>2 ps) requires inclusion of diffusive reorientation or slow structural reorganization of the liquid. Our major point is once the short Kerr response can be understood via Brownian oscillators, the response of the solvent to a step function change in the electronic nature of a solute can be understood on the same basis.

2. Our model is based on a weak coupling rapid dephasing approximation. In this limit I believe it is possible to distinguish between direct (superexchange) and sequential mechanisms. When the coupling matrix elements and relaxation times are better characterized (e.g. by non–linear spectroscopy) it will be very interesting to see over what range the Golden rule approach is applicable.

Dynamics and Mechanisms of
Photoinduced Transfer and Related Phenomena
N. Mataga, T. Okada and H. Masuhara (Editors)

Picosecond Solvation Dynamics of Electrons in Several Alcohols

Yoshinori Hirata and Noboru Mataga

Department of Chemistry,
Faculty of Engineering Science and Research Center for Extreme Materials,
Osaka University, Toyonaka, Osaka 560, Japan

Abstract

Dynamic behavior of the electron ejected by two–photon ionization of p–phenylenediamine in several alcohols and cellosolves has been investigated by picosecond transient absorption measurements in the temperature range from −90 to 23°C. The solvation process of the electron in alcohols is characterized by two time constants. One is similar to τ_{D2}, the dielectric relaxation time for rotation of the monomer alcohol and the other is well correlated to τ_{L1}, the longitudinal relaxation time accompanying hydrogen bond breaking. These results indicate that the formation process of the solvated electron, e_s^-, consists of at least two steps, and the simple two state model, $e_t^- \rightarrow e_s^-$, cannot hold. Partially solvated state, e_{ps}^-, should be involved in this process. The solvation of electrons is initiated by the reorientation of alcohol monomer in the first solvation shell and the polarization of the outer shell is established later.

1. Introduction

There have been many studies concerned with the dynamics and physical properties of electrons in liquids. Electrons ejected into liquids will interact with solvent molecules and, in polar solvents, most of them will be trapped in a solvent cage to form solvated electron, e_s^-. The first direct observation of the formation of e_s^- in liquid alcohol was made by the pulse radiolysis of 1–propanol[1]. At low temperature (6° above the freezing point), the decay of the transient absorbance at 1.3 μm was observed in hundreds of nanoseconds and the lifetime was the same as the rise time at 500 nm. They also observed the longer lived component in 600~1100 nm.

Chase and Hunt[2] studied the formation of e_s^- in water, methanol, ethanol, 1–propanol, and 1–butanol by using a stroboscopic pulse radiolysis techniques. They measured the solvation time of electrons at 600 nm and the decay time of the trapped electron, e_t^-, at 1.3 μm in the temperature range from −65 to 60°C. Since these time constants were the same and well correlated to the dielectric relaxation time τ_{D2}, they concluded that intermolecular hydrogen bond breaking does not limit the rate of the electron solvation but rotation of monomer alcohol is important for the stabilization of electron. Although the slower decay of the transient absorbance was also observed at 1050 nm, they did not discuss it in detail.

Kenny–Wallace and Jonah[3] proposed a cluster model as the mechanism of the electron solvation. They measured the time dependence of the transient absorbance in the visible region (450~ 750 nm) and concluded that the electron solvation observed in the picosecond time regime can be described by a single physical step, $e_t^- \rightarrow e_s^-$.

In the previous papers[4,5], we reported the formation process of e_s^- in alcohols at room temperature by using picosecond laser photolysis techniques. Gradual blue shift of the transient absorption spectra was observed in a time region longer than τ_{D2}. The slow decay was observed only in the longer wavelength region than 800 nm. The formation time of e_s^- deduced from the decay rate around 900 nm was rather similar to τ_{L1}, the longitudinal dielectric relaxation time of alcohols accompanying the hydrogen bond breaking. We also observed the solvation time similar to τ_{D2} around 650 nm. These findings suggest that the solvation of e_t^- cannot be a single step but it consists of two steps. Our model for the localization of electron in alcohols is the four stage process:

e_{qf}^-	$\rightarrow$	e_t^-	$\rightarrow$	e_{ps}^-	$\rightarrow$	e_s^- .
extended state		trapped state		partially solvated state		fully solvated state

In order to clarify the relation between the formation time of e_s^- and τ_{L1} and to elucidate the solvation process of electron in alcohols, we measured the picosecond time resolved absorption spectrum of electrons in alcohol solutions where p-phenylenediamine (PD) was ionized by picosecond laser pulse, in the temperature range between −90 and 23 ℃ . In the following, details of the formation process of e_s^- will be discussed in conjunction with the theoretical model[6] of e_s^-.

2. Experimental

Picosecond transient absorption spectra were measured by using a rhodamine-6G dye laser photolysis system pumped by the second harmonics of a mode-locked Nd^{3+}: YAG laser (Quantel, Picochrome). The sample was excited with the second harmonics of the dye laser (295 nm, about 10 ps FWHM, 0.5 mJ), while the transient absorption spectra were monitored by the picosecond white light generated in H_2O/D_2O mixture from the fundamental pulse of YAG laser. The details of this system was described elsewhere[7].

For the measurements of the temperature effect, sample cell was put in the quartz dewar with flat windows. Temperature was monitored by the thermocoulpe contacted with the cell holder and was controlled by changing the flow rate of cold N_2 gas evaporated from liquid N_2.

PD was purified by repeated recrystallization from ethanol followed by sublimation in vacuo. 1-Propanol (GR grade) was treated with sodium borohydride and then distilled twice. Spectrograde ethanol was kept in contact with molecular sieves over night and then distilled. GR-grade methylcellosolve and ethylcellosolve were distilled twice. Spectrograde methanol, 2-propanol and 1-butanol were used without further purification. The sample was degassed by several freeze-pump-thaw cycles and sealed in quartz cell of 1 cm path length.

3. Results and Discussion

3.1. Time Resolved Absorption Spectra

Figure 1 shows the picosecond transient absorption spectra of PD in (a) isooctane, (b) acetonitrile, and (c) methanol. The spectrum in isooctane at 100 ps after the excitation can be assigned to the $S_n \leftarrow S_1$ transition of PD. The spectrum in acetonitrile was a superposition of $S_n \leftarrow S_1$ absorption, cation band of PD, and the anionic species. The peaks of PD^+ band appear at 480 and 510 nm. The broad band observed in acetonitrile at longer wavelengths should be due to the dimer anion of acetonitrile.

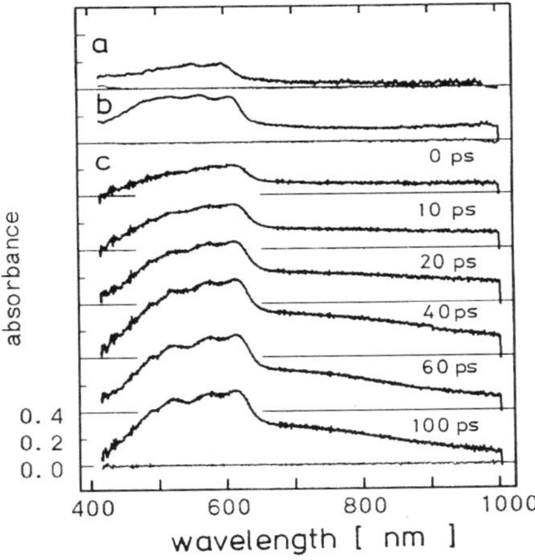

Figure 1. Picosecond time-resolved absorption spectra of PD in n-hexane (a), acetonitrile (b), and methanol (c).

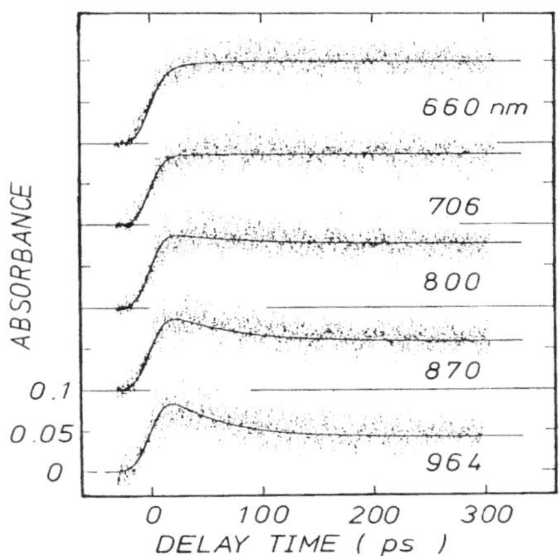

Figure 2. Time dependence of the transient absorbance of PD in 1-propanol at room temperature.

No strong bands appeared in isooctane or acetonitrile in the wavelength region longer than 650 nm, while the broad band was observed in methanol in that wavelength region. At delay times longer than 60 ps after excitation, the spectrum of methanol solution in the longer wavelength region did not change significantly, whereas a blue-shift of the spectrum was observed in earlier times. Similar behavior is observed also when indole was ionized biphotonically in normal alcohols as reported previously[4]. In other alcohols, the similar spectral changes are observed on a different time scale. In water, we observed the well-known spectrum of the hydrated electron even immediately after laser pulse excitation and undergoes no change over a period of tens of picoseconds. Since the formation time of the hydrated electron was reported to be less than 1 ps, the solvation process of the electron in water is not expected to be observed by our measurements with a 10-ps time resolution.

As reported previously[5], the spectral change observed in alcohols can be ascribed to the solvation of the ejected electron. The spectrum at short delay times was assigned to the partially solvated electron, e_{ps}^-, while the spectrum at longer delay times was due to the fully solvated electron, e_s^-[5]. Similar changes were observed in other alcohols on a different time scale. In cellosolves the relaxed spectra seems to be broader than those in alcohols. The extra intensity in near–IR region may be due to the ethereal region of the solvent. Although such effect was known to be evident for cellosolves with longer alkyl chain in the low temperature glass, we could not measure the transient absorption spectra in butylcellosolve because of the decomposition of sample.

3.2. Kinetics of the Electron Solvation Process

Typical time dependences of the transient absorbance of PD in 1–propanol at several wavelengths are shown in Figure 2. At all wavelengths a portion of the transient absorbance grows up within a response time of the apparatus and further growth or decay depending on the monitor wavelength follows. At shorter wavelengths than 700 nm the rapid rise is followed by a further growth with a time constant of about 20 – 30 ps. The additional rise should be due to the formation of e_{ps}^-,

$$e_t^- \xrightarrow{k_{ps}} e_{ps}^- .$$

Although the slow decay is observed at wavelengths longer than 800 nm, no decay corresponding to the formation of e_{ps}^- can be observed in the wavelength region shorter than 1 μm. Such decay was observed at 1.08 μm as reported previously[5]. The transient absorbance does not decay completely in the time range of the observation and has a long lived component due to e_s^-. According to Chase and Hunt[2], single exponential decay ascribed to the solvation of e_t^- was observed at 1.3 μm in pure alcohols. The slow decay in 800~ 1000 nm observed in the present work may be attributable to the solvation process of e_{ps}^-,

$$e_{ps}^- \xrightarrow{k_s} e_s^- .$$

In alcohols, the rise corresponding to the formation of e_{ps}^- can be hardly observed with our experimental accuracy at room temperature. However, we have observed it in methanol at low temperatures. In cellosolves the slow decay was observed only at longer than 900 nm and the rise was clearly observed at 660 nm. The rate constants kps and k_s can be obtained from the rise time at 660 nm and from the decay time at 960 nm, respectively. Smooth lines in Fig. 2 are the convolution curves calculated by assuming the kinetics of eq(1),

$$OD(t) = A \cdot \{1 - \alpha \cdot \exp(-k_{ps} \cdot t)\} \qquad < 700\ \text{nm} \qquad (1a)$$
$$= A \cdot \{1 - \alpha \cdot (1 - \exp(-k_s \cdot t))\} \qquad < 800\ \text{nm}. \qquad (1b)$$

The values obtained from the best fit are listed in Table 1. Since at 660 nm, in Eq. 1a, α is estimated to be about 0.4, the short wavelength tail of e_t^- band has a rather large absorbance even at 660 nm.

Table 1
Observed Rise and Decay Times of The solvated Electron at Various Wavelength in Several Alcohols.

	Observed results (ps)			Dielectric Relaxation Time	
	This Work				
	$1/k_s$	$1/k_{ps}$	τ_s	τ_{D2}	τ_{L1}
Methanol	10.7±1.7	<10	10.7,[a] 10[c]	12,[d] 7.09[e]	9,[f] 9.4[e]
Ethanol	24.±5.2	11	23,[a] 18,[b] 18[c]	16,[d] 8.97[e]	30,[f] 30.1[e]
1–Propanol	50.1±3.5	21	34,[a] 22,[b] 24[c]	22,[d] 15.1[e]	81,[d] 60.2[e]
2–Propanol	66.1±7.0	22	23,[b] 25[c]	14.5[e]	64.2[e]
1–Butanol	77.6±4.9	26	39,[a] 21,[b], 30[c]	27[d]	127[d]
methylcellosolve	45	15			
ethylcellosolve	120	15			

a: Ref. 2,
b: D. Huppert, Keny–Wallace, and P. M. Rentzepis, J. Chem. Phys., 75 (1981) 2265.
c: Y. Wang, M. K. Crawford, M. J. McAuliffe, and K. B. Eisenthal, Chem. Phys. Lett., 74 (1980) 160.
d: S. K. Garg and C. P. Smyth, J. Phys. Chem., 69 (1965) 1294.
e: J. Barthel, K. Bachhuber, R. Buchner, and H. Hetzenauer, Chem. Phys. Lett., 165 (1990) 369.
f: J. A. Saxton, R. A. Bond, G. T. Coats, and R. M. Dickinson, J. Chem. Phys., 37 (1962) 2132.

The formation time of e_{ps}^- is in a good agreement with the literature values of the rise time in the visible region 600 nm or the decay time in the IR region. As reported by Chase and Hunt[2], these values can be well correlated to the dielectric relaxation time for the rotation of alcohol monomer, τ_{D2}. On the other hand, the decay time at 960 nm seems to correlate with the longitudinal relaxation time, τ_{L1}, obtained from the dielectric relaxation time for hydrogen bond breaking, although the decay times obtained in the present work are slightly shorter for alcohols other than methanol.

3.3. Temperature Dependence of the Electron Solvation Process

Figure 3 shows the time dependence of the transient absorbance of PD in methanol at various temperatures. The decay time in the longer wavelength region increased with decreasing temperature. The fraction of the decaying component at 960 nm also increased at low temperatures. This should be due to the blue shift of the e_s^- band at the low temperatures. Such spectral shifts of e_s^- are well known for many alcohols and according to Sauer et al[8] the band maxima of e_s^- in methanol are 1.97 and 2.20 eV at 298 and 195 K, respectively.

Figure 4 shows the temperature dependence of k_s which was determined by ignoring the initially rising part of decay curve monitored at 960 nm. Broken lines in the figure show the temperature dependence of $1/\tau_{L1}$[9]. In methanol, k_s seems to be smaller than $1/\tau_{L1}$ at all temperatures, while k_s of other alcohols at higher temperatures appears to be larger than $1/\tau_{L1}$ and at lower temperatures k_s and τ_{L1} are quite close to each other. As shown in the next section, k_s depends on the monitor wavelength and the plotted values in Figure 4 may be the upper limit.

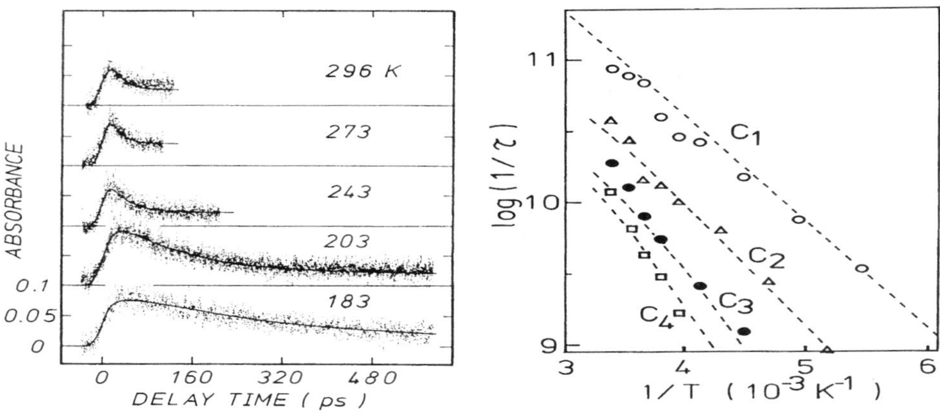

Figure 3. Time dependence of the transient absorbance of PD in methanol at several temperatures.

Figure 4. Arrhenius plots of k_s and $1/\tau_{L1}$ (broken lines) in several alcohols.

3.4. Dependence of $1/k_s$ on the Monitor Wavelength

Figure 5 shows the dependence of $1/k_s$ on the monitor wavelength in methanol and 1–propanol. As we reported previously, the triple–exponential decay was observed at 1.08 μm[5], while in the wavelength region below 1 μm, the decay was biexponential. The longer the observation wavelength, the shorter $1/k_s$ was obtained. Similar behavior was observed for other normal alcohols in the wide range of temperature. If e_{ps}^- is a chemically identifiable species and the solvation is described by a single step process, k_s should be independent on the probing wavelength. Our present findings are consistent with our previous results[5] that we could not observe the isosbestic point for the spectra of electrons in the course of the solvation of e_{ps}^-. The partially solvated state should not be a chemically identifiable species but an ensemble of the many species. The last step of the solvated electron formation should be a continuous step including many intermediate states with a different grade of solvation.

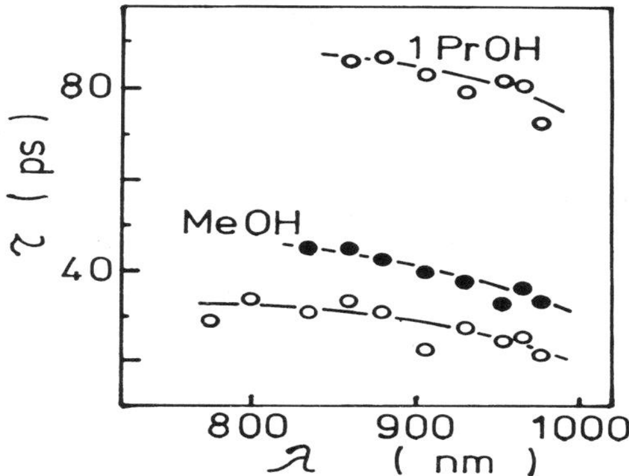

Figure 5. Monitor wavelength dependence of $1/k_s$ in methanol (−10°C (○), −30°C (●)) and 1−propanol (10°C).

3.5. Solvation Mechanism

The structure of e_s^- has been studied extensively[10−12] and, by using the sophisticated magnetic resonance techniques, Narayana and Kevan[10] determined the structure of e_s^- in the low temperature ethanol glass. Only four equivalent ethanol molecules are in the first solvation shell and the molecular dipole (not the O−H bond) is oriented toward the electron. Although the alcohol radical disturbed the signal significantly, similar structure was expected also for e_s^- in methanol glass[11].

According to Fueki et al.[6] the properties of e_s^- can be explained by using semicontinuum model which includes the short− and long−range interactions between the electron and the solvent molecules. The short−range charge−dipole interaction acts between the electron and a fixed number (4 or 6) of symmetrically situated solvent molecules in the first solvation shell, while the solvent molecules beyond the first shell are treated as a dielectric continuum. They predicted that about 60 % of the stabilization energy of the ground state of e_s^- in methanol is due to the short−range interaction.

Our observation clearly shows that the solvation process of electron can be divided into two parts. The formation time of e_{ps}^- is well correlated with τ_{D2}, which suggests that the orientation of the monomer alcohols in the first solvation shell and the inner shell structure or the core of the solvated electron should be established in this stage. On the other hand, the formation of e_s^- is determined by τ_{L1}. Therefore, it is better to consider that the solvation of e_{ps}^- accompanies the hydrogen bond breaking in the outer shell. The peak position of the absorption band shifts from 1.5 μm (e_t^-) to around 900nm (e_{ps}^-) and the solvation of e_{ps}^- brings the band maximum to 650 nm. Although the band maxima do not show the stabilization energy exactly, the amount of the spectral shift seems to be in a good agreement with the theoretical prediction.

We can safely conclude that the solvation of the electron is initiated by the reorientation of the solvent molecules in the first solvation shell and the polarization of the outer shells is established later. The polarization of outer shell may not be an additional small adjustment but may bring almost 50 % of the stabilization energy. It seems reasonable that the alcohols in the hydrogen bonding network contribute to the formation of the solvated electron because the major part of the alcohol molecules are connected each other by hydrogen bond even at room temperature. In the e_{ps}^- formation, the monomer alcohol plays an important role, which may suggests that the site happen to be rich in the monomer alcohol can trap the photo–ejected dry electron.

Acknowledgement: Present work was partially supported by the Grant–in–Aid for Special Promoted Research No. 6265006 to N. M. from the Ministry of Education, Science and Culture of Japan.

1. J. H. Baxendale and P. Wardman, Nature, 230 (1971) 450.
2. W. J. Chase and J. W. Hunt, J. Phys. Chem., **79** (1975) 2835.
3. G. A. Kenney–Wallace and C. D. Jonah, J. Phys. Chem., **86** (1982) 2572.
4. Y. Hirata, N. Murata, Y. Tanioka, and N. Mataga, J. Phys. Chem, **93** (1989) 4527.
5. Y. Hirata and N. Mataga, J. Phys. Chem., **94** (1990) 8503.
6. K. Fueki, D.–F. Feng, and L. Kevan, J. Am. Chem. Soc., **95** (1973) 1398.
7. Y. Hirata and N. Mataga, J. Phys. Chem., **95** (1991) 1640.
8. M. C. Sauer Jr., S. Arai, and M. Dorfman, J. Chem. Phys., **42** (1965) 708, S. Arai and M. C. Sauer Jr., J. Chem. Phys., **44** (1966) 2297.
9. E. W. Castner Jr., B. Bagchi, M. Maroncelli, S. P. Webb, A. J. Ruggiero, and G. R. Fleming, Ber. Bunsenges. Phys. Chem., **92** (1988) 363.
10. M. Narayana and L. Kevan, J. Chem. Phys. **72** (1980) 2891.
11. L. Kevan, Chem. Phys. Lett., **66** (1979) 578.
12. L. Kevan, Acc. Chem. Res., **14** (1981) 138.

Dynamics and Mechanisms of
Photoinduced Transfer and Related Phenomena
N. Mataga, T. Okada and H. Masuhara (Editors)

Monte Carlo simulation study on solvation energy of ionic molecule, reorganization energy of electron transfer reactions and mean ion-pair potential

M.Saito[a], T.Kakitani[b], and Y.Hatano[c]

[a]Protein Engineering Research Institute, Furuedai Suita, Osaka565, Japan

[b]Department of Physics, Faculty of Science, NagoyaUniversity, Furocho, Chikusa-ku, Nagoya 464-01, Japan

[c]Faculty of Liberal Arts, Chukyo University Syowa-ku, Nagoya 466, Japan

Abstract

We conducted Monte Carlo simulations for solutions assuming spherical hard core models for solute and solvent molecules. By means of Bennett's acceptance ratio method, we calculated the solvation energy for an ionic molecule. Adopting the umbrella sampling method, we calculated the free energy curves of neutral pair and ionic pair of solute molecules as a function of the reaction coordinate. From them, we obtained the reorganization energy for some values of the distance of the ion pair. We also calculated the mean ion-pair potential by means of Bennett's acceptance ratio method. Those results are compared with the theories which are based on the dielectric continuum model and/or the linear response approximation. As results, it was found that the theories based on the dielectric continuum model overestimate the simulation data nearly twice. The MSA theory is rather good for the reorganization energy but underestimates the solvation energy. The nonlinear effect is rather significant for the reorganization energy.

1. INTRODUCTION

Electron transfers (ET) are one of the most fundamental reactions in chemistry and biology. Among them, the photoinduced charge separation (CS) and the charge recombination (CR) in polar solutions

$$A^{*} \cdots B \longrightarrow A^{\pm} \cdots B^{\mp} \qquad (CS) \tag{1}$$
$$A^{\pm} \cdots B^{\mp} \longrightarrow A \cdots B \qquad (CR) \tag{2}$$

are the most basic reactions. This ET proceeds essentially by an electron tunneling mechanism. It is important to notify that the reaction proceeds only at the time when the initial state energy becomes equal to the final state energy which is brought about by the fluctuation of polar media. The crossing of energy surfaces of the initial and final states in the multidimensional configuration space of solvent molecules forms a hypersurface of the transition state. Thus, the ET rate is obtained by averaging over the probability of passing this hypersurface. At the present time, this reaction rate is definitely formulated by means of the reaction coordinate theory [1-4].

The ET rate is expressed as a product of the frequency factor and the thermally averaged Franck-Condon factor. The latter factor depends on the free energy gap $-\Delta G$ between the initial and final states and this relation is called the energy gap law. According to the Marcus theory [5] which is based on the linear response approximation, the activation energy $\Delta G^{\ddagger}$ is expressed as

$$\Delta G^{\ddagger} = \frac{(\lambda + \Delta G)^2}{4k_B T \lambda} \tag{3}$$

where λ is the reorganization energy and k_B Boltzmann factor. Equation (3) indicates that $\Delta G^{\ddagger}$ has a parabolic dependence upon $-\Delta G$ with a minimum at $-\Delta G=\lambda$. The reorganization energy is one of the most important parameters in the ET reaction. The other important parameter is the mean ion-pair potential U^{ip} due to the Coulombic attraction between ion pair in solution. That is, the energy gap is expressed as

$$-\Delta G = -\Delta G^0 - U^{ip} + U^{np} \qquad for\ CS \tag{4}$$
$$-\Delta G = -\Delta G^0 + U^{ip} - U^{np} \qquad for\ CR \tag{5}$$

where $-\Delta G^0$ is the standard free energy gap which is defined at an infinit distance between the reactants and U^{np} is the mean potential between neutral pair. The contribution of Unp is usually much smaller than that of U^{ip}. As we see in Eqs. (4) and (5), U^{ip}-U^{np} affects the energy gap in an opposite direction between the CS and CR reactions.

The reorganization energy and mean potentials are the function of the distance d between the reactants. Determination of their distance dependences are the central theme of this paper. On the other hand, there remains a famous theoretical problem that whether nonlinear response due to the dielectric saturation of solvent molecules adjacent to the charged molecule affects appreciably the ET process or not. We investigate those problems by means of the Monte Carlo simulation method using hard core spherical models of molecules. Quantitative examination of the reorganization energy and mean ion-pair potential and evaluation of the magnitude of nonlinearity are ultimately important at the present stage of the investigation of the ET. Based on those data, we can proceed to analyze in much detail why the energy gap law as obtained by the experiment is so much different between the CS and CR reactions [6,7].

2. SOLVATION ENERGY

We start our study by calculating the solvation energy of a single charged molecule. The method of Monte Carlo simulations is the same as before [8-10]. A solute molecule with a spherical hard core (radius r_0) and a point charge ze at its center is fixed at the center of a spherical vessel of radius R. N solvent molecules with a spherical hard core (radius a) and a point dipole moment μ at its center are put into this vessel and are allowed to move under the condition that the center of the solvent molecules is prohibited from going beyond the vessel wall. The packing fraction η is calculated by a formula $\eta=(Na^3+r_0^3)/R^3$

In calculating the solvation energy, we adopt Bennett's acceptance ratio method [11]. For this purpose, se conduct Monte Carlo simulations for 8 states with various solute valencies z_i (i=1,...,8) between $z_1^2=0$ and $z_8^2=1.0$. Free energy difference $-\Delta\Delta G(i,i+1)$ between two states with z_i and z_{i+1} is calculated for i=1,...,7 using the Bennett acceptance ratio method [11]. Then, the solvation energy $-\Delta G_s$ of a solute with charge $z_j e$ is calculated from

$$-\Delta G_s(z_j e) = -\sum_{i=1}^{j-1} \Delta\Delta G(i, i+1) \tag{6}$$

The parameter values are chosen as follows;

μ=2D, a=2.2A, r_0=2.2A, R=30.2A, N=1000, η=0.387 and T=300K. (7)

We conduct 10^4 Monte Carlo steps (10^7 configurations), and we use the last 5000 steps for the statistics.

The simulation values of -$\Delta\Delta G$ and -ΔG_S are listed in the second and third columns of Table 1.

Table 1. The simulation values of the difference free energy -$\Delta\Delta G_S$ between the two successive valence states, and the solvation energy -ΔG_S of a single molecule with a charge ze. The corrected values due to the contribution from the solvent outside the vessel are also shown.

z^2 (ev)	-$\Delta\Delta G_S$ (ev)	-ΔG_S (ev) Simulation	-ΔG_S (ev) Corrected
0			
	0.177		
0.1		0.177	0.20
	0.173		
0.2		0.350	0.39
	0.173		
0.3		0.523	0.59
	0.182		
0.4		0.705	0.79
	0.334		
0.6		1.039	1.17
	0.330		
0.8		1.369	1.54
	0.330		
1.0		1.699	1.91

The solvation energy which was obtained above is only for the solvent molecules inside the spherical vessel. We try to evaluate the contribution from the solvent outside the vessel by assuming the dielectric continuum model with a linear response as follows:

$$-\Delta G' = (1 - \frac{1}{\varepsilon})\frac{z^2 e^2}{2R} \tag{8}$$

where ε is the dielectric constant. We estimate the value of ε by the mean spherical approximation (MSA) theory [12] as

$$\varepsilon = \frac{(1+4\xi)^2(1+\xi)^4}{(1-2\xi)^6} \tag{9}$$

where ξ is a solution of

$$\frac{(1+4\xi)^2}{(1-2\xi)^4} - \frac{(1-2\xi)^2}{(1+\xi)^4} = \frac{\eta\mu^2}{a^3 k_B T} \tag{10}$$

Using Eqs. (9) and (10), we obtain ε=7.97. Using this value, the correction free energy -$\Delta G'$ is, for example, 0.21 eV for z=1. The corrected solvation energy is listed in the fourth columns of Table 1. In Figure 1, we draw the corrected solvation energy as a function of z^2 by black circles.

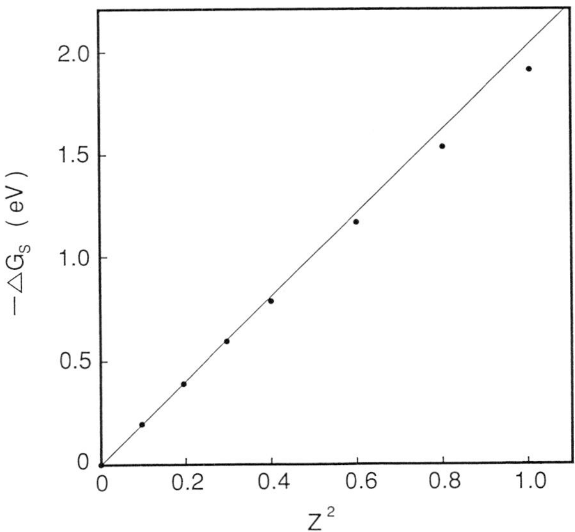

Figure 1. Solvation energy -ΔG_S as a function of squre of valence z^2. The black circles denote the simulation data and the straight line represents the correlation line corresponding to the linear response.

The straight line is the solvation energy in the linear response case which has the same slope as that of z^2=0.1. The simulated data are slightly below the straight line, representing the contribution of the non-linear response. But, its contribution is not so large; e.g. the decrease of the solvation energy due to the nonlinear response is only 6 % for z=1.

The solvation energy for z=1 calculated by the Born formula [13] which bases on the dielectric continuum model is 2.87 eV, which is much larger than the simulation value 1.91 eV. The solvation energy calculated by the MSA theory [14] is 1.50 eV, which is appreciably smaller than the simulation value.

3. REORGANIZATION ENERGY

Next, we calculate the reorganization energy. For this purpose, we rely on the reaction coordinate theory. We define the reaction coordinate operator f as [4]

$$f = H^{np} - H^{ip} \tag{11}$$

where H^{np} and H^{ip} are Hamiltonians of the neural and ion pair states exept the electronic energy, respectively. Free energy curves of the neutral and ion pair states as a function of a reaction coordinate x are defined as [4]

$$G^{np}(x,d) = -k_B T ln(k_B T \int \delta(f-x) e^{-H^{np}/k_B T} d\Gamma) \tag{12}$$

$$G^{ip}(x,d) = -k_B T ln(k_B T \int \delta(f-x) e^{-H^{ip}/k_B T} d\Gamma) \tag{13}$$

where Γ is the normalized configuration space of solvent molecules and d the distance between donor and acceptor molecules. From the definition of $G^{np}(x,d)$ and $G^{ip}(x,d)$ in Eqs. (12) and (13), we easily obtain a relation

$$G^{ip}(x,d) = G^{np}(x,d) - x \tag{14}$$

In Figure 2, we schematically draw $G^{np}(x,d)$ and $G^{ip}(x,d)-\varepsilon$, where ε is the electronic energy difference between the neutral pair and ion pair states. The reorganization energy for the CS and CR reactions are, then, defined as

$$\lambda^{CS}(d) = G^{ip}(x_0^{np}, d) - G^{ip}(x_0^{ip}, d) \tag{15}$$

$$\lambda^{CR}(d) = G^{np}(x_0^{ip}, d) - G^{np}(x_0^{np}, d) \tag{16}$$

where x_0^{np} and x_0^{ip} are coordinates corresponding to the minima of $G^{np}(x,d)$ and $G^{ip}(x,d)$, respectively. Using Eqs. (14)-(16), we obtain a relation

$$\lambda^{CS}(d) + \lambda^{CR}(d) = x_0^{ip} - x_0^{np} \tag{17}$$

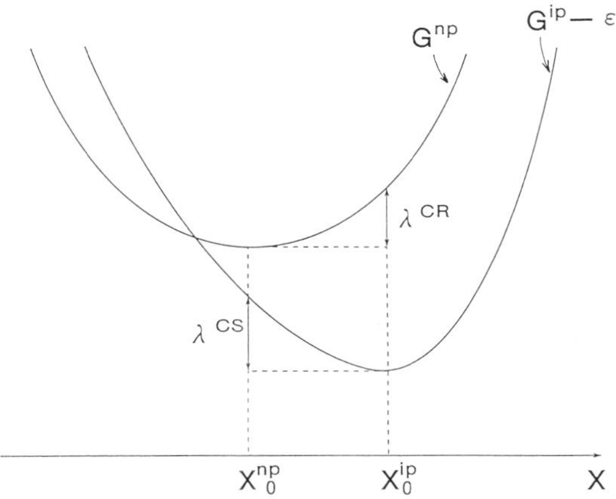

Figure 2. Schematic free energy curves of neutral pair G^{np} and ion pair U^{ip}-ε as a function of the reaction coordinate x. ε is the electronic energy difference between the neutral pair and ion pair states. $x_0{}^{np}$ and $x_0{}^{ip}$ are the reaction coordinates corresponding to the minima of U^{np} and U^{ip}, respectively. λ^{CS} and λ^{CR} are the reorganization energies for the CS and CR reactions.

Since $G^{ip}(x,d)$ is related to $G^{np}(x,d)$ by means of Eq.(14), we only need to calculate $G^{np}(x,d)$, $x_0{}^{np}$ and $G_0{}^{ip}$. In order to accumulate sufficient configurations at a larger free energy level of $G^{np}(x,d)$ in the Monte Carlo simulation, we adopt the umbrella sampling method as follows [15].

$$G^{np}(x,d) = -k_B T ln(k_B T \int \delta(H^{np} - H^{ip} - x) e^{-H^{ip(m)}/k_B T} d\Gamma) + mx \tag{18}$$

where m is a fraction of the charge of an ion pair state and $H^{ip}(m)$ is the Hamiltonian of its state. We smoothly connect free energy curves for m=0, 0.3, 0.6 and 1.0. The Monte Carlo simulations are done for a system where two solute molecules (donor and acceptor) with spherical hard cores (radii a_D and a_A) with a mutual distance d are fixed inside a spherical vessel of radius R. N solvent molecules with a spherical hard core and a point dipole moment. We choose the following parameter values

$$a_D = a_A = 2.2A, \; N=500, \; R=22.9A, \; \eta=0.45 \tag{19}$$

The other parameters are the same as Eq.(7). The different parameter values for N and η between Eqs.(7) and (19) has no physical meaning but is only for convenience of calculations.

The simulation values of λ^{CS} and λ^{CR} for some values of d as obtained by the above method are listed in Table 2. It is seen that λ^{CS} and λ^{CR} increases considerably with increase of d and that λ^{CS} is always substantially larger than λ^{CR}. The latter fact indicates that the nonlinear effect is much more appreciabe in the reorganization energy than in the solvation energy. In the same table, we have also listed the corrected reorganization energies due to contribution from the solvent outside the vessel. This correction was made as follows: The solvent outside the vessel is assumed to be a dielectric continuum. The ion pair is simplified as a dipole moment with a magnitude ed.

Table 2. The simulation values of the reorganization energies λ^{CS} and λ^{CR} for some values of d. The corrected values due to the contribution from the solvent outside the vessel are also shown.

d (A)	λ^{CS} (ev) Simulation	λ^{CS} (ev) Corrected	λ^{CR} (ev) Simulation	λ^{CR} (ev) Corrected
4.4	1.14	1.15	0.93	0.94
5.4	1.46	1.48	1.21	1.23
6.5	1.81	1.83	1.48	1.50
7.6	2.13	2.16	1.74	1.77
8.8	2.35	2.39	1.94	1.98
9.9	2.52	2.58	2.11	2.17
11.0	2.66	2.73	2.22	2.29
12.1	2.79	2.87	2.36	2.44

The correction is

$$\Delta\lambda = \frac{e^2 d^2}{R^3} \frac{\varepsilon - 1}{2\varepsilon + 1} \tag{20}$$

The bulk dielectric constant ε is evaluated as 10.1 by the MSA theory using parameter values in Eq.(19). We find that the correction is very small (0.01-0.08eV for 4.4-12.1 A) in contrast to the case of the solvation energy.

The corrected reorganization energies λ^{CS} and λ^{CR} as a funcition of the ion pair distance are drawn in Figure 3. For comparison, we have also drawn the curve calculated by the Marcus formula [5]

$$\lambda = e^2(1 - \frac{1}{\varepsilon})(\frac{1}{a} - \frac{1}{d}) \tag{21}$$

where a is a radius of donor and acceptor molecules. This theoretical curve is almost twice as large as the simulation data. Simularly, we drew the curve calculated by an approximate MSA formula

$$\lambda = e^2(1 - \frac{1}{\varepsilon})(\frac{1}{a+\Delta} - \frac{1}{d}) \tag{22}$$

where

$$\Delta = (1 - \frac{3\xi}{2+8\xi})a \tag{23}$$

The MSA curve is just intermediate between λ^{CS} and λ^{CR} for d>5.5A. So, we can say that the Marcus theory is improved very much by introducing the molecular picture of MSA theory.

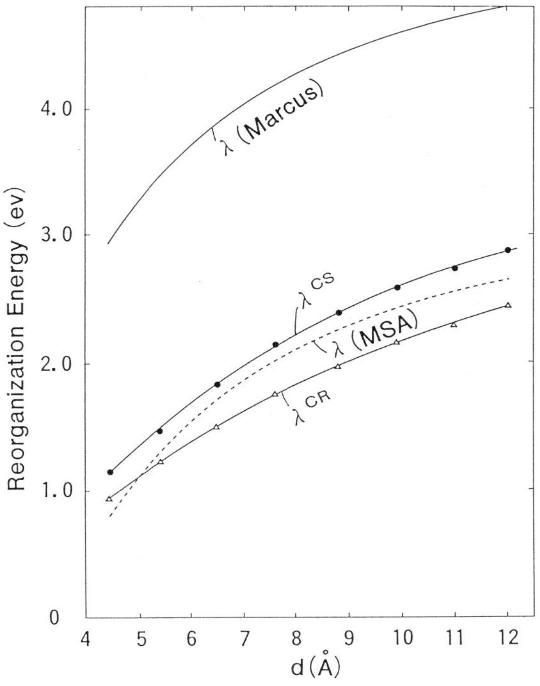

Figure 3. Reorganization energy as a function of the distance d between the two reactants. λ^{CS} and λ^{CR} are the simulation data. λ(Marcus) is obtained using Eq.(21). λ(MSA) is obtained using Eq.(22).

4. MEAN ION-PAIR POTENTIAL

Next we go to the problem of calculating the mean potential U^{ip}-U^{np} which is defined as

$$e^{-U^{ip}(d)/k_B T} = \frac{\int e^{-[H^{ip}(d)-\frac{e^2}{r}]/k_B T} d\Gamma}{\int e^{-[H^{ip}(\infty)/k_B T} d\Gamma} \quad (24)$$

$$e^{-U^{np}(d)/k_B T} = \frac{\int e^{-H^{np}(d)/k_B T} d\Gamma}{\int e^{-H^{np}(\infty)/k_B T} d\Gamma} \quad (25)$$

Using Eqs.(12) and (13), we can rewrite Eqs.(24) and (25) as

$$U^{ip}(d) = G_{ip}(d) - G_{ip}(\infty) - \frac{e^2}{r} \quad (26)$$

$$U^{np}(d) = G_{np}(d) - G_{np}(\infty) \quad (27)$$

where

$$G_{ip}(d) = \frac{1}{k_B T} \int_{-\infty}^{\infty} G^{ip}(x, d) dx \quad (28)$$

$$G_{np}(d) = \frac{1}{k_B T} \int_{-\infty}^{\infty} G^{np}(x, d) dx \quad (29)$$

Then, we obtain

$$U^{ip}(d) - U^{np}(d) = \Delta G_s(d) - \Delta G_s(\infty) - \frac{e^2}{r} \quad (30)$$

where $-\Delta G_S(d)$ is the solvation energy of the ion pair at distance d,

$$\Delta G_s(d) = G_{ip}(d) - G_{np}(d) \quad (31)$$

In the following, we evaluate $-\Delta G_S(d)$ and $-\Delta G_S(\infty)$ by the Monte Carlo simulation. In this, $-\Delta G_S(\infty)$ is the sum of the solvation energy of each single ion. We can read it as 2×1.91eV = 3.82eV in Table 1. For $-\Delta G_S(r)$, we adopt the Bennett acceptance ratio method by assuming 5 states with various valencies of the ion pair between 0 and 1.0 for each value of d. The calculation method and parameter values are the same as section 2.

The difference free energy $-\Delta\Delta G$ between the two successive valence states as obtained by the Monte Carlo simulation for some values of d are listed in Table 3. The solvation energy $-\Delta G_S(d)$ is obtained by the summation in the similar way to Eq.(6).

We make correction due to the solvent outside the vessel by the formula of Eq.(20). We find that this correction is much smaller than that for the solvation energy in a single ion molecule. In Table 4, we list the values of corrections to $-\Delta G_S(d)$, corrected $-\Delta G_S(d)$, -e2/d and $U^{ip}(d)$-$U^{np}(d)$ for some values of d. The mean ion-pair potential so obtained are plotted in Figure 4, as well as the curve obtained by the dielectric continuum model (U^{ip}-U^{np}=$-e^2/\varepsilon d$).

Table 3. The simulation values of the difference free energy $-\Delta\Delta G$ between the two successive valence states and the solvation energy at z=1 for some values of the distance d between the ion pair.

z^2	$-\Delta\Delta G$ (ev) d=4.4A	d=5.5A	d=6.6A	d=7.7A	d=8.8A	d=11.0A
0.0						
	0.233	0.308	0.356	0.447	0.503	0.551
0.2						
	0.223	0.299	0.356	0.421	0.477	0.542
0.4						
	0.216	0.299	0.360	0.425	0.477	0.529
0.6						
	0.212	0.299	0.360	0.430	0.477	0.525
0.8						
	0.216	0.286	0.351	0.421	0.460	0.516
1.0						
$-\Delta G_s$ at z=1.0 (ev)	1.10	1.49	1.78	2.14	2.39	2.66

Table 4. Correction to ΔG^{ip}, corrected ΔG^{ip}, Coulomb attraction energy and $U^{ip}-U^{np}$ as a function of the distance d between the ion pair. We have used the value $\Delta G^{ip}(\infty)=3.82$ev.

d (A)	Correction to $\Delta G^{ip}(d)$ (ev)	Corrected $\Delta G^{ip}(d)$ (ev)	$-e^2/d$ (ev)	$U^{ip}-U^{np}$ (ev)
4.4	-0.004	-1.10	-3.27	-0.55
5.5	-0.007	-1.50	-2.61	-0.29
6.6	-0.010	-1.79	-2.18	-0.15
7.7	-0.013	-2.15	-1.86	-0.19
8.8	-0.016	-2.41	-1.64	-0.23
11.0	-0.026	-2.69	-1.31	-0.18

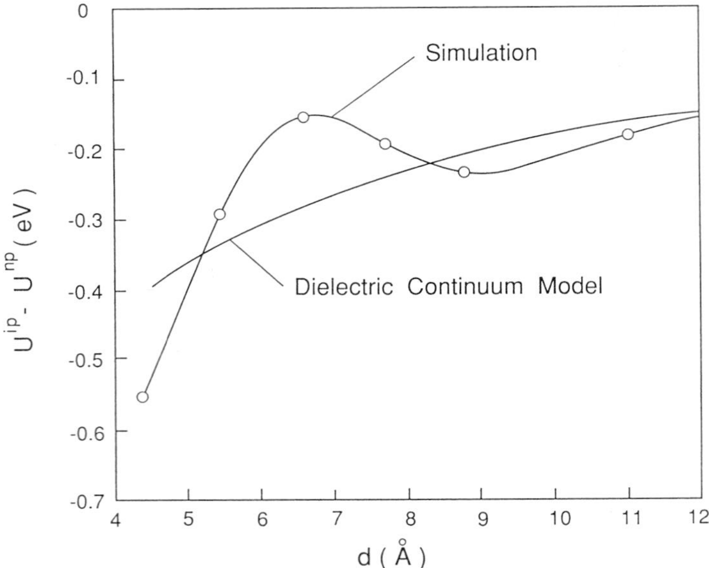

Figure 4. Mean ion-pair potential U^{ip}-U^{np} as a function of the distance d between the two reactants. Open circles are the simulation data. The solid line is obtained by the dielectric continuum model.

We see that the curve of U^{ip}-U^{np} which is deep at the contact ion pair distance (d=4.4A) increases rapidly until d=6.6A and decreases down to the minimum at the solvent separated ion pair distance (d=8.8A) and again goes up by approaching to the curve of the dielectric continuum model. A great oscillatory behavior is due to the molecular characteristic of the ion pair and this feature is much more evidently seen than the previous simulation result by Patey and Valleau [16]. Therefore, the oscillatory property of U^{ip}-U^{np} apprears to depend somewhat upon the molecular parameters. Its detailed analysis will be made in the forthcoming paper.

5. CONCLUSION

In the present simulation studies, we adopted spherical hard core models for solute and solvent molecules. It is only this case that the validity of primitive theories such as Born formula for the solvation energy, Marcus formula for the reorganization energy and the dielectric continuum theory for the mean ion-pair potential is strictly checked by the simulation calculations. The validity of the MSA theory can also be checked. As results, the dielectric continuum model overestimates nearly twice the solvation energy of an ion molecule and reorganizaiton energy of an ion pair. The MSA theory rather underestimates the solvation energy of an ion molecule and well reproduce the reorganization energy. Other results we obtained by the simulation study are that the nonlinear effect for the solvation energy for an ion molecule is small but that the nonlinear effect for the reorganization energy is large ($100(\lambda^{CS}-\lambda^{CR})/\lambda^{CR}$=21% at d=8.8A). We also observed that the mean ion-pair potential Uip-Unp as a function of the distance between donor and acceptor molecules greatly deviates from the dielectric continuum model, representing a remarkable oscillatory behavior. This property must substantially affect the energy gap law by means of the modification of the local energy gap. In the forthcoming paper, we will discuss on the relation between the reorganization energy and the mean ion-pair potential.

Acknowledgement

T.K. was supported by Grants-in-Aid(0130009) from the Japanese Ministry of Education, Science and Culture. The Monte Carlo simulations were done using FACOM VP200 in the Computation Center in Nagoya University, HITAC S-820/80 of Institute for Molecular Science and FACOM VP400E in Protein Engineering Research Institute.

REFERENCES

1 R.A.Marcus, Discuss.Faraday Soc. 29 (1960) 21.
2 A.Warshel, J.Phys.Chem. 86 (1982) 2218.
3 D.F.Calef and P.G.Wolynes, J.Chem.Phys. 78 (1983) 470.
4 A.Yoshimori, T.Kakitani, Y.Enomoto, and N.Mataga, J.Phys.Chem. 93 (1989) 8316.
5 R.A.Marcus, J.Chem.Phys. 24 (1956) 966, 979.
6 T.Kakitani, A.Yoshimori, and N.Mataga, in Electron Transfer in Inorganic, Organic, and Biological Systems, Advances in Chemistry Series, J.Bolton et al.(eds.), Am.Chem.Soc., Chapt.4, 1991.
7 T.Kakitani, papers in this proceedings.
8 Y.Hatano, T.Kakitani, A.Yoshimori, M.Satio, and N.Mataga, J.Phys.Soc.Jpn, 59 (1990) 1104.

9 M.Saito and T.Kakitani, Chem.Phys.Letters, 172 (1990) 169.
10 Y.Enomoto, T.Kakitani, A.Yoshimori, Y.Hatano, and M.Saito, Chem.Phys.Letters 178 (1991) 235.
11 C.H.Bennett, J.Comput.Phys. 22 (1976) 245.
12 M.S.Wertheim, J.Chem.Phys. 55 (1971) 4291.
13 H.L.Friedman and C.V.Krishnan, in Water: A Comprehensive Treatise, F.Franks (ed), Plenum Press, vol. 3, 1973.
14 D.Y.C.Chan, D.J.Mitchell, and B.W.Ninham, J.Chem.Phys. 70 (1979) 2946.
15 R.A.Kuharshi, J.S.Bader, D.Chandler, M.Sprik, M.L.Klein, and R.W.Impey, J.Chem.Phys. 89 (1988) 3248.
16 G.N.Patey and J.P.Valleau, J.Chem.Phys. 63 (1975) 2334.

Dynamics and Mechanisms of
Photoinduced Transfer and Related Phenomena
N. Mataga, T. Okada and H. Masuhara (Editors)

TICT MOLECULES IN A FREE JET

Okitsugu Kajimoto

Department of Chemistry, Faculty of Science, Kyoto University, Kitashirakawa-Oiwakecho, Sakyo-Ku, Kyoto 606 Japan

Abstract

Several kinds of molecules which form the twisted intramolecular charge transfer (TICT) state were investigated in a free jet. The information on the torsional potential was deduced based on the LIF excitation spectra. The formation of the TICT state was observed in a jet-cooled condition for the van der Waals complexes with polar molecules.

1. Introduction

Electron donor-acceptor interaction plays an important role in various chemical systems. when a molecule possesses both electron-donating and electron-accepting parts within a single molecular frame, intramolecular charge transfer may occur on electronic excitation. Its efficiency depends on the strength of electron-releasing and electron-withdrawing ability of the component parts, as well as the distance and the orientation between them. When a typical electron-donor-acceptor molecule, N,N-dimethylaminobenzonitrile is photoexcited in polar solvent, anomalously red-shifted emission is observed in addition to the normal fluorescence [1]. The origin of this "anomalous" emission had been a subject of controversy during 60's and 70's. Lippert [1] first interpreted this as a result of solvent-induced reversal of the first excited state, namely from the ^1L_b-type to the more polar ^1L_a-type. Rotkiewicz, Grellmann and Grabowski proposed that the emission comes from a charge-transfer state where the dimethylamino group twists by 90° from the benzene plane [2]. Although several other suggestions appeared, the concept of "twisted intramolecular charge transfer (TICT)" state seems to be established now.

Although a great number of studies have been carried out on the compounds giving the TICT state in solution, detailed structural and dynamical studies by means of supersonic jet have just started and are still quite few. The free-jet technique has great ability for clarifying the dynamics of electronically excited state and hence the application of this technique to the TICT state-formation is very promising. In the present paper, such kind of information obtained recently in our laboratory will be summarized.

2. Potential Energy Curve for the Twisting Motion

In the case of TICT molecules, the key motion for the formation of charge-transfer state is the internal rotation around the bond which connects the donor and the acceptor moieties of the molecule. Therefore, the experimental determination of the potential energy curve along such a torsional angle is of essential importance. The detailed spectroscopic data obtained from the jet-cooled molecules now provides unambiguous information on this subject.

The torsional potential can be represented usually in the following forms. The trigonometric potential function

$$V(\vartheta) = \sum (1/2)\, V_n\, (1 - \cos n\vartheta)$$

is the simplest and most frequently used form. Sometimes, the Gaussian perturbed harmonic oscillator approximation is more suited to represent the double minimum potential. The potential function is expressed as,

$$V(\alpha) = -\alpha^2 + A \exp(-B\alpha^2)$$

$$\alpha = (8\pi^2 V_2 I_\vartheta / h^2)^{1/4}\, \vartheta$$

From the diagonalized wavefunctions of each vibrational level, one can easily calculate the Franck-Condon factors and simulate both the LIF excitation and the dispersed fluorescence spectra.

3. Typical TICT Molecules

3.1 Bianthryl(BA)

9,9'-Bianthryl has been one of the favorite molecules in the research of charge transfer (CT) complexes. Schneider and Lippert [3] first observed the emission from the charge-transfer (CT) state of BA and studied the dependence of the emission intensity on the polarity and viscosity of solvent. The CT emission was detectable only in polar solvent. High viscosity as well as low temperature generally prevented BA from forming the CT state. From this fact they suggested that the torsional angle between two anthracene moieties, being 90° in the ground state, must be relaxed from 90° in the excited state so that the electron can transfer from one moiety to the other. From simple empirical molecular orbital calculations they estimated the angle to be 78° [4]. Rettig and Zander also performed a quantum mechanical force field calculation of 9,9'- bianthryl and concluded that the torsional angle would be around 70° [5]. Subsequent investigations confirmed the interpretation of the emission by Schneider and Lippert. The emission most probably comes out from a TICT state [6].

The first spectroscopic determination of the torsional potential for TICT compounds was carried out independently by Yamasaki et al. [7] and Zewail et al. [8] for bianthryl (BA). The LIF excitation spectrum of jet-cooled BA, as well as those of cyclohexane and acetone complexes, is shown in Fig. 1. The 0-0 band (except for the torsional vibration) consists of a train of sharp peaks separated by 11-16 cm^{-1} with each other. Yamasaki et al. used the trigonometric potential to calculate the ground and excited state energy levels (and hence to simulate the spectrum) and concluded that the most stable conformation in the excited state deviates from the perpendicular form by 23 degree. They assumed that the 0-0 transition is hardly observed due to the small Franck-Condon factors and estimated the position of the genuin 0-0 band so as to best reproduce the total features of the observed spectrum including the hot band intensities. On the other hand, Zewail et al. assumed the

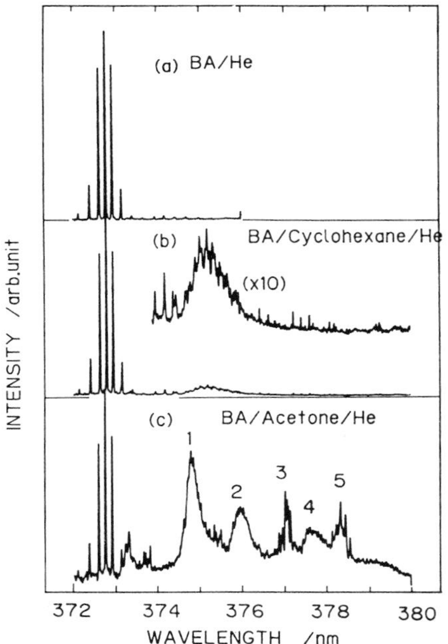

Figure 1. The LIF excitation spectra of 9,9'-bianthryl and its complexes around 373 nm.

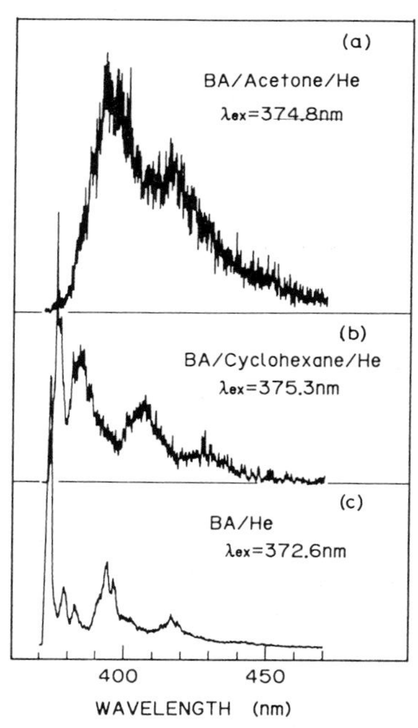

Figure 2. The dispersed fluorescence spectra of (a) bare BA, (b) BA-cyclohexane and (c) BA-acetone complexes.

peak appearing at the longest wavelength as the 0-0 band and derived the excited state angle to be 78°.

Very recently, Rettig et al. [9] determined the torsional potential by simulating the single vibronic emission spectra and established the position of the 0-0 transition. The angle and the barrier of the upper electronic state they obtained are 69° and 200 cm^{-1} respectively, which are just in between those of the previous two research groups.

Figure 2 shows the dispersed fluorescence of BA and its complexes. The fluorescence from the BA-cyclohexane complex is similar to that of bare BA, showing that nonpolar molecule does not help BA to form the CT state. On the other hand, the acetone-BA complex gives a totaly different red—shifted fluorescence spectrum, probably indicating the formation of the CT state.

To confirm this, the lifetime of the acetone-BA complex was determined at the individual peak of the complex appearing in Fig. 1. From the mass-selected 2-photon ionization spectra, these peaks are assigned to the BA-$(acetone)_n$ complexes (No. 1 as 1:1, No. 2 and 3 as 1:2 and No. 4 and 5 as 1:3 or higher complexes). The results are shown in Fig. 3. Comparing with 10 ns for bare BA, even the 1:1 complex shows much longer lifetime; 1:2 and 1:3 have still longer lifetimes. This fact, together with the red-shifted emission from these complexes, indicates that even the 1:1 complex can form the CT state and the additional solvation further helps to stabilize the polar CT state.

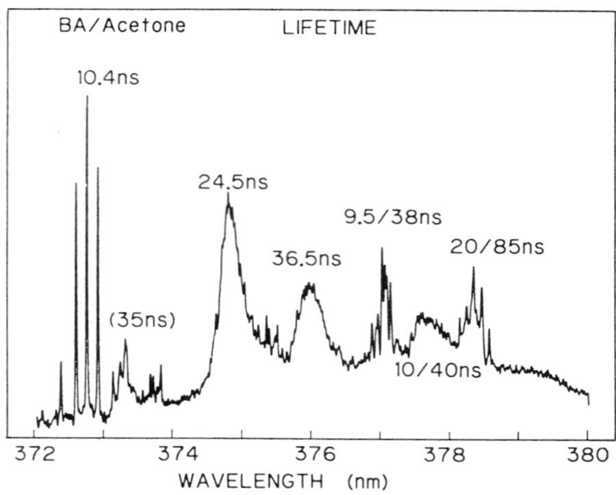

Figure 3. The lifetimes of the BA-acetone complexes. The heavier is the solvation, the longer the lifetime.

3.2 4-(9-Anthryl)-dimethylaniline (ADMA)

ADMA is another typical compound which forms a TICT state and hence its structure and dynamics has been extensively studied in the liquid phase by many research groups [6,10]. It seems to be established that in polar solvent the charge transfer occurs from the dimethylamino group to the anthracene moiety and the internal rotation between these two moieties plays a key role in the formation of the TICT state.

The torsional potential of ADMA was first analized by Kajimoto et al. [11]. The LIF spectrum of ADMA taken in a free jet is given in Fig. 4 which shows very complicated features due to the presence of four rotors and one inversion motion. However, a rough shape of the torsional potential can be drawn as shown in Fig. 5. In the excited state, the angle considerably deviates from perpendicular position by about 30° and the barrier is fairly large (900 cm^{-1}). The presence of small splitting in the bottom level of the ground state is confirmed by the temperature dependence of the spectral shape. Such splitting most probably indicates the double minimum character of the torsional potential with very small barrier at 90°.

The torsional potentials of several substituted anthracenes were reported by Gentry et al. [12,13] and Lim et al.[14]. The torsional angle of the most stable form and the barrier of the double minimum potential are listed for several relevant compounds.

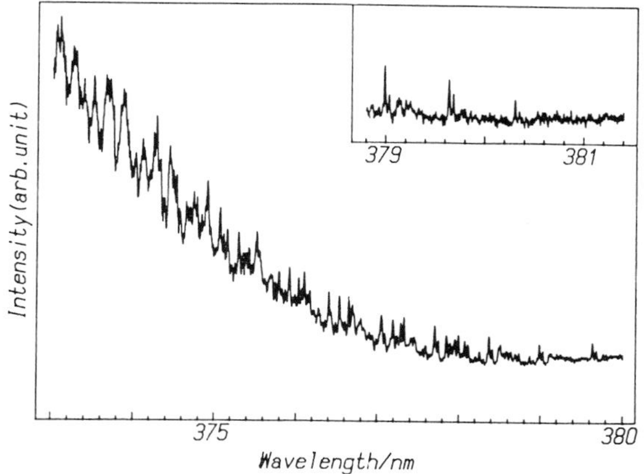

Figure 4. The LIF excitation spectra of the 0-0 transition region of ADMA. The presence of four internal rotors and one inversion motion makes the spectrum very complicated. However, the same pattern repeats every 44 cm^{-1}.

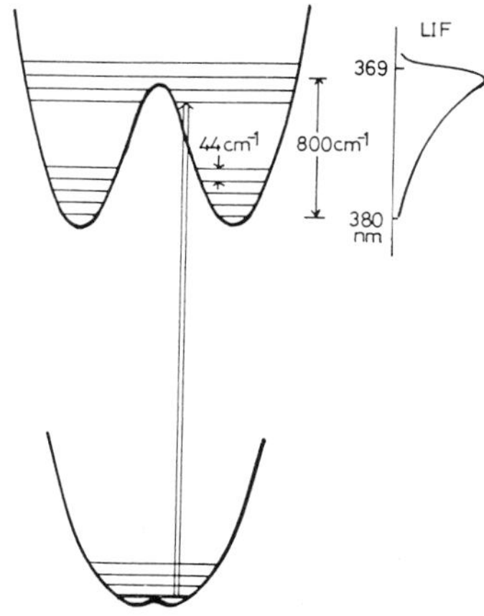

Figure 5. The schematic representation of the torsional potentials for the S_0 and S_1 states of ADMA.

Table 1
The parameters available for the torsional vibration of various polynuclear aromatic molecules

molecule	state	$\Delta\nu$ /cm^{-1}	ϑ /deg.	Vp^a/cm^{-1}	Ref.
biphenyl	S_0	-	42	-	15
	S_1	67	0	-	16
binaphthyl	S_0	-	88	-	17
	S_1	30	80-90	150	17
9-phenylanthracene	S_0	13-101	90	-	12
	S_1	25-49	60	243	12
9,9'-bianthryl	S_0	14	90	-	9
	S_1	11-16	69	200	9
ADMA	S_0	-	>85	very small	11
	S_1	44	70	900	11

[a] Barrier height at 90°.

3.3 4-(9-anthryl)-3,5,N,N-tetramethylaniline (ATMA)

Figure 6 shows the LIF excitation spectrum of ATMA around 371-374 nm. Although the detailed analysis has not yet completed, the general features suggest that the steric hindrance of 3,5-dimethyl group sharpens the ground state potential curve with respect to the torsional angle and thereby narrows the Franck-Condon region of the upper potential curve accessible from the ground state.

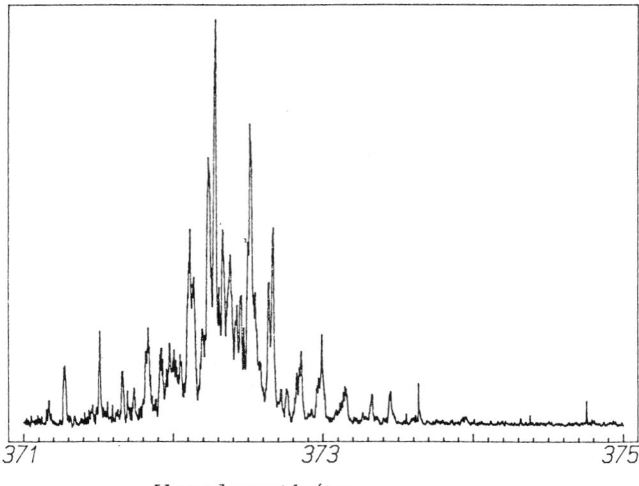

Figure 6. The LIF excitation spectrum of ATMA.

3.4 4-(9-anthryl)-aniline (AA)

AA, synthesized for the first time, was found to show the CT fluorescence in polar liquid solvent. Figure 7 displays the LIF excitation spectrum of jet-cooled AA near its 0-0 transition. In contrast to ADMA and ATMA, the congestion of peaks is much reduced and the sharp peaks separate with each other; the substitution of H atoms for the CH_3 groups reduces the number of internal rotors and thereby the spectrum becomes much simpler.

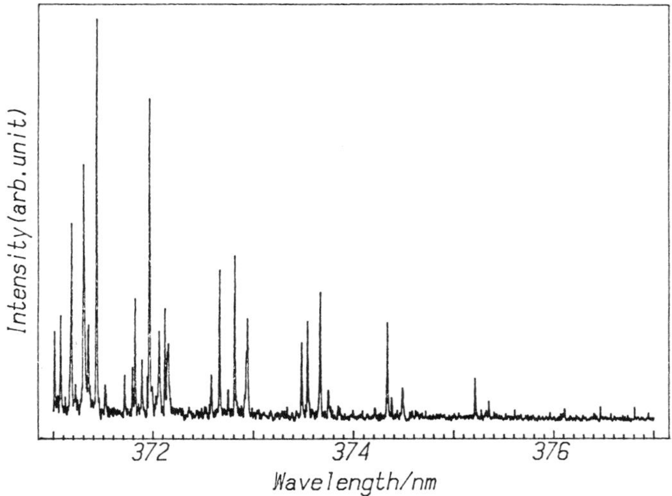

Figure 7. The LIF excitation spectrum of AA.

It is interesting here to consider the relation between the width of the 0-0 band region and the steric hindrance against the internal rotation. As shown in Fig. 8, the 0-0 band region is very narrow for AB and ATMA, less than 3 nm, whereas the 0-0 band covers more than 8 nm for AA and ADMA. In the former case, the steric hindrance against the internal rotation is very large because of the presence of the ortho-substituent to the anthryl group, as depicted in Fig. 8. Such large hindrance probably raises the barrier height of the potential for the internal rotation and hence the ground state potential has a sharp well at the perpendicular position. This considerably limits the Franck-Condon region on the S_1 potential which is accessible from the bottom of the ground state, resulting in the narrow 0-0 band. The reverse is true for ADMA and AA where no substituent is present at the ortho position to the anthryl group.

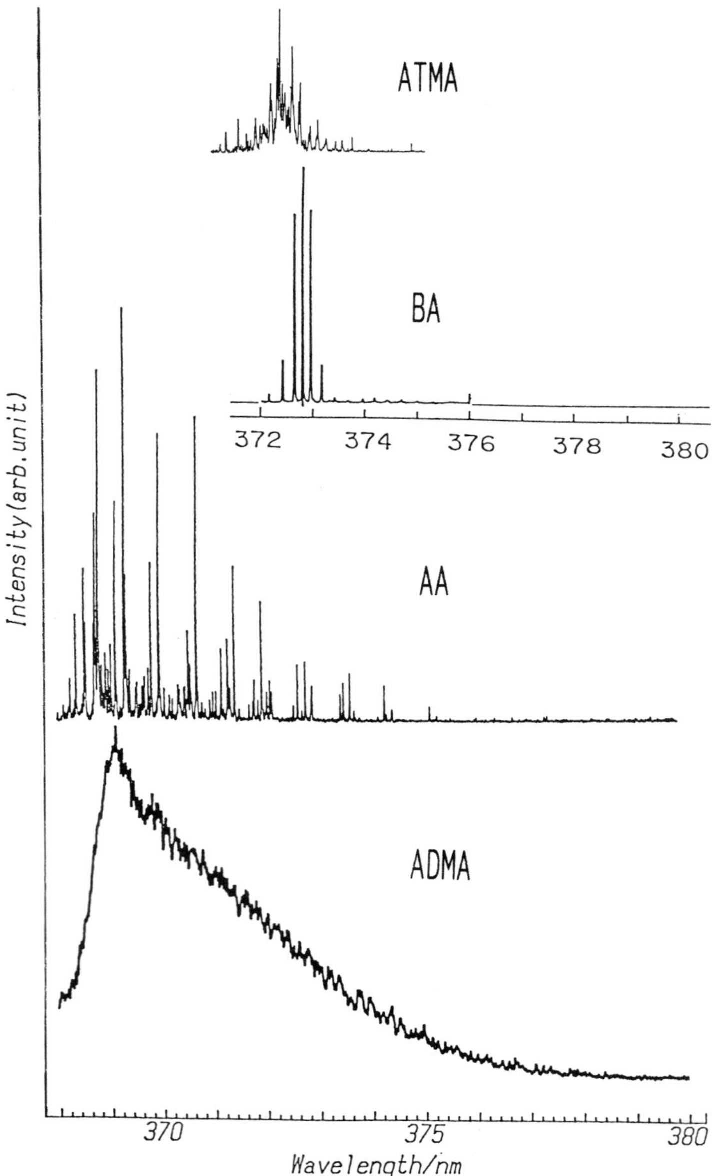

Figure 8. The comparison of the features among the LIF spectra of various TICT molecules. The 0-0 bands of BA and ATMA appear within very narrow range whereas those of AA and ADMA spread over wide spectral range.

3.5 Dimethylaminobenzonitrile (DMABN)

DMABN, the simplest compound yielding the TICT state, has also been a subject of the study of torsional potential. The LIF spectrum of DMABN near the origin exhibits the characteristic pattern ascribable to the internal rotation. However, the analysis is more complicated than that of the anthryl compound [18].

The structure of DMABN was first analyzed spectroscopically by Yokoyama et al.[19] using a combination of FTMW and high resolution LIF excitation spectra. They first determined structure of the ground state using the moments of inertia derived from the MW spectra. Then, the rotational envelope of the single vibronic transition was utilized to estimate the moments of inertia of the electronically excited state. Fig. 9 shows the high resolution LIF spectrum of DMABN at one of the 0-0 transitions.

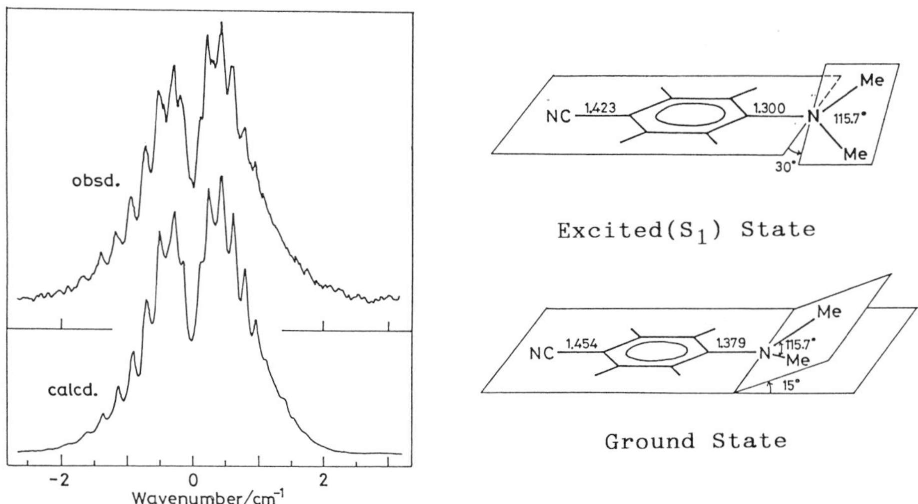

Figure 9. The rotational envelope of the vibronic transition of DMABN at the 0-0 region. The best simulated spectrum and the resulting geometry of DMABN are also shown.

The structure of the electronically excited state was determined so that the simulated spectrum best reproduced the observed rotational contour as shown in Fig. 9. The results shows that the torsional potential has a minimum at 0° in the ground state whereas in the excited state the lone-pair electron of the N atom in the $N-(CH_3)_2$ group tilts by about 30°. Concerning the inversion potential, the most stable angles were found to be 15° and 0° for the ground and excited states, respectively.

Acknowledgment

The author is grateful to Professor H. Shizuka for the sample of ADMA and also to Dr. K. Tominaga for supplying ATMA to us. The author also thank Professor K. Tamao for his valuable help in synthesizing AA. The collaboration of Prof. K. Honma, Dr. K. Yamasaka and Messrs. K. Arita, S. Hayami and H. Yokoyama is gratefully acknowledged.

References

1 E. Lippert, W. Luder and H. Boos, in Advan. Mol. Spectrosc. (A. Mangini, Ed.), Pergamon Press, Oxford, 1962, p. 443.
2 K. Rotkiewicz, K.H. Grellmann, Z.R. Grabowski, Chem. phys. Lett. 19 (1973) 315; ibid. 21 (1973) 212.
3 F. Schneider and E. Lippert, Ber. Bunsenges. Phys. Chem. 72 (1968) 1155.
4 F. Schneider and E. Lippert, Ber. Bunsenges. Phys. Chem. 74 (1970) 624.
5 W. Rettig and M. Zander, Ber. Bunsenges. Phys. Chem. 87 (1983) 1143.
6 Z.R. Grabowski, K. Rotkiewicz, A. Siemiarczuk, D.J. Cowley and W. Baumann, Nouv. J. Chim., 3 (1979) 443.
7 K. Yamasaki, K. Arita, O. Kajimoto and K. hara, Chem. Phys. Lett. 123 (1986) 277.
8 L.R. Khundkar and A.H. Zewail, J. Chem. Phys. 84 (1986) 1302.
9 A. Subaric-Leitis, Ch. Monte, A. Roggan, W. Rettig, P. Zimmermann and J. Heinze, J. Chem. Phys. 93 (1990) 4543.
10 T. Okada, T. Fujita, M. Kubota, S. Masaki, N. Mataga, R. Ide, Y. Sakata and S. Misumi, Chem. Phys. Lett. 14 (1972) 563.
11 O. Kajimoto, S. Hayami and H. Shizuka, Chem. Phys. Lett. 177 (1991) 219.
12 D.W. Werst, W.R. Gentry and P. Barbara, J. Phys. Chem. 89 (1985) 729.
13 D.W. Werst, A.M. Brealey, W. Ronald Gentry ana P.F. Barbara, J. Am. Chem. Soc. 109 (1987) 32.
14 V. Swayambunathan and E.C. Lim, J. Phys. Chem. 91 (1987) 6359.
15 J. Murakami, M. Ito, and K. Kaya, J. Chem. Phys. 74 (1984) 6505.
16 J.D. Lewis, T.B. Malloy, Jr., T.H. Chao, and J. Laane, J. Mol. Struct. 12 (1972) 427.
17 H.T. Jonkman and D.A. Wiersma, J. Chem. Phys. 81 (1984) 1573.
18 V.H. Grassian, J.A. Warren and E.R. Bernstein, J. Chem. Phys. 90 (1989) 3994; See also the comment by R.D. Gorden, J. Chem. phys. 93 (1990) 6908.
19 H. Yokoyama and O. kajimoto, Chem. phys. Lett. in press.

Dynamics and Mechanisms of
Photoinduced Transfer and Related Phenomena
N. Mataga, T. Okada and H. Masuhara (Editors)

Femtosecond–Picosecond Laser Photolysis Studies on Proton Transfer Dynamics in Solutions

Hiroshi MIYASAKA and Noboru MATAGA

Department of Chemistry, Faculty of Engineering Science, Osaka University, Toyonaka, Osaka 560, Japan

ABSTRACT

Proton transfer (PT) processes in the excited hydrogen-bonding (HB) complex as well as in radical ion pair states were investigated by means of femtosecond–picosecond laser photolysis. From these results, it has been revealed that the PT rate constants in the excited HB complex (*ca.*1ps) were almost independent of the polarity of the solvents. In addition, the deuteration effect was found to be very small at room temperature. In the PT process in the radical ion pair states, it was strongly suggested that the orientation and the distance between two radical ions determined at the electron transfer (ET) reaction regulated the subsequent PT process. On the basis of the experimental results and theoretical considerations, the relation between the ET and the PT processes was discussed.

1. PREFACE

Proton transfer (PT) in the excited state as well as the photo-induced electron transfer (ET)[1,2] are processes playing the most important and fundamental roles in a number of photochemical reactions. Compared to the intermolecular ET undergoing through the overlap of the mutual electronic wave-functions which are widely spread, the PT process seems much more sensitive to the mutual distance and the orientation between the proton donor and the acceptor. In the present paper, we report pico- and femtosecond laser photolysis studies on the following intermolecular proton transfer processes in solutions:

(1) PT process of the excited hydrogen-bonding complexes,

(2) PT process in the radical ion pair states.

By integrating these experimental results on PT processes, the effects of the solvent and the relation between the structure of the complex or the ion pair and the rate of the PT process will be discussed.

2. EXPERIMENTAL

A microcomputer-controlled picosecond laser photolysis system with a repetitive mode-locked Nd^{3+}:YAG laser was used for transient absorption spectral measurements in 10ps to nanosecond region[3]. In the measurements in shorter time region, a femtosecond laser photolysis system was used[2b,4]. The output of a CW mode-locked Nd^{3+}:YAG laser operated at 82MHz (12W output-power) was compressed by a fiber-grating system and the SHG pulse (FWHM 4ps and output-power 1.2W) synchronously pumped a pyridine-1 dye laser (at 710nm and FWHM:300fs). The output of the dye laser was amplified to 0.4mJ/pulse by a three-stage dye amplification system pumped by a frequency-doubled Q-switched Nd^{3+}:YAG laser operating at 10Hz. The pulse duration of the amplified pulse was typically 500fs. In the case where the saturable absorber was used in the amplifier, the width of the pulse was 230fs and the output-power was 0.25mJ/pulse. The amplified pulse was frequency doubled (355nm) and used for the exciting light. The rest of the fundamental pulse was focussed into D_2O to generate a white continuum probe pulse. Two set of multichannel photodiode detectors were used to observe transient absorption spectra. All the measurements were performed at 22$\pm$2°C.

3. RESULTS AND DISCUSSION

3.1. Proton Transfer Process in the Excited Hydrogen-Bonding Complex.

A number of investigations have been carried out on the role of the hydrogen-bonding interaction in the fluorescence quenching processes. From these results, the following mechanisms have been established [5,6,7].

(1) Charge-transfer interaction through the hydrogen bond plays an important role for the fluorescence quenching when two conjugate π-electron systems are directly connected, such as 1-aminopyrene-pyridine and 1-pyrenol-pyridine.

$$\mathbf{(D\text{-}H\cdots A)^* \rightarrow D\text{-}H^{+}\cdots A^{-}}$$

(2) Ultrafast proton transfer leading to the formation of the ion pair in the hydrogen-bonding complex between aromatics with hydroxyl group and aliphatic amines.

$$\mathbf{(D\text{-}H\cdots A)^* \rightarrow (D^{-}\cdots H\text{-}A^{+})^*}$$

In this study, we have directly observed the proton transfer process of the pyrenol(PYOH)-triethylamine(TEA) hydrogen-bonding complexes in various solvents by means of the femtosecond laser photolysis.

1-Pyrenol (PYOH) forms stable 1:1 hydrogen-bonding (HB) complex with triethylamine (TEA) in the ground state. The equilibrium constant of the HB complex formation is more than 100 in nonpolar solvents such as n-hexane and benzene. In Figure 1 exhibited are the fluorescence spectra of PYOH-TEA system in various solvents, which were obtained by the excitation of the HB complex formed in the ground state. The spectrum in n-hexane shows rather sharp structure, which is ascribed to the emission of the excited HB complex (HB^*). This fluorescence spectrum is very similar to that of free PYOH in this solvent but shifted to longer wavelength region by 3-4nm.

In TEA solution, broad band in longer wavelength region was observed in addition to the fluorescence of HB^* appeared in shorter wavelength region. The former structureless fluorescence is attributed to the ion pair (IP^*) produced by the PT reaction in the HB^* complex. On the basis of the coincidence of the fluorescence lifetime of IP^* with that of HB^*, it was concluded that rapid equilibrium was established between HB^* and IP^* in the excited state in TEA solution. With increase in the polarity of the solvent, the fluorescence due to IP^* shifts toward longer wavelength region. No emission due to HB^* was observed in more polar solvents than TEA, indicating that the equilibrium shifted dominantly to IP^* in polar solvents.

In the fluorescence spectrum in acetonitrile solution, a shoulder around 470nm was observed. This new fluorescence band was ascribed to the excited state of the dissociated PYO^- produced by the PT reaction followed by the ionic dissociation(ID) on the basis of measurements of photoconductivity, time-resolved fluorescence spectrum, and the spectral shape of $(PYO^-)^*$ produced in alkaline solution.

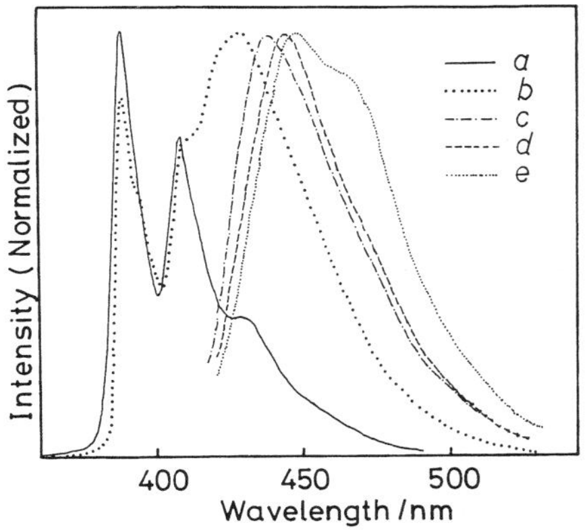

Figure 1. Fluorescence spectrum of PYOH-TEA system in various solvents. (a) n-hexane, (b) TEA, (c) benzene, (d) acetone, and (e) acetonitrile.

Summarizing above results, total reaction scheme of the excited PYOH-TEA HB complex is depicted as follows:

$$(\mathrm{PYOH}\cdots\mathrm{TEA})^* \xrightarrow{\mathrm{PT}} (\mathrm{PYO}^-\cdots\mathrm{H}\text{-}\mathrm{TEA}^+)^* \xrightarrow{\mathrm{ID}} (\mathrm{PYO}^-)_s^*$$

The main species observed in the steady state fluorescence spectrum is dependent on the polarity of the solvent. In this paper, we concentrate our discussion on the initial step of the PT process.

In Figure 2, we show time-resolved transient absorption spectra of PYOH - TEA(2M) in benzene solution excited with a femtosecond laser pulse. With increase of the delay time after the excitation, an absorption maximum at 475nm, which is ascribable to the ion pair state produced by the PT reaction, increases together with the decrease of the absorbance around 500nm (due to HB^*). From the time profiles of the absorbance, the time constant for the PT reaction was obtained to be 850fs. Time constants for the PT in other solvents are summarized in Table 1, where no clear dependence of the PT rate on the solvent can be observed. In addition, the deuteration effect on the time constant was found to be very small.

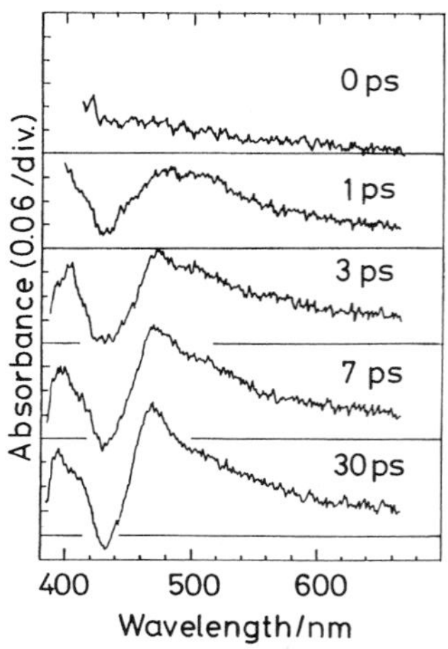

Figure 2. Transient absorption spectra of PYOH–TEA(2M) in benzene excited with a 355nm femtosecond laser pulse.

Table 1. Proton Transfer Time Constants of Excited PYOH–TEA Hydrogen Bonding Complex in Solutions.

Solvent	Time Constant / ps	
	PYOH	PYOD
Benzene	0.85±0.15	0.9±0.2
TEA	0.8±0.07	0.8±0.1
THF	0.7±0.1	---
Acetone	1.0±0.1	1.15±0.15
Methanol	1.5	----
Acetonitrile	1.05±0.15	1.15±0.15

The PT process of PYOH–TEA hydrogen bonding complex at 77K in the glass matrix of TEA solution was also investigated by means of picosecond single-photon-counting method. It should be noted here that there exist at least two kinds of HB complexes in the TEA glassy solvent at 77K, which were identified by the steady state fluorescence excitation spectrum. From the results of the excitation spectrum, it was concluded that one of these two HB complexes undergoes the PT process and the other cannot undergo the PT reaction during the lifetime of the excited state. In anyway, it has been revealed that the PT time constant of the HB complex which can undergo the PT reaction was extremely short (<<3ps) even in the glass matrix at 77K, where the large molecular motion seems to be restricted.

Summarizing these experimental results on the effects of solvents, deuteration, and the temperature on the PT time constants, it is suggested that the small motion involving a little larger part of the HB complex than the H–bond itself such as the torsional motion around carbon–oxygen bond might regulate the PT process *at room temperature.*

The intermolecular PT rates obtained in the present study are smaller than those of the intramolecular PT process; the time constant for the PT of excited 2-(2-hydroxide)-benzothiazole in methylene chloride solution was reported to be (170±20)fs[8]. The difference between the rates of the inter- and intramolecular processes may be attributed to the difference in the degree of the coupling between the change of the electronic structure by excitation and the PT process. While the former intermolecular process is the formation of the ion pair state, the latter intramolecular PT is the enol-keto isomerisation and the change of the electronic structure by excitation may more directly assist the PT process.

3.2. Proton Transfer Process in the Radical Ion Pair States.

Hydrogen abstraction reaction of excited benzophenone (BP^*) has been widely investigated for a long time. Especially, much attention has been paid for the photoreduction process of BP by amines (AH), since the production yields of the benzophenone ketyl radical (BPH) are usually very high in the reaction with AH and the reaction rate constants are very close to that of the diffusion-controlled reaction. On the basis of these experimental results, CT interaction has been supposed to play important roles in the hydrogen abstraction reaction of BP^* from AH since amines usually have rather low oxidation potential.

Although a number of investigations have been performed in order to clarify the role of CT interaction in the reaction mechanism, clear conclusion on this problem has not been established. In order to elucidate the roles of the CT interaction and the ion pair state(s) in the reaction and the relation between the ET reaction and the PT process, the effects of the solvent, the oxidation potential of the amine, and the structure of the ion pair (IP) on the reaction mechanism were investigated by means of femtosecond-picosecond laser photolysis and transient absorption spectroscopy.

3.2.1 Reaction Mechanism of Hydrogen Abstraction of $^3BP^*$.

Since the $S_1 \rightarrow T_1$ intersystem crossing (ISC) time of BP is very short (*ca.*10ps), much attention has been concentrated on the photoreduction of BP in T_1 state ($^3BP^*$). From our investigations on the dynamics of the photoreduction of $^3BP^*$ with a number of amines in various solvents[9-13], it has been revealed for secondary amines that the hydrogen abstraction and CT or IP state formation by ET are competing at encounter collision between $^3BP^*$ and AH, and the CT or IP state relaxed with respect to D-A configurations and solvation does not contribute to ketyl radical (BPH) formation.

$$^3BP^* + AH \rightarrow BPH + A\cdot \quad ; \quad \searrow (BP^{-\cdot}AH^{+}) \qquad \textbf{Scheme 1}$$

On the other hand, for some tertiary amines with low oxidation potential such as MDPA, DMA, and DEA (see Table 2)in acetonitrile, it has been revealed that the hydrogen abstraction of $^3BP^*$ is described by such mechanism as

$$^3BP^* + AH \rightarrow (BP^{-\cdot}AH^{+}) \rightarrow BPH + A\cdot \quad ; \quad \searrow BP^{-} + AH^{+} \qquad \textbf{Scheme 2}$$

The first step of the photoreduction is the ET at encounter collision leading to the stable IP formation $^3(BP^-AH^+)_{enc}$, followed by the proton transfer. This difference of the mechanism depending on the nature of amines and solvent might be related to the difference of the energy level of the IP state and the mutual configuration of BP and AH at encounter collision.

3.2.2 Dependence of Production Yields of BPH radical on Solvents and Amines.

The production yields of BPH radical in the reaction between $^3BP^*$ and various amines were determined. As will be discussed later, the contribution from the reaction between $^1BP^*$ (S_1 state of BP) and AH and that from the excited state of CT complex between BP and AH formed in the ground state increase with increase of the concentration of AH. Therefore, the reaction yields listed in table 2 were obtained in the condition where the concentration of AH was kept low enough to avoid such contributions.

Table 2. Production Yields of BPH Radical in the Reaction of $^3BP^*$ and Various AH in Solutions.

AMINE		OXIDATION POTENTIAL$^{(a)}$	Yields of BPH	
			CH_3CN	isooctane
ANILINE	(AN)	0.98	0.78	0.78
TRIETHYLAMINE	(TEA)	0.97	0.73	0.72
DIPHENYLAMINE	(DPA)	0.96	0.76	0.75
N-METHYLDIPHENYLAMINE	(MDPA)	0.86	0.74	0.73
N,N-DIMETHYLANILINE	(DMA)	0.76	0.74	0.79
N,N-DIETHYLANILINE	(DEA)	0.72	0.23	0.73
N,N-DIETHYL-P-TOLUIDINE	(DET)	0.69	0	0.7
N-ETHYL-P-TOLUIDINE	(NET)	0.68	0.85	0.77
1,4-DIAZABICYCLO -[2,2,2]OCTANE	(DABCO)	0.68	0	0.7

(a) vs. S.C.E. in acetonitrile.

In isooctane solution, the yields of BPH radical are almost independent of the structure and the oxidation potential of AH. On the other hand, the decrease of the yield of BPH radical with decrease of the oxidation potential of AH was observed for the reaction with tertiary amines in acetonitrile, while such a dependence on the oxidation potential of AH was not observed and the yield of BPH was independent of the oxidation potential of AH for the reaction with secondary amines. The difference in the BPH yield depending on the nature of AH in acetonitrile is related to the mechanism of the reaction as described in the previous section. The abstraction mechanism for BP–tertiary AH in acetonitrile was the consecutive one and, hence, the yield of BPH radical was dependent not only on the PT rate constant in the ion pair state but also on other competitive processes such as the ionic dissociation and the charge recombination, which will be discussed later in detail.

3.2.3 Dependence of the Proton Transfer Reactivity of IP on its Production Process.

In addition to the ET in T_1 state, the excitation of the CT complex in the ground state as well as the ET reaction in S_1 state also produce IP states.

$$^3BP^* + AH \longrightarrow {}^3(BP^- \cdot AH^+)_{enc} \qquad {}^3IP_{enc}$$

$$^1BP^* + AH \longrightarrow {}^1(BP^- \cdot AH^+)_{enc} \qquad {}^1IP_{enc}$$

$$(BP \cdot AH) \xrightarrow{h\nu} {}^1(BP^- \cdot AH^+)_{com} \qquad {}^1IP_{com}$$

It should be noted that $^1IP_{enc}$ and $^1IP_{com}$ are produced efficiently in the concentrated solution of AH (>0.3M); the production yield of $^1IP_{com}$ in BP–DMA(1.0M) in acetonitrile solution is 30%, that of $^1IP_{enc}$ is 35%, and only 35% of the absorbed energy of the incident excitation light is terminated in the production of T_1 state of BP[10,11].

From the recent investigation on the singlet IP between various donors (D) and acceptors (A)[14], it was indicated that the structure of IP such as the mutual configuration of ions and the inter–ionic distance in the pair including the surrounding solvents was different depending on the mode of its production. Hence, the elucidation of the reactivity of each IP, which consists of the same D–A pair and has different pathway of production, may provide important information on the relation between the geometry and the reactivity of the IP in the hydrogen abstraction. In the following, we will discuss the dependence of the proton transfer reactivity of these three kinds of IP's.

Informations on dynamic behaviors and reaction profiles of $^3IP_{enc}$ can be obtained by the analysis of the experimental results in the dilute solutions of AH (<0.1M). In this section, we will discuss the ET process in the S_1 state of BP.

In Figure 3, we show the time–resolved transient absorption spectra of BP–DEA (0.6M) in acetonitrile solution (Figure 3). The absorption spectra of $^1BP^*$ and $^3BP^*$ in this solvent are exhibited by dotted lines. S_1->T_1 ISC time constant of BP in acetonitrile has been obtained to be 9.0ps[10,11]. A transient absorption spectrum at 1ps after the excitation with fs 355nm laser pulse in Figure 3 shows absorption maxima at 470nm (DEA^+) and 575nm ($^1BP^*$).

In addition to these two peaks, broad absorption whose intensity increases in wavelength region longer than 600nm (BP^-) is observed. Since the rapid increase of absorption signal due to (BP^-DEA^+) was almost identical with the response of the apparatus and BP and DEA in acetonitrile solution forms stable weak CT complex in the ground state, the rapid appearance of IP was concluded to be due to the excitation of the ground state CT complex. From the equilibrium constant of the complex formation and the extinction coefficient of the complex, ca.30% of the excitation light at 355nm was absorbed by the CT complex.

By the laser photolysis at 397nm where the CT complex can be selectively excited, it was revealed that the production of BPH was not observed from the IP produced by the excitation of the CT complex, $^1(BP^- DEA^+)_{com}$, and rapid disappearance of the absorption mainly due to the charge recombination with the time constant of 93ps was observed.

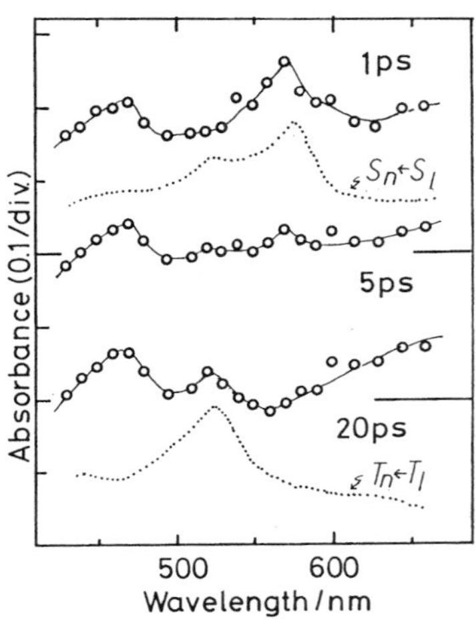

Figure 3. Transient absorption spectra of BP–DEA(0.6M) in acetonitrile excited with a 355nm fs laser pulse.

In Figure 3, we can also see that the absorption due to (BP^- DEA^+) increases together with the decrease of $^1BP^*$ absorbance and increase in the absorbance at 525nm ($^3BP^*$) with the increase of the delay time after the excitation. The rise times of $^3BP^*$ and IP were 5.5ps and 5.4ps, respectively, and the decay time constant at 575nm ($^1BP^*$) was 5.5ps. From the coincidence of the time constants and the time evolution of the transient absorption spectrum, it may be concluded that ET reaction between $^1BP^*$ and DEA competes with the ISC process. Since the lifetime of $^1BP^*$ in amine free solution was 9.0ps, almost half of the $^1BP^*$ was quenched by DEA, which resulted in the production of $^1(BP^- DEA^+)_{enc}$, and the rest in the triplet state formation (which also resulted in the IP formation as stated above). Summarizing above results, it is concluded that three kinds of IP are produced in the acetonitrile solution of BP–DEA (0.6M) by 355nm excitation.

By changing the concentration of DEA and the excitation wavelength, we have determined the lifetime and reaction rate constants of each IP (Table3). For comparison, included in this table are the results on BP–DMA system[10,11]. As is clear from this table, the rate constant of each process is quite different depending on the mode of the production. It should be noted that the structure of each IP may be considered to be maintained at least during several hundreds of picoseconds, since the rate constants are different depending on the nature of each IP. The difference in k_{CR} between singlet and triplet IP's may be explained by the difference of the spin multiplicity. However that in k_{PT} cannot be accounted for only by this difference. According to the results of recent investigation[14], the compact IP probably with no intervening solvent molecule between A^- and D^+ would be produced in the case of the excitation of the CT complex, while more loose IP with intervening solvent molecules between A^- and D^+ would be formed by the ET at encounter. Along with this interpretation, it may be concluded from the present results that the factor controlling the hydrogen abstraction process is not only the mutual distance between the pair but also the initial geometry, especially the orientation, for the interaction. More details concerning the relation between the PT rate and the inter–ionic distance will be discussed in the next section.

Table 3. Dependence of the Reaction Rate Constants of the Ion Pairs in Acetonitrile Solution upon the Mode of their Production.

	k_{PT}/s^{-1}	k_{ID}/s^{-1}	k_{CR}/s^{-1}
$^3(BP^-\ DEA^+)_{enc}$	7.3×10^8	2.1×10^9	——
$^1(BP^-\ DEA^+)_{enc}$	4.3×10^8	1.4×10^9	1.1×10^9
$^1(BP^-\ DEA^+)_{com}$	$<<10^8$	1.4×10^9	9.3×10^9
$^3(BP^-\ DMA^+)_{enc}$	5.4×10^9	1.4×10^9	——
$^1(BP^-\ DMA^+)_{enc}$	6.6×10^8	9.5×10^8	5.8×10^8
$^1(BP^-\ DMA^+)_{com}$	$<<10^8$	$<4\times10^8$	1.1×10^{10}

(PT : proton transfer, ID : ionic dissociation, and CR : charge recombination.)

3.2.4 Difference of the Reactivity of Ion Pairs Depending on the Oxidation Potential of the Amine.

In addition to the difference of the reactivity of the IP depending on the mode of its production, a large difference of the reactivity of $^3IP_{enc}$ was also observed between BP–DMA and BP–DEA systems. In order to clarify factors causing this difference, we have investigated photoreduction processes of BP– N,N–diethyl–p–toluidine (DET) and BP– N–methyldiphenylamine (MDPA) systems. We have selected these tertiary amines on the following reasons. First, the reaction mechanisms for these tertiary amines are all consecutive one of Scheme 2. Second, molecular structures of these tertiary amines are rather close to one another and have different oxidation potentials (listed in Table 4). Although the difference of the reactivity depending on the mode of the IP production, i.e. the difference among $^3IP_{enc}$, $^1IP_{enc}$ and $^1IP_{com}$, was observed also in BP–DET and –MDPA systems just as in BP–DEA and BP–DMA systems, we concentrate our discussion on the reduction process of $^3BP^*$.

In the case of the reaction between $^3BP^*$ and DET, the yield of the dissociated free BP^-_s and DET^+_s was 0.95±0.05, and neither decay process of the IP nor the formation of the ketyl radical was observed, which made impossible the determination of the lifetime of $^3IP_{enc}$ by the present experiment. Hence, we estimate the lifetime and the reaction rate constants in the following manner. The rate constant of the ionic dissociation process of IP's of various donors and acceptors produced by ET at encounter collision in acetonitrile at room temperature (ca.20 C) has been determined to be $0.5–2\times10^9 s^{-1}$[14,15]. As listed in Table 3, k_{ID} of $^3(BP^-\ DEA^+)_{enc}$ and that of $^3(BP^-\ DMA^+)_{enc}$ were also in this range. Since the yield of the ionic dissociation of the $^3(BP^-\ DET^+)_{enc}$ was 0.95±0.05, the lifetime of the IP was estimated to be ≥500ps. Since the ketyl radical formation was not recognized, the rate constant for the proton transfer is estimated to be $<<2\times10^8 s^{-1}$. Anyhow, summarizing above results on the reduction process of $^3BP^*$–DET system, it may be concluded that the decrease of the yield of BPH is attributable to the decrease of k_{PT}. This result indicates that, k_{PT} of $^3IP_{enc}$ between BP and the tertiary aromatic amine decreases with the decrease of the oxidation potential of the amine. In order to confirm the effect of the oxidation potential of the amine on k_{PT} of $^3IP_{enc}$, the reduction process of $^3BP^*$–MDPA system was also investigated by means of the picosecond transient absorption spectroscopy. The lifetime of $^3(BP^-\ MDPA^+)_{enc}$ was obtained to be 90ps and the reaction yields of the proton transfer and the ionic dissociation were obtained to be 0.74 and 0.10, respectively.

In Table 4, collected are the values of the lifetime and the rate constants for the PT and the ID processes of $^3IP_{enc}$ of BP–tertiary aromatic amine systems in acetonitrile. This table shows clear dependence of each rate constant, especially k_{PT}, on the oxidation potential of the amine. This dependence of k_{PT} on the oxidation potential seems to indicate the difference in the structure of the IP depending on the energy of the IP state, and seems to be closely related to the problems of the energy gap dependence of the ET reaction.

The rate of the photoinduced ET reaction in polar solutions is regulated by[16]:

(a) Franck–Condon factor for the ET process related to the energy gap ($-\Delta G^o$) between the initial and final state of ET,
(b) the reorganization energy (λ) including the intramolecular vibrational modes and the polarization of the solvent surrounding D and A,
(c) the solvent dynamics,
(d) the electronic interaction responsible for ET which depends on mutual distance and orientations between D and A.

When the factor (d) is not very strong and is not much different throughout a series of D, A systems, the ET rate may be determined mainly by the factor (a) and (b).

The CR decay rate constant, k_{CR}, of loose IP (LIP) or solvent separated IP (SSIP) formed by CS at encounter in the fluorescence quenching reaction [1b,2a,15b] as well as the phosphorescence quenching reaction[17] has been confirmed to show bell-shaped energy gap dependence. These energy gap dependence were rather easy to be explained by the usual theories on the ET reactions. On the other hand, the charge separation (CS) rate constant, k_{CS}, at encounter in the luminescence quenching reaction rises steeply around zero energy gap and becomes diffusion-limited at favorable $-\Delta G^o$ regions and no inverted effect has been experimentally confirmed yet. In order to obtain the intrinsic CS rate constant which cannot be obtained by the stationary measurement due to the diffusion limit, we have evaluated the k_{CS} values in the diffusion-limited $-\Delta G^o$ regions by measuring the transient effect in the fluorescence quenching reaction of a series of D, A systems in acetonitrile solutions[18]. From this experiment, it has been demonstrated that the $-\Delta G^o$ dependence of k_{CS} in this top region with large $-\Delta G^o$ is rather flat and does not show such a clear bell-shaped $-\Delta G^o$ dependence as observed in the CR reaction. On this peculiar $-\Delta G^o$ dependence of k_{CS} in the top region, we have made quantitative theoretical studies and have shown that the observed $-\Delta G^o$ dependence of k_{CS} and k_{CR} can be systematically and consistently interpreted by assuming D, A distance distributions depending on $-\Delta G^o$ values in the CS reaction and also nonlinear polarization of solvent around ions[19,20]. Briefly speaking, such $-\Delta G^o$ dependence of k_{CS} can be interpreted by assuming the change of the ET distance depending on the change of $-\Delta G^{o}$ value. That is, for larger $-\Delta G^{o}$ value, CS at a little larger distance is favorable because λ becomes larger at larger D, A distance, which makes the $-\Delta G^{o}$ dependence of k_{CS} broader than that expected from the simple theoretical consideration[21].

The present result of the dependence of k_{PT} on the oxidation potential of the amine is well interpreted along with the above considerations on the energy gap dependence of k_{CS} in the following manner. The inter–ionic distance in those BP–amine ^{3}LIP will become longer with decrease of the oxidation potential of the amine to some extent. Since the PT reaction rate is much more severely dependent on the mutual distance than those of ET reactions, small change of the inter–ionic distance in the ^{3}LIP arising from the difference of $-\Delta G^{o}$ in the ET reaction drastically affects the subsequent PT processes in the condition where the orientation of two molecules are similar among various D–A pairs and the averaged structure is kept until the proton transfer taking place in competition with the ionic dissociation. Consequently, the k_{PT} will decrease with decrease of the amine oxidation potential due to the increase of the proton transfer distance.

Table 4. Dependence of the Reaction Rate Constants of Triplet Ion Pairs between BP and Tertiary Amines upon the Oxidation Potential of the Amine in Acetonitrile.

	Oxid. Potential Vs. S.C.E.(V)	Lifetime	k_{PT}/s^{-1}	k_{ID}/s^{-1}
$^{3}(BP^{-}MDPA^{+})_{enc}$	0.86	80ps	9.5×10^{9}	1.0×10^{9}
$^{3}(BP^{-}DMA^{+})_{enc}$	0.76	140ps	5.4×10^{9}	1.4×10^{9}
$^{3}(BP^{-}DEA^{+})_{enc}$	0.72	290ps	7.3×10^{8}	2.1×10^{9}
$^{3}(BP^{-}DET^{+})_{enc}$	0.69	>500ps	$<<2\times10^{8}$	(2×10^{9})

(MDPA, N-methyldiphenylamine, DEA, N,N-diethylaniline, and DET, N,N-diethyl-p-toluidine)

3.2.5 Concluding Remarks on the PT Process in the Radical Ion Pairs.

From the experimental results on the k_{PT} of the radical ion pair, it was revealed that the PT processes are regulated severely by the geometrical structure which were determined at the initial interaction between the BP^{*} and the tertiary aromatic AH in the ET process. On the other hand, the hydrogen abstraction process of the BP^{*} from primary and secondary AH is described by Scheme 1 even in acetonitrile. This difference of the mechanism depending on the structure of AH might be related to weak HB interaction between π–electrons on oxygen of the BP^{*} and primary or secondary AH at encounter. Consequently, the mutual configuration and distance between BP^{*} and AH at encounter may be appropriate for the rapid direct hydrogen transfer processes in these amines.

In relation to this, we have studied also the dynamic behaviors of chained molecules such as BP–O–$(CH_2)_n$–O–AH. In this case, the reaction mechanism for the hydrogen abstraction (scheme 1 or 2) changed depending on the chain length. This results also suggests that the mutual geometry for the interaction is important in the reaction mechanism[22].

ACKNOWLEDGEMENT

The present work was partially supported by a Grant-in-Aid for Specially Promoted Research (No.6265006) from Ministry of Education, Science, and Culture of Japan to N.M.

REFERENCES

1.a **N.Mataga**, *Pure. Appl. Chem.*, **56**, 1255 (1984).
1.b **N.Mataga**, *Acta. Phys. Polon.*, **A71**, 767 (1987).
2.a **N.Mataga**, *"Photochemical Energy Conversion"*, Elsevier, Amsterdam (1988), p.32.
2.b **N.Mataga, H.Miyasaka, T.Asahi, S.Ojima, and T.Okada**, *"Ultrafast Phenomena VI"*, Springer-Verlag, Berlin, (1988), p.511.
3. **H.Miyasaka, H.Masuhara, and N.Mataga**, *Laser Chem.*, **1**, 357 (1983).
4. **H.Miyasaka, S.Ojima, and N.Mataga**, *J.Phys.Chem.*, **93**, 3380 (1990).
5. **N.Mataga, Y.Kaifu, and M.Koizumi**, *Nature*, **175**, 731 (1955).
6. **N.Mataga and T.Kubota**, *"Molecular Interactions and Electronic Spectra"*,Marcel Dekker, New York, 1970.
7. **N.Ikeda, H.Miyasaka, T.Okada, and N.Mataga**, *J.Am.Chem.Soc.*, **105**, 5206, (1983); **M.M.Martin, D.Grand, N.Ikeda, T.Okada, and N.Mataga**, *J.Phys.Chem.*, **88**, 167 (1984); and references cited therein.
8. **F.Laermer, T.Elsaesser, and W.Kaiser**, *Chem. Phys. Lett.*, **148**, 119 (1988).
9. **H.Miyasaka and N.Mataga**, *Bull. Chem. Soc., Jpn.*, **63**, 131, (1990).
10. **H.Miyasaka, K. Morita, M.Kiri, and N.Mataga**, *"Ultrafast Phenomena VII"*, Springer-Verlag, Berlin, (1990) p.552.
11. **H.Miyasaka, K.Morita, K.Kamada, and N.Mataga**, *Bull. Chem. Soc. Jpn.*, **63**, 3368, (1990).
12. **H.Miyasaka, K.Morita, K.Kamada, and N.Mataga**, *Chem.Phys.Lett.*, **178**, 504, (1991).
13. **H.Miyasaka, K.Morita, K.Kamada, T.Nagata, M.Kiri, and N.Mataga**, *Bull.Chem.Soc.Jpn.*, submitted.
14.a **T.Asahi and N.Mataga**, *J.Phys.Chem.*, **93**, 6575 (1989).
14.b **T.Asahi and N.Mataga**, *ibid*, **95**, 1956 (1991).
15.a **N.Mataga, Y.Kanda,T.Asahi, H.Miyasaka, T.Okada, and T.Kakitani**, *Chem.Phys.*, **127**, 239(1988).
15.b **N.Mataga, T.Asahi, Y.Kanda, T.Okada, and T.Kakitani**, *Chem.Phys.*, **127**, 249 (1988), and references cited therein.
16. **N.Mataga**, "*Perspective in Photosynthesis*", ed. by J.Jortner and B.Pullman, Kluwer Academic, Dordrecht (1990) p.227.
17. **T.Ohno, A.Yoshimura, and N.Mataga**, *J.Phys.Chem.*, **94**, 4871 (1990).
18. **S.Nishikawa, T.Asahi, T.Okada, N.Mataga, and T.Kakitani**, submitted to *Chem.Phys.Lett.*.
19. **T.Kakitani, A.Yoshimori, and N.Mataga**, *Advances in Chemistry Series, ACS, "Electron Transfer in Inorganic Organic and Biological Systems"*, ed. by J.R.Bolton, N.Mataga, and G.McLenden (1991), Chap.4.
20. **A.Yoshimori, T.Kakitani, Y.Enomoto, and N.Mataga**, *J.Phys.Chem.*, **93**, 8316 (1989), and references cited threrin.
21. **R.A.Marcus**, *J.Chem.Phys.*, **24**, 966 (1956); R.A.Marcus and N.Sutin, *Biochim.Biophys.Acta.*, **811**, 265 (1985).
22. **H.Miyasaka, M.Kiri, K.Morita, N.Mataga, and Y.Tanimoto**, to be submitted for publication.

Dynamics and Mechanisms of
Photoinduced Transfer and Related Phenomena
N. Mataga, T. Okada and H. Masuhara (Editors)

Dynamics of double proton transfer reaction in the excited state of hydrogen bonded dimers as studied in a supersonic jet

K. Fuke[a], A. Nakajima[b], and K. Kaya[b]

[a] Institute for Molecular Science, Myoudaiji, Okazaki 444, Japan
[b] Department of Chemistry, Faculty of Science and Technology, Keio University, 3-14-1 Hiyoshi, Kohoku-Ku, Yokohama 223, Japan

Abstract

Double proton transfer reaction in the lowest singlet excited state of hydrogen bonded dimers of 1-azacarbazole, and 7-azaindole was studied by the laser induced fluorescence method, mass selected multiphoton ionization spectroscopy, and picosecond time resolved spectroscopy. Intermolecular N-H--N H-bond stretching vibration was found to stimulate the reaction. Deuterium substitution experiment also supports the coupling of proton transfer reaction with the intermolecular vibration.

1.Introduction

Proton transfer reaction is one of the most fundamental chemical reactions and plays a crucial role in a variety of chemical and biological processes. Photoinduced proton transfer in neighbouring two H-bondings named as double proton transfer (DPTR) is important in biological systems such as hydrogen bonded base pairs of DNA. As model compounds of these base pairs, H-bonded dimers of 7-azaindole(7-AI) and 1-azacarbazole(1-AC) have been studied extensively in condensed phases (solid and solution). Excited state DPTR of 7-AI dimer has been studied first by El-Bayoumi and his coworkers in solution. According to their work, 7-AI dimers after absorbing UV photons(-300 nm) emit both UV and visible fluorescenc. UV one comes from the nonreacted normal dimers and green emission is emitted from the tautomers after DPTR in the excited state. Eisenthal et al. measured the DPTR rate of 7-AI dimers by the picosecond pulsed laser and determined the rate to be faster than 6 ps. Similarly to 7-AI dimers, dimers of 1-azacarbazole(1-AC) tautomerize in the lowest excited state. Even though extensive studies have been conducted in the elucidation of the DPTR process, microscopic mechanism cannot be unveiled until recently because the reaction is taken place in condenced phase where the electronic spectrum has no fine structure.

Supersonically expanded free jet combined with laser spectroscopy enables one to obtain well resolved vibrational and rotational fine structure in the electronic spectrum of large molecules and microscopic information on DPTR will be obtained under the supersonic jet condition. For the past several years, our group has been studying DPTR in the excited state of 7-AI and 1-AC dimers in a supersonic jet using both wavelength and time resolved spectroscopies.

In the present paper, summarized results of the studies for DPTR in the excited states of these dimers will be discussed.

2.Experimental

Most of the experiments were conducted under a supersonic free jet condition using a pulsed nozzle.

(Wavelength resolved spectroscopy)

In order to assign the location of the electronic transitions in repective clusters, mass selected resonance enhanced multiphoton ionization(REMPI) spectroscopy, and laser induced fluorescence method were utilized. As is known from the solution studies, the tautomer emitts visible fluorescence, while the normal dimer emitts UV fluorescence. An excitation spectrum monitoring the visible emission is the action spectrum of DPTR and UV excitation spectrum corresponds to the absorption spectrum of normal form dimer.

(Time resolved spectroscopy)

For time resolved measurements, excitation was provided by a synchronously pumped, mode-locked cavity-dumped picosecond dye laser system. The second harmonic of a mode-locked Nd:YAG laser at 1.5 W average power pumped pyridine-1 dye to generate 12 ps pulses with the repetition rate of 4 MHz. The dye pyridine-1(exciton) covers wavelength region from 680 to 720 nm. The second harmonic of the dye laser was generated by KDP crystal with the bandwidth of 0.1 nm. Time resolved measurement was made using single-photon counting techniques.

3.Results and Discussion

(1) 7-Azaindole dimer

Figure 1 stands for the excitation spectra of 7-AI dimer monitoring UV (fig. 1a) and visible emission(fig.1b), respectively. 0-0 band of the S_0-S_1 transition is red shifted by about 2,500 cm^{-1} from that of the monomer. As is clear from the figures, former spectrum exhibits sharp fine structure, while the latter one shows diffuse bandwidth in the respective vibrational bands even under the supercooled jet condition.It was found that the variation of stagnation pressure in the nozzle affects the relaive intensity of UV and visible monitoring spectra. This implies that a series of sharp bansds and that of diffuse ones are in thermal equilibrium each other,i.e., there exist two types of isomers in the 7-AI dimers. The former one which emits UV emission is suspected to have a non-coplanar structure while the latter isomer which emits visible emission has a planar structure. It is concluded that 7-AI dimer with a planar structure tautomerizes promptly after excited to the lowest singlet state. One finds variation of the linewidths of the individual vibrational bands in the excitation spectra. In the case of the UV monitoring one,which corresponds to the absorption spectrum of non-tautomerized dimer, linewidth does not exhibits appreciable change and intensity of the fluorecscence excitation spectrum decreases as the excitation energy is increased. This can be explained by the non-radiative transition of the dimer including internal conversion and intersystem crossing. The linewith of the laser used in the experiments (1 cm^{-1})does not resolve the increment of non-radiative rate in this case. On the other hand, in the case of visible monitoring spectrum, which corresponds to the action spectrum of DPTR, linewidth is appreciably large as compared to the that of the excitation laser, and it also exhibits unusual variation as the excess energy is increased. The spectrum consists of the progressions and combinations of 98- and 120-cm^{-1} vibrations which are assigned to the intermolecular H-bond bending and stretching vibrations, respectively. The bandwidth of the 0-0 band at 32,552 cm^{-1} is 5 cm^{-1}. The v'=1 and 2 bands of the 120 cm^{-1} vibration have the bandwidths of 10 and 30 cm^{-1} which correspond to the lifetime of these bands of 0.5 and 0.2 ps. On the other hand, the linewidth of the v'=1 band of the bending vibration(98 cm^{-1}) shows the

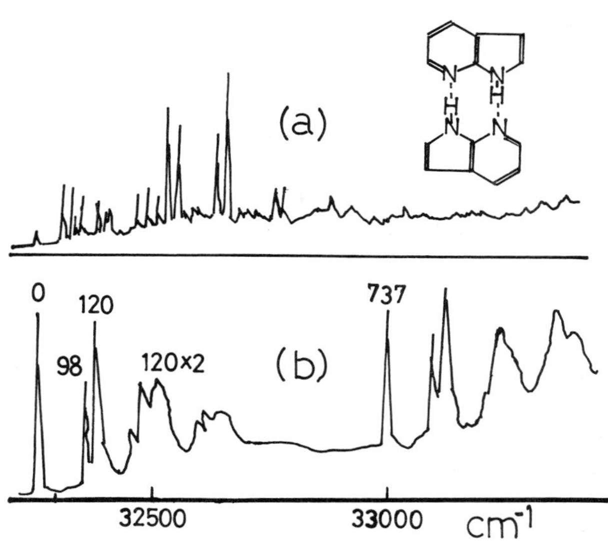

Fig. 1 UV(a) and visible excitation spectra of 7-AI dimer.

value 3 cm^{-1} and the combination band with 120 cm^{-1} vibration has the increased bandwidth of 7 cm^{-1}. This implies that symmetric H-bond stretching vibration selectively enhances the DPTR in contrast the role of deaccelerator of bending mode. More remarkable fact is that when one excites intramolecular vibration of the 7-AI moiety at 737 cm^{-1}, DPTR rate is as fast as that of the 0-0 band. Even though excess energy is large compared to the intermolecular modes, no enhancement of the reaction takes place. However, exciting the combination bands of the H-bond stretching mode with the 737 cm^{-1} mode, remarkable broadening of the linewidth is seen again. All of these results indicate that the intermolecular H-bond stretching vibration plays the role of the promoting mode of DPTR.The role of N-H--N stretching vibration as a promoting mode of DPTR can be explained qualitatively as following. Figure 2 schematically illustrates the potential energy surface (abbreviated in two dimensions) of proton transfer coordinate(r_{N-H}) and the stretching vibration (r_{N-H--N}). Here, D and T denote normal dimer before DPTR and the tautomer after DPTR, respectively. At the zero vibrational level of the normal dimer, DPTR proceeds by a tunneling as shown in the broken line in the figure. When N-H--N vibration is excited, N-H--N distance is reduced and barrier height to the tautomer surface is also reduced, which results into the appreciable increase of DPTR rate.

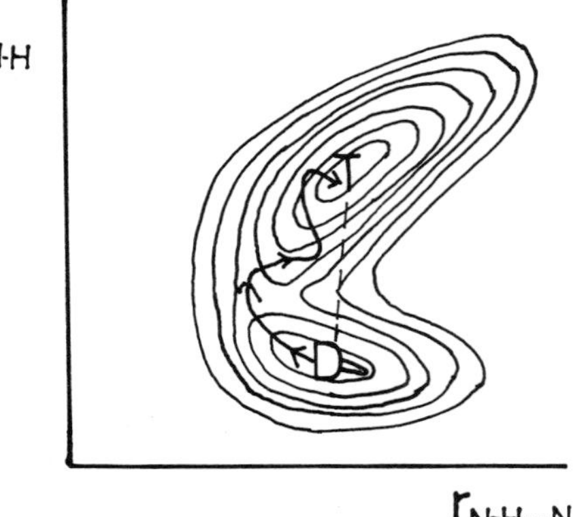

Fig. 2 Schematic potential energy surface for DPTR(r_{N-H}) and intermolecular N-H--N mode (r_{N-H--N}).

In order to confirm the coupling of the stretching vibration with DPTR, effect of the deuterium substitution for the hydrogen bonded protons was examined. The reduction rate of the reaction was observed to be on the order of 0.5, which is by far smaller than expected in pure proton transfer reaction. When the coupling works, effective mass of the DPTR is influenced by the the mass of surrounding heavy nuclei. Deuterium substitution can not change so dramatically the DPTR rate.

(2)1-Azacarbazole dimer

Figure 3 represents the fluorescence excitation spectra of jet-cooled 1-AC dimer monitoring UV (fig. 3a) and visible (fig. 3b) emission, respectively. The origin band for S_0-S_1 transition is located at 28,556 cm^{-1}. As is similar to 7-AI dimer, there exist conformational isomers for 1-AC dimer whose 0-0 bands are separated by 80 cm^{-1} with eachother. As seen in the figure, the isomer whose origin is at 28,556 cm^{-1} emitts visible fluorescence indicating that the isomer can be tautomerized. The isomer whose origin band is 80cm^{-1} above that of the reactive isomer does not emitt visible tautomer emission. In contrast to 7-AI dimer, even the dimer species of 1-AC which tautomerizes by DPTR also emits UV fluorescence and linewidth of the respective vibrational bands of reactive dimer is as sharp as that of non-reactive isomer. This means that the reaction rate for DPTR in 1-AC dimer is by far slower than that of 7-AI dimer.The excitation spectrum of DPTR (fig.3b) consists of progression and combinations of 55-, 67- and 109 cm^{-1} vibrations among which the 109 cm^{-1} one is assigned to the N-H--N stretching mode.

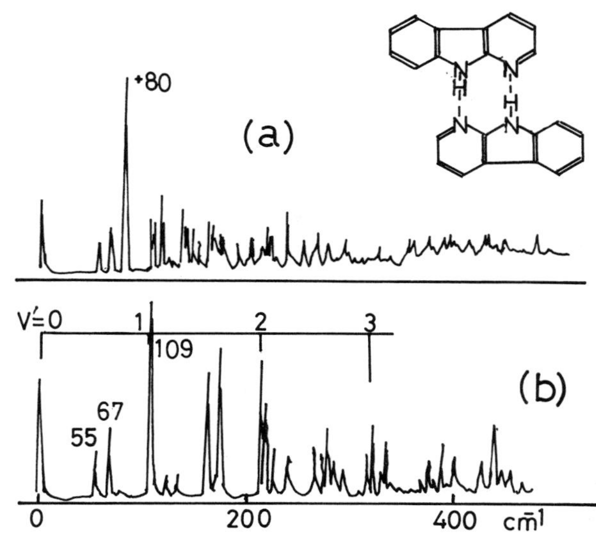

Fig. 3 UV(a) and visible excitation(b) spectra of 1-AC dimer.

Because of the relatively slow rate of DPTR , DPTR rate for 1-AC dimer is expected to be determined by the real time resolved spectroscopy. We have conducted fluorescence rise-and-decay time measurement for the reactive 1-AC dimer in the various vibronic bands using ps UV laser as an excitation source. Figure 4 is the temporal data for the decays of the UV and visible emissions excited at 0-0 + 109 cm^{-1} band. The deacay time of the UV fluorescence was simulated to be 130 ps. While the rise time and decay time of the visible fluorescence emitted from the tautomer is 130 ps and 2.2 ns, respectively. The coincedence of decay time of the fluorescence from the normal dimer with the rise time of the tautomer emisssion is the direct evidence of the DPTR.

In Table I, observed decay rates of the normal dimer fluorescence which corresponds to the reciprocal of the DPTR rate, at various bands are tabulated. It is clear from the table that 109 cm^{-1} vibration promotes the DPTR while the 55 and 67 cm^{-1} vibrations never enhances the reaction. The results is in complete agreement with those depicted from the static spectroscopic means for 7-AI dimer.

Table 1
Observed decay time of 1-AC dimer for individual vibrational bands.

$\Delta E(cm^{-1})$	τ(ps)
0	330
109	130
2x109	65
55	365
55+109	150
67	355
67+109	130

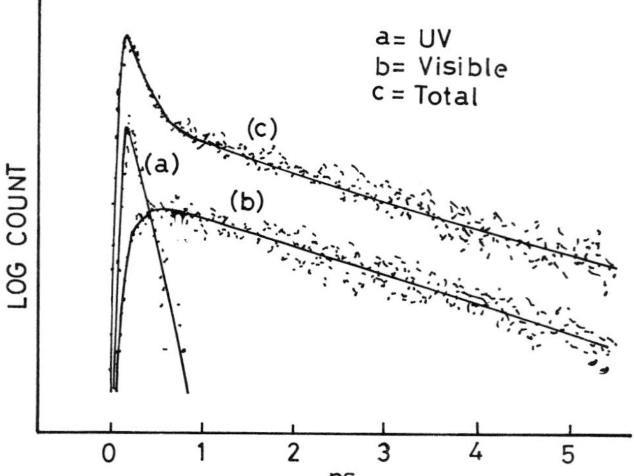

Fig. 4 Temporal data for the decays of UV(a) and visible(b) emission from 0-0 + 109 cm^{-1} band of 1-AC dimer.

We have observed pronounced decrease in the DPTR rate by the deuterium substitution for the H-bonded protons, which is in contrast to the case of 7-AI dimer where deuterium substitution has small influence on the DPTR. The difference comes from the fact that even 109 cm^{-1} mode is excited to shorten the N-H--N distance, still 1-AC dimer has large barrier for the DPTR to occur through tunnel process.

4. Conclusions

Both static laser spectroscopic means and real-time ps spectroscopy have been applied to investigate the DPTR of the H-bonded dimes of 7-AI and 1-AC under supercooled jet condition. N-H--N intermolecular stretching vibation was found to be the key vibration of the DPTR which lowers the effective barrier height of the reaction. Deuterium substitution experiments also supports the coupling mechanism of intermolecular mode with the reaction coordinate.

5. Acknowledgement

We wish to thank Prof. S. Iwata for his theoretical advices for the interpretation of the experimental results. The part of this research is supported by the Grant-in-aid from the Ministry of Education, Science, and Culture for the Priority Area.

6. References

(1)K. C. Ingham, and M. A. El-Bayoumi, J. Am. Chem. Soc. 96, 1674 (1974)
(2)W. M. Heterington, R. H. Micheels, and K. B. Eisenthal, Chem. Phys. Lett. 66, 230 (1979)
(3)K. Fuke, H. Yoshiuchi, and K. Kaya, J. Phys. Chem. 88, 5840 (1984)
(4)K. Fuke, T. Yabe, T. Chiba, T. Kohida, and K. Kaya, J. Phys. Chem. 90, 2309 (1986)
(5)K. Fuke, and K. Kaya, J. Phys. Chem. 93, 514 (1989)
(6)J. Waluk, B. Grabowska, B. Pakula, and J. Sepiol, J. Phys. Chem. 88, 1160 (1984)
(7)J. Waluk, J. Herbich, D. Oelkrug, and S. Uhl, J. Phys. Chem. 90, 3866 (1986)
(8)B. H. Meier, F. Graf, and R. R. Ernst, J. Chem. Phys. 76, 767 (1984)
(9)N. Sato, and S. Iwata, J. Chem. Phys. 89, 2932 (1989)

Part 2: Fundamental Aspects of Electron Transfer and Related Processes II

Dynamics and Mechanisms of
Photoinduced Transfer and Related Phenomena
N. Mataga, T. Okada and H. Masuhara (Editors)

Effects of fast intramolecular vibrations on rates of nonthermalized reactions dependent on solvation dynamics through viscosity

Hitoshi Sumi

Institute of Materials Science, University of Tsukuba, Tsukuba, 305 Japan

Abstract

It has been interesting many people that intramolecular electron, proton, or atom-group transfers in solvents are observed to have often rate constants decreasing with increasing the solvent viscosity. In these reactions,rate constants depend on dynamics of solvation of a solute molecule where these reactions take place, and the reactions proceed with a nonthermal-equilibrium distribution of molecular arrangements in the solvated state. These reactions have been treated so far by the Kramers theory or by its recent modifications, in which reactions are modeled as diffusive surmounting of a potential barrier by a Brownian particle. The Brownian motion describes diffusive fluctuations of molecular arrangements in the solvated state and the strength of friction retarding its speed is regarded as proportional to the solvent viscosity η. It cannot be understood by these theories, however, that observed rate constants decrease as a fractional power of η^{-1}. In actual systems, not only molecular-arrangement fluctuations, but also intramolecular atomic vibrations contribute to form the transition state for reaction. The latter is much faster than the former, with a characteristic time of the order of atomic-vibration periods. When both of them are taken into account, it is shown that the rate constant has a form of $1/(k_e^{-1} + k_f^{-1})$, where k_e represents the rate constant given by the transition-state theory and does not depend on η, while k_f (> 0) represents a part which decreases with increasing η as $\eta^{-\alpha}$ with $0 < \alpha < 1$. The degree of decrease of α from unity is determined by the degree of contribution of the intramolecular atomic vibrations in the reaction. The rate constant approaches the transition-state value k_e in fast solvents with small η's for $k_f \ll k_e$. On the other hand, it approaches k_f in slow solvents with large η's for $k_f \gg k_e$, and it becomes decreasing as $\eta^{-\alpha}$ with increasing η with a fractional-power dependence, in agreement with the recent observations.

1. SLOW FLUCTUATIONS OF MOLECULAR ARRANGEMENTS IN THE SOLVATED STATE

Recently much interest has been aroused in chemical reactions whose rates decrease with increasing the viscosity of solvents [1~6]. The viscosity retards the speed of thermal fluctuations of molecular arrangements in the solvated state of a solute molecule. Therefore, decrease of the reaction rates with the viscosity shows that they are limited by a slow speed of these fluctuations. The speed of these thermal fluctuations corresponds to the speed of thermalization among molecular arrangements in the solvated state. Therefore, the observed viscosity dependence of the reaction rates means that thermal equilibrium among molecular arrangements in the solvated state cannot be maintained during reaction. In this situation, traditional theories such as the transition-state theory cannot be applied to describe the reaction rates. In fact, it is tacitly assumed in these theories that thermal equilibrium is always maintained during reaction in the initial reactant state, which is composed of many substates of solvation different in molecular arrangement little by little and fluctuates among them in the present problem. Therefore, a new theory has been required which describes these reaction rates by taking full account of dynamics of slow thermal fluctuations in the solvated state, not by simply assuming thermal equilibrium among them. As the rate limiting step of these reactions we consider electron, proton or atom-group transfers taking place in solute molecules in solvents hereafter.

We should first notice how slow thermal fluctuations of molecular arrangements are in the solvated state. As an example, let us take a biological macromolecule in water in the physiological condition. Figure 1 shows the tertiary structure of about one quarter of an enzyme called carboxypeptidase A (abbreviated as CPA) [7]. The enzyme is secreted from the pancreas and catalyzes hydrolysis of proteins in the small intestine. The enzyme itself is one of proteins. The tertiary structure of a protein is usually composed of comparatively-rigid large domains with deep clefts among them and small groups of atoms protruding into water from the domain surface. For CPA we notice in Fig.1 a deep cleft entering the protein interior from above and bifurcating on its way. The active site of an enzyme is usually located around a cleft in the protein structure. That in CPA is composed of the 145th arginine, the 248th tyrosine and the 270th glutaminic-acid residues (shown by solid discs in Fig.1) and the Zn^{2+} ion, and they are in fact placed around the deep cleft mentioned above. Some of the amino-acid residues comprising the active site usually protrude into water from the protein surface, as the 145th arginine and the 248th tyrosine residues do in CPA shown in Fig.1. It has been considered that the rate-limiting step in protein hydrolysis by CPA is proton transfer from a water molecule bound by Zn^{2+} ion to a carboxylate oxygen ion of the 270th glutaminic-acid residue [8].

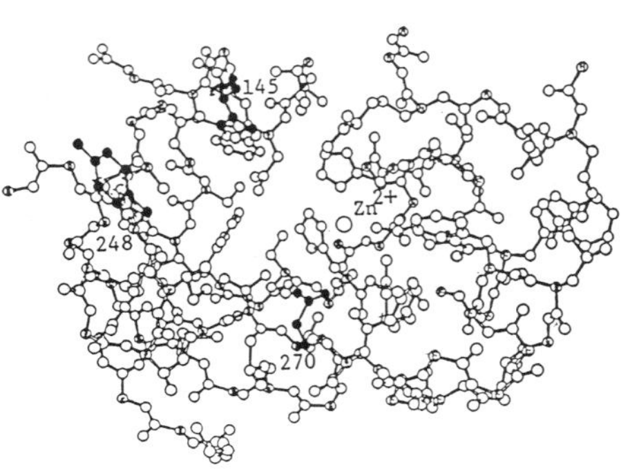

Figure 1. Tertiary structure of about one quarter of carboxypeptidase A.

The tertiary structure of a protein thermally fluctuate slowly, when it is hydrated in water. The speed of these fluctuations can directly be measured by Doppler broadening of the Rayleigh scattering of Mössbauer radiation, with which Gol'danskii and his coworkers have observed two kinds of slow fluctuations in the hydrated state of a protein [9]. The faster one of them has a relaxation time of the order of $10^{-11} \sim 10^{-9}$ sec, while the slower one of the order of $10^{-8} \sim 10^{-7}$ sec. The former has been considered to represent rotational fluctuations of small groups of atoms protruding into water from the protein surface, while the latter to represent mutual (hinge-bending) fluctuations of comparatively-rigid large domains in the protein structure. These slow thermal fluctuations of the tertiary structure of a protein can be observed only when the protein is well hydrated in water around room temperature. Therefore, they can be regarded as Brownian motions excited and/or damped solely by small perturbations due to microscopic thermal motions of water molecules. In this situation, we can consider that the Brownian motions representing the slow thermal fluctuations of the tertiary structure of a protein in solvents have a relaxation time proportional to the solvent viscosity.

2. FRACTIONAL-POWER DEPENDENCE OF THE RATE CONSTANT ON THE SOLVENT VISCOSITY

The slow thermal fluctuations of the tertiary structure of biological macromolecules in the hydrated state have been considered to play an important role in facilitating functions of these molecules. In other words, these fluctuations are utilized to find a conformation most favorable for reaction. In fact, it has been observed that the rate constant of biochemical processes in these molecules depends on the solvent viscosity proportional to the relaxation time of these fluctuations. To be more exact, the rate constant k_s decreases as a fractional power of the solvent viscosity η, that is, of the relaxation time τ of these fluctuations, as

$$k_s \propto \eta^{-\alpha} \propto \tau^{-\alpha}, \qquad \text{for} \quad 0 < \alpha < 1 . \tag{2.1}$$

For the catalytic process by CPA, the value of α in (2.1) is about 0.8 [1]. The dependence of (2.1) has been observed in various reactions in solvents, for example, in the O_2 or CO binding reaction in hemoglobin or myoglobin [2], the photocycle in bacteriorhodopsin [3], the proton-exchange reaction in lysozyme [4], intramolecular electron-transfer reactions in polar solvents [5], and the trans-cis isomerization reaction of stilbene [6]. Since the dependence of (2.1) is observed not only in biochemical reactions but also in various other kinds of reactions in solvents, τ in (2.1) should be described, in more general terms, as the relaxation time of thermal fluctuations of molecular arrangements in the solvated state.

Theories on reaction rates based on Brownian motions, which can represent these thermal fluctuations in solvents, can be traced back to the pioneering work by Kramers [10] presented in 1940. He modeled a chemical reaction as diffusive surmounting by a Brownian particle over a potential barrier in a double-well potential, as shown in Fig.2. When the coordinate of the Brownian particle with unit mass is written as X(t) at time t, it obeys the Langevin eq. written as

$$d^2X/dt^2 = - dW(X)/dX - \zeta\, dX/dt + R(t) , \tag{2.2}$$

in the Markovian limit, where W(X) represents the double-well potential, ζ the friction coefficient, and R(t) the random force arising from random perturbations due to microscopic thermal motions of solvent molecules. R(t) is related to ζ through the fluctuation-dissipation theorem giving $\langle R(t)R(t') \rangle = 2\zeta\, k_B T\, \delta(t-t')$

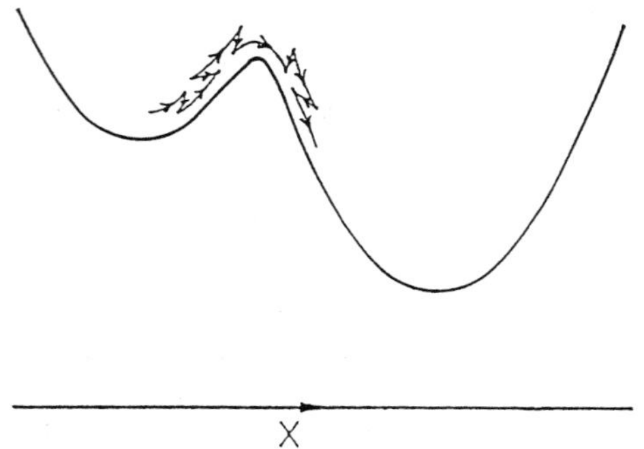

Figure 2. Potential curve for reaction in the Kramers scheme and an example of reactive trajectories there.

where ⟨•••⟩ represents the statistical average. Dynamics of Brownian motions in (2.2) is characterized by the friction coefficient ζ, and it is proportional to the solvent viscosity η. If the transition-state theory is applicable in (2.2), the rate constant should tend to

$$k_e = (\omega_o/2\pi) \exp(-\Delta G^*/k_BT) , \qquad (2.3)$$

where ΔG^* represents the height of the potential barrier measured from the bottom of the initial well and ω_o^2 the curvature of this well when the double-well potential W(X) is approximated as

$$W(X) \approx \frac{1}{2}\, \omega_o^2\, X^2 , \text{ in the initial well.} \qquad (2.4)$$

In this case, the rate constant becomes determined only by the static quantities characterizing the double-well potential also in the Kramers reaction scheme, and it does not depend on the quantities such as the friction coefficient ζ characterizing dynamics of Brownian motions in the double-well potential.

If the friction coefficient ζ is infinite in (2.2), dX/dt should vanish and the Brownian particle stops somewhere. Then, its coordinate value becomes a constant of motion in (2.2) and the reaction does not take place. In the limit of large ζ, therefore, (2.2) should describe Brownian motions only of the coordinate X of the Brownian particle with its momentum being thermalized more rapidly. In this limit, the rate constant should become proportional to ζ^{-1}, since it is proportional to the speed of the Brownian particle passing through the potential-barrier region from the initial to the final wells [11]. (In the limit of small ζ, too, the rate constant becomes dependent on ζ. This limit, however, cannot be realized in the liquid-phase reaction, but in the gas-phase one [6].) When ζ proportional to the solvent viscosity η is very large, therefore, (2.2) enables us to obtain the rate constant decreasing with increasing η as observed experimentally.

In the limit of large ζ, it is known [12] that (2.2) is equivalent to the diffusion equation, called the Smoluchowski eq., for the distribution function Q(X;t) for the coordinate value X of the Brownian particle at time t in the potential W(X), as

$$\partial Q(X;t)/\partial t = \zeta^{-1} (\partial/\partial X) [\, k_BT (\partial/\partial X) + dW(X)/dX \,] Q(X;t) . \qquad (2.5)$$

The relaxation time τ of fluctuations of the coordinate X in the initial well is

$$\tau = \zeta / \omega_o^2 . \qquad (2.6)$$

In fact, it is easy to show that the average value of X at time t given by $\bar{X}(t) \equiv \int X\, Q(X;t)\, dX$ satisfies $d\bar{X}(t)/dt = -\bar{X}(t)/\tau$ in the initial well described as (2.4) and hence it decays with the time constant τ given by (2.6). Kramers solved this equation under the condition of steady decay of the population of the Brownian particle in the initial well, and obtained the rate constant given by

$$k_o = (\omega_b/\zeta)\, k_e = \tau^{-1}(\omega_b/2\pi\omega_o)\exp(-\Delta G^*/k_BT)\,, \tag{2.7}$$

where k_e represents the rate constant in (2.3) given by the transition-state theory, and $\omega_b{}^2$ represents the absolute value of the curvature of the potential at the top of the barrier between the initial and the final wells. (2.7) is for the potential barrier of a round shape. But, when it is of a sharp (cusp-like) shape, k_o tends to

$$k_o = \tau^{-1}\sqrt{\Delta G^*/\pi k_BT}\,\exp(-\Delta G^*/k_BT)\,. \tag{2.8}$$

Both (2.7) and (2.8) are obtianed when the barrier height satisfies $\Delta G^* \gg k_BT$.

Since τ is proportional to the solvent viscosity η and hence to the friction coefficient ζ, the rate constant k_o in the high-friction limit in the Kramers reaction scheme decreases as a linear function of $\eta^{-1} \propto \tau^{-1}$. This η dependence, however, does not reproduce the observed one shown in (2.1). To understand the observed fractional-power dependence of the rate constant on η, Grote and Hynes argued that the speed μ of the Brownian particle passing through the potential-barrier region between the initial and the final wells might not be much smaller than the speed μ_c of microscopic thermal motions of solvent molecules [13]. If this is the case, the friction coefficient retarding the speed μ must be that effective at the speed μ, written as $\zeta(\mu)$. Therefore the $\mu \propto \zeta(\mu)^{-1}$, and (2.7) must be replaced by

$$k_o \propto \mu\, k_e \propto \zeta(\mu)^{-1}\, k_e\ . \tag{2.9}$$

On the other hand, the friction coefficient ζ determining the Kramers rate constant of (2.7) or (2.8) is that effective at a speed $\ll \mu_c$. Since $\zeta(\mu) < \zeta$, the rate constant of (2.9) given by Grote and Hynes is not so small as that of (2.7) or (2.8).

To justify their argument, it must be checked that the speed of the Brownian motions is in fact not much smaller than that of microscopic thermal motions of solvent molecules. The former should have a magnitude of the order of the inverse of the relaxation time of thermal motions of molecular arrangements in the solvated state, which is about $10^{-11} \sim 10^{-9}$ sec for the rotational Brownian motions of small groups of atoms or about $10^{-8} \sim 10^{-7}$ sec for the mutual (hinge-bending) Brownian motions between large domains in the tertiary structure of a protein. On the other hand, the latter can be considered to have a magnitude of the order of 10^{12} sec^{-1} or larger since it should have a magnitude of the order of the lowest frequency of localized vibrations in liquids. In contradict to their argument, therefore, the former is much smaller than the latter. Moreover, as long as the rate constant of (2.9) obtained by them decreases with increasing the viscosity η in the high-friction limit it decreases more rapidly than the Kramers rate constant does as η^{-1}: This arises from that the former is always larger than the latter as mentioned above and simultaneously the former approaches the latter as η approaches ∞ since μ proportional to the rate constant in (2.9) approaches zero. Contrary, observed rate constants of (2.1) decrease more slowly than η^{-1}.

3. EFFECTS OF FAST VIBRATIONAL FLUCTUATIONS OF INTRAMOLECULAR ATOMIC ARRANGEMENTS IN THE SOLVATED STATE

In the Kramers scheme described by Fig.2, the reaction occurs as a result of interaction with Brownian motions along the coordinate X. This means in the reaction taking place in a protein in solvents explained in Sec.1, for example, that there exist, between the initial and the final states of the reaction, differences in molecular arrangements composed of mutual positions between large domains in the tertiary structure of the protein and/or of orientations of small groups of atoms protruding from the protein surface. Between the initial and the final states of the reaction, however, there exist, in general, not only differences in molecular arrangements, but also those in intramolecular atomic arrangements within large domains and/or small groups of atoms regarded as rigid above. Fluctuations of these atomic arrangements can be described by phonons. Their speed has a magnitude of the order of their characteristic frequency ($\sim 10^{13}$ sec^{-1}), and it is much larger than that of thermal fluctuations of molecular arrangements in the solvated state. In actual systems, therefore, the reaction occurs as a result of interaction not only with slow diffusive fluctuations of molecular arrangements described as Brownian motions, but also with fast vibrational fluctuations of atomic arrangements due to intramolecular phonons. The structural distortion taking place in the final state of the reaction can be measured by an excess distortion energy there seen from the structure at the initial state. We can consider a distortion energy associated with each of the two kinds of fluctuations participating in the reaction, and denote the part due to intramolecular phonons by λ_i and that due to intermolecular Brownian motions by λ_o. They have been called the (partial) reorganization energies associated with the reaction, with $\lambda_i + \lambda_o$ called the total reorganization energy.

Importance of taking into account both of these two kinds of reorganization energies in calculating the reaction rate has well been recognized in electron-transfer reactions in polar solvents, although in the frame of the transition-state theory. In these reactions, λ_o represents the energy due to rearrangement of polar solvent molecules occurring associated with the electron transfer, and has been called the outersphere reorganization energy. On the other hand, λ_i represents the energy due to rearrangements of atoms within donor and/or acceptor molecules occurring associated with the electron transfer, and has been called the innersphere reorganization energy. In the Kramers theory and also in its recent modifications [14] including the Grote and Hynes theory mentioned above, however, no account is taken of effects of fast vibrational fluctuations of intramolecular atomic arrangements on reaction rates dependent on the solvent viscosity. Let us call λ_i and λ_o by the same name as used in the electron-transfer reaction in polar solvents, for more general reactions in solvents, too.

Let us consider that contributing in reaching the transition state for reaction are not only thermal diffusive fluctuations of molecular arrangements in the solvated state, but also thermal vibrational fluctuations of intramolecular atomic arrangements. The contribution from the latter is constructed with a duration time of the order of the phonon period ($\sim 10^{-13}$ sec), which is much shorter than the relaxation time of the former. In this situation, the reaction takes place by thermal activation due to the latter at each molecular arrangement in the solvated state during its thermal diffusive fluctuations in the initial well. Then, we can introduce a rate constant due to thermal vibrational fluctuations of intramolecular atomic arrangements, denoted by k(X) with a value different at each molecular arrangement specified by the coordinate X in the solvated state. Let us represent by P(X;t) dX the population of reactants with molecular arrangements at the coordinate values from X to X + dX at time t in the initial well V(X). Then, P(X;t) should satisfy the following diffusion reaction equation

$$\partial P(X;t)/\partial t = \zeta^{-1} (\partial/\partial X) [k_BT (\partial/\partial X) + dV(X)/dX] P(X;t) - k(X) P(X;t) . \qquad (3.1)$$

The first term on the right-hand side describes diffusion as in (2.5), while the last term works as a sink of reactants due to the reaction. It is reasonable to assume for the sink term the reflective boundary condition that $k(X) \to 0$ for $X \to \pm\infty$. Differences between (2.5) and (3.1) should be noted: W(X) in (2.5) represents the double-well potential shown in Fig.2, while V(X) in (3.1) represents a single-well potential for the initial well of the reaction. Moreover, reaction in (2.5) takes place as a result of diffusive surmounting of a potential barrier in W(X), while in (3.1) it takes place as a result of operation of the last term on the right-hand side.

The reaction scheme described by (3.1) can be visualized by Fig.3, where the abscissa denotes the coordinate X of molecular arrangements in the solvated state, while the ordinate denotes the coordinate q of intramolecular atomic arrangements. Equi-energy contours drawn there describe the potential surface for reaction, in which the parts below and above the line C represent respectively the reactant and the product surfaces for reaction. These two parts cross each other on the line C. Therefore, the line C represents the position of the transition state for reaction in the (X,q) plane. The potential energy becomes lowest at the point S on the line C. Therefore, the height of the point S measured from the lowest point O of the reactant surface gives the thermal-activation energy of the rate constant in the transition-state theory. It is written by ΔG^* as before.

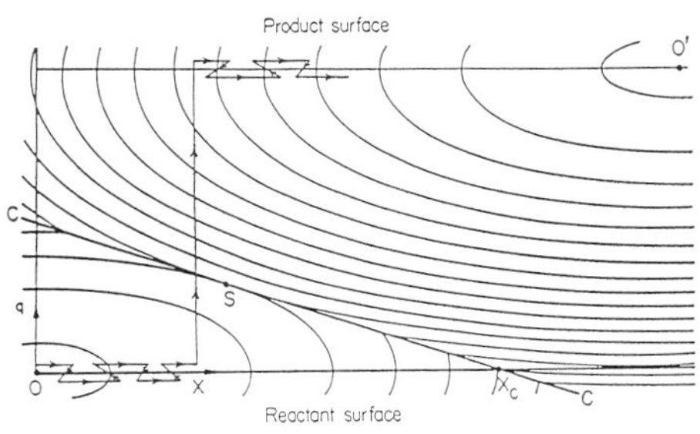

Figure 3. Potential curve for reaction in the two-mode model and an example of reactive trajectories there.

In Fig.3, reaction takes place as described below: First, on the reactant surface below the line C, the coordinate X is fluctuating as Brownian motions, shown as a zigzag line along the potential V(X) which represents the cut of the reactant surface along the X axis passing through its lowest point O. Then, at some point of the coordinate X during its Brownian motions, it occurs that the amplitude of intramolecular atomic arrangements becomes suddenly so large as to reach the line C as a result of its thermal vibrational fluctuations, as shown by a vertical line in Fig.3, and the reactant changes to the product, that is, the reaction takes place. Afterwards, the coordinate X starts Brownian motions, now, on the product surface, shown as a zigzag line along the X axis passing through its lowest point O', and approaches gradually the point O'.

If the transition-state theory is applicable in (3.1), the rate constant tends to

$$k_e \equiv \nu \exp(-\Delta G^*/k_B T) = \langle k(X) \rangle_e , \tag{3.2}$$

$$\langle f(X) \rangle_e \equiv \int f(X) \exp[-V(X)/k_B T]\, dX \,/ \int \exp[-V(X)/k_B T]\, dX , \tag{3.3}$$

given by the average of k(X) over the thermal-equilibrium distribution in the initial well V(X) of the reaction, written explicitly as (3.3). The pre-exponential factor of k_e was written as ν in (3.2). This rate constant k_e depends only on the static quantities characterizing the potential V(X), and is independent of dynamics of Brownian motions in the potential V(X) characterized by the friction coefficient ζ or their relaxation time τ proportional to ζ by (2.6) where $\omega_0{}^2$ represents the curvature of the initial well V(X) as in (2.4). Because of the presence of the last term proportional to k(X) on the right-hand side of (3.1), the distribution function P(X;t) decreases with time t. When the transition-state theory is applicable, it should show a single-exponential decay determined by k_e of (3.2), as

$$P(X;t) = P_e(X) \exp(-k_e t) , \quad \text{in the transition-state theory,} \tag{3.4}$$

where $P_e(X)$ is given by the thermal-equilibrium distribution function in the initial well V(X) of the reaction, written as

$$P_e(X) = \exp[-V(X)/k_B T] \,/ \int \exp[-V(X)/k_B T]\, dX , \quad \text{with } \int P_e(X)\, dX = 1 . \tag{3.5}$$

In general, neither the rate constant derived from (3.1) coincides with k_e of (3.2), nor the distribution function P(X;t) shows a single-exponential decay. When P(X;t) does not necessarily show a single-exponential decay, it is convenient to use, as a measure of the rate constant, the inverse of the first passage time for the time decay of the total population of reactants $\int P(X;t)\, dX$, written as

$$k_s = 1 \,/ \int_0^\infty dt \int dX\, P(X;t) , \quad \text{with } \int P(X;0)\, dX = 1 , \quad \text{in the general case.} \tag{3.6}$$

The value of k_s depends, in general, on the form of the initial distribution of reactants at t = 0, written as P(X;0). When the steady state is realized, P(X;t) should become decaying as a single-exponential function in t with the decay rate k_s, as

$$P(X;t) = P_s(X) \exp(-k_s t) , \quad \text{in the steady state,} \tag{3.7}$$

maintaining a steady form of the distribution of X given by $P_s(X)$. Only in this case, we can define unambiguously the rate constant given by k_s since it can be determined irrespective of what form the initial distribution of reactants is. In this case, the inverse of the first passage time given by the right-hand side of (3.6) becomes the true rate constant. Even in this case, k_s does not necessarily coincide with k_e of (3.2), and $P_s(X)$ with the thermal-equilibrium distribution function $P_e(X)$ of (3.5). The steady state mentioned above is realized only when

$$k_s \ll \tau^{-1}; \quad \text{the condition for the steady state,} \tag{3.8}$$

where τ represents the relaxation time of Brownian motions in the initial well, given by (2.6). In fact, (3.8) ensures that internal agitations due to Brownian motions in the initial well (with a frequency of the order of τ^{-1}) are so often compared with the reaction with a frequency of k_s that the distribution of reactants in the initial well has enough time to be recovered to a steady form and maintained there during the reaction. Since the steady state is not necessarily realized, k_s determined by (3.6) is calculated in the present work under the condition that the initial distribution function of reactants P(X;0) coincides with the thermal-equilibrium one in the initial well V(X), given by $P_e(X)$ of (3.5), as

$$P(X;0) \Rightarrow P_e(X) . \tag{3.9}$$

The k_s thus calculated is called the rate constant hereafter for simplicity. Similarly, let us call a function calculated below, the distribution function

$$P_s(X) = k_s \int_0^\infty P(X;t)\, dt , \quad \text{with} \quad \int P_s(X)\, dX = 1 , \quad \text{in the general case.} \tag{3.10}$$

When (3.8) is satisfied for k_s calculated by (3.6), the k_s and $P_s(X)$ of (3.10) should have become respectively the true rate constant and the true steady-state distribution function, both of which become independent of the initial condition of (3.9).

It can be shown rigorously [15] that irrespective of the form of k(X), the rate constant k_s of (3.6) derived from the diffusion reaction equation (3.1) has a form

$$k_s = 1 / (k_e^{-1} + k_f^{-1}) , \quad \text{with} \quad k_f > 0 . \tag{3.11}$$

In this formula, k_e does not depend on the relaxation time τ of Brownian motions in the initial well V(X), while k_f depends on τ, decreasing with increasing τ. In the small-τ limit where τ is so small that $k_f \gg k_e$ is satisfied, k_s of (3.11) tends to k_e and the result of the transition-state theory is recovered. In this limit, k_f becomes proportional to τ^{-1}, although it does not influence the value of k_s in (3.11). In the opposite large-τ limit where τ is so large that $k_f \ll k_e$ is satisfied, on the other hand, k_s of (3.11) tends to k_f, giving the rate constant decreasing with increasing τ, and the traditional theories such as the transition-state theory break down. The k_f can be called the fluctuation-limited rate constant, since in (3.11) k_f describes the part of the rate constant k_s limited by the slow speed of thermal diffusive fluctuations of molecular arrangements in the solvated state.

The boundary beween these two limits is determined by a quantity given by

$$k_c = \tau^{-1} (k_BT/\omega_o^2)\, k_e^2 \Big/ \int_{-\infty}^{\infty} P_e(X)^{-1} \{\int_{-\infty}^{X} [k(Y) - k_e]\, P_e(Y)\, dY\}^2\, dX\ , \tag{3.12}$$

with a dimension of sec^{-1}, where $P_e(X)$ represents the thermal-equilibrium distribution function in the initial well $V(X)$ written as (3.5). This k_c is proportional to τ^{-1}. In terms of k_c, the small-τ limit mentioned above corresponds to

$$k_s \approx k_e\ , \text{and}\ \ k_f \approx k_c\ , \quad \text{when} \quad k_c \gg k_e\ , \quad \text{in the small-}\tau\text{ limit.} \tag{3.13}$$

Therefore, k_f becomes proportional to τ^{-1} in this limit, as mentioned above. Similarly, the large-τ limit mentioned above can be described as

$$k_s \approx k_f\ , (\text{but}\ \ k_f \not\approx k_c)\ , \quad \text{when} \quad k_c \ll k_e\ , \quad \text{in the large-}\tau\text{ limit.} \tag{3.14}$$

It is not easy to directly calculate k_f in (3.11), but k_c ($\propto \tau^{-1}$) can easily be calculated by (3.12) instead of k_f. Then, comparison of k_c to k_e enables us to determine whether the traditional theories such as the transition-state theory are still applicable in a situation of small τ, or, they break down in a situation of large τ.

Let us first investigate a limit that the participation of thermal vibrational fluctuations of intramolecular atomic arrangements in the reaction approaches zero. The limit can be described as $\lambda_i/\lambda_o \to 0$, where λ_i and λ_o represent respectively the innersphere and the outersphere reorganization energies explained before. In this limit, the lowest point O' of the product surface in Fig.3 should be located on the same X axis as passing through the lowest point O of the reactant surface. Therefore, the transition-state line C should be perpendicular to the X axis, and the saddle point S on the line C should move to the point X_c shown in Fig.3. In this limit, the transition-state line C cannot be reached by thermal vibrational fluctuations of intramolecular atomic arrangements along the coordinate q except at $X = X_c$, and the X-dependent rate constant $k(X)$ becomes proportional to a delta function at $X = X_c$. In this limit, we can show that k_f in (3.11) tends to the rate constant of (2.8) in the Kramers reaction scheme, as

$$k_f \to \tau^{-1} \sqrt{\Delta G^*/\pi k_BT}\ \exp(-\Delta G^*/k_BT)\ , \quad \text{when} \quad \lambda_i/\lambda_o \to 0\ . \tag{3.15}$$

Therefore, k_f becomes proportional to τ^{-1} irrespective of the magnitude of τ in this limit. Moreover, k_c defined by (3.12) tends also to the rate constant of (2.8) in this limit. Then, k_f coincides with k_c irrespective of τ in this limit, although in general cases k_f tends to k_c only in the small-τ limit, as noted in (3.13). Many authors [14] obtained k_f in (3.11) equal to the Kramers rate constant of (2.8), but it should be emphasized that it can be justified only for $\lambda_i/\lambda_o \to 0$ as in (3.15).

Since k_s given by (3.11) approaches k_f in the large-τ limit, the rate constant becomes proportional to τ^{-1} in this limit when $\lambda_i/\lambda_o \to 0$. Therefore, the experimental observation shown in (2.1) cannot be reproduced in the limit of $\lambda_i/\lambda_o \to 0$.

Except for the limit of $\lambda_i/\lambda_o \to 0$, the transition-state line C is not perpendicular to the X axis, and it crosses the X axis passing through the lowest point O of the reactant surface at $X=X_c$, as shown in Fig.3. Then,the line C can be reached at various values of X by thermal vibrational fluctuations of intramolecular atomic arrangements along the q axis. It means that the X-dependent rate constant k(X) has a nonvanishing width around $X = X_c$. In this case, concerning the asymptotic behavior of k_f in the large-τ limit, it was shown [15] that if the thermal average $\langle [dk(X)^{-1}/dX]^2 \rangle_e$ calculated by (3.3) diverges, k_f behaves as

$$k_f \sim \tau^{-\alpha} \nu^{1-\alpha} \exp(-\gamma \Delta G^*/k_BT), \quad \text{with } 0 < \alpha < 1, \text{ and } 0 < \gamma < 1, \text{ in the large-}\tau \text{ limit,} \tag{3.16}$$

where ΔG^* and ν represent respectively the thermal-activation energy and the pre-exponential factor of the rate constant k_e of (3.2) given by the transition-state theory. Therefore, the rate constant k_s becomes proportional to $\tau^{-\alpha}$ in the large-τ limit of (3.14). This reproduces the experimental observation in (2.1) since τ is proportional to the solvent viscosity η. It should be noted here that the relation of (2.1) was obtained as an analytic behavior of the rate constant k_s of (3.11) in the large-τ limit. The fractional-power dependence on τ as that in (3.16) cannot be obtained by perturbational expansions in τ^{-1} even if it is realized in the large-τ limit. If the thermal average $\langle [dk(X)^{-1}/dX]^2 \rangle_e$ is finite, on the other hand, α in (3.16) tends to unity, although γ remains smaller than unity even in this case.

When the value of the coordinate X at the saddle point S on the line C in Fig.3 is denoted by X_s, it can be shown that the thermal-activation energy $\gamma\Delta G^*$ of k_f in (3.16) is equal to $V(X_s)$ where V(X) represents the initial well along the coordinate X. Therefore, k_f represents approximately a probability per unit time with which the situation of $X = X_s$ is realized by thermal Brownian motions of the coordinate X in the potential V(X). Then, it turns out that the reaction takes place as a consecutive two-step process composed of the first step described by k_f and the subsequent step of a transition to the product state at an X around X_s due to k(X). This interpretation is consistent with (3.11) for the rate constant since it has a general form in the consecutive two-step mechanism for reaction.

Deviation of α from unity in (3.16) originates in the distribution of the position of X for the transition mentioned above around X_s due to the nonvanishing width of k(X), which is realized except for the limit of $\lambda_i/\lambda_o \to 0$. The concrete value of α can be obtained when a concrete form of the X dependence of k(X) is given.

4. SUMI-MARCUS MODEL FOR k(X)

As the most simplified model [16], we can assume that the reactant and the product surfaces in Fig.3 are the same in shape, being quadratic both in variables X and q, but different only in the position of their lowest point. In this model, the height of the transition-state line C at an X measured from the initial well V(X) (equal to $\omega_o^2 X^2/2$) at the X is given by $(\lambda_o/\lambda_i)\,\omega_o^2\,(X-X_c)^2/2$, where $\lambda_o+\lambda_i$ representing the energy of the reactant surface at the lowest point O'of the product surface, λ_o and λ_i represent respectively its component along the X and q axes. This height gives the activation energy of the rate constant k(X) due to thermal vibrational fluctuations of intramolecular atomic arrangements, as

$$k(X) = \nu_q \exp[-(\lambda_o/\lambda_i)\,\omega_o^2\,(X-X_c)^2 / 2k_BT] , \tag{4.1}$$

where ν_q represents an appropriate pre-factor. For this k(X) the thermal average $\langle\, [dk(X)^{-1}/dX]^2 \,\rangle_e$ calculated by (3.3) diverges when $\lambda_i/\lambda_o \leq 2$. Then, we get

$$0 < \alpha < 1 , \quad \text{when} \quad 0 < \lambda_i/\lambda_o < 2 , \quad \text{and} \quad \alpha = 1 , \quad \text{otherwise.} \tag{4.2}$$

The analytic formula for α when $0 < \alpha < 1$ can be obtained only around the boundary of the region of λ_i/λ_o for such α's, as

$$\alpha \approx |1 - \lambda_i/\lambda_o| , \quad \text{when} \quad \lambda_i/\lambda_o \text{ or } 2 - \lambda_i/\lambda_o \ll 1 . \tag{4.3}$$

Numerical calculations of k_f shown later enable us to see that α decreases to about 0.6 for $\alpha \approx 1$. The concrete value of γ can be obtained as

$$\gamma \approx 1/(1 + \lambda_i/\lambda_o) , \tag{4.4}$$

which is applicable for any value of λ_i/λ_o. The formulas given in (4.3) and (4.4) can be applied also in the limit of $\lambda_i/\lambda_o \to 0$. In fact, these equations give $\alpha = \gamma = 1$ for $\lambda_i/\lambda_o \to 0$, and this result is consistent with the expression (3.15) for k_f obtained in this limit, when α and γ are obtained in comparison with (3.16).

k_f in (3.11) can be obtained by solving (3.1) and then by calculating k_s of (3.6). The τ dependence of k_f for $\Delta G^*/k_BT = 10$ is shown in Fig.4, where its ratio to the transition-state value k_e of the rate constant of (3.2) is plotted as a function of $\tau\nu$ for various values of λ_i/λ_o. We see therein that k_f is proportional to τ^{-1} for $\lambda_i/\lambda_o \to 0$ in the entire region of τ, as

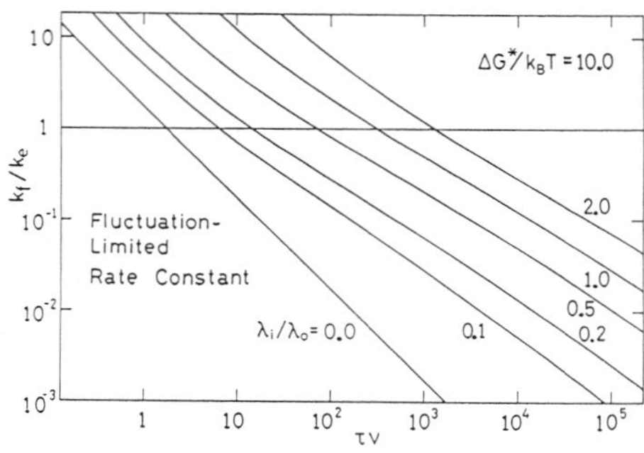

Figure 4. $\tau\nu$ dependence of k_f/k_e for $\Delta G^*/k_BT = 10$ and various λ_i/λ_o's.

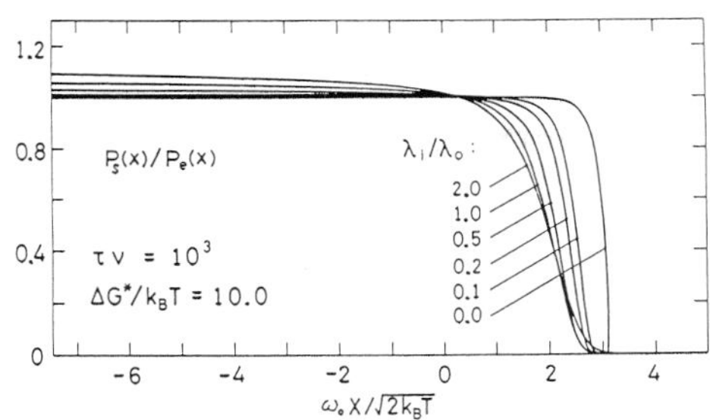

Figure 5. Distribution function $P_s(X)$ scaled by $P_e(X)$ for $\tau v = 10^3$, $\Delta G^*/k_BT = 10$, and various values of λ_i/λ_o.

shown in (3.15). For $\lambda_i/\lambda_o \neq 0$, however, it is only in the small-τ limit of $k_f \ll k_e$ that k_f is proportional to τ^{-1}, as shown in (3.13). As k_f becomes smaller than k_e in the large-τv region, the linear dependence of k_f on τ^{-1} tends to a weaker one. From the approximate linearity between log(k_f) and log(τv) in this region in Fig.4, we see that the τ dependence of k_f changes, in fact, to a fractional one as shown in (3.16) in this region. We also see that the value of α in (3.16) decreases to about 0.6 for $\lambda_i/\lambda_o \approx 1$.

The distribution function $P_s(X)$ calculated by (3.10) for $\Delta G^*/k_BT = 10$ and $\tau v = 10^3$ is shown in Fig.5, where its ratio to the thermal-equilibrium distribution function $P_e(X)$ of (3.5) is plotted as a function of $\omega_o X/\sqrt{2k_BT}$ for various values of λ_i/λ_o. If fluctuations of the coordinate X in the initial well were maintained in thermal equilibrium during reaction, as assumed in the transition-state theory, the ratio of $P_s(X)/P_e(X)$ would be constant over the entire region of X. The actual value of $P_s(X)/P_e(X)$, however, drops sharply to zero in a certain region of X, although it becomes almost constant for X much lower than this region. Therefore, Fig.5 shows explicitly that fluctuations of the coordinate X in the initial well are not maintained in thermal equilibrium during reaction, and hence the transition-state theory is inapplicable. This can be inferred also from Fig.4 showing that k_f at $\tau v = 10^3$ is smaller than k_e for any value of $\lambda_i/\lambda_o \leq 2$, then, k_f^{-1} cannot be neglected in comparison with k_e^{-1} in the formula for the rate constant k_s in (3.11), and the transition-state theory cannot be applied. The X value around which the sharp drop of $P_s(X)/P_e(X)$ occurs in Fig.5 corresponds to X_s representing the X value at the saddle point S in Fig.3. It means, therefore, that reactants reaching a region of X around X_s as a result of Brownian motions make a transition to the product state very rapidly due to the rate constant k(X). In other words, the reaction takes place by a consecutive two-step mechanism composed of fluctuations of X up to the region around X_s and the subsequent transition to the product state there. It has already been mentioned in the previous section that the formula (3.11) with (3.16) for the rate constant is consistent with this picture for reaction.

5. DISCUSSION

The pre-exponential factor v of the rate constant k_e of (3.2) can be estimated at a value from about 10^{13} sec^{-1} (in the adiabatic limit) to about 10^{11} sec^{-1} (in the nonadiabatic limit), since in general ν must be smaller than the average frequency of intramolecular atomic vibrations participating in the reaction [17], and it is about 10^{13} sec^{-1}. On the other hand, as mentioned in Sec.1, the relaxation time τ of molecular arrangements in the solvated state of a protein, for example, has been estimated experimentally to be about $10^{-11} \sim 10^{-9}$ sec for rotational Brownian motions of a small group of atoms protruding into water from the protein surface, or about $10^{-8} \sim 10^{-7}$ sec for mutual (hinge-bending) Brownian motions of large domains in the protein structure. Therefore, $\tau\nu$ has a magnitude of about $1 \sim 10^4$ for the former, or about $10^3 \sim 10^6$ for the latter. We see in Fig.3 that k_f at these values of $\tau\nu$ is comparable to or smaller than k_e, and hence k_f^{-1} cannot be neglected in comparison with k_e^{-1} in (3.11) for the rate constant k_s. Therefore, biological reactions taking place in proteins are in a nonthermal-equilibrium-reaction regime where traditional theories such as the transition-state theory are inapplicable and the rate constant should be written as (3.11).

We have thus seen that experimental observations described by (2.1) can be understood, in principle, by the recent theory giving the rate constant k_s of (3.11) with k_f of (3.16). In the isomerization reaction of stilbene in solvents, however, it has been observed [6] that the value of α in (2.1) can decrease to about 0.2, and such a small value of α cannot be reproduced in the large-τ limit of the present calculation of this theory with the Sumi-Marcus model explained in Sec.4. It might originate in an apparent intermediate τ dependence of k_s seen in the transition region between the small-τ limit of (3.13) with $k_s \propto \tau^{-0}$ and the large-τ limit of (3.14) with $k_s \propto \tau^{-\alpha}$, or in multi-dimensional Brownian motions in the initial well for reaction [18] which gives rise to multi-dimensional broadening of k(X). Let us remember that distribution of X around X_s for reaction is important in giving $\alpha < 1$. Anyhow, further investigations must be made on this point,

The rate constant k_s given by (3.11) contains a term k_f dependent on τ which corresponds to the thermalization time of fluctuations in the initial well. Then, the reaction proceeds with a nonthermal-equilibrium distribution of fluctuations there. Even in this case it can be shown that thermal equilibrium between the reactant and the product populations can be attained as a result of this nonthermal-equilibrium reaction, that is, the law of chemical equilibrium still holds. In fact, we can extend the diffusion reaction equation (3.1) so as to incorporate not only the forward reaction but also the backward one [15]. Solving the equation thus extended, we can prove that both the forward rate constant $k_s^{(f)}$ and the backward one $k_s^{(b)}$ can be written in the same form as (3.11), as

$$k_s^{(f)} = [1/k_e^{(f)} + 1/k_f^{(f)}]^{-1}, \quad \text{and} \quad k_s^{(b)} = [1/k_e^{(b)} + 1/k_f^{(b)}]^{-1}, \tag{5.1}$$

where both $k_e^{(f)}$ and $k_e^{(b)}$ are the rate constants given by the transition-state theory and independent of τ, while both $k_f^{(f)}$ and $k_f^{(b)}$ depend on τ, decreasing with increasing τ. We can show, however, that $k_f^{(f)}/k_f^{(b)}$ is independent of τ and equal to the equilibrium constant K of the reaction equal to $k_e^{(f)}/k_e^{(b)}$, as

$$k_e^{(f)}/k_e^{(b)} = K = k_f^{(f)}/k_f^{(b)} . \tag{5.2}$$

In the transition-state theory, the law of chemical equilibrium is ensured to hold by the first equality in (5.2). Since combination of (5.1) and (5.2) gives $k_s^{(f)}/k_s^{(b)} = K$, the same law is ensured to hold also in the present nonthermal-equilibrium reaction scheme. In other words, the chemical reaction proceeds toward thermal equilibrium between the reactant and the product populations, even if the reactant and the product distributions respectively in the initial and the final wells are maintained steadily in a nonthermal-equilibrium form during the reaction. In other words, the nonthermal-equilibrium distribution in the reactant and the product wells let both the forward and the backward rate constants deviate from the transition-state-theory value, but leaves the ratio between them unchanged from that determined by the transition-state theory.

6. REFERENCES

1 B. Gavish, *The Fluctuating Enzyme,* ed. G.R. Welch (Wiley, New York, 1986) p.263; and N.G. Goguadze, J.M. Hammerstad-Pedersen, D.E. Khoshtariya & J. Ulstrup, Eur. J. Biochem. (1991) in press.
2 D. Beece, L. Eisenstein, H. Frauenfelder, D. Good, M.C. Marden, L. Reinish, A.H. Reynolds, L.B. Sorensen, & K.T. Yue, Biochem. **19** (1980) 5147.
3 D. Beece, S.F. Bowne, J. Czégé, L. Eisenstein, H. Frauenfelder, D. Good, M.C. Marden, J. Marque & K.T. Yue, Photochem. Photobio. **33** (1981) 517.
4 B. Somogyi, J.A. Norman & A. Rosenberg, Biophys. Chem. **32** (1988) 1.
5 X. Zhang, J. Leddy & A.L. Bard, J. Am. Chem. Soc. **107** (1985) 3719; M. McGuire & G. McLendon, J. Phys. Chem. **90** (1986) 2549; R.M. Nielson, G.E. McManis, M.N. Golovin & M.J. Weaver, J. Phys. Chem. **92** (1988) 3441; R.M. Nielson & M.J. Weaver, J. Electroanal. Chem. **260** (1989) 15; and as a review, M.J. Weaver & G.E. McManis, Acc. Chem. Res. **23** (1990) 294.
6 G.R. Fleming, S.H. Courtney & M.W. Balk, J. Stat. Phys. **42** (1986) 83; and N.S. Park & D.H. Waldeck, J. Chem. Phys. **91** (1989) 943.
7 F.A. Quiocho & W.N. Lipscomb, *Advances in Protein Chemistry,* Vol.25, ed. J.T. Edsall, C.B.Anfinsen & F.M. Richards (Academic, New York, 1971) p.1.
8 B. Breslow & A. Schepartz, Chem. Lett. **1987** (1987) 1.
9 As a review, V.I. Gol'danskii & Yu.F. Krupyanskii, Quart. Rev. Biophys. **22** (1989) 39.
10 H.A. Kramers, Physica **7** (1940) 284.
11 H. Frauenfelder & P.G. Wolynes, Science **229** (1985) 337.
12 For example, H. Risken, *The Fokker-Planck Equations* (Springer, Berlin, 1984).
13 R.F. Grote & J.T. Hynes, J. Chem. Phys. **73** (1980) 2715; *ibid.* **75** (1981) 2191, and as a review, J.T. Hynes, J. Stat. Phys. **42** (1986) 149.
14 L.D. Zusman, Chem. Phys. **49** (1980) 295; *ibid.* **80** (1983) 29; I. Rips & J. Jortner, J. Chem. Phys. **87** (1987) 2090 & 6513; *ibid.* **88** (1988) 818; M. Sparpaglione & S. Mukamel, J. Chem. Phys. **88** (1988) 3263 & 4300; Y.J. Yan, M. Sparpaglione & S. Mukamel, J. Phys. Chem. **92** (1988) 4842; M. Morillo & R.I. Cukier, J. Chem. Phys. **89** (1988) 6736; and D.Y. Yang & R.I. Cukier, J. Chem. Phys. **91** (1989) 281.
15 H. Sumi, J. Phys. Chem. **95** (1991) 3334.
16 H. Sumi & R.A. Marcus, J. Chem. Phys. **84** (1986) 4894; see also, W. Nadler & R.A. Marcus, J. Chem. Phys. **86** (1987) 3906.
17 For example, H. Sumi, J. Phys. Soc. Jpn. **49** (1980) 1701.
18 For example, N. Agmon & R. Kosloff, J. Phys. Chem. **91** (1987) 1988.

DISCUSSION

Jortner

Your work which extends the Kramers formalism to incorporate the role of atomic rearrangement modes is timely and significant. A central issue is the classical treatment of these modes. Barbara has already pointed out in his talk that this approach may require revision when applied to electron transfer. Such an approach may be applicable to treat the effect of some internal modes of a protein at room temperature. For instance, we estimated that characteristic frequencies of the protein medium of the photosynthetic reaction center which couples to electron transfer are 80–100 cm^{-1}, so that for these modes the classical treatment is adequate at room temperature. On the other hand, the intramolecular modes of stilbene (for which extensive spectroscopic information is available) cannot be treated this way. It will be important to extend your treatment to incorporate high-frequency quantum modes.

Sumi

Thank you very much for your encouragement. Quantum–mechanical treatments of intramolecular vibrational modes can easily be incorporated in the present theory, since the theory can be applied to any form of k(X) which represents the rate constant of a reaction brought about by intramolecular vibrational modes at various molecular arrangements specified by the coordinate X. Only as an example, I showed in my talk a calculation of the rate constant k_s obtained by using the most simplified Sumi–Marcus model in which k(X) is calculated with the classical treatment of intramolecular vibrational modes. When high–frequency vibration modes take part in the reaction, they must be treated quantum mechanically. In this case, k(X) should have an X dependence asymmetric between the normal and the inverted regions, being greatly enhanced in the inverted region due to the quantum effect of phonon tunneling.

Fleming

I do not believe that your comments on the viscosity dependence of the Grote-Hynes theory are correct. Kramers and Grote–Hynes equations both go to the transition–state limit as the friction goes to zero. As Bagchi and Oxtoby showed, when estimates for the frequency dependence of the viscosity of real liquids (eg. alkanes) are made, the Grote–Hynes expression does, in fact, show a fractional dependence on the macroscopic skew viscosity.

Secondly, I think it is very important to distinguish between fractional dependence on friction and fractional dependence on viscosity. For example, the rotational diffusion time of trans–stilbene shows a fractional dependence on viscosity if the viscosity is changed (at constant temperature) by changing from one hydrocarbon solvent to another. However, if the viscosity is varied by changing the temperature in a single solvent a completely linear dependence on viscosity is observed. But the slope of the straight line changes from solvent to solvent. In other words, the coupling constant changes with solvent (Kim and Fleming; J. Phys. Chem.). Thus, when comparing experimental data with theory it is important to consider the way in which the viscosity was varied.

Sumi

Concerning the first comment, I think that you misunderstood my talk. What I pointed out is concerned not with the limit of low friction, but with the limit of high friction. I pointed out that when the rate constant given by the Grote and Hynes theory decreases with increasing the solvent viscosity as observed experimentally it should decrease more rapidly than that given by the Kramers theory (decreasing as η^{-1}) in the high-friction limit, while the observed rate constant decreases more slowly than η^{-1}. This occurs since the Grote-Hynes rate constant is always larger than the Kramers one and *simultaneously the former approaches the latter as approaches infinity with the speed of the Brownian particle proportional to the rate constant approaching zero*. This property of the Grote-Hynes rate constant is, of course, concerned only with the high-friction limit. In the intermediate friction region, it is possible for the Grote-Hynes rate constant to show a milder viscosity dependence as you point out. In order to reproduce the milder viscosity dependence in the high-friction limit in the Grote-Hynes theory, however, the rate constant must be smaller than the Kramers one as long as it approaches the Kramers one as the viscosity approaches infinity, and this can be obtained only in an unphysical situation that the high-frequency friction is stronger than the low-frequency one.

Concerning your second comment, I completely agree with you. Calculated in the dynamical theories for reaction rates is the rate constant as a function of the friction coefficient. I assumed simply in my talk that the friction coefficient was proportional to the solvent viscosity. It is quite true that the relationship between them must be checked more carefully. Moreover, since the exponential factor of the rate constant is very sensitive to a small change in the coupling with solvent fluctuations, only the viscosity dependence of the pre-exponential factor must be extracted experimentally in order to discuss the dynamical role of solvent fluctuations in the reaction.

Hynes

I have a number of difficulties with a number of your statements, particularly regarding Grote-Hynes (GH) theory and its predictions. But let me focus only on one issue. As Professor Fleming has just pointed out, examination of predictions of GH with realistic models of solvent friction indeed gives power law dependence over a very wide range of solvent viscosities. However, again as Professor Fleming pointed out, there can be uncertainties about the connection of the friction to the viscosity. It is therefore very important to have unambiguous tests of GH and other rate theories. In this connection, there are now a number of molecular dynamics simulations tests for realistic reactions, which have the key ingredients of, loosely stated, a nuclear coordinate and a solvent coordinate. These include S_N2, S_N1, ion pair interconversion and ion transport in membrane channels (References are given in my paper). In all of these cases (and several others), the simulation results and GH theory agree to within the error bars. I think it would be useful and instructive to examine the applicability of your theory to these well-defined, realistic simulation results.

Sumi

As mentioned in my answer to Professor Fleming concerning the viscosity dependence of the rate constant given by the Grote–Hynes theory, as long as it decreases with the increase of the viscosity toward infinity in the high-friction limit, it should decrease more rapidly than the Kramers rate constant does. I do not deny that it is possible for the Grote–Hynes rate constant to show a milder viscosity dependence in the intermediate friction region. However, if the observed fractional power dependence of the rate constant written as $\eta^{-\alpha}$ with $0<\alpha<1$ is ascribed to the behavior of their rate constant in the intermediate friction region, no analytical formula would exist for , since it behaves differently both in the high-friction limit as mentioned above and in the low-friction limit where the rate constant recovers the transition-state value becoming independent of η. Then, the fractional power dependence would tend to an apparent phenomenon in the intermediate friction region arising from mutual cancellation between the two types of viscosity dependence in the high- and the low-friction limits.

I consider that the frequency dependence of friction is important in some systems, but inoperative in others: The speed of microscopic thermal motions of solvent molecules has a magnitude of the order of 10^{12} sec^{-1} or larger, as pointed out also in your talk. It must be compared with the speed of the Brownian motions coupling to a reaction. In electron-transfer reactions in polar solvents, for example, the speed of solvent fluctuations as a whole in the field of a solute charge is given by the inverse of the longitudinal dielectric-relaxation time which is about 0.3 psec in water while larger than 10 psec in higher alcohols at room temperature, as noted also by Barbara and by Fleming in their talks. Therefore, the former is smaller than the latter in water, and we must take into account the frequency dependence of friction there, while the former is much larger than the latter in higher alcohols, and the frequency dependence of friction is inoperative there. It has been observed that the tertiary structure of a protein in water fluctuates as Brownian motions with speed slower than 10^{11} sec^{-1}, as mentioned in my talk. Therefore, the frequency dependence of friction is inoperative in biochemical reactions taking place in proteins.

Dynamics and Mechanisms of
Photoinduced Transfer and Related Phenomena
N. Mataga, T. Okada and H. Masuhara (Editors)

SOLVATION DYNAMICS IN ELECTRON-TRANSFER AND FEMTOSECOND NONLINEAR SPECTROSCOPY

Shaul Mukamel and Wayne Bosma

Department of Chemistry, University of Rochester, Rochester, NY 14627

Abstract

A unified theory for electron-transfer rate and hole-burning spectroscopy is presented using the density matrix and its evolution in Liouville space. Both solvent modes and high frequency intramolecular modes are incorporated in the same manner using a multimode Brownian oscillator model for the nonlinear response function.

I. Introduction

Solvent-solute interactions play an important role in chemical processes and optical spectroscopy in condensed-phases. Laser pulses with durations as short as ~ 10 fs may be used in nonlinear optical experiments to probe the solute and solvent degrees of freedom separately.[1-2] Additionally, a solvated molecule's interaction with its solvent environment directly affects the rates of electron-transfer processes.[3-6]

We have developed a microscopic theory based on the evolution of the density matrix, which establishes a general fundamental connection between electron-transfer rates and nonlinear optical processes in solution.[8] Solvent correlation functions and dephasing rates obtained from nonlinear optical measurements may thus be directly used in the calculation of molecular rate processes. This theory provides an insight on the transition from nonadiabatic to adiabatic rates, and the relevant solvent timescale which controls the adiabaticity is precisely defined. In addition, the incorporation of internal vibrational modes of the solute can be made in the current theory in a straightforward way. We shall discuss here the application of this density-matrix approach to a multimode solute-solvent system, where we consider specifically the vibronic states of one or more solute modes. We shall present a semiclassical theory which provides a unified description of molecular rate processes and nonlinear optical lineshapes. We will furthermore develop in parallel formulas for the electron-transfer rate and the ultrafast hole-burning signal, to reveal the exact relationship between the two quantities and demonstrate how they depend on the same dynamical correlation functions.

The connection between electron-transfer rates and optical lineshapes may be understood as follows: Reaction rates may be calculated by starting with a nonadiabatic (two-state) model and expanding the rate perturbatively in the nonadiabatic coupling V. Optical lineshapes are usually calculated by expanding the optical polarization in powers of the electric field E. Both expansions are expressed in terms of correlation functions. To lowest order (V^2), the nonadiabatic rate is given by the Fermi golden rule; the optical response to first order in E (e.g., the absorption lineshape) is given by the linear susceptibility $\chi^{(1)}$. Both quantities are related to a two-time correlation function of the solvent. In the next order (V^4), the rate is related to a four-point correlation function. The same correlation function enters in the calculation of the third order nonlinear susceptibility (to order E^3), $\chi^{(3)}$. Numerous nonlinear optical measurements can be interpreted in terms of $\chi^{(3)}$. Fluorescence, coherent and spontaneous Raman, hole-burning, pump-probe, and four wave mixing are a few examples of optical measurements related to $\chi^{(3)}$. The expansions can be carried out to higher orders and, in general, the rate to order V^{2n} is related to $\chi^{(2n-1)}$. This connection establishes a fundamental link between the dynamics of rate processes and nonlinear optical measurements.

Åkesson, et al.,[4] have recently performed electron-transfer experiments on Betaine-30, which is frequently used as a probe of solvent polarity. They find a significant contribution of both solute and solvent modes to the electron-transfer rate. In modelling their experiments, they find that a model proposed by Jortner and Bixon,[7] which explicitly includes solute vibrations in a quantum mechanical fashion, does rather well at predicting the electron-transfer rate for solvents such as acetone which have fast relaxation timescales (~ 1 ps). However, for more slowly relaxing solvents, such as triacetin (relaxation timescales ~ 100 ps - 10 ns, depending on temperature), that theory gives reaction rates which are orders of magnitude too slow. This suggests that when the solvent timescale is very slow it may become irrelevant to the electron-transfer, and some other nuclear relaxation time (e.g. that of a low frequency intramolecular vibration) may take over. The present theory, which incorporates both solvent and intramolecular relaxation, shows this crossover very clearly. Furthermore, we suggest that the solute and solvent dynamics observed in these experiments can be understood in the same way as those which are probed directly in nonlinear optical experiments, such as hole-burning spectroscopy.

II. Response Functions in Nonlinear Optical Spectroscopy and Electron-Transfer

Consider a solute-solvent system with two relevant electronic states, $|A\rangle$ and $|B\rangle$. $|A\rangle$ and $|B\rangle$ can represent the ground and excited electronic states in an optical experiment, or the initial and final states in an electron-transfer experiment. In addition, there is a coupling between the states. We may write the Hamiltonian for this system as

$$H = |A\rangle H_A \langle A| + |B\rangle (H_B + \hbar\omega_{BA}) \langle B| + V(|A\rangle\langle B| + |B\rangle\langle A|) \quad . \tag{1}$$

H_A is the nuclear Hamiltonian while the system is in the $|A\rangle$ state, and H_B is the nuclear Hamiltonian for state $|B\rangle$. $\hbar\omega_{BA}$ is the difference in ground-state energies between the two states. For an optical experiment, V is equal to $\underset{\sim}{\mu} \cdot \underset{\sim}{E}$, where $\underset{\sim}{\mu}$ is the transition dipole moment between the electronic states and $\underset{\sim}{E}$ is the applied electric field. In an electron-transfer experiment, V is the nonadiabatic coupling between the initial and final states.

In an electron-transfer experiment, the quantity of interest is the rate, K; in an optical measurement, the experimental signal may generally be expressed in terms of the polarization of the system. Both of these quantities may be expressed as perturbation series in the interstate coupling:[8]

$$K = V^2 C_2 + V^4 C_4 \;\; \ldots \tag{2a}$$

$$P(\underset{\sim}{r},t) = P^{(1)}(\underset{\sim}{r},t) + P^{(3)}(\underset{\sim}{r},t) \; + \; \ldots \tag{2b}$$

For typical experiments on isotropic condensed-phase systems, the odd-order terms in Eq. (2a) vanish, as do the even-order terms in Eq. (2b). We shall focus on the two lowest order nonvanishing terms in each series. For the optical experiments, these are:[8]

$$P^{(1)}(\underset{\sim}{r},t) = -\,2\,|\mu|^2 \,\mathrm{Im} \int_0^{\infty} dt_1 \; J(t_1) \; E(\underset{\sim}{r},t-t_1) \; . \tag{3a}$$

and

$$P^{(3)}(\underset{\sim}{r},t) = 2\,|\mu|^4\,\mathrm{Im}\int_0^{\infty} dt_1 \int_0^{\infty} dt_2 \int_0^{\infty} dt_3$$

$$R(t_3,t_2,t_1)\, E(\underset{\sim}{r},t-t_1-t_2-t_3)\, E(\underset{\sim}{r},t-t_2-t_3)\, E(\underset{\sim}{r},t-t_3) \quad . \tag{3b}$$

Here $J(t_1)$ and $R(t_3, t_2, t_1)$ are the linear and third-order nonlinear response functions, respectively. Im (Re) denotes the imaginary (real) part. $P^{(1)}$ is used to calculate linear optical properties, such as the absorption lineshape. $P^{(3)}$ is relevant to a variety of nonlinear optical experiments, such as hole-burning, coherent and spontaneous Raman spectroscopies, and photon echo. For the case of electron-transfer experiments, the two lowest-order terms in Eq. (2a) are given by:[8]

$$C_2 = 2\,\mathrm{Re}\int_0^{\infty} dt_1\, J(t_1) \quad , \tag{4a}$$

$$C_4 = 2\,\mathrm{Re}\int_0^{\infty} dt_1 \int_0^{\infty} dt_2 \int_0^{\infty} dt_3\, [R(t_3,t_2,t_1) - R(t_3,t_2{=}\infty,t_1)] \quad . \tag{4b}$$

In the limit of weak coupling between the initial and final states of the electron-transfer process, the reaction is in the nonadiabatic limit, and the rate is simply given by V^2C_2. This expression for the rate can be derived using the Fermi golden rule, and the electron-transfer process is then analogous to ordinary (linear) absorption. In general, however, we need all orders in V to adequately calculate the rate. To that end, we approximately resum the series in Eq. (2a) with a Padé approximant:[8]

$$K = \frac{V^2C_2}{1 + V^2(C_4/C_2)} \quad . \tag{5}$$

In the nonadiabatic limit, the electron-transfer rate is given by $K_{NA} = V^2C_2$. In the other extreme (the adiabatic limit), we have $K_{AD} = 1 / \langle\tau\rangle$, where $\langle\tau\rangle = C_4 / (C_2)^2$ is the relaxation timescale (solvent or intramolecular) which controls the adiabatic rate.

Eqs. (3) and (4) show the fundamental connection between electron-transfer and optical experiments: Each observable is related to the linear and nonlinear response functions

$J(t_1)$ and $R(t_3, t_2, t_1)$. While the specific states $|A>$ and $|B>$ may be different for the two types of experiments, we shall show in the following that the relevant response functions can be calculated in the same manner.

III. A Multimode Brownian Oscillator Model for Hole-Burning Spectroscopy

In order to directly compare the experimental observables for electron-transfer and optical experiments, we should focus on a specific four wave mixing technique. To that end, we present formulas for the ordinary (linear) absorption lineshape, and for the pump-probe difference absorption spectrum, for a direct comparison to C_2 and C_4. In a pump-probe absorption experiment, the system is subjected to a pump pulse, centered at time $t=-\tau$, and a probe pulse, centered at $t=0$. The applied electric field is:

$$E(\underset{\sim}{r},t) = E_p(t+\tau)e^{i\Omega_P t + i \underset{\sim}{k}_P \cdot \underset{\sim}{r}} + E_T(t)e^{i\Omega_T t + i\underset{\sim}{k}_T \cdot \underset{\sim}{r}} + c.c. \tag{6}$$

Here $E_p(t)$ and $E_T(t)$ are the envelopes of the pump and probe pulses, with center frequencies Ω_P and Ω_T, respectively. The transmitted probe pulse is then frequency-dispersed through a monochromator, and subtracted from the transmitted probe in the absence of the pump pulse. The resulting difference absorption signal, measured at frequency ω_T, is given by:[9]

$$S(\Omega_p, \Omega_T; \omega_T, \tau) = -2\,\mathrm{Im}\,E_T^*(\omega_T) \int_{-\infty}^{\infty} dt\, \exp[i(\omega_T-\Omega_T)t]\, P^{(3)}(\underset{\sim}{k}_T, t) \quad . \tag{7}$$

We now consider a specific model for a solute-solute system which can be applied to both nonlinear optics and electron-transfer. The solute may have several harmonic coordinates, representing the intramolecular vibrations which are strongly coupled to the electronic system. For each mode, the equilibrium position of the oscillator representing the nuclear dynamics of the $|A>$ state is displaced from that of the $|B>$ state. We follow the convention that $|a>$ and $|c>$ represent vibronic eigenstates of the $|A>$ state, and $|b>$ and $|d>$ correspond to levels in the $|B>$ electronic state (Fig. 1).

For the solvent, we use a single Brownian oscillator mode in the overdamped limit.[9] Representing the solvation coordinate by $U \equiv H_B - H_A$,[8] we can express the solvent contribution to the response functions $J(t_1)$ and $R(t_3, t_2, t_1)$ by the first two moments of U and its two-time correlation function. We define:

$$\lambda \equiv \langle U \rangle \tag{8}$$

$$\Delta^2 \equiv \langle U^2 \rangle - \langle U \rangle^2 \tag{9}$$

and

$$M(t) \equiv \frac{1}{\Delta^2} \mathrm{Re}\left[\langle U(t)\, U \rangle - \langle U \rangle^2\right] = e^{-\Lambda t} \quad . \tag{10}$$

For a Debye solvent, Λ^{-1} is the longitudinal relaxation time. This solute-solvent system, for a single solute mode, is illustrated schematically in Figure 1:

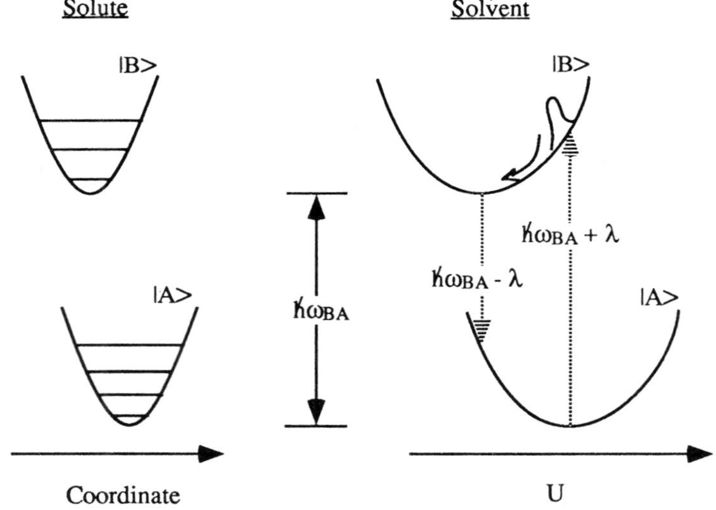

Figure 1: Two-Mode Model for Hole-Burning Spectroscopy

For the solute-solvent system described above, the frequency-dispersed linear absorption is given by:

$$\sigma(\omega_T) = J'(\omega_T - \omega_{BA}) \tag{11}$$

where

$$J'(\omega) = 2\mathrm{Re} \sum_{a,b} P(a)\, |f_{ba}|^2 \int_0^\infty dt_1 \exp[\, i(\omega_0 - \omega_{ba})t_1 - g(t_1)\,] \tag{12}$$

$$g(t) = i\lambda \int_0^t dt' M(t') + \Delta^2 \int_0^t dt' \int_0^{t'} dt'' M(t'') \tag{13}$$

In Eq. (12), f_{ba} represents the Franck-Condon overlap between vibronic states |b> and |a>. P(a) is the probability for the solute to be in state |a> at thermal equilibrium.

If we take the solvent nuclear motions to be slow compared with the electronic dephasing timescale, we may perform the integral in Eq. (12), to obtain:[9]

$$J'(\omega) = \sqrt{\frac{2\pi}{\Delta^2}} \sum_{a,b} |f_{ba}|^2 \exp\left[-\frac{1}{2\Delta^2}(\omega - \omega_{ba} - \lambda)^2\right] \tag{14}$$

We next consider an ideal pump-probe experiment in the hole-burning limit, where the pump pulse is long compared to the timescale of the electronic dephasing processes and the pump and probe pulses are well-separated in time. Additionally, we take the probe pulse to be a δ-function in time; since the frequency resolution of the experiment is determined by the monochromator, rather than the probe pulse, this ideal limit does not cause any difficulties. Under these conditions, and further invoking the rotating-wave approximation, the probe difference absorption is given by:

$$S(\Omega_p; \omega_T, \tau) = |\mu|^4 \int_0^\infty dt_2\; I_P(\tau - t_2)\, [R'(\omega_T - \omega_{BA}, t_2, \Omega_P - \omega_{BA}) + R'(\omega_T - \omega_{BA}, t_2, \omega_{BA} - \Omega_P)] \;. \tag{15}$$

Here $I_P(t)$ is the pump pulse temporal intensity profile, and

$$R'(\omega_3, t_2, \omega_1) = 2\,\mathrm{Re} \int_0^\infty dt_1 \int_0^\infty dt_3 \;\exp[i\omega_3 t_3 + i\omega_1 t_1]\; R(t_3, t_2, t_1) \tag{16}$$

For the present model, we have[10]

$$\begin{aligned} R'(\omega_3, t_2, \omega_1) = 2\mathrm{Re}\, |\mu^{(B)}|^4 \int_0^\infty dt_1 \int_0^\infty dt_3 \sum_{a,b,c,d} P(a)\, f_{ab} f_{bc} f_{cd} f_{da} \exp[i\omega_3 t_3 + i\omega_1 t_1] \\ \times \{ \exp[- i\omega_{dc}t_3 - i\omega_{db}t_2 - i\omega_{da}t_1 - g^*(t_3) - g(t_1) - (z^* - zM(t_1))\, M(t_2)\, (1 - M(t_3))] \\ + \exp[- i\omega_{dc}t_3 - i\omega_{db}t_2 - i\omega_{da}t_1 - g^*(t_3) - g^*(t_1) - (z - z^*M(t_1))\, M(t_2)\, (1 - M(t_3))] \\ + \exp[- i\omega_{dc}t_3 - i\omega_{ac}t_2 + i\omega_{ba}t_1 - g(t_3) - g^*(t_1) - z^*(1 - M(t_1))\, M(t_2)\, (1 - M(t_3))] \\ + \exp[- i\omega_{ba}t_3 - i\omega_{ca}t_2 - i\omega_{da}t_1 - g(t_3) - g(t_1) - z(1 - M(t_1))\, M(t_2)\, (1 - M(t_3))] \} . \end{aligned} \tag{17}$$

Here

$$z = \frac{\Delta^2}{\Lambda^2} - \frac{2\lambda}{\Lambda} \quad .$$

The third - order response function for the system, given by the integrand in eq. (17) is the result of considering the system's dynamics in the following way: Initially, the system is in thermal equilibrium, with electronic density matrix $\rho = | A >< A |$. The system then interacts with the pump pulse, leaving ρ in an electronic coherence (either $| A >< B |$ or $| B >< A |$), where it evolves for a time period t_1. The system then interacts again with the pump pulse, bringing the system back to an electronic coherence (note, however, that the system may then be in a vibrational coherence, e.g. $|b><d|$), where it evolves for a period t_2. The first two terms in eq. (17) correspond to the system propagating on the $| B >$ electronic state during t_2, while the last two terms involve propagation in the $| A >$ state. The system then interacts with the probe pulse, after which it evolves in an electronic coherence during the time period t_3. Following that time period, the detection of the signal takes place. It should be noted that in the current model we have chosen to represent the solute modes using the vibronic state representation. Alternatively, we can use the Wigner phase space representation to model the solute modes[9]; the choice of representation is made for convenience of calculation; if the solute modes are low-frequency, the phase space representation may become more convenient.

Assuming the solvent nuclear motions to be slow compared to the timescale of the electronic dephasing, and making the additional assumption that the system evolves in a vibrational population during the t_2 time period (i.e. considering only terms with b = d in the first two terms of Eq. (17) and a = c in the last two terms), we obtain:

$$R'(\omega_3, t_2, \omega_1) = \sum_{a,\,b,\,d} P(a)\,|f_{ab}|^2\,|f_{ad}|^2 \frac{2\pi}{\Delta^2[1 - M^2(t_2)]^{1/2}}$$

$$\times \exp\{\frac{1}{\Delta^2 - \Delta^2 M^2(t_2)}[-\frac{1}{2}(\omega_3 - \omega_{da} - \lambda)^2 - \frac{1}{2}(\omega_1 - \omega_{ba} - \lambda)^2$$

$$+ M(t_2)(\omega_3 - \omega_{da} - \lambda)(\omega_1 - \omega_{ba} - \lambda)]\}$$

$$+ \sum_{a,\,b,\,c} P(a)\,|f_{ab}|^2\,|f_{cb}|^2 \frac{2\pi}{\Delta^2[1 - M^2(t_2)]^{1/2}}$$

$$\times \exp\{\frac{1}{\Delta^2 - \Delta^2 M^2(t_2)}[-\frac{1}{2}(\omega_3 - \omega_{bc} - \lambda(t_2))^2 - \frac{1}{2}(\omega_1 - \omega_{ba} - \lambda)^2$$

$$+ M(t_2)(\omega_3 - \omega_{bc} - \lambda(t_2))\,(\omega_1 - \omega_{ba} - \lambda)]\} \qquad (18)$$

with

$$\lambda(t) \equiv \lambda[2M(t) - 1] \quad .$$

The neglect of the vibrational coherence terms (propagation with $b \neq d$ or $a \neq c$ during t_2) may be justified for high frequency vibrations. The fast oscillations during the t_2 period make their contribution small. This is not true for vibrations of lower frequency.

We have calculated the hole-burning spectrum for a two-mode system, with the following parameters: The solute mode has frequency $\omega = 600\ cm^{-1}$, and the oscillators are displaced by a dimensionless displacement d = 1.2. The solute mode has $\lambda = 625\ cm^{-1}$, $\Delta = 510\ cm^{-1}$, and $\Lambda^{-1} = 3.5$ picoseconds. The pump pulse is 62 fs fwhm, and the probe pulse is taken to be a δ-function. Figure 2 illustrates the frequency-resolved hole-burning signal for three delay times: $\tau = 200$ fs, 600 fs, and 8 ps:

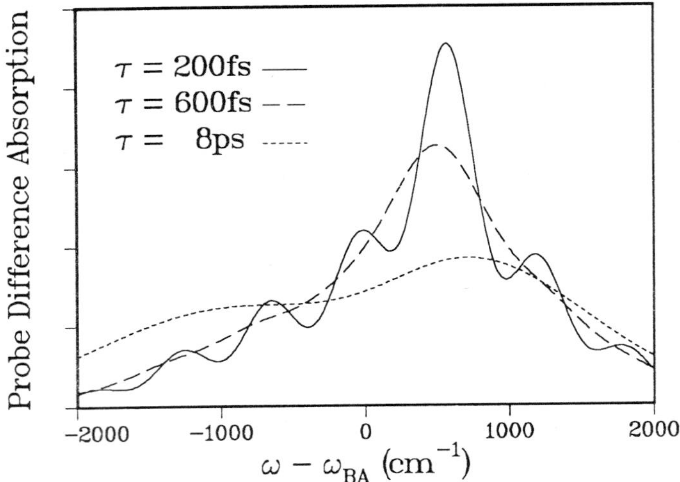

Figure 2: Hole-Burning Spectra for 3 Delay Times.

In the shortest of the delay times, we see peaks due to vibronic transitions in the solute mode. This narrowing results from the spectral selectivity of the initial excitation, whereby the inhomogeneous broadening of the solvent is eliminated, and the underlying vibronic structure is then visible. At longer times, the solvent mode broadens the spectrum, causing the vibronic peaks to be smeared out in the $\tau = 600$ fs curve. At still longer times, the solvent mode causes a Stokes shift in the difference absorption signal, whereby the contributions from excited-state propagation during the t_2 time period are separated from ground-state the ground-state propagation part.[11]

IV. The Multimode Electron-Transfer Rate

For electron-transfer experiments, we may use a model virtually identical to the one described in the previous section, as illustrated in Figure 3:

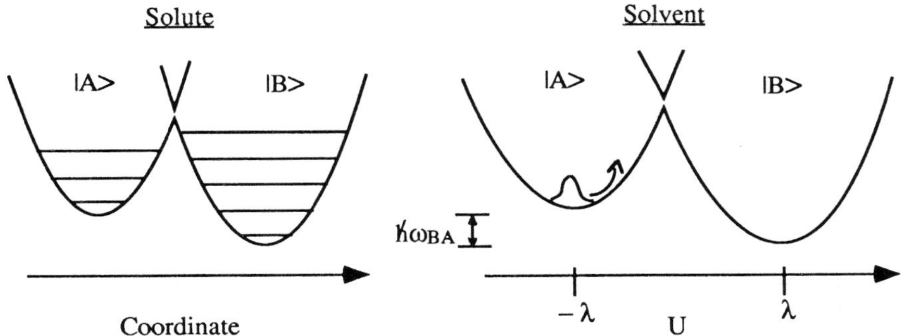

Figure 3: Two-Mode Model for Electron-Transfer

For this solute-solute system, eqs. (4) become[8]

$$C_2 = J'(-\omega_{BA}) \tag{19a}$$

and

$$C_4 = \int_0^{\infty} dt_2 \; [R'(-\omega_{BA}, t_2, -\omega_{BA}) - R'(-\omega_{BA}, t_2 = \infty, -\omega_{BA})] \quad , \tag{19b}$$

where J' and R' are given by Eqs. (14) and (17), respectively. Inserting Eqs. (19) into Eq. (5) gives us the electron-transfer rate for the multimode Brownian oscillator model. In general, the parameters used in calculating J' and R' for the two types of experiments could be different, as the two experiments probe different electronic states. However, the physical significance of the parameters are the same, and we can thus consider the two types of experiments using the same theoretical framework. Eq. (19b) allows us to identify the relevant molecular dynamics timescale which controls the adiabatic electron-transfer rate. In many cases, this is due to the solvent relaxation time. However, if that timescale is very slow, other processes (such as the vibrational relaxation of the low frequency modes) may take over and dominate the relaxation of R' to its $t_2 = \infty$ value. Since R' is a multimode quantity which depends on solvent dynamics as well as intramolecular vibrations, Eq. (19b) can be used in either case.

The expression for the electron-transfer rate, given by Eq. (5) is similar, but not identical, to one presented recently by Jortner and Bixon.[7] The main difference is that Jortner and Bixon assume the rate to be the sum of contributions from pairs of vibrational levels in the initial and final state. In resumming eq. (2a) to obtain eq. (5), we could have made the same assumption. Rather than expressing the electron-transfer rate as the sum of contributions from separate "reaction channels", we have used a different resummation of the perturbation series in V. Furthermore, the analogy with nonlinear optics allows us to explicitly consider the intermediate vibrational states involve in the reaction process, rather than just the initial and final vibrational states. In general, the contributions of the various modes to the rate are nonadditive, just as the various contributions to the hole-burning spectra are not additive. However, if we neglect the contributions of vibrational coherences (i.e. terms with $b \neq d$ and $a \neq c$) to the rate, as we have done here, we could rewrite the series in Eq. (2a) in terms of a sum over reaction channels; thus, a rearrangement of that series, followed by a Padé resummation, would produce a similar result to that of Jortner and Bixon.

In conclusion, we have shown how a unified treatment of nonlinear optics and electron-transfer kinetics may be used to clarify the different solute and solvent contributions to the electron-transfer rate. It is thus, in principle, possible to used the results of nonlinear optical experiments, which probe the specific characters of the solute and solvent modes, to obtain parameters necessary for the calculation of electron-transfer rates.

Acknowledgements

We would like to thank Prof. P. Barbara for useful discussions. The support of the Center for Photoinduced Charge Transfer sponsored by the National Science Foundation, the Air Force Office of Scientific Research, and the Petroleum research fund, administered by the American Chemical Society, is gratefully acknowledged.

References

1. C. H. Brito-Cruz, R. L. Fork, W. H. Knox, and C. V. Shank, Chem. Phys. Lett., **132**, 341 (1986).
2. P. C. Becker, H. L. Fragnito, J. Y. Bigot, C. H. Brito-Cruz, R. L. Fork, and C. V. Shank, Phys. Rev. Lett., **63**, 505 (1989).
3. N. Mataga, Bull. Chem. Soc. Japan, **36**, 654 (1963); N. Mataga, T. Okada, Y. Kanda, and H. Shioyama, Tetrahedron, **42**, 6143 (1986); N. Mataga, Pure and Applied Chemistry, **56**, 1255 (1984); H. Masuhara and N. Mataga, Acc. Chem. Res., **14**, 312 (1981).
4. E. Åkesson, G. C. Walker, and P. F. Barbara, J. Chem. Phys (submitted).

5. M. Maroncelli, J. MacInnis, and G. R. Fleming, Science, **243**, 1674 (1989).
6. J. D. Simon, Acc. Chem. Res., **21**, 128 (1988).
7. J. Jortner and M. Bixon, J. Chem. Phys. **88**, 167 (1988).
8. Y.J. Yan, M, Sparaglione, and S. Mukamel, J. Phys. Chem., **92**, 4842 (1988); S. Mukamel and Y. J. Yan, Acc. Chem. Res., **22**, 301 (1989).
9. Y.J. Yan and S. Mukamel, Phys. Rev. A, **41**, 6485 (1990).
10. S. Mukamel and Y.J. Yan in Recent Trends in Raman Spectroscopy, S. B. Banerjee and S. S. Jha, Eds., World Scientific, Singapore (1989), pp. 160-191.
11. W. B. Bosma, Y.J. Yan, and S. Mukamel, J. Chem. Phys., **93**, 3863 (1990).

DISCUSSION

Fleming

Could you comment on the Non-Markovian interpretation of their recent 3-pulse photon echo experiments by Shank and coworkers?

Mukamel

The three pulse echo experiments cannot be adequately interpreted using the stochastic model. That model neglects completely the effect of the chromophore on the solvent dynamics and therefore misses the Stokes shift which occurs during the t_2 period (delay between the second and the third pulses). The model result may be obtained by setting $\lambda=0$ in eq (13) thereby taking g(t) to be real. The Brownian oscillator model does incorporate the Stokes shift properly. Three pulse echo measurements provide therefore very valuable dynamical information which is not contained in two pulse echoes. Apart from showing the Stokes shift and establishing a connection with fluorescence, three pulse echo measurements can also have a simple Fourier transform relation with hole burning when the delay time t_2 is sufficiently large [Y. J. Yan and S. Mukamel, J. Chem. Phys. **94** *179* (1991)].

Hynes

Is it not true that the oscillator model that you use would predict that, for example, the time dependent fluorescence spectral width is either constant or proportional to the square of what you call M(t) [depending on initial conditions]? If so, this seems to be inconsistent with experimental results from the Fleming group and computer simulations [Carter and Hynes, JPC 1991], where widths narrow with a time dependence more like that of the first power of M(t). Can you comment on this?

Mukamel

This is true if you assume a single solvation coordinate. By corporating several overdamped (solvent and molecular) coordinates with different relaxation times and incorporating vibrational relaxation, the time dependence of the width does not have to satisfy a simple relation with the Stokes shift time scale (R. F. Loning, Y. J. Yan and S. Mukamel, J. Chem. Phys. *87*, 5840 (1987)).

Sumi

In your treatment of the rate constant for electron transfer, you assembled a perturbation series by using the Pade approximation to get a compact formula. In the adiabatic limit for a large electron-transfer matrix element V, your formula approaches a dependence proportional to $1/\tau$, where τ represents the relaxation time of Brownian motions. I solved the same problem in which both fast vibrational and slow Brownian motions take part in the reaction, but by a nonperturbational treatment. I could show rigorously that the rate constant approaches, in the adiabatic limit, to a dependence proportional to a fractional (less than unity) power of $1/\tau$. This means that the Pade approximation cannot be applied in the adiabatic limit for large V. Therefore, the correspondence between the problem of nonlinear spectroscopy and that of the rate constant you pointed out can be obtained only in the nonadiabatic limit for small V, but fails to hold as V increases.

Mukamel

It is true that we have used an approximate resummation procedure in our solution. However, our simulation can give a fractional dependence on the viscosity. [M. Sparpaglione and S. Mukamel, J. Chem. Phys., 88, 4300 (1988)]. Our τ is a complex function of the viscosity which for large viscosities scales as $1/\eta$ but for intermediate viscosities we do get the fractional power dependence of $\eta^{-\alpha}$.

Jortner

I would like to comment briefly on the Zusman limit alluded by you and by Hynes. From a general formalism he derived the expression $k_{ET}=k_{NA}/H$, where k_{NA} is the nonadiabatic rate while $H=V^2C_4/C_2$ in your notation or $H=(4\pi V^2\tau_L/\hbar E_m)$ (where E_m is the medium reorganization energy) as derived by us for a Debye solvent. This equation incorporates the dissipative properties of the solvent in the frequency factor for the rate. What is significant is that the more general expression $k_{ET}=k_{NA}/(1+H)$, which was derived by several groups, bridges between the nonadiabatic limit and the Zusman limit. This last expression should be confronted with the results of simulations and utilized for the interpretation of experimental data.

Mukamel

Our expression agrees with yours and with Zusman's expression for a Debye solvent. Its main advantage is that it provides a simple generalization to any dielectric function $\varepsilon(\omega)$ as well as an arbitrary number of high frequency molecular vibrations. A 29 mode calculation for hole burning illustrates that point (S. Mukamel, Adv. Chem. Phys. **70** *165* (1988)).

Hynes

I believe that there is some confusion arising in the meaning of "adiabatic" in the discussion between Sumi and Mukamel. In the electronically adiabatic limit (i.e., large enough electronic coupling), the barrier height itself is modified (decreased) by the coupling (I think this is also Sumi's sense of "adiabatic"); I do not think that this is included in Mukamel's development. Further, I believe – as indicated in my talk – the solvent dynamical effects arise from events occurring in the neighborhood of the parabolic barrier top. The formula which Mukamel showed – which also appears in J. Phys. Chem. 90, 3701 (1986) – involving M(t), and also the more general formula connecting ET and nonlinear spectroscopy, can only refer to solvent motion in stable solvent wells, and not on the unstable barrier top of relevance for adiabatic reactions (at least for fast solvents).

Mukamel

Our expression is based on a resummation of a perturbative expression for the rate. C_2 and C_4 contain the dynamics of both wells in the absence of nonadiabatic coupling. The full adiabatic expression will require an extension of the present approach.

Jortner

There are two parameters which specify electron transfer dynamics, the Landau Zenner parameter $\gamma_{LZ}=2v^2/\hbar \mid \vec{F}_1-\vec{F}_2 \mid v$ (where F_1 and F_2 are the slopes of the potential curves at the intersection and v the velocity at this point) and the solvent dynamic parameter $H=4\pi V^2\tau_L/\hbar E_m$ (which was referred by us as the "nonadiabaticity parameter"). The electron transfer rate can be described in the γ_{LZ} and plane so that three limits are realized.

(i) Nonadiabatic limit. Both γ_{LZ} and H are small.
(ii) Solvent dynamic controlled limit. Large H.
(iii) Adiabatic transition–state theory limit. Large γ_{LZ} and moderate H.

When we describe the "transition" between limits (i) and (ii) the now familiar expression $k_{ET}=k_{NA}/(1+H)$ is appropriate, while Hynes focused in his presentation at his conference on limit (iii), Rips and myself have attempted to provide an interpolation formulae between the three limits [I. Rips and J. Jortner in "Perspectives in Photosynthesis" Ed. J. Jortner and B. Pullman, Dochecht Publishing Co. (1989)]

Mukamel

I do not need to reply.

Dynamics and Mechanisms of
Photoinduced Transfer and Related Phenomena
N. Mataga, T. Okada and H. Masuhara (Editors)
1992 Elsevier Science Publishers B.V.

Intramolecular Charge Transfer as Revealed by Results from the Measurement of Ground and Excited State Dipole Moments

W. Baumann, Z. Nagy, A. K. Maiti, H. Reis, Silvana Vianna Rodrigues and N. Detzer

Institute of Physical Chemistry, University of Mainz, D 6500 Mainz, Germany

Abstract

The principles of solvent shift measurements, electro optical absorption and emission measurements and of time resolved conductivity measurements as tools for the determination of ground and excited state dipole moments are discussed. Applications to photon induced charge transfer in a non-conjugated bichromophoric molecule, in two coumarin dyes and in 6-cyanobenzoquinuclidine are presented.

1. INTRODUCTION

The electric dipole moment μ_0 of an isolated molecule is an observable measure of its charge distribution. This holds for all quantum states and different states may have different dipole moments. Then an observed change $\Delta\mu$ of the electric dipole moment with the absorption of light (or the emission of light) will supply us with a quantitative measure for the photoinduced intramolecular charge transfer in the considered molecular system. This may be the reason that scientists thought out new methods and adopted all improvements of scientific instrumentation to determine ground and excited state dipole moments or the change of the dipole moment with the excitation or emission process as precisely as possible. Regrettably, even to-day there are only a few methods available [1]. The situation is perhaps worse if the direction of the transition moment involved in the considered excitation or emission process is to be determined and it becomes yet worse if this information has to be gathered from solute molecules. This communication will only consider solute molecules and the following methods will be discussed briefly in a first part:

1. Solvent shift measurements
2. Electro optical absorption and emission measurements
3. Time resolved conductivity measurements.

In a second part, ground and excited state dipole moments of solute molecules are presented and discussed with respect to photoinduced charge transfer.

2. THEORETICAL

2.1 Electric field effects on solute molecules

It is well known that a molecule with dipole moment μ_0 and polarizability α_0 solved in a liquid with the relative permittivity ε and the refractive index n polarizes the surrounding dielectric. A so-called internal electric field E_i is related to this polarization and induces an additional dipole moment in the solute molecule. This interaction of the solute dipole and the dielectric was first described quantitatively by Onsager [2] whose model is preferentially used to-day although other models have been developed [1]. This may be due to its clear conception. Mataga [3] and Lippert [4] adopted this model for the early development of the solvent shift method and Liptay [5] extended it for the description of the effect of an externally applied electric field E_a on electronic transitions.

The Onsager model describes the interaction of a solute molecule in a given state with dipole moment μ_o and polarizability α_o as a respective polarizable point dipole in the center of an empty sphere in the surrounding homogeneous dielectric. The size of the sphere about that of the solute molecule and its radius a_o is called the Onsager radius. More elaborate models are known [1] that extend this model to non-spherical cavities or that take into account the solute's real charge distribution, known from quantum mechanical model calculations [6]. According to Onsager, the internal electric field E_i is

$$E_i = f\,\mu \qquad (1)$$

where

$$\mu = \mu_o + \alpha_o E_i \qquad (2)$$

is the total dipole moment that is the sum of the permanent and the solvent induced part.

$$f = (2\pi\varepsilon_o a_o^3)^{-1} g \qquad (3)$$

where ε_o is the permittivity of the vacuum and

$$g = (\varepsilon - 1)/(2\varepsilon + 1) \qquad (4)$$

Combining the eqs. (1) and (2) yields

$$E_i = (1 - f\alpha_o)^{-1} f\,\mu_o \qquad (5)$$

and

$$\mu = \mu_o\,(1 - f\,\alpha_o)^{-1} \qquad (6)$$

It is very important to note already here that all methods that probe the dipole moment of a solute in a state thermally equilibrated with its dielectric surrounding will yield μ not μ_o of the solute in the considered state. In non-polar solvents the difference is about 10 to 20%.

The situation becomes much more complex if molecular states are considered that are not in thermal equilibrium like Franck Condon states, since the orientational part of the total polarization remains frozen during the electronic transition to a Franck Condon state. Liptay has studied this effect in detail with the theoretical model of electro optical absorption measurements [5,7].

A simple estimation following to eq.(5) makes clear that all external electric fields that can be applied to solutions without dielectric breakdown are some orders of magnitude lower than is the internal electric field E_i. That means that the effect of the external electric field E_e to the solute can be calculated independently and both effects can be summed up.

Due to the difference in the relative permittivity of the dielectric and the empty sphere, the electric field E_e in the center of the sphere is different from E_a. This problem has been solved by basic electrostatics and it is

$$E_e = f_e\,E_a \qquad (7)$$

where

$$f_e = 3\varepsilon/(2\varepsilon + 1) \qquad (8)$$

Combining eqs.(5) and (7) yields the total internal electric field E_{ei} affecting a solute molecule

$E_{ei} = (1 - f\alpha_o)^{-1} (f\mu_o + f_e E_a)$ (9)

This extended Onsager model is a well-defined model and therefore E_{ei} eq.(9) should be preferred over the often used Lorentz field E_{eiL}

$E_{eiL} = (\varepsilon + 2)/3$ (10)

which is poorly defined in solutions and which fails completely in polar solvents [1].

2.2 Solvent shift measurements

With solvent shift measurements the solvent induced shift of the maximum (or center) wavenumber $\tilde{\nu}_{max}$ of an absorption or fluorescence band is observed as a function of the solvent polarity. The theoretical model [8] connects $\tilde{\nu}_{max}$ with the relative permittivity ε and the dipole moment and polarizability of the ground and excited state involved in the considered transition. For this purpose, it calculates the interaction energy W_g and W_e of the solute polarizable dipole in the ground and excited state with the internal electric field E_i. The difference of W_e and W_g is thus directly related to the ground and excited state dipole moment (and polarizability) and on the other side can be measured by the solvent induced shift of the considered transition. Neglecting minor terms, the following equations describe the observed solvent induced shift of $\tilde{\nu}_{a,max}$ and $\tilde{\nu}_{f,max}$ with sufficient accuracy if solvents are used that differ in their relative permittivity but have very similar refractive indices:

$h\, c_o \tilde{\nu}_{a,max} = \text{constant} - (\mu_{eo}^{FC} - \mu_{go})\, f\, (1 - f\, \alpha_{go})^{-1}\, \mu_{go}$ (11)

and

$h\, c_o\, \tilde{\nu}_{f,max} = \text{constant}' - (\mu_{eo} - \mu_{go}^{FC})\, f\, (1 - f\, \alpha_{eo})^{-1}\, \mu_{eo}$ (12)

h is Planck's constant and c_o is the velocity of light. FC denotes a Franck Condon state and subscripts g and e a ground and an excited state. Vectors are printed italic. The two constants comprise a dispersion interaction term and the gas phase wavenumber maximum. For more details see for example [8,1]. Obviously, these two equations can be simplified even further if terms $(1 - f\alpha)$ can be considered ≈ 1. Then with eqs. (3) and (4) a linear regression of observed values $\tilde{\nu}_{max}$ on g yields values of $(\mu_e - \mu_g)\mu_g / a_o^3$. Hence a_o^3 has to be estimated in order to get values for dipole moments which therefore depend strongly on the relatively arbitrary choice of a_o^3 . Nevertheless, all important phenomena that are paralleled by strong photoinduced charge transfer have been observed through this method first.

2.3 Electro optical measurements

Based on early work of Kuhn [9], Czekalla [10] and Labhart [11], Liptay [5,7] has presented the most elaborate theoretical description of the external electric field effect on the absorption of solute molecules. There are two main effects that have to be considered:

1. Electric dipole moments are preferentially oriented towards the direction of the total internal electric field E_{ei}. The orientation distribution can be described by a Boltzmann distribution function if the system is in thermal equilibrium. This is the

case for solutes in their ground state or in their fluorescent excited state if their life time is considerable longer than all reorientation times.
This distribution is then probed by photo selection.

2. The external electric field induces an unsymmetric peak broadening of observed absorption or fluorescence bands of solute molecules. Basically, the explanation is like with solvent shift measurements. But opposite to solvent shift measurements, the solute dipoles are not parallel to the electric field - there is only a small preferential orientation in the direction of the external electric field, as mentioned above. Hence blue and red shifted transitions will be observed simultaneously from the solution which means peak broadening that is unsymmetric due to the preferential orientation of the solute molecules.
At a given wavenumber, the electric field induced wavenumber shift is observed as a change of the absorption coeffficient or the fluorescence intensity.
Minor effects like the explicite dependence of the transition moment on the external electric field [7] or the effect of fluctuation of E_{ei} are not considered here; the reader is referred to [5,7].
Since the detailed theoretical treatment of the effect of an external electric field on the absorption process [7] and the emission process [12,1] have been presented, only the final equations will be given and explicite polarizability terms and the field dependence of the transition moment are neglected. Studying the former effect is called electro optical absorption measurements (EAOM) and the latter electro optical emission measurements (EOEM).
Using Liptay´s formalism, the following formulae describe the electric field effect on the absorption $A(\tilde{\nu},\chi)$:

$$A^E(\tilde{\nu},\chi) = A(\tilde{\nu},\chi)\,[\,1 + L(\tilde{\nu},\chi)\,E_a^2\,] \quad (13)$$

$A^E(\tilde{\nu},\chi)$ is the absorption of the solution in the external electric field and χ is the angle between the electric field vector of the analyzing linearly polarized light and the direction of the external electric field E_a.

$$L(\tilde{\nu},\chi) = {}^aD\,r(\chi) + ({}^aE/30)\,s(\chi) + 3\,[\,{}^aF\,r(\chi) + {}^aG\,s(\chi)]\,{}^at(\tilde{\nu}) + 3\,[\,{}^aH\,r(\chi) + {}^aI\,s(\chi)]\,{}^au(\tilde{\nu}) \quad (14)$$

where

$$r(\chi) = (2 - \cos^2\chi) \quad (15)$$

$$s(\chi) = (3\cos^2\chi - 1) \quad (16)$$

and

$${}^at(\nu) = \frac{1}{15hc_o(\kappa/\tilde{\nu})}\left(\frac{\partial}{\partial\tilde{\nu}'}\,\frac{\kappa}{\tilde{\nu}'}\right)_{\tilde{\nu}=\tilde{\nu}'} \quad (17)$$

$${}^au(\tilde{\nu}) = \frac{1}{30h^2c_o^2(\kappa/\tilde{\nu})}\left(\frac{\partial^2}{\partial\tilde{\nu}'^2}\,\frac{\kappa}{\tilde{\nu}}\right)_{\tilde{\nu}=\tilde{\nu}'} \quad (18)$$

κ is the absorption coefficient at wavenumber $\tilde{\nu}$.

${}^aD \approx 0$, within the approximation used here.

$${}^aE = \beta^2 f_e^2\,[3(m_a\mu_g)^2 - \mu_g] \quad (19)$$

$$^{a}F = \beta\, f_e^2\, \mu_g\, \Delta^a\mu \tag{20}$$
$$^{a}G = \beta\, f_e^2\, (m_a\, \mu_g)\, (m_a\, \Delta^a\mu) \tag{21}$$
$$^{a}H = f_e^2\, (\Delta^a\mu)^2 \tag{22}$$
$$^{a}I = f_e^2\, (m_a\, \Delta^a\mu)^2 \tag{23}$$

$\beta = 1/(kT)$ with the temperature T and Boltzmann´s constant k. m_a is a unit vector in the direction of the transition moment involved in the observed absorption process and the following definitions have been used:

$$\mu_g = \mu_{go}\, (1 - f\, \alpha_{go})^{-1} \tag{24}$$
$$\Delta^a\mu = (1 - f\, \alpha_{go})^{-1}(1 - f'\alpha_{go})\, (1 - f\, \alpha_{eo}^{FC})\, (1 - f'\alpha_{eo}^{FC})^{-1}\, (\mu_e^{FC} - \mu_g) \tag{25}$$

with

$$\mu_e^{FC} = \mu_{eo}^{FC}\, (1 - f\, \alpha_{eo}^{FC})^{-1} \tag{26}$$
$$f' = (2\pi\varepsilon_o a_o^3)^{-1}\, g' \tag{27}$$

with

$$g' = (n^2 - 1)/(2n^2 + 1) \tag{28}$$

In sufficiently non-polar solvents f is approximately equal to f´ and eq.(25) reduces to

$$\Delta^a\mu = (\mu_e^{FC} - \mu_g) \tag{29}$$

Following eq.(13) the quantity $L(\tilde{\nu},\chi)$ can be determined experimentally by measuring the relative field induced change of the absorption as a function of the wavenumber $\tilde{\nu}$ and the angle χ. Since the effect is very small, it is modulated by modulating the electric field and the small modulated component of the light passing the absorbing solution must be detected by a phase sensitive rectifier, as is the essential part of a lock-in amplifier. Eq.(14) is used for further evaluation. This is acomplished by a multilinear regression of of the measured values of $L(\tilde{\nu},\chi)$ on the variables $r(\chi)$, $s(\chi)$, $r(\chi)\, {}^{a}t(\tilde{\nu})$, $s(\chi)\, {}^{a}t(\tilde{\nu})$, $r(\chi)\, {}^{a}u(\tilde{\nu})$ and $s(\chi)\, {}^{a}u(\tilde{\nu})$. The first and second derivative ${}^{a}t(\tilde{\nu})$ and ${}^{a}u(\tilde{\nu})$ are determned by stepwise fitting a third order polynomial to a region of e.g. 4nm of the spectrum and calculating the derivatives at the center wavelength of the fitted range. This procedure yields terms ${}^{a}D$ to ${}^{a}I$. Further evaluation towards dipole moments depends on the investigated molecular system.

A consistent formalism has been developed in [12] for the description of the field effect on the fluorescence quantum intensity $q(\tilde{\nu},\phi)$ of solute molecules, founding on the early communication of Liptay [13]. It is

$$q^E(\tilde{\nu},\phi) = q^{E=0}(\tilde{\nu},\phi)\, [\, 1 + X(\tilde{\nu},\phi)\, E_a^2\,] \tag{30}$$

$q^E(\tilde{\nu},\chi)$ is the fluorescence quantum flux from the solution in the external electric field and ϕ is the angle between the electric field vector determined by an analyzing linear polarizer and the direction of the external electric field E_a.

$$\begin{aligned} X(\tilde{\nu},\phi) = {} & L(\tilde{\nu}_a,\chi_a) + {}^{e}D/3 + ({}^{e}E/30)\, s(\phi) \\ & +3\, [\, {}^{e}F\, r(\phi) + {}^{e}G\, s(\phi)]\, {}^{e}t(\tilde{\nu}) \\ & +3\, [\, {}^{e}H\, r(\phi) + {}^{e}I\, s(\phi)]\, {}^{e}u(\tilde{\nu}) \end{aligned} \tag{31}$$

where $L(\tilde{\nu}_a,\chi_a)$ is the effect introduced by the field effect on the absorption at the excitation conditions $\tilde{\nu}_a$ and χ_a and where $r(\phi)$ and $s(\phi)$ are defined analoguous to eqs.(15) and (16)

$$^{e}t(\tilde{\nu}) = \frac{1}{15hc_o(q/\tilde{\nu}^3)}\left(\frac{\partial}{\partial\tilde{\nu}'}\frac{q}{\tilde{\nu}'^3}\right)_{\tilde{\nu}=\tilde{\nu}'} \qquad (32)$$

$$^{e}u(\tilde{\nu}) = \frac{1}{30h^2c_o^2(q/\tilde{\nu}^3)}\left(\frac{\partial^2}{\partial\tilde{\nu}'^2}\frac{q}{\tilde{\nu}'^3}\right)_{\tilde{\nu}=\tilde{\nu}'} \qquad (33)$$

$^{e}D \approx 0$, within the approximation used here.

$$^{e}E = \beta^2\, f_e^2\, [3(m_e\,\mu_e)^2 - \mu_e^2] \qquad (34)$$

$$^{e}F = \beta\, f_e^2\, \mu_e\, \Delta^f\mu \qquad (35)$$

$$^{e}G = \beta\, f_e^2\, (m_e\,\mu_e)\,(m_e\,\Delta^f\mu) \qquad (36)$$

$$^{e}H = f_e^2\, (\Delta^f\mu)^2 \qquad (37)$$

$$^{e}I = f_e^2\, (m_e\Delta^f\mu)^2 \qquad (38)$$

with the definitions

$$\mu_e = \mu_{eo}\,(1 - f\,\alpha_{eo})^{-1} \qquad (39)$$

$$\Delta^f\mu = (1 - f\,\alpha_{eo})^{-1}(1 - f'\alpha_{eo})\,(1 - f\,\alpha_{go}^{FC})\,(1 - f'\alpha_{go}^{FC})^{-1}\,(\mu_e - \mu_g^{FC}) \qquad (40)$$

where

$$\mu_g^{FC} = \mu_{go}^{FC}\,(1 - f\,\alpha_{go}^{FC})^{-1} \qquad (41)$$

Again, eq. (40) reduces in non-polar solvents to

$$\Delta^f\mu = (\mu_e - \mu_g^{FC}) \qquad (42)$$

Following eq.(30) the quantity $X(\tilde{\nu},\chi)$ can be determined experimentally by measuring the relative field induced change of the fluorescence intensity as a function of the wavenumber and the angle ϕ. Since the effect is very small, the electric field effect is modulated for this purpose and the small modulated component of the fluorescence light must be detected by a phase sensitive rectifier. The evaluation procedure is completely similar to that described with electro optical absorption measurements. Fig. 1 shows the experimental set-up used for EOAM and EOEM. It is selfexplaining.

Especially when investigating compounds that exhibit low quantum yield, the signal-to-noise ratio is so small that poor photon statistics set a limit to the accuracy of the measurement. To overcome this drawback, integral electro optical emission measurements (IEOEM) have been developed [14,15]. Here, the total photon flux is optically integrated over the whole fluorescence band. As a result, only the excited state orientation distribution is probed and eqs. (30) and (31) reduce to

$$q^{E}(\phi) = q^{E=0}(\phi)\,[\,1 + {}^{i}X(\phi)\,E_a^2\,] \qquad (43)$$

and

$${}^{i}X(\phi) = L(\tilde{\nu}_a,\chi_a) + {}^{e}D/3 + ({}^{e}E/30)\,s(\phi) + O(PM) \qquad (44)$$

where O(PM) are small correction terms related to the spectral characteristic of the photomultiplier used to detect the fluorescence light [15]. In this case of integral electro optical emission measurements only two values for ${}^{i}X(\phi)$ at two different angles ϕ must be determined. This is especially desirable with photochemically

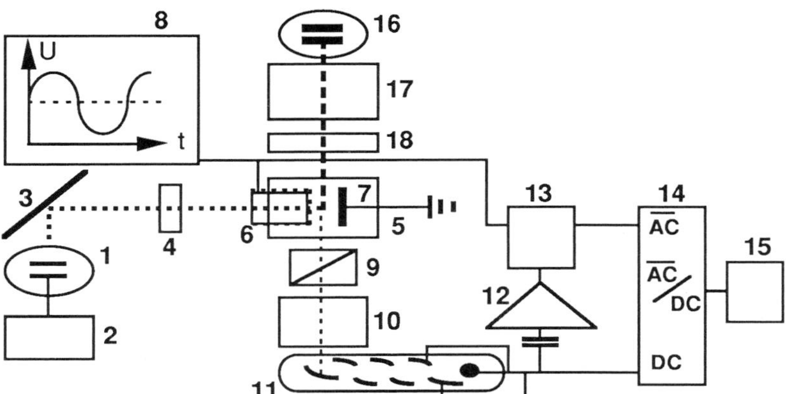

Figure 1 Block diagram of the experimental set-up used with EOAM and EOEM
1: 500W highest pressure mercury lamp; 2: lamp power supply; 3: mirror; 4: excitation filter; 5: measuring cell; 6: quartz window covered with SnO_2; 7: counter electrode; 8: modulated DC-high voltage supply; 9: polarizer; 10: monochromator; 11: photomultiplier tube; 12: AC-preamplifier; 13: lock-in amplifier; 14: electronic devider; 15: data processor; 16: Xe-highest pressure lamp; 17: monochromator; 18: filter.

instable compounds. The drawback is that considerable information is lost compared to the spectrally resolved electro optical emission measurements, since only the essential term ^{e}E can be determined here and hence the direction m_e of the transition moment must be known from symmetry in order to be able to calculate μ_e using eq. (34). The experimental procedure is as has been discussed with EOEM with the exception that the monochromator 10, fig. 1, is taken out from the fluorescence channel (see fig.1).

2.4 Determination of the direction of the transition moment

By photoselection in frozen solutions (or in sufficiently viscous solutions) the relative direction of the transition moments involved in electronic absorption or emission spectra can be determined. The situation over the whole UV/Vis spectral range usually is visualized by plotting the measured degree of anisotropy or the degree of polarization of the fluorescence observed at a fixed fluorescence (excitation) wavelength as a function of the excitation (fluorescence) wavelength. A severe drawback with this method is that such measurements cannot be performed in liquid solutions, generally.

From eqs. (19) to (23) or (34) to (38) it is obvious that with electro optical measurements the relative directions of the transition moment and the ground and excited state dipole moment can be determined. On the other hand, it is difficult to visualize complex situations which are revealed during the regression analysis process according to eq. (14) or (31) whenever this regression is performed throughout a non-homogeneous absorption or fluorescence spectrum. On the other hand, often molecules are investigated which have C_{2v} symmetry, at least in sufficient approximation. Then m can only be parallel or perpendicular to the C_2-axis and thus to all states′ dipole moments. In this case, using the approximations already introduced it follows in the case of EOAM directly from the defining eqs. (14)

to (23) that

$$\Delta_{a,||}(\tilde{\nu}) = L(\tilde{\nu},\chi=0^o) \; - 3\, L(\tilde{\nu},\chi=90^o) = \frac{f_e^2\,\mu_g^2}{3(kT)^2} \tag{45}$$

if the transition moment m_a and the ground state dipole moment μ_g are parallel, and

$$\Delta_{a,\perp}(\tilde{\nu}) = 2\, L(\tilde{\nu},\chi=0^o) \; - \; L(\tilde{\nu},\chi=90^o) = \frac{-\,f_e^2\,\mu_g^2}{6(kT)^2} \tag{46}$$

if the transition moment m_a and the ground state dipole moment μ_g are perpendicular.

In the case of EOEM it follows analoguously from eqs. (31) to (38) if $L(\tilde{\nu}_a,\chi_a)$ can be neglected that

$$\Delta_{e,||}(\tilde{\nu}) = X(\tilde{\nu},\phi=0^o) \; - 3\, X(\tilde{\nu},\phi=90^o) = \frac{f_e^2\,\mu_e^2}{3(kT)^2} \tag{47}$$

if the transition moment m_e and the fluorescent excited state dipole moment μ_e are parallel, and

$$\Delta_{e,\perp}(\tilde{\nu}) = 2\, X(\tilde{\nu},\phi=0^o) \; - \; X(\tilde{\nu},\phi=90^o) = \frac{-\,f_e^2\,\mu_e^2}{6(kT)^2} \tag{48}$$

if the transition moment m_e and the fluorescent excited state dipole moment μ_e are perpendicular.

Hence, in a homogeneous part of the fluorescence or the absorption spectrum that quantity Δ from eq.(45) to eq.(48) should be constant and have the right sign for which the assumed symmetry condition holds true.

Examples will be given in the results section.

2.5 Dielectric loss measurements

Although dielectric loss experiments are not available in the authors´ laboratory, this method is described here since it is an independent method to determine excited state dipole moments. It is based on the complex dielectric constant ε^*

$$\varepsilon^* = \varepsilon - i\,\varepsilon'' \tag{49}$$

ε is the relative permittivity (as above) and ε'' is the dielectric loss factor which can be determined measuring the conductivity σ which is related to ε'' through

$$\sigma = \omega\,\varepsilon_o\,\varepsilon'' \tag{50}$$

Usually, the dielectric loss tangent $\tan\Theta$ is considered which is defined as

$$\tan\Theta = \varepsilon''/\varepsilon \tag{51}$$

For a dilute solution of one solute species in the given state g with dipole moment μ_g the loss tangent is given by Debye´s theory [16]

$$\tan\Theta = \frac{(\varepsilon + 2)^2}{\varepsilon\,\varepsilon_o}\,\frac{1}{27}\,\beta\,\mu_{go}^2\,\frac{\omega\,\tau_r}{1 + (\omega\,\tau_r)^2}\,N_g. \tag{52}$$

N_g is the number density of the solute in the ground state. If part of the solutes are pumped from the ground state g to one excited state e by strong laser excitation [17] the number density of molecules in the ground state decreases by the number density N_e of the excited state molecules. Hence the total loss tangent will be a sum of that due to the remaining ground state species and to the new excited state species. Hence, the difference $\Delta\tan\Theta$ of the loss tangent observed with the pumping process will be

$$\Delta\tan\Theta = \frac{(\varepsilon + 2)^2}{\varepsilon\,\varepsilon_o}\frac{1}{27}\,\beta\,(\mu_{eo}^2 - \mu_{go}^2)\,\frac{\omega\tau_r}{1 + (\omega\tau_r)^2}\,N_e \qquad (53)$$

Here it was assumed that the Debye relaxation time τ_r has not changed with the excitation and the Lorentz field is used. N_e can be determined from the known absorbed laser power. Warman and de Haas [17,18] have worked out the method for the determination of excited state dipole moments even when the systems under consideration are not simple two state systems. They observe the time dependence of the microwave conductivity change after a short laser excitation pulse and therefore have called the method time resolved microwave conductivity technique (TRMC). Nice results from this method have been reported for rigid molecules [19] and for systems related to the TICT phenomenon [20,21]. The importance of this method for the present communication is that it is a method completely independent from EOEM (or EOAM) by which excited state dipole moments can be determined, too, not only estimated as it is the case with solvent shift measurements.

3. EXPERIMENTAL RESULTS AND DISCUSSION

3.1 4-Dimethylamino-4′-nitrostilbene (DMANS)

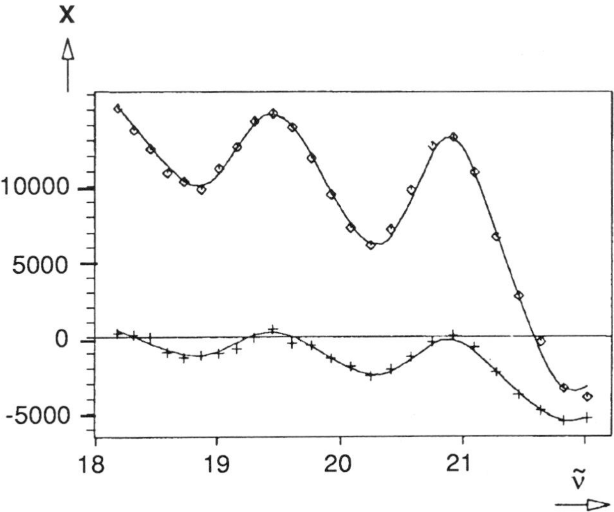

Figure 2 Values $X(\tilde{\nu},\phi)/10^{-20}V^{-2}m^2$ determined from EOEM on DMANS in cyclohexane at 298 K, plotted against the wavenumber $\tilde{\nu}$ $/10^5m^{-1}$. Solid line : fit according to eq. (31)

This molecule has been often investigated and thus is well suited for the evaluation of new methods or for an overall check of the used instrumentation. Here new results are presented from spectrally resolved EOEM of DMANS in cyclohexane. The experimental procedure was as described in the previous section and a regression analysis according to eq. (31) yielded values for the coefficients ^eE to ^eI, given in table 1. Fig.2 shows the original data $X(\tilde{\nu},\phi)$ and the fit according to eq. (31).
Since ^eF = ^eG and ^eH = ^eI, and since ^eE - 2^eD is large and positive, the conclusion can be drawn that $m_e \,//\, \mu_e \,//\, \Delta^f\mu$, which is in agreement with symmetry demands. Then with the dipole approximation discussed in the section before the dipole moment values listed in the lower part of table 1 result. They agree well with literature values [22,17].

Table 1
Results from EOEM on DMANS in cyclohexane at 298 K, λ_{exc} = 405 nm

^eE / $10^{-20}V^{-2}m^2$		122000 ± 1300
^eF / $10^{-40}CV^{-1}m^2$		17400 ± 500
^eG / $10^{-40}CV^{-1}m^2$		17700 ± 500
^eH / 10^{-60} C^2m^2		6300 ± 600
^eI /10^{-60} C^2m^2		6000 ± 600
μ_e / 10^{-30}Cm	(^eE)	84.7 ± 0.4
$\mu_e \Delta^f\mu$ / 10^{-30}Cm	(^eF,^eG)	5000 ± 150
$\Delta^f\mu$ / 10^{-30}Cm	(^eE, ^eF, ^eG)	59 ± 2
$\Delta^f\mu$ / 10^{-30}Cm	(^eH, ^eI)	65.6 ± 3
μ_g^{FC} / 10^{-30}Cm	(^eE, ^eF, ^eG)	25.7 ± 2.5
μ_g^{FC} / 10^{-30}Cm	(^eE, ^eH, ^eI)	19.1 ± 3.5

To show the effect of the solvent on the excited state dipole moment μ_e of DMANS in its first excited singlet state, IEOEM have been performed in dioxane and fluorobenzene. The results are given in table 2 and show some dependence of the excited state dipole moment from the solvent polarity. The results are in sufficient agreement with the results from TRMC investigations on DMANS in dioxane [17] where 79 10^{-30}Cm have been reported.

Table 2
Results from IEOEM on DMANS in three solvents at 298 K (* values taken from table 1)

	cyclohexane*	dioxane	fluorobenzene
^eE / $10^{-20}V^{-2}m^2$	122000 ± 1300	151000 ± 3000	180000 ± 3000
μ_e / 10^{-30}Cm	84.7 ± 0.4	92.6 ± 1.0	89.9 ± 0.7

3.2 A non-conjugated bichromophoric donor-acceptor-molecule

Although the results from EOEM on DMANS already show that this method yields reliable results, a compound was choosen that had been studied before by TRMC measurements by Mes et al. [23] and which consists of two non-conjugated chromophors. This compound is shown in fig.3. A strong photo-induced charge transfer had been observed from solvent shift measurements [23,24] and was confirmed in agreement with results from the TRMC technique.

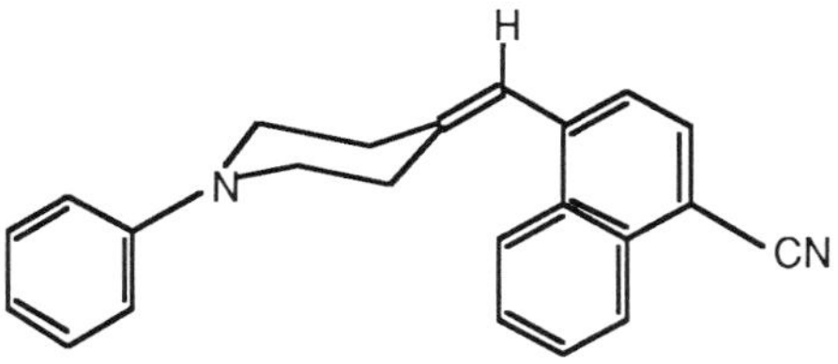

Figure 3. The bichromophoric compound

EOEM have been performed in cyclohexane, benzene and dioxane and IEOEM in cyclohexane and dioxane. The resulting values for ${}^eE$ to ${}^eG$ are presented in table 3; values for ${}^eH$ and ${}^eI$ could not be determined since they only amount to some percent to the measured effect.

Table 3
Results from EOEM and IEOEM on the bichromophoric compound shown in figure 2, in three solvents, at 298 K. * values from IEOEM; ** values from ref. [23]

	cyclohexane	benzene	dioxane
${}^eE$ / $10^{-20}V^{-2}m^2$	70000 ± 2000	122000 ± 2000	115000 ± 3000
${}^eE$ / $10^{-20}V^{-2}m^2$ *	69000 ± 2000		115000 ± 2000
${}^eF$ / 10^{-40} $CV^{-1}m^2$	16000 ± 1000	21000 ± 1000	20000 ± 1000
${}^eG$ / 10^{-40} $CV^{-1}m^2$	10500 ± 1000	23000 ± 1000	21500 ± 1000
μ_e/ 10^{-30}Cm	67.5 ± 2.1	79.5 ± 1.8	74.3 ± 1.9
μ_e / 10^{-30}Cm **	103.4	80.1	83.4
$m_e\mu_e$ / 10^{-30}Cm	65.2 ± 2.4	80.7 ± 1.1	78.7 ± 1.4
$m_e\Delta^f\mu$ / 10^{-30}Cm	45.9 ± 2.2	78.2 ± 3.4	75.5 ± 3.6

It is assumed for further evaluation that $\mu_e = \Delta^f\mu$, since μ_g^{FC} is small and moreover includes an angle with $\Delta\mu$ considerable different from zero. Then eqs.(34) and (36) can be rearranged to yield

$$\mu_e = \frac{1}{f_e}\sqrt{\frac{F}{\beta}}, \tag{54}$$

$$m_e\mu_e = \frac{1}{f_e}\sqrt{\frac{1}{3}\left[\frac{E}{\beta^2} + \frac{F}{\beta}\right]} \tag{55}$$

and

$$m_e\Delta^f\mu = \frac{1}{f_e} \frac{\sqrt{3}\ G}{\sqrt{E + \beta F}} \tag{56}$$

With these equations the values for the dipole moment terms in table 3 have been calculated. From the fact that in dioxane and benzene solutions μ_e, $m_e\mu_e$ and $m_e\Delta^f\mu$ are in very good agreement follows that m_e, μ_e and $\Delta^f\mu$ are parallel and supports the assumption $\mu_e = \Delta^f\mu$ used to derive these terms. The value found for the excited state dipole moment μ_e in dioxane and benzene is in very good agreement with the results presented by Warman et al. [23] determined from TRMC measurements. Since they used the Lorentz field their values have been recalculated in order to be comparable. The values determined in cyclohexane are not in agreement and furthermore $m_e\Delta^f\mu$ is not equal to $m_e\mu_e$. The reasons may be various. They are discussed in detail in [25].
As a final remark, the overall good agreement between the results from TRMC and EOEM on this non-conjugated bichromophoric compound is quite encouraging, since it shows that with uncomplecated systems at least reliable results can be achieved from both methods.

3.3 Coumarin 47 and coumarin 102

Some coumarins have been object of investigations as potential candidates of spontaneous TICT state formation after excitation [26,27,28]. In the present study, the Laser dyes Coumarin 47 and 102 have been investigated by electro optical absorption and emission measurements. In the coumarin 102 the diethylamino group is forced to a near co-planar structure whereas in coumarin 47 the diethylamino group is free. Table 4 shows the results from EOAM, EOEM and IEOEM.

Table 4
Results from EOAM, EOEM and IEOEM on the coumarin dyes 47 and 102 determined in dioxane at 298 K. * from IEOEM

	47	102
${}^eE$ / $10^{-18}V^{-2}m^2$	290 ± 60	290 ± 60
${}^eE$ / $10^{-18}V^{-2}m^2$ *	350 ± 10	350 ± 10
${}^eF$ / $10^{-38}CV^{-1}m^2$	27 ± 2	25 ± 1.5
${}^eG$ / $10^{-38}CV^{-1}m^2$	25 ± 2	25 ± 1.5
${}^aE$ / $10^{-18}V^{-2}m^2$		73 ± 1
${}^aF$/ $10^{-38}CV^{-1}m^2$		22.4 ± 0.4
${}^aG$ / $10^{-38}CV^{-1}m^2$		22.0 ± 0.4
μ_e / 10^{-30} Cm	**40.6** ± 4	**40.6** ± 4
$\Delta^f\mu$ / 10^{-30}Cm	17.7 ±1.5	17 ± 2
μ_g^{FC} / 10^{-30}Cm	**22.9** ± 3	**23.6** ± 3
μ_g / 10^{-30} Cm		**20.4** ± 0.1
$\Delta^a\mu$ / 10^{-30}Cm		30 ± 0.6
μ_a^{FC} / 10^{-30}Cm		**50.4** ± 1

In [26] Rettig and Klock have claimed that in 6-aminocoumarin a TICT state emits in alcohols, and even part of the fluorescence in non-polar solvents is ascribed to TICT state fluorescence, quite different from 7-aminocoumarin. The 7-aminocoumarin derivates investigated her by EOEM both show μ_e = 40.6 10^{-30} Cm and both show the same value for μ_g^{FC}, which is roughly the same as μ_g. Hence, these results completely rule out a TIICT state formation after excitation in coumarin 47, in agreement with arguments of Rettig and Klock [26]. The FC-excited state dipole moment seems to be larger than that of the equilibrated excited state. Although this might happen if the FC-excited state is far from equilibrium, one must have in mind that the discussion here is on dipole moment terms μ that might differ from μ_o values through solvent induced parts in dioxane (seefore eqs. (39) to (41)).

3.4 6-propionyl-2-(dimethylamino)naphthalene (PRODAN)

PRODAN was studied by electro optical measurements in order to elucidate the nature of the fluorescence of this compound the dimethylamino group of which was assumed to undergo twisting after excitation [29]. The results from EOEM and EOAM on PRODAN in cyclohexane and dioxane are shown in table 5.

Table 5
Results from EOAM, EOEM and IEOEM on PRODAN, determined in cyclohexane and in dioxane at 298 K. * from IEOEM

	cyclohexane	dioxane
${}^eE$ / $10^{-18}V^{-2}m^2$		260 ± 20
${}^eE$ / $10^{-18}V^{-2}m^2$ *	280 ± 30	330 ± 20
${}^eF$ / $10^{-38}CV^{-1}m^2$		41 ± 2
${}^eG$ / $10^{-38}CV^{-1}m^2$		37 ± 2
${}^aE$ / $10^{-18}V^{-2}m^2$		44 ± 1
${}^aF$/ $10^{-38}CV^{-1}m^2$		20 ±0.7
${}^aG$ / $10^{-38}CV^{-1}m^2$		20 ±0.7
μ_e / 10^{-30} Cm	40.6 ± 1*	**41.3** ± 2
$\Delta^f\mu$ / 10^{-30}Cm		26.1 ± 3
μ_g^{FC} / 10^{-30}Cm		**15.2** ±3.5
μ_g / 10^{-30}Cm		**15.8** ±0.2
$\Delta^a\mu$ / 10^{-30}Cm		35 ±1.5
μ_a^{FC} / 10^{-30}Cm		**50.8** ±1.5

Since ${}^eF \approx {}^eG$, it may be assumed that $m_e // \mu_e // \Delta^f\mu$ and it follows the value for μ_e from ${}^eE$ according to eq.(34) and with this μ_e and the average value of ${}^eF$ and ${}^eG$ a value for $\Delta^f\mu$. Combining both yields μ_g^{FC}. A completely analoguous procedure applies to the results from EOAM. Here, the spectral range from 376 to 324 nm was found to represent a transition with uniform direction of the transition moment, whereas the red edge of the absorption spectrum down to 400 nm shows decreasing values for $\Delta_{||}(\tilde{\nu})$ with encreasing wavelength: its value falls from 730 10^{-20} $V^{-2}m^2$ in the uniform part of the absorption to 155 10^{-20} $V^{-2}m^2$ at 400 nm. From $\Delta_{a||}(\tilde{\nu})$ = 730 10^{-20} $V^{-2}m^2$ follows with eq.(45) μ_g = (15.8 ± 0.1) 10^{-30}Cm, as resulted from the regression analysis (see table 5). The terms ^eH and ^eI as well as ^aH and ^aI did not amount siginificantly to the electro optical effect. Both the measured dipole moments and the finding of a superposition of transitions at the red edge of the absorption of PRODAN will be studied and compared with results from [30] in more detail in a forthcoming paper [31].

3.5 4-cyanobenzoquinuclidine (CBQ)

This compound is perhaps the key compound to the discussion on dipole moments of twisted compounds in the class of derivates of p-dimethylaminobenzonitril. The structure shows that the dimethylamino group is twisted out of the phenyl plane and is rigidly fixed in a perpendicular position with respect to the phenyl plane. With this model compound for a TICT state geometry all results on dipole moments and the direction of the transition moments should be highly valuable.

The compound was synthesized basically following the procedure described in [32]. So far, solvent shift measurements of the fluorescence and EOAM have been performed. Fig. 4 shows the absorption spectrum of CBQ, its fluorescence spectrum and its absorption anisotropy spectrum induced by an external electric field in cyclohexane.

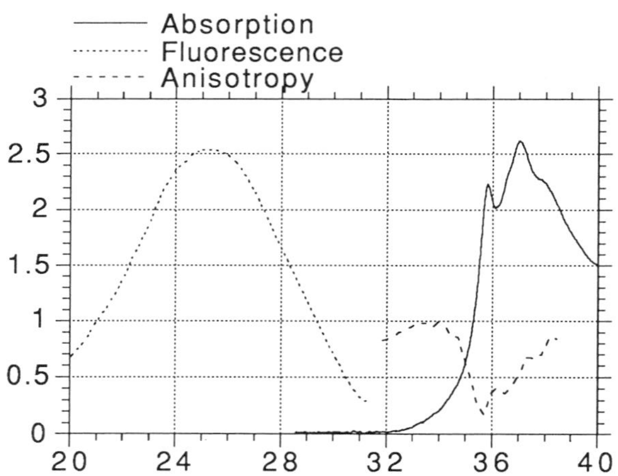

Figure 4. Absorption, Fluorescence and Absorption anisotropy ($= \Delta_{a||}(\tilde{\nu})$) spectra (all arbitrary units) of CBQ in cyclohexane, at 298 K, plotted against the wavenumber $\tilde{\nu}$ / $10^5 m^{-1}$

The absorption spectrum does not change its shape very much nor shifts much with solvent polarity. The plot of $\Delta_{a||}(\tilde{\nu})$ clearly reveals that there is a homogeneous spectral range at the red edge of the absorption where $\Delta_{a||}(\tilde{\nu})$ is found constant and positiv, indicating $m_a \parallel \mu_g$. The same constant value is achieved around the maximum of the absorption spectrum, but around the small peak at 278 nm $\Delta_{a||}(\tilde{\nu})$ drops considerably, indicating a transition moment perpendicular to the dipole moments. The results from a quantitative analysis of $\Delta_{a||}(\tilde{\nu})$ at the red edge of the absorption in cyclohexane and in dioxane are presented in table 6.

Table 6
Results from EOAM in the red edge of the absorption band of CBQ in cyclohexane and in dioxane at 298 K.

	cyclohexane	dioxane
$\Delta_{a\|\|}(\tilde{\nu})$ / 10^{-20} $V^{-2}m^2$	420 ± 20	540 ± 30
μ_g / 10^{-30}Cm	12.2 ± 0.3	13.6 ± 0.3

The ground state dipole moment is found to be much less than that of N,N-dimethylaminobenzonitril, the structure of which is assumed not to be far away from planarity. This red edge absorption polarized parallel to the ground state dipole moment may be assigned as to the forbidden transition to a TICT state which could well be the energetically lowest excited state in CBQ, even in non-polar solvents. This assignment is further supported by the fact that the fluorescence band in fig. 4 shows a slight overlap with the red edge slope of the absorption spectrum, in the nonpolar cyclohexane. It is also in agreement with results from standard fluorescence anisotropy measurements in sufficiently viscous low temperature solutions [33].

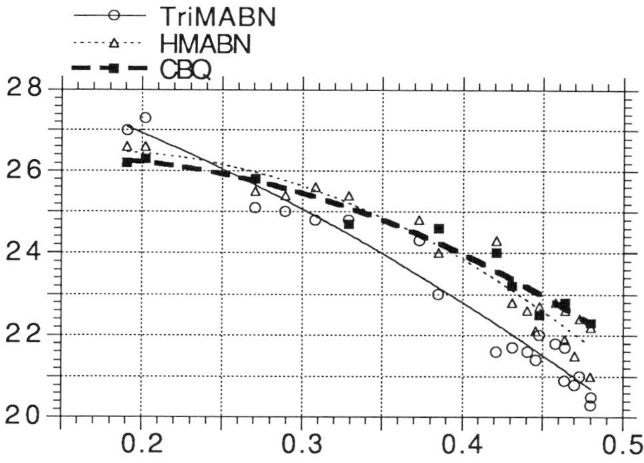

Figure 5. The fluorescence wavenumber $\tilde{\nu}_{f,max}$ plotted versus the solvent polarity parameter g, for CBQ compared with TriMABN and HMABN

Fig. 5 shows the results from solvent shift measurments on CBQ compared to the solvent induced shift of the fluorescence of 2,N,N-trimethylaminobenzonitril (TriMABN) and 2,3,5,6,N,N-hexamethylaminobenzonitril (HMABN). All three plots are siginificantly non-linear and show an encreasing overall slope with encreasing flexibility, but all three show the same near linear slope of the fitted curves in the solvent polarity range g >0.35 consistent with solvent assisted stabilization of the emitting TICT state.

Detailed electro optical investigations on the fluorescence of CBQ are under way, but turn out to be difficult due to its low fluorescence quantum efficiency.

4. REFERENCES

1 W. Baumann in: Physical Methods of Chemistry, second edition, Vol III part B, eds. B.W. Rossiter and J.F. Hamilton, J. Wiley & Sons, New York 1989

2 L. Onsager, J. Amer. Chem. Soc. **58**, 1486 (1936)

3 N. Mataga, Y. Kaifu, and M. Koizumi, Bull. Chem. Soc. Jpn., **29**, 465 (1956)

4 E. Lippert, Z. Naturforsch., **10a**, 541 (1955)

5 W. Liptay, Z. Naturforsch., **20a**, 272 (1965)

6 W. Rettig, J. Mol. Struct., **84**, 303 (1982)

7 W. Liptay in: Excited States, Vol.1, ed. E.C. Lim, Academic Press, New York 1974

8 W. Liptay, Z. Naturforsch. **20a**, 1441 (1965)

9 W. Kuhn, H. Dührkop, and H. Martin, Z. Phys. Chem. Abt. B, **45**, 121 (1940)

10 J. Czekalla, Chimia, **15**, 26 (1961)

11 H. Labhart, Chimia, **15**, 20 (1961)

12 W. Baumann and H. Deckers, Ber. Bunsenges. phys. Chem. **81**, 786 (1977)

13 W. Liptay, Z. Naturforsch., **18a**, 705 (1963)

14 W. Baumann and H. Bischof, J. Mol. Struct., **84**, 181 (1982)

15 W. Baumann and H. Bischof, J. Mol. Struct., **129**, 125 (1985)

16 P. Debye, Physik. Z., **35**, 101 (1934)

17 M.P. de Haas and J.M. Warman, Chem. Phys., **73**, 35 (1982)

18 J.M. Warman, M.P. de Haas, A. Hummel, C.A.G.O. Varma, and P.H.M. van Zeyl, Chem. Phys. Lett., **87**, 83 (1982)

19 J.M. Warman, M.P. de Haas, M. N. Paddon-Row, E. Cotsaris, N. S. Hush, H. Oevering, and J. W. Verhoeven, Nature (London), **320**, 615 (1986)

20 R. J. Visser, P. C. M. Weisenborn, C. A. G. O. Varma, M. P. de Haas, and J. M. Warman, Chem. Phys. Lett., **104**, 38 (1984)

21 R. J. Visser, P. C. M. Weisenborn, P. J. M. van Kan, B. H. Huizer, C. A. G. O. Varma, J.M. Warman, and M.P. de Haas, J. Che. Soc. Faraday Trans. 2, **81**, 689 (1985)

22 W. Baumann, H. Deckers, K.-D. Loosen, and F. Petzke, Ber. Bunsenges. Phys. Chem., **81**, 799 (1977)

23 G.F. Mes, B. de Jong, H.J. van Ramesdonk, J.W. Verhoeven, J.M. Warman, M.P. de Haas and L.E.W. Horsman-van den Dool, J. Am. Chem. Soc **106**, 6524 (1984)

24 R. M. Hermant, N. A. C. Bakker, T. Scherer, B. Krijnen and J. W. Verhoeven, J. Amer. Chem. Soc. **112,** 1214 (1990)

25 Silvana V. Rodrigues , A. K. Maiti, H. Reis, and W. Baumann. Mol. Phys., accepted for publication

26 W. Rettig and A. Klock, Can. J. Chem., **63**, 1649 (1985)

27 G. Jones II, W. R. Jackson, and A. M. Halpern, Chem. Phys. Lett., **72**, 391 (1980)

28 G. Jones II, W. R. Jackson, C. Choi, and W. R. Bergmark, J. Phys. Chem., **89**, 294 (1985)

29 P. Ilich and F. G. Prendergast, J. Phys. Chem., **93**, 4441 (1989)

30 W. Nowak, P. Adamczak, A. Balter, A. Sygula, J. Mol. Struct., **139**, 13 (1986)

31 W. Baumann, Z. Nagy, and W. Nowak, to be published

32 A. Krówczyński, Pol. J. Chem. **58**, 933 (1984)

33 K. Rotkiewicz and W. Rubaszewska, Chem. Phys. Lett., **70**, 444 (1980)

DISCUSSION

Barbara

It is interesting that the emission and absorption properties of CBQ are analogous to that of ADMA. In particular, in polar solvents only a CT fluorescence is observed and the absorption has a red "tail" which is a consequence of the state ordering, i.e. S_1 is primarily of CT character, while S_2 is primarily of LE character. Our recent experiments on ADMA show that the S_1->S_2 process is very rapid for ADMA and irreversible since the S_2/S_1 gap > k_BT. We would predict that CBQ should have similar photodynamics to ADMA, i.e. a very rapid <100 fs electron transfer component due to vibrationally promoted internal conversion. It would be interesting to test this prediction experimentally.

Baumann

I follow your speculations and think that it is worth testing your prediction, as far as CBQ is concerned. However, the situation seems to be more complicated with CBQ, since our electric field induced anisotropy measurements seem to indicate three partially overlapping transitions within the first absorption band from 32 to 40×10^3 cm^{-1}.

Hynes

For coumarin 102, you indicated that the equilibrium μ_e was less than the μ_e^{FC}, and suggested some possible reasons for this. I would like to suggest another possibility, connected to some theoretical efforts with Dr. H. Kin in our group. This is a change in the electronic structure of the excited state as solvation proceeds from the FC state to the equilibrium state. Another consequence of this, according to the theory, is that the oscillator strength will deviate significantly from the conventional first power frequency dependence. Have you measured the oscillator strength frequency dependence?

Baumann

In my talk I made a short remark that we neither know anything about possible differences of μ_e and μ_e^{FC} (better say μ_{e0}^{FC}) nor what would oppose such differences. I also mentioned that the absolute error of all our μ is around 6%. Adding the given statistical error our experimental values for μ_e and μ_e^{FC} differ just outside the error bar. Hence I take our results as a hint (and no more) to such differences and therefore think that your mentioned efforts on the theoretical side of view are of high interest.

What concerns the oscillator strength we did not measure it. Accepting your remark one could do this with coumarin 102 with the absorption spectrum in different solvents.

Kakitani

It seems to me very nice that the absolute values of dipole moments in the ground and excited states of dye were determined by carefully considering the electronic polarizability effect. Now, I want to clarify one point. How much are the contributions of electronic polarizabilities in the experimentally obtained dipole moments? If those contributions are significant, one should be careful in discussing the magnitude of charge shift in the excited state. That is, one should derive the dipole moments in free states by extracting contributions of the electronic polarizabilities from the observed dipole moments.

Baumann

The complete theory of EOEM (and Liptay's EOAM as well) includes not only $(1-f\alpha_{e0})$ terms, which I discussed here, but also explicit α–terms which I omitted here for the sake of clarity. From electro optical measurements in frozen solution and from such measurements on compounds which do not have a permanent dipole moment in liquid solutions we know that electronic polarizabilities may be pretty large. For example, we published results from such measurements in frozen solutions and in liquid solutions on DMANS already in 1977. Therefore, our experience shows that with such very large dipole moments explicit polarizability terms may be well neglected. A quite different question is that for the terms $f\alpha_{e0}$. In some of our papers I have shown that the observed solvent dependent value $\mu_e=\mu_{e0}\ (1-f\alpha_{e0})^{-1}$ can be interpreted by large polarizability densities α_{e0}/a^3 (a=Onsager radius). But at the same time we have often said that this polarizability must not be considered as electronic polarizability in the case of non–rigid molecules, in particular not with TICT forming species, where the whole electronic structure is changed due to drastic solvent induced changes of the nuclear framework.

Jortner

A crucial ingredient for obtaining microscopic information from these significant experiments rests on a proper description of the inner field corrections. These are still described in terms of continuum dielectric models, which require modification because of (at least) two reasons: (1) The microscopic description of the short range solute–solvent interactions cannot rest on a description of the solute molecule in terms of a point dipole. (2) The Onsager cavity picture may be oversimplified. It is an open question whether a real or a "virtual" cavity should be used to accommodate the solute molecule. This problem was considered recently in the context of medium effects on radiative lifetimes of a guest molecule in condensed phase and in clusters, where it was found that the "virtual" cavity description seems to be appropriate. It will be extremely important to quantify the description of inner filed effects.

Baumann

In my talk I emphasized on the necessity of applying an appropriate model for the description of the internal field. Therefore I completely agree. In addition, I should like to draw your attention to a proposed procedure just in line with your recommendation.

Some years ago, Rettig introduced a model in which he treated the interaction of the solutes charge distribution (calculated by quantum mechanical model calculations) with the surrounding solvent as a homogeneous dielectric. The interesting result is that his treatment comes out with a final formula that precisely resembles the Onsager reaction field description, with the improvement that the a^3 (a=Onsager radius) in his model is replaced by a constant, calculated through the procedure sketched above. Its advantage obviously is that a^3 in Rettig's treatment has got more physical life and that the choice of an appropriate value for a^3 has got less arbitrary.

Dynamics and Mechanisms of
Photoinduced Transfer and Related Phenomena
N. Mataga, T. Okada and H. Masuhara (Editors)

Inter- and Intramolecular Exciplexes studied by Single Photon Timing and Laser Induced Optoaccoustic Spectroscopy

Frans C. De Schryver*, Mark Van der Auweraer, Noël Boens, Mostafa M.H. Khalil, Nicole Helsen, Philippe Van Haver, Kaoru Iwai#

Chemistry Department KULeuven, Celestijnenlaan 200 F, B-3001 Heverlee, Belgium and Nara Women University, Japan.

Abstract

Although the intermolecular formation of exciplexes between pyrene and aliphatic amines was investigated by stationary fluorescence experiments, the low efficiency of this quenching process did not allow determination of the different kinetic parameters. Using global analysis and compartimental analysis, which increased the accuracy and model testing capacity of single photon timing, it was possible to determine the kinetic parameters for the intermolecular exciplex formation between 1-methylpyrene and triethylamine in toluene at room temperature.

Also for the intramolecular exciplex formation of 3-(1-pyrenyl)-(2-indolyl)propane in isooctane global and compartimental analysis of the fluorescence decays at different wavelengths allowed an accurate determination of the rate constants of the processes involved in the exciplex formation. While the single curve analysis of the fluorescence decays of 3-(1-pyrenyl)-(3-indolyl)propane at different wavelengths suggested a biexponential decay, global analysis of the decay data surface indicates the presence of three species on the potential energy surface of the excited singlet state. The wavelength dependence of the contribution of the three components of the decay suggests the formation of two different exciplexes.

The use of laser induced optoaccoustic spectroscopy allowed to determine in a quantitative way the relative contribution of internal conversion and intersystem crossing in the nonradiative decay of intramolecular exciplexes between pyrene and indole. The relative increase of the internal conversion observed upon increasing the solvent polarity could be interpreted in the framework of the Marcus theory.

Introduction

Exciplex formation between pyrene and aliphatic[1,2] or aromatic amines[3,4,5] has received substantial attention. Intermolecular exciplex formation, involving two singlet excited species, can be described by following kinetic scheme[6,7].

$$A \xrightarrow{h\nu} A^* \qquad \text{excitation of } A^*$$

A^*	$\xrightarrow{k_{fle}}$	$A + h\nu_{le}$	fluorescence of the locally excited state
A^*	$\xrightarrow{k_{iscle}}$	A^3	intersystem crossing of the locally excited state
A^*	$\xrightarrow{k_{icle}}$	A	internal conversion of the locally excited state
$A^* + D$	$\xrightarrow{k_{21}[D]}$	$(AD)^*$	exciplex formation
AD	$\xrightarrow{h\nu}$	$(AD)^*$	excitation of AD
$(AD)^*$	$\xrightarrow{k_{12}}$	$A^* + D$	exciplex dissociation
$(AD)^*$	$\xrightarrow{k_{fe}}$	$A + D + h\nu_e$	exciplex fluorescence
$(AD)^*$	$\xrightarrow{k_{isce}}$	$A^3 + D$	intersystem crossing for the exciplex
$(AD)^*$	$\xrightarrow{k_{ice}}$	$A + D$	internal conversion for the exciplex

$$k_{01} = k_{fle} + k_{iscle} + k_{icle} \qquad (1)$$

$$k_{02} = k_{fe} + k_{isce} + k_{ice} \qquad (2)$$

$$r = k_{ice}/(k_{isce} + k_{ice}) \qquad (3)$$

The time dependence of the concentration of A^* and $(AD)^*$ is given by:

$$[A^*(t)] = \beta_{11}\exp(\gamma_1 t) + \beta_{12}\exp(\gamma_2 t) \qquad (4)$$

$$[(AD)^*(t)] = \beta_{21}\exp(\gamma_1 t) + \beta_{22}\exp(\gamma_2 t)] \qquad (5)$$

with

$$\beta_{11} = \{[A^*(0)]\,(\gamma_2 + X_1) - [(AD)^*(0)]\,k_{12}\}/(\gamma_2 - \gamma_1) \qquad (6)$$

$$\beta_{12} = -\{[A^*(0)]\,(\gamma_1 + X_1) - [(AD)^*(0)]\,k_{12}\}/(\gamma_2 - \gamma_1) \qquad (7)$$

$$\beta_{21} = \{[(AD)^*(0)]\,(\gamma_2 + X_2) - [A^*(0)]\,k_{21}[D]\}/(\gamma_2 - \gamma_1) \qquad (8)$$

$$\beta_{22} = - \{[(AD)^*(0)] (\gamma_1 + X_2) - [A^*(0)] k_{21}[D]\} / (\gamma_2 - \gamma_1) \quad (9)$$

$$\gamma_{1,2} = - ½ \{(X_1 + X_2) \mp [(X_1 - X_2)^2 + 4[D]k_{21}k_{12}]^{1/2}\} \quad (10)$$

$$X_1 = k_{01} + k_{21}[D] \quad (11)$$

$$X_2 = k_{02} + k_{12} \quad (12)$$

$[A^*(0)]$ and $[(AD)^*(0)]$ denote the concentration of A^* and $(AD)^*$ at time zero, respectively. When $[(AD)^*(0)]$ can be neglected, equations (6) -(9) can be simplified to

$$\beta_{11} = [A^*(0)](\gamma_2+X_1)/(\gamma_2 - \gamma_1) \quad (13)$$

$$\beta_{12} = -[A^*(0)](\gamma_1+X_1)/(\gamma_2 - \gamma_1) \quad (14)$$

$$\beta_{21} = -[A^*(0)]k_{21}[D]/(\gamma_2 - \gamma_1) \quad (15)$$

$$\beta_{22} = [A^*(0)]k_{21}[D]/(\gamma_2 - \gamma_1) \quad (16)$$

Due to the overlap between the emission[9] of the exciplex and the locally excited state, the fluorescence δ-response function at an emission wavelength λ_{em} will be a linear combination of equations (4) and (5).

$$f(\lambda_{em},t) = c_1[A^*(t)] + c_2[(AD)^*(t)]$$

$$f(\lambda_{em},t) = [c_1\beta_{11}+c_2\beta_{21}]\exp(\gamma_1 t)+[c_1\beta_{12}+c_2\beta_{22}]\exp(\gamma_2 t) \quad (17)$$

or

$$f(\lambda_{em},t) = \alpha_1\exp(\gamma_1 t)+\alpha_2]\exp(\gamma_2 t) \quad (18)$$

In equation (17) c_1 and c_2 are a function of the emission wavelength λ_{em}. The pre-exponential factors α_1 and α_2 have contributions from the rate constants, the factors c_1 and c_2 describing the overlap of the emission spectra of the exciplex and the locally excited state and the initial concentrations $[A^*(0)]$ and $[(AD)^*(0)]$.

For intermolecular exciplex formation the rate constants k_{01}, k_{12}, k_{21} and k_{02} can be determined from γ_1 and γ_2 as a function of concentration using equation (19) and (20).

$$- (\gamma_1 + \gamma_2) = k_{01} + k_{02} + k_{12} + k_{21}[D] \quad (19)$$

$$\gamma_1\gamma_2 = k_{01}(k_{02} + k_{12}) + k_{02}k_{21}[D] \quad (20)$$

When - ($\gamma_1 + \gamma_2$) and $\gamma_1\gamma_2$ are plotted versus the donor concentration a linear relationship must be obtained. From the slopes values for k_{21} and $k_{02}k_{21}$ can be determined. The ratio of the slopes yields k_{12}. From the intercept of equation (19) the sum ($k_{12} + k_{01}$) can be calculated. The intercept of equation (20) yields a nonlinear expression in $k_{01}k_{12}$ and therefore it is possible to obtain two physically acceptable sets of values for k_{12} and k_{01}. Therefore, it is recommended to derive k_{01} independently from the decay of A^* in the absence of D.

The correlation[8] between the decay times and preexponential factors observed in the analysis of the fluorescence decay of A* limits severely the accuracy and precision with which the decay parameters and therefore the rate constants can be obtained. The accuracy and precision with which the decay parameters can be determined also reduces the ability to discriminate between the kinetic scheme shown here and kinetic schemes that are more complex due to e.g. the formation of more than one complex[10]. This problem can be reduced by global analysis[11] according to equation 17 where fluorescence decays observed at different wavelengths and/or different time scales and characterized by the same decay times are analyzed simultaneously. As this procedure reduces the number of unknown decay parameters that have to be determined, it allows a more accurate determination of the decay parameters γ_1 and γ_2 and therefore also of the rate constants. This more accurate determination of the decay parameters will also lead to a better capacity to discriminate between different models.

Global analysis according to equation 17 does not allow to link fluorescence decays obtained at different concentrations of the donor D because those decays are characterized not only by a different ratio of the preexponential factors but also by different decay times. They are, however, characterized by the same rate constants (k_{01}, k_{12}, k_{21}, and k_{02}). Using compartmental analysis[12,13] which allows to fit directly for the decay rate constants it is possible to analyse simultaneously fluorescence decays obtained at different emission wavelengths and concentrations of D. Using this method it becomes possible to determine accurately the different decay rates and the contributions of both excited species to the emission intensity at each wavelength.

Although the intermolecular formation of exciplexes between pyrene and aliphatic amines was investigated by stationary fluorescence experiments[14], the low efficiency of this quenching process did not allow determination of the different kinetic parameters using single curve analysis. In the present contribution it will be demonstrated that those kinetic parameters can be obtained with increased accuracy and precision using global analysis[11] and compartmental analysis[12].

The kinetic scheme used for intermolecular exciplex formation can also be used for intramolecular exciplex formation when the second order rate constant k_{21} is replaced by a first order rate constant k'_{21}. In that case the rate constant k_{01} can be determined from the fluorescence decay of a model compound without D present. When the ratios $[(AD)^*(0)]/[A^*(0)]$ and c_1/c_2 at a certain wavelength are known the rate constants k_{12}, k'_{21} and k_{02} can be obtained. When the ratio c_1/c_2 is known at two different wavelengths it is possible to determine the rate constants k_{12}, k'_{21} and k_{02} and the ratio $[(AD)^*(0)]/[A^*(0)]$.

It has been demonstrated that the formation of an intramolecular exciplex between aromatic moieties and aliphatic amines is determined[15,16,17] to a large extent by the strength of the Coulomb attraction between the radical ions. To investigate the relevance of this effect, the intramolecular exciplex formation of three isomers of 1-(1-pyrenyl)-3-(indolyl)-propane, where the trimethylene chain was linked to a different position of the indole moiety, was investigated. Our interest in those compounds was increased further by the fact that molecular models suggest that the formation of different intramolecular exciplexes is in principle possible between pyrene and indole linked by a trimethylene chain. Although the formation of two different excimers has been observed on several occasions in bichromophoric systems[10,18], no similar information exists concerning intramolecular exciplex formation.

Although the use of stationary and time-resolved fluorescence spectroscopy allows to analyze the exciplex formation, it is generally difficult to separate the contributions of internal conversion and intersystem crossing to the radiationless decay process of those exciplexes. For some exciplexes intersystem crossing (ISC) from the charge transfer state to the locally excited triplet state has been studied by means of laser photolysis[19-20]. While some experiments suggested that no triplet formation occurred[21], other experiments indicated efficient intersystem crossing[22]. However, Laser Induced Optoacoustic Spectroscopy (LIOAS) is a method suitable to study the radiationless decay processes of excited states[23-27]. The time resolution of the optoacoustic experiment is limited by the response time of the piezoelectric detector, that determines (depending upon the experimental conditions) the heat that is released within about 250 ns after excitation. The amplitude of the optoacoustic deflection is directly proportional to the amount of energy released within these 250 ns. This so called "prompt heat" differs from the "slow heat" which has its origin in the radiationless decay of long living species (triplets, radicals...). The first optoacoustic experiments on systems where intermolecular exciplexes and radical ion pairs were formed were executed by Gould et al.[28] who used LIOAS to study the electron transfer reactions in the Marcus inverted region.[29], In the present contribution the results of optoacoustic and transient absorption measurements of 1-(1-pyrenyl)-3-(2-(N-methyl)indolyl)-propane (1Py2In), 1-(1-pyrenyl)-3-(3-(Nmethyl)indolyl)-propane (1Py3In) and in solvents of different polarity are compared. Those molecules are characterized by an efficient quenching of the locally excited state in different solvents which allows an accurate determination of the relative importance of the different decay channels of the exciplex.

Experimental

Triethylamine (Janssen, p.a.) was purified by distillation under reduced pressure. Methylpyrene was purified by preparative thin layer chromatography on silica using a 1/1 mixture of hexane and methylene chloride as eluent. The synthesis and purification of 1-(1-pyrenyl)-3-(2-(N-methyl)indolyl)-propane (1Py2In) and 1-(1-pyrenyl)-3-(3-(N-methyl)indolyl)-propane (1Py3In) will be described elsewhere[30]. The solvents were of spectroscopic or fluorescence grade and used as received. All solutions used for stationary and time-resolved fluorescence measurements were degassed by freeze-pump-thaw cycles. Argon was bubbled through the solutions used for the LIOAS experiments.

Absorption spectra were recorded with a Perkin Elmer Lambda 6 UV/VIS spectrometer. Corrected fluorescence and excitation spectra were obtained with a SPEX Fluorolog 212. The fluorescence decays were recorded by the single photon timing technique[31]. The global and compartmental analysis of the fluorescence decays were performed as described[11,12,13]. For the LIOAS experiments the experimental set-up described earlier[32] was used. The fraction α of the absorbed laser energy converted into prompt heat is determined by comparing the optoacoustic signal of the sample with that of a solution of a reference compound (2-hydroxy benzophenone) in the same solvent.

Results and Discussion

Intermolecular exciplex formation between 1-methylpyrene and triethylamine.

Upon addition of triethylamine to a solution of 1-methylpyrene ($5.0x10^{-6}$ $molL^{-1}$) in toluene quenching of the structured pyrene fluorescence occurs and the a new structureless fluorescence with a maximum at 520 nm are observed (figure 1). Even at a concentration of 0.107 $molL^{-1}$ triethylamine no change of the absorption spectrum is observed and the excitation spectrum of the structured pyrene emission and the structureless emission are identical. A linear Stern-Volmer plot, characterized by a Stern-Volmer constant of 23.0 $Lmol^{-1}$, is obtained. As the lifetime of the excited singlet state of 1-methylpyrene in toluene amounts to 167 ns[33] in the absence of triethylamine a value of $1.4x10^8$ $Lmol^{-1}s^{-1}$ can be obtained for the quenching rate constant k_q. Global analysis of fluorescence decays obtained at different emission wavelengths allowed to obtain γ_1 and γ_2. By plotting $\gamma_1+\gamma_2$ and $\gamma_1\gamma_2$ versus the concentration of triethylamine the values of k_{21}, k_{12} and k_{02} represented in table I were obtained according to equation (19) and (20).

Table I: Rate constants for exciplex formation between 1-methylpyrene and triethylamine in toluene at 298 K determined by global analysis

$k_{01} = (5.99\pm0.01)x10^6 s^{-1}$ $k_{02} = (2.50\pm0.90)x10^7 s^{-1}$
$k_{21} = (1.77\pm0.45)x10^9 Lmol^{-1}s^{-1}$ $k_{12} = (1.08\pm0.30)x10^8 s^{-1}$

From the values of the rate constants in table 1 a value of $2.0x10^8$ $Lmol^{-1}s^{-1}$ could be calculated for k_q according to equation (21).

$$k_q = (k_{21}k_{02})/(k_{12} + k_{02}) \qquad (21)$$

By simultaneous analysis of 31 decays obtained at different wavelengths and four triethylamine concentrations - including zero - the rate constants shown in table II could be obtained using compartmental analysis. From the rate constants in table II a value of $1.33\pm0.04x10^8$ $Lmol^{-1}s^{-1}$ could be determined for k_q. This value agrees well with the value obtained from the Stern-Volmer plot. The larger precision and the better agreement with the stationary quenching experiments obtained for the results displayed in table II indicate the advantages of compartmental analysis to study intermolecular quenching.

Table II: Rate constants for exciplex formation between 1-methylpyrene and triethylamine in toluene at 298 K determined by compartmental analysis

$k_{01} = (5.97\pm0.01)x10^6 s^{-1}$ $k_{02} = (4.78\pm0.06)x10^7 s^{-1}$

$k_{21} = (6.54\pm0.19)x10^8 Lmol^{-1}s^{-1}$ $k_{12} = (1.86\pm0.06)x10^8 s^{-1}$

Intramolecular exciplex formation of 1-(1-pyrenyl)-3-(indolyl)propanes

As observed for 1-(1-pyrenyl)-(1-indolyl)propane[4] (1Py1In) the emission spectra of 1Py2In and 1Py3In at 298 K consist of a structured band like the emission spectrum of 1-methylpyrene and a bathochromic structureless band (figure 1). The excitation spectra of the structured and structureless emission are identical and resemble at wavelengths larger than 310 nm that of 1-methylpyrene. When the solvent polarity is increased the maximum of the emission (λ_{max}) of the structureless band is shifted to longer wavelengths (table III and IV). This indicates that the structureless emission at longer wavelengths is due to an exciplex. The relative intensity of the exciplex emission increased from 1Py1In over 1Py2In to 1Py3In (figure 1). When the solvent polarity is increased, a more important quenching of the emission of the locally excited state (Φ_{le}) is observed for 1Py1In and 1Py2In. Contrary to observations made for exciplexes with dimethylaniline[21,34] or intramolecular exciplexes with aliphatic amines[1,35,36], no drastic decrease of the fluorescence quantum yield of the exciplex (Φ_{ex}) is observed in solvents of high polarity. In this aspect the data in table III and IV indicate that for 1Py2In and 1Py3In even in polar solvents the formation of a radical ion pair is not an important decay channel of the exciplex. In this aspect the behaviour of 1Py2In and 1Py3In resembles that of 1Py1In[4], indicating that even in polar solvents the exciplex is not converted into a solvent separated ion pair.

By plotting the emission energy at the maximum versus solvent polarity according to the Lippert-Mataga equation[37,38,39] a value of 11500 cm^{-1} and 10000 cm^{-1} could be obtained for the ratio μ^2/ρ^3 for 1Py2In and 1Py3In respectively; μ and ρ are respectively the exciplex dipole moment and the radius of the solvent cavity. Those values are significantly larger than the value of 8800 cm^{-1} obtained for 1Py1In. As the distance between the planes of the pyrene moiety and the indole moiety is the same in the three exciplexes, this would indicate that in 1Py1In a more extensive mixing with the locally excited state occurs, leading to a decrease of the charge transfer character of the exciplex.

Table III: Stationary fluorescence data of 1Py2In at 298 K

solvent	Φ_{le}	Φ_{ex}	λ_{max}
isooctane	0.15	0.43	425 nm
diethylether	0.06	0.48	447 nm
ethylacetate	0.07	0.60	471 nm
acetone	0.02	0.51	487 nm
acetonitrile	0.03	0.28	496 nm

Table IV: Stationary fluorescence data of 1Py3In at 298 K

solvent	Φ_{le}	Φ_{ex}	λ_{max}
isooctane	0.01	0.57	442 nm
diethylether	0.02	0.57	463 nm
ethylacetate	0.06	0.58	474 nm
tetrahydrofuran	0.05	0.50	475 nm
acetone	0.02	0.42	493 nm
acetonitrile	0.01	0.49	515 nm

Using single curve analysis, the fluorescence decay of the emission of 1Py2In in isooctane could be analysed at 298 K as a sum of two exponentials. Fluorescence decays obtained between 378 nm and 510 nm could be analysed simultaneously with common values of the parameters γ_1 and γ_2. On the basis of this analysis, the rate constants in table V could be obtained. Using compartmental analysis similar values (table VI) were obtained for the different rate constants. This suggests that the kinetic schema displayed in the introduction, where the second order rate constant k_{21} is replaced by a first order rate constant k'_{21} can be applied.

Table V: Rate constants for exciplex formation of 1Py2In in isooctane at 298 K determined by global analysis.

$k_{01} = (3.77 \pm 0.01) \times 10^6 s^{-1}$ $\quad$ $k_{02} = (2.4 \pm 7.5) \times 10^7 s^{-1}$

$k'_{21} = (1.56 \pm 0.08) \times 10^8 s^{-1}$ $\quad$ $k_{12} = (4.24 \pm 0.60) \times 10^8 s^{-1}$

$Z\chi_2 = 2.634$

Table VI: Rate constants for the intramolecular exciplex formation in 1Py2In in isooctane at 298 K determined by compartmental analysis assuming that $[(AD)^*(0)]$ and c_2/c_1 at 378 nm equal zero.

$k_{01} = (3.77 \pm 0.01) \times 10^6 s^{-1}$ $\quad$ $k_{02} = (2.39 \pm 0.04) \times 10^7 s^{-1}$

$k'_{21} = (1.62 \pm 0.04) \times 10^8 s^{-1}$ $\quad$ $k_{12} = (4.24 \pm 0.05) \times 10^8 s^{-1}$

$Z\chi_2 = 2.385$

Table VII: Rate constants for the intramolecular exciplex formation in 1Py2In in isooctane at 298 K determined by compartmental analysis assuming that c_2/c_1 at 378 nm and c_1/c_2 at 500 nm equal zero.

$k_{01} = (3.77 \pm 0.01) \times 10^6\ s^{-1}$ $\quad$ $k_{02} = (8.9 \pm 0.85) \times 10^7\ s^{-1}$

$k'_{21} = (3.74 \pm 0.37) \times 10^7\ s^{-1}$ $\quad$ $k_{12} = (4.7 \pm 0.07) \times 10^8\ s^{-1}$

$Z\chi_2 = 2.723$ $\quad$ $[(AD)^*(0)]/[A^*(0)] = -0.208 \pm 0.011$

According to equations 15, 16 and 17 the ratio of the preexponential factors of the decay of the exciplex emission should equal -1 at wavelengths where no overlap with the emission of the locally excited state occurs (c_1 equals zero). For 1Py2In deviations from -1 were observed. It could be possible that those deviations were due to the fact that already in the ground state a fraction of the molecules is in the $g^{\pm}g^{\mp}$ conformation[40,41,42] which would make $[(AD)^*(0)]$ different from zero. Application of the bicompartmental analysis allows to calculate the fraction of the molecules that are already in the ground state in the $g^{\pm}g^{\mp}$ conformation when c_1/c_2 is known at two different emission wavelengths. Assuming that at 378 nm no emission of the exciplex is observed and that at 500 nm no emission of locally excited is observed the ratio $[(AD)^*(0)]/[A^*(0)]$ amounts to -0.208±0.011 (Table VII). This value has no physical meaning and therefore it is more likely that the assumptions concerning the ratio's of c_1/c_2 are incorrect.

Using single curve analysis, the decay of the emission of 1Py3In at 378 nm in isooctane can be described at 209 K by a monoexponential with a decay time of 56.44 ns. Between 410 and 510 nm the decay of the emission can be analyzed as a difference of two exponentials and the ratio of the two preexponential factors changes gradually from -0.62 at 410 nm to -0.97 at 510 nm. The two decay times change gradually from 53.5 ns and 20.9 ns at 410 nm to 55.6 ns and 29.7 ns at 510 nm. If the results would fit the kinetic scheme one would expect the invariance of the decay times as a function of the emission wavelength. However, using single curve analysis the change of the decay times could be due to the limited resolving power of single curve analysis. Using global analysis, however, it is not possible to obtain an acceptable fit to a biexponential function. When decays between 378 nm and 510 nm are analysed simultaneously as a sum of three exponentials with common decay times, a good fit is obtained and the three decay times amount respectively to 20.8 ns, 26.7 ns and 55.9 ns. The ratio of the sum of the negative preexponential factor to the sum of positive preexponential factors changes gradually from -0.717 at 410 nm to -0.975 at 510 nm. Between 430 nm and 480 nm the ratio of the negative preexponential factors changes from 5.5 to 0.125. At 378 nm the decay remains monoexponential indicating that the exciplex formation is not reversible. The monoexponential decay at 378 nm also excludes a modification of the kinetic schema due to the presence of different starting conformations that form the exciplex with a different rate[43]. These results can be rationalized by the assumption that two different complexes[10,18] are formed, one emitting around 430-450 nm and the other emitting around 460-480 nm. Although analogous results were obtained at higher temperatures, the situation is now more complex due to dissociation of the exciplexes leading to a triple exponential decay at 378 nm.

The spectra in figure I indicate that the the tendency to form an exciplex increases from 1Py1In over 1Py2In to 1PY3In. In the exciplex of 1Py1In the small value of the dipole moment indicates that important mixing with the locally excited state occurs. In this case the overlap between the orbitals of donor and acceptor are important for the stabilisation of the exciplex. For 1Py2In and 1Py3In the exciplex is characterized by a considerably larger dipole moment, indicating that this exciplex has a very large charge transfer character. In this case the exciplex energy is determined rather by the Coulomb attraction between the positive charge on the indole and the negative charge on the methylpyrene. The larger stability of the exciplex of 1Py3In is also suggested by molecular models indicating that for one of the rotamers of 1Py3In the indole nitrogen can approach very close to the centre of the methylpyrene in the $g^{\pm}g^{\mp}$ conformation. Molecular models also suggest that for this compound there is a second rotamer with a larger distance between the centre of the methylpyrene and the indole nitrogen but characterized by a better overlap

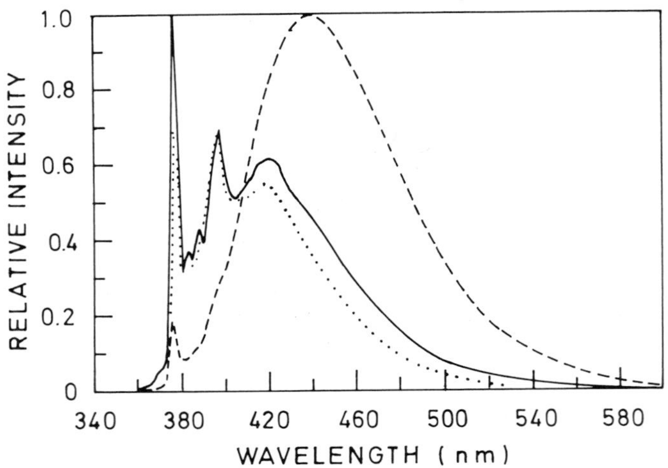

Figure 1: Corrected emission spectra of 1Py1In (·······), 1Py2In (————) and 1Py3In (--------) in isooctane at 298 K. The spectra are normalized at the maximum. Excitation occurred at 345 nm.

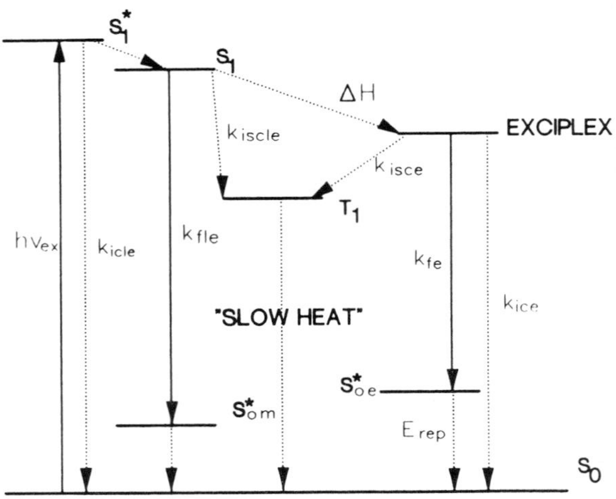

Figure 2: Kinetic scheme of the photophysics of 1Py3In in isooctane.

between orbitals of the methylpyrene and the indole moiety in the $g^{\pm}g^{\mp}$ conformation.

The decay of intramolecular exciplexes with indole chromophores

For 1Py2In and 1Py3In the amplitude of the optoacoustic signal is linear with the incident laser energy indicating that no biphotonic processes or ground state depletion occurs[44]. Because the fluorescence decay times of the bichromophores in the different solvents used are considerably shorter than the response time of the optoacoustic set up (250 ns) all heat emitted by the system, except that generated by the triplet decay, can be considered as "prompt" heat. Since there is, except for the locally excited triplet state, no long living species formed all absorbed energy must be trapped in this triplet state or be dissipated as fluorescence or heat. The heat generated by the decay of the triplet state A^3 ($\tau \approx$ several μs) will not contribute to the amount of prompt heat release, but will be dissipated as slow heat. Therefore for 1Py2In in isooctane the amount of prompt heat generated per absorbed photon can be calculated according to the scheme in figure 2:

$$\alpha \times h\nu_{ex} = 4650\ \text{cm}^{-1} + \Phi_{fle} \times 900\ \text{cm}^{-1} + \Phi_{iscle} \times 9800\ \text{cm}^{-1} + \Phi_{fe} \times 3000\ \text{cm}^{-1} + (\Phi_{ice} + \Phi_{isce}) \times r \times 26600\ \text{cm}^{-1} + (\Phi_{ice} + \Phi_{isce})\ (1 - r) \times 9800\ \text{cm}^{-1} \tag{22}$$

The energy parameters in equation (22) are obtained as follows:

4650 cm^{-1} = The energy difference between the Franck-Condon (FC) excited state A^*_{FC} and the relaxed locally excited state A^* of methylpyrene. Since this conversion is very fast (10^{-13} sec) it is within the time resolution of the experiment and will contribute in total to the amount of prompt heat release.

900 cm^{-1} = The energy difference between the FC ground state reached by fluorescence, A_{FC}, and the relaxed ground state, A. This heat is also released promptly (10^{-13}s). This quantity corresponds to the difference between the first moment of the fluorescence spectrum of 1-methylpyrene and the 0-0 transition of the fluorescence spectrum of 1-methylpyrene, which are both experimentally accessible.

9800 cm^{-1} = The energy difference between the relaxed locally excited state A^* and the locally excited triplet state A^3 of methylpyrene[45].

3000 cm^{-1} = The sum of the ground state repulsion and the stabilization enthalpy of the exciplex for 1Py2In. This amount is calculated from the difference of the fluorescence maxima of the locally excited state and the exciplex $(AD)^*$ augmented with 900 cm^{-1} (see above). This energy difference depends on the polarity of the solvent and equals 3000 cm^{-1} in isooctane, 5000 cm^{-1} in diethylether, and 6840 cm^{-1} in acetonitrile (for 1Py3In see table VIII).

26600 cm^{-1} = The energy difference between the relaxed locally excited state A^* and the relaxed ground state A. This quantity corresponds to the energy of the 0-0 transition in the fluorescence spectrum of 1-methylpyrene.

$h\nu_{ex}$: The energy of an absorbed photon.

The amplitude of the optoacoustic signal is plotted as a function of the incident laser energy for both the reference (characterized by $\alpha = 1$) and the sample. The ratio of the two slopes equals α (table IX). Using equation (22) and the experimentally determined values of the emission maxima and quantum yields of the exciplex (tables III and IV) and α (table IX) it is possible to calculate r (table X). Tables IX and X suggest an important influence of the solvent polarity on the values of α and r. The large propagating errors of the r values are inherently due to a 5% error of the α values.

With increasing polarity an increase of both α and r is observed. The increase of α correlates with the lower quantum yields of fluorescence of the locally excited state and of the exciplex observed in more polar solvents. The increase in r, giving direct information about the non-radiative behaviour of the intramolecular exciplex, suggests that internal conversion is favoured in more polar solvents. The values of r obtained by LIOAS for 1Py3In and 1Py2In correspond, within the experimental error, to those obtained by transient absorption[46].

Table VIII: Energies of the different excited states involved in the photophysics of intramolecular exciplexes between methylpyrene and indole (in cm^{-1}). ΔH°, E_{rep}, $E(A^{*}_{FC})$ and $E(A_{FC})$ correspond respectively to the enthalpy of exciplex formation, the repulsion energy in the ground state, the energy of the Franck-Condon excited state and the energy of the Franck-Condon ground state of 1-methylpyrene.

	1Py2In	1Py3In
$E(A^{*})$	26600	26600
$E(A^{*}_{FC})$	31250	31250
$E(A_{FC})$	900	900
$-\Delta H^{\circ} + E_{rep}$ (isooctane)	3000	3980
$-\Delta H^{\circ} + E_{rep}$ (diethylether)	4230	5000
$-\Delta H^{\circ} + E_{rep}$ (acetonitrile)	6840	7180
$E(A^{3})$	16800	16800

Table IX: Experimental values of α for 1Py2In and 1Py3In.

	1Py2In	1Py3In
Isooctane	0.34 ± 0.02	0.40 ± 0.02
Diethylether	0.46 ± 0.02	0.47 ± 0.02
Acetonitrile	0.64 ± 0.03	0.68 ± 0.03

Table X: Influence of the solvent polarity on r for the intramolecular exciplexes of 1Py2In and 1Py3In

	1Py2In	1Py3In
Isooctane	0.11 ± 0.13	0.21 ± 0.09
Diethylether	0.46 ± 0.11	0.48 ± 0.16
Acetonitrile	0.58 ± 0.09	1.00 ± 0.12

Both sets of r values suggest that, contrary to former interpretations[1,36,43,47] of the radiationless decay of intramolecular exciplexes, internal conversion is an important decay process of the intramolecular exciplex and that it becomes more important with increasing solvent polarity. This was already observed for intermolecular exciplexes[21,48]. The difference in the solvent dependence of k_{ice} and k_{isce} can be rationalized in the framework of the current theory describing electron-transfer[29,49,50]. The change in solvent polarity has a different influence on both reaction rates k_{ice} and k_{isce}. Assuming that the solvent is a uniform structureless dielectric, the electron transfer rate k_{et} can be written as:

$$k_{et} = A\exp(-\Delta G^{\#}/RT) \qquad (23)$$

with:

$$\Delta G^{\#} = (\lambda/4) \times (1 + \Delta G^{\circ}/\lambda)^2 = \lambda/4 + \Delta G^{\circ}/2 + \Delta G^{\circ 2}/4\lambda \qquad (24)$$

$$\lambda = \lambda_i + \lambda_s \qquad (25)$$

ΔG° and $\Delta G^{\#}$ correspond to the overall free energy change and the free energy of activation respectively. λ_i and λ_s are the intramolecular vibrational reorganization energy and the solvent contribution to the reorganization energy, λ.

Increasing the solvent polarity will make ΔG° less negative and increases λ_s for charge recombination from both the solvated radical ion pair and the exciplex to the ground state ($(AD)^*$ - S_0 transition) or the locally excited triplet state ($(AD)^*$ - A^3 transition). For both internal conversion and intersystem crossing ΔG° is much smaller than $-\lambda$ (Marcus inverted region[29]) and increasing the solvent polarity will increase both k_{ice} and k_{isce}. However, the latter process is characterized by values of ΔG° close to the maximum of the Marcus parabola and will therefore be less dependent on ΔG°. Therefore, increasing the solvent polarity will lead to a more pronounced increase of k_{ice} than of k_{isce}. If it is argued that a transition to a more polar[34] exciplex occurs upon increasing the solvent polarity, this would, according to Mataga, decrease the rate of intersystem crossing compared to that of internal conversion and therefore enhance the observed effect.

The rate of the intersystem crossing, which can be expected to be a non-adiabatic electron transfer process, will be proportional to the square of the electronic part of the matrix element between the initial and the final state. The analysis of the LIOAS and transient absorption data indicates that the radiationless decay of the exciplex occurs, even in solvents of intermediate polarity, for a large part by internal conversion. This apparently contradicts some results indicating a small (or even negative) activation energy and preexponential factor for the radiationless decay of exciplexes[1,36,43,47]. A small activation energy and preexponential factor can however be obtained for electron transfer in the inverted region if important nuclear tunnelling occurs[51,52,53]. The small preexponential factor is in this case due to a small Franck-Condon factor between the vibrational ground state of the exciplex and the vibrationally excited electronic ground state. This Franck-Condon factor will also modulate the matrix element between both states which can become several orders of magnitude smaller than in the absence of electron phonon coupling[54]. Although in the exciplex the undressed matrix element between the charge transfer configuration and the ground state can be of the order of magnitude of 0.01 to 0.1 eV[15,16,17,39], the decrease[54] of the matrix elements by "phonon dressing" will make the "dressed" matrix element small enough to make the internal conversion nonadiabatic. This can explain the small preexponential factors obtained experimentally. Following this line of thought increasing the solvent polarity will make the electron transfer less exothermic leading to the population of lower vibrational levels of the electronic ground state characterized[48] by a larger Franck-Condon factor. Furthermore, increasing the solvent polarity will increase the reorganization energy which also leads to the population of lower vibrational levels. Both factors will increase the rate of the internal conversion when the solvent polarity is increased.

Acknowledgments

M.V.d.A. is a research associate of the F.K.F.O.; N.B. is a research associate of the F.G.W.O.; Ph.V.H. and N.H. thank the I.W.O.N.L. and the K.U.Leuven for financial support. M.M.H.K. and K.I. thank the K.U.Leuven for financial support. The authors thank the Belgian Ministry of Scientific Programming, the F.K.F.O. and the F.G.W.O. for financial support to the laboratory.

References

1) A.M. Swinnen, M. Van der Auweraer, F.C. De Schryver, K.Nakatani, T.Okada, N. Mataga, J. Am. Chem. Soc., 109 (1987) 321.
2) K. Nakatani, T. Okada, N. Mataga, F.C. De Schryver, M. Van der Auweraer, Chem. Phys. Lett., 145 (1988) 81.
3) J. Hinatu, H. Masuhara, N. Mataga, Y. Sakata, S. Misumi, Bull., Chem. Soc. Jpn., 51 (1978) 1032.
4) J.P. Palmans, A.M. Swinnen, G. Desie, M. Van der Auweraer, J. Vandendriessche, F.C. De Schryver, J. of Photochem., 28 (1985) 419.
5) K. Nakatani, T. Okada, N. Mataga, F.C. De Schryver, Chem. Phys., 121 (1988) 87.
6) N. Mataga, K. Ezumi, Bull. Chem. Soc. Jpn., 40 (1967) 1355.
7) W.R. Ware, D. Watt, D. Holmes, J. Am. Chem. Soc., 96 (1974) 7853.
8) A.E.W. Knight, B. Selinger, Austr. J. of Chemistry, 26 (1973) 1.
9) L. Brand, W.R. Laws in "Time-Resolved Fluorescence Spectroscopy in Biochemistry and Biology", NATO ASI Series A, eds. R.B. Cundall and R.E. Dale, Plenum Press, New York, 1983 pp. 319
10) K.A. Zachariasse, G. Duveneck, W. Kuhnle, Chem. Phys. Lett., 113 (1985) 337.
11) N. Boens, L. Janssens, M. Ameloot en F.C. De Schryver in Proceedings SPIE Conference on "Time-resolved Laser Spectroscopy in Biochemistry II", p 456, R.Lakowitz editor, Los Angeles 1990, SPIE, Washington 1990
12) M. Ameloot, N. Boens, R. Andriessen, V. Van den Bergh and F.C. De Schryver, J. Phys. Chem., 95 (1991) 2041.
13) R. Andriessen, N. Boens, M. Ameloot and F.C. De Schryver, J. Phys. Chem. 95 (1991) 2048.
14) N. Nakashima, N. Mataga, F. Ushio, C. Yamanaka, Z. Phys. Chem. N.F., 78 (1972) 153.
15) M. Van der Auweraer, A.M. Swinnen, F.C. De Schryver, J. Chem. Phys. (1982) 77.

16) A.M. Swinnen, M. Van der Auweraer, F.C. De Schryver, Chem. Phys. Lett. 109 (1983) 574.
17) A.M. Swinnen, M. Van der Auweraer, F.C. De Schryver, J. Photochemistry 28 (1985) 315.
18) F.C. De Schryver, K. Demeyer, M. Van der Auweraer, E. Quanten, Annals of the New York Academy of Sciences, 23 (1981) 93.
19) T. Okada, I. Karaki, E. Matsuzawa, N. Mataga, Y. Sakata, S. Misumi, J. Phys. Chem. 85 (1981) 3957.
20) T. Nishimura, N. Nakashima, N. Mataga, Chem. Phys. Lett. 46 (1977) 334.
21) N. Mataga, T. Okada, N. Yamamoto, Chem. Phys. Lett. 1 (1967) 119.
22) H. Leonhardt, A. Weller, Ber. Bunsenges. Phys. Chem. 67 (1963) 791.
23) A.C. Tam, C.K.N. Patel,. Appl. Opt. 18 (1979) 3348.
24) J. Lavilla, J. Goodman, J. Am. Chem. Soc. 111 (1989) 712.
25) K. Heihoff, S. Braslavsky, Chem. Phys. Lett. 131 (1986) 183.
26) K. Heihoff, S. Braslavsky, Chem. Phys. Lett. 134 (1987) 335.
27) M. Bilmes, S. Braslavsky, O. Tocho, Chem Phys. Lett. 134 (1987) 335.
28) I. Gould, J. Moser, B. Armitage, S. Farid, J. Goodman, M. Herman, J. Am. Chem. Soc. 111 (1989) 1917.
29) P. Siders, R.A. Marcus, J. Am. Chem. Soc. 103 (1981) 741.
30) N. Helsen, Ph. Van Haver, M. Van der Auweraer, F.C. De Schryver, to be published.
31) N. Boens, L. Janssens, F.C. De Schryver, Biophys. Chem., 33 (1989) 77.
32) Ph. Van Haver, N. Helsen, S. Depaemelaere, M. Van der Auweraer and F.C. De Schryver, J.Am.Chem.Soc., in press.
33) J.P. Palmans, M. Van der Auweraer, A.M. Swinnen, F.C. De Schryver, J. Am. Chem. Soc. 106 (1984) 7721.
34) N. Mataga, T. Okada, N. Yamamoto, Bull. Chem. Soc. Jpn., 39 (1966) 2562.
35) H. Beens, H. Knibbe, A. Weller, J. Chem. Phys. 47 (1967) 1183.
36) M. Van der Auweraer, Academiae Analecta, 48 (1986) 29.
37) E.Z. Lippert, Naturf. Phys. Chem. 10a (1955) 541.
38) N. Mataga, Y. Kaifu, M. Koizumi, Bull. Chem. Soc. Jpn. 28 (1955) 690.
39) H. Beens, A. Weller, Organic Molecular Photophysics, J. B. Birks: Wiley: London, 1975, Vol. 2, p.159.
40) M. Goldenberg, J. Emert, H. Morawetz., J. Am. Chem. Soc., 100 (1978) 7171.
41) J. Vandendriessche, J.-P. Palmans, S. Toppet, N.Boens, F.C. De Schryver and H. Masuhara, J. Am. Chem. Soc., 106 (1984) 8057.
42) P. Reynders, W. Kuhnle, K.A. Zachariasse, J. Phys.Chem., 94 (1990) 4073.
43) M. Van der Auweraer, A. Gilbert, F.C. De Schryver, J. Am. Chem. Soc., 102 (1980) 4007.
44) Ph. Van Haver, M. Van der Auweraer, F.C. De Schryver, to be published.
45) L. Peter, G. Vaubel, Chem. Phys. Letters., 21 (1973) 158.
46) Ph. Van Haver, M. Van der Auweraer, F.C. De Schryver, to be published.
47) M. Van der Auweraer, A. Gilbert, F.C. De Schryver, J. Phys. Chem., 85 (1981) 3198.
48) I. Desparasinska, E. Gaweda, M. Mandziuk, J. Prochorov, Advances in Molecular Relaxation and Interaction Processes 1982, 23, 45.
49) R.A. Marcus, J. Chem. Phys., 81 (1984) 4494.
50) N.S.Hush in Supramolecular Photochemistry NATO ASI Series C, Vol 214, V.Balzani eds., Reidel Publishing Company, Dordrecht 1987, p.53.
51) S. Efrima, M. Bixon, Chem. Phys. Lett., 25 (1974) 341.
52) N.R. Kestner, J. Logan, J. Jortner, J. Phys. Chem., 78 (1974) 2148.
53) R. Hermant, Ph.D. Thesis, Universiteit Amsterdam, 1990.
54) J. Jortner, Bixon M., J. Chem. Phys., 88 (1988) 167.

DISCUSSION

Rettig

Is it possible to analyze systems using global or compartmental analysis, where one (or more) of the components shows a continuous fluorescence red shift?

De Schryver

In compartmental analysis they will show a poor fit. In global analysis one can include a time dependent rate constant to elucidate such kinetic.

Baumann

You showed us clearly how important it is to take into account all necessary reaction channels when extracting kinetic data from single photon timing and you have introduced compartmental analysis around a 4 stage model system. Could you comment on how does phase fluorimetry compare with respect to the necessity of putting into the evaluation procedure the assumed (maximum) kinetic scheme?

De Schryver

The information obtained by phase fluorimetry is of course nothing but the Fourier transform of the data obtained by single photon timing. Therefore the modeling should be identical.

Steiner

You said no model is necessary if global compartmental analysis is applied. What did you mean exactly, since global compartmental analysis itself is based on a kinetic model?

De Schryver

What I wanted to say is that it is possible to solve intramolecular complex formation without using "model lifetimes" if one does quenching studies. The compartmental system is of a kinetic model.

Sakata

Previously we prepared intramolecular exciplex systems, where N,N–dimethylaniline and anthracene are linked with a trimethylene chain (**1**, **2**, **3**). Among these three compounds only **3**, in which nitrogen atom can overlap well at sandwiched conformation, shows exciplex emission at longer wavelength.

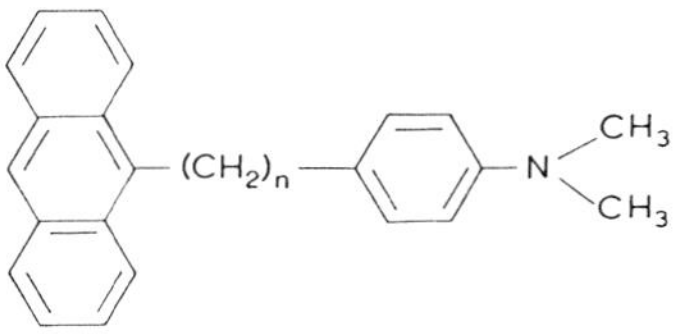

$n = 0,1,2,3$

<u>De Schryver</u>

Thank you for your comment confirming our results.

Dynamics and Mechanisms of
Photoinduced Transfer and Related Phenomena
N. Mataga, T. Okada and H. Masuhara (Editors)

On the Non-exponential Behavior of Intramolecular Electron Transfer Processes in Polar Solvents

Tadashi Okada

Department of Chemistry, Faculty of Engineering Science and Research Center for Extreme Materials, Osaka University, Toyonaka, Osaka 560, Japan

Abstract

The dynamics of photoinduced intramolecular electron transfer in polar solvents have been investigated using femtosecond-picosecond time resolved absorption and fluorescence spectroscopy. The model compounds studied here with different degrees of electronic interactions between combined donor-acceptor groups show a simple one electron transfer from the excited state localized in the acceptor to the intramolecular ion-pair state. Time dependences of the electron transfer reaction were expressed by an extended exponential function, although the main part of the reaction can be approximated by an exponential function, from which we can estimated the approximate electron transfer time of the compounds. In some compounds, the approximate electron transfer times were found to be faster than the solvation time and the longitudinal dielectric relaxation time of the solvents. The possible mechanisms responsible for the extended exponential behavior were discussed.

1.INTRODUCTION

When the electronic interaction between an electron donor(D) and an acceptor(A) is sufficiently strong and the energy gap relations are also favorable, the activation barriers in electron transfer reaction will become very small. In the case of the small barrier electron transfer reaction, it is believed that the orientational motions of polar solvent molecules or polar groups in the surrounding donor and acceptor molecules, of which the time dependences are characterized by the longitudinal dielectric relaxation time τ_L under a continuum model of solvent, play an important role in the course of electron transfer process[1]. In this work, we focus on measuring the dynamics of electron transfer in excited intramolecular D-A systems using femtosecond-picosecond time resolved absorption and fluorescence spectroscopy. The combination of the absorption and emission measurements provide us an precise time profile for electron transfer process affected by solvent orientation dynamics in large dynamic range and wide time scale.

Time resolved fluorescence Stokes shift of polar fluorescent probe molecule has recently been extensively studied for understanding of the microscopic relaxation processes of the solvent in the environment of the probe[2-4]. The important experimental results for the solvation dynamics show that the time scales observed for solvation are usually larger than τ_L and the solvent relaxation function is not predicted using bulk dielectric relaxation times of the neat solvent and the continuum solvent description model. However, the systematic studies on the photoinduced electron transfer affected by solvent relaxation dynamics are very few and these studies have been investigated mainly by means of time resolved fluorescence measurements[2,5].

In the present report, we have examined the following systems with different degrees of electronic interactions between combined D-A groups which seem to be appropriate for the elucidation of the polar solvent dynamics associated with the electron transfer reaction. In these compounds, there is no appreciable change of absorption spectra due to the electronic interaction between donor and acceptor groups both in the ground state and in the excited state immediately after laser excitation. The transient absorption spectra of these systems show the simple electron transfer from the state localized in the pyrenyl or anthryl part(LE state) to the intramolecular ion-pair state(IP state). The electron transfer in the excited state in polar solvent seems to take place easily when the surrounding solvent orientations are favorable.

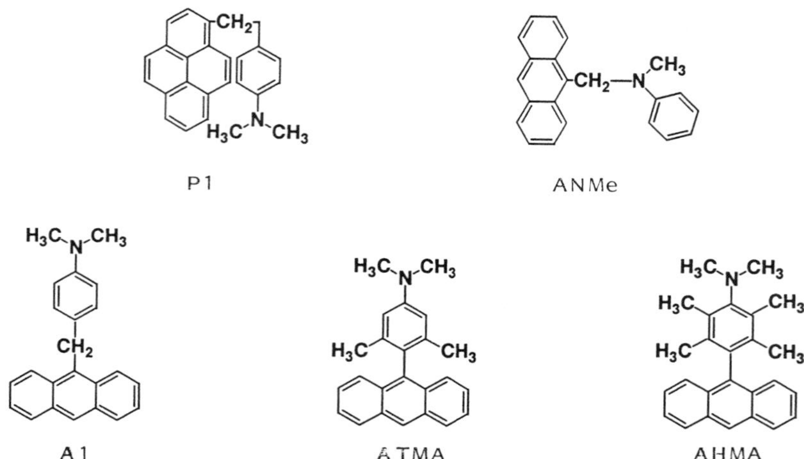

Figure 1. Molecular structures for the intramolecular D-A compounds studied here.

2.EXPERIMENTAL

A femtosecond laser photolysis system was used for the measurement of time resolved absorption spectra in subpicosecond to tens of picosecond regions. The out put of a synchronously pumped Pyridine-1(710nm) or Rhodamine 6G(590nm) dye laser was amplified to several hundreds of μJ/pulse.The amplified pulse with several hundreds of fs was frequency doubled and used for exciting the sample. The rest of the fundamental pulse was focused into D_2O to generate a white light probe pulse. Two set of multichannel photodiode detectors were used to observe wide band transient absorption spectra. When it is necessary, the observed spectra were corrected for the chirping of the monitoring probe pulse.

Time resolved fluorescence was measured by using a correlated single photon counting technique with a synchronously pumped, cavity dumped dye laser. The observed pulse width of the exciting laser pulse was about 40 ps(FWHM). Fluorescence was detected under the optics setting at the magic angle. Fluorescence decay curves were analyzed by using ACOS 1000 system of the Computation Center, Osaka University. Sample solutions were deaerated by freeze-pump-thaw cycles or by flushing N_2 stream.

3. TIME RESOLVED ABSORPTION SPECTRA

Time resolved absorption spectra of P1 in hexanenitrile(A) and A1 in 1-butanol(B) are indicated in Figure 2. In the case of P1, one can see the rapid rise of the characteristic sharp absorption band at 500nm(pyrene anion) and around 460nm(N,N-dimethylaniline cation and pyrene anion) due to the intramolecular ion pair state. Time resolved absorption spectra of A1 in 1-butanol show the rise of a cation(480nm) and an anion(>630nm) bands and the decay of absorbance at 600nm which is ascribed to the $S_n \leftarrow S_1$ transition localized in the anthracene part. The photoinduced charge separation process in polar solvents for every system indicated in Figure 1 can be analyzed by a simple one electron transfer reaction as a two-state model of LE to IP[6,7]. The rise curves of ion pair state are also indicated in the lower panels in Figure 2. The main part of the rise curves of the ion-pair state absorbance can be expressed approximately by an exponential function, though the rise curve obtained by more precise measurements shows an non-exponential behavior even in the case of P1 in hexanenitrile as described later. The obtained approximate rise times in alkanenitriles are listed in Table 1 together with τ_L of the solvent and solvation time τ_S determined by the measurement of the time dependent Stokes shift of fluorescence[8]. τ_L and τ_S values of hexanenitrile are estimated by logarithmic plot of these values for alkanenitriles against the ratio of solvent viscosity to dielectric constant, η/ε.

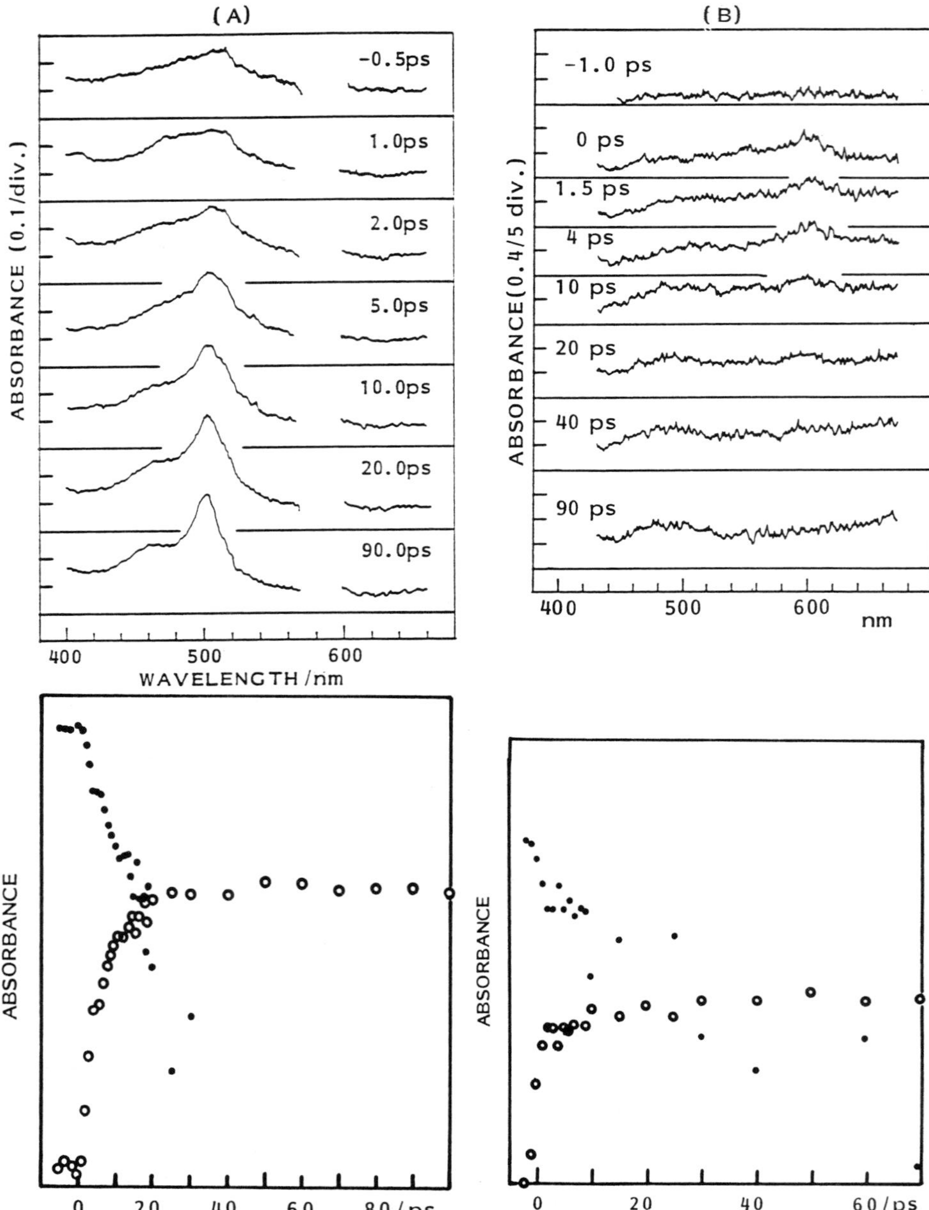

Figure 2. Time resolved absorption spectra and the rise curves of IP state at room temperature. Closed circles denote logarithmic plot. (A):P1 in hexanenitrile.(B):A1 in 1-butanol.

Table 1. Approximate rise time of the ion pair state in alkanenitriles at room temperature and τ_L and τ_S of the solvents.

solvent	acetonitrile	butyronitrile	hexanenitrile
A1	0.65 ps	1.0 ps	1.4 ps
			1.6[a)]
P1	1.7	2.5	4.5
			4.5[a)]
ANMe			0.7
ATMA	0.6		1.5
AHMA	1.5		12.5
τ_L	0.19	0.53	0.98 - 1.1
τ_S	0.4 - 0.9	1.5 - 2.1	3.5 - 4.5

a) Rise time obtained by excitation at 295nm.

The values listed in Table 1 were obtained by the excitation at 355nm which corresponds to the excitation up to the acceptor LE state with excess vibrational energy of thousands of wavenumbers. We have also examined using 295nm laser pulse as an excitation light as indicated in the Table. Although the irradiation at 295nm corresponds to the excitation of the donor part, we could not detect the rise of the S_n<–S_1 absorption of anthracene or pyrene moiety and the electron transfer times obtained were same to those obtained by 355nm excitation within experimental error.

The approximate electron transfer times of D–A systems studied here are rather close to the solvation time τ_S but longer than the longitudinal dielectric relaxation time τ_L of solvents except in the case of ANMe. It is noted here that the electron transfer time of ANMe in hexanenitrile is faster than τ_L. The observed results suggests that the photoinduced electron transfer process is possibly controlled by the solvent orientation dynamics, especially in the systems of A1, ATMA, and ANMe. The electronic interaction responsible for the electron transfer reaction between D and A in ANMe seems to be stronger than that in the case of A1, since the charge density on N. atom of dimethylaniline is much larger than that on C atom at p-position. It is plausible for the reasons of the rapid electron transfer time of ANMe in hexanenitrile that the contribution of intramolecular vibrations to electron transfer process will enhance the reaction rate. It is also plausible that some presolvated solvent molecules closely to N atom exist in the case of ANMe, because the bulky chromophores connected to N atom may prevent random fluctuational motion of solvent molecules, leading to the increase of electron transfer rate. The results might be ascribed to the fact that the solvent dielectric response on the electron transfer dynamics for the compound ANMe can not be treated with a homogeneous continuum model of solvent, though the detailed mechanism is not clear at the present stage of investigation.

4.DECAY CURVES OF THE LE FLUORESCENCE

Time resolved absorption method has a high time resolution but the dynamic range of this method is rather poor. More precise informations on the time dependences of the electron transfer rate were obtained by time resolved fluorescence measurements by means of a single photon counting method[9]. The fluorescence decay curves monitored at the fluorescence band of pyrene or anthracene moiety were non-exponential and could not beanalyzed by two or three exponential functions. We have analyzed the decay curves by means of an extended exponential function,

$$F(t) = A\exp[-(t/\tau)^{\beta}] + B\exp(-kt) \qquad (1)$$

where the second term is due to the contribution from the photoproducts,the amount of which was about 0.1% in a fresh sample.

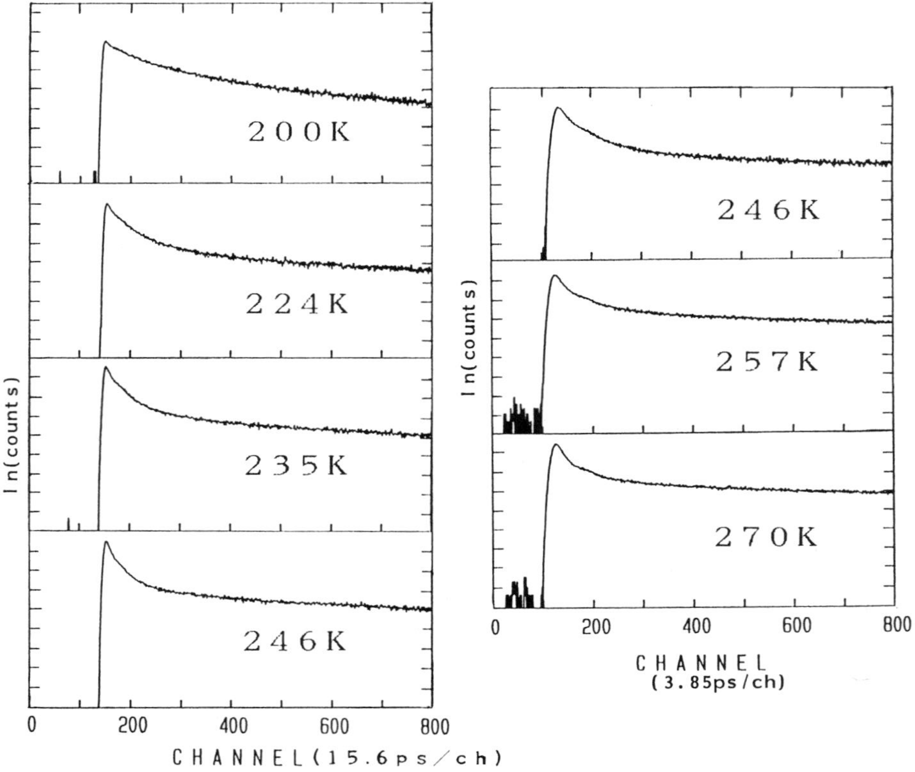

Figure 3. Temperature dependence of the LE fluorescence decay curve of A1 in 1-butanol.

Table 2. β and τ values of the intramolecular D–A systems in some solvents.

	solvent	temperature/K	β	τ/ps
A1	1-butanol	298	0.62	8.2
		244	0.46	43.5
		235	0.41	44.2
		224	0.35	38.2
	2-methylTHF	298	0.79	9.2
P1	1-butanol	298	0.95	86.9
		254	0.82	192
		248	0.75	243
		226	0.69	292
		206	0.74	692
	hexanenitrile	298	0.8	5.8
ATMA	1-butanol	298	0.52	3.6
AHMA	1-butanol	298	1.0	86.9
		254	0.90	590
		196	0.73	2270

The temperature dependences of the decay curve of LE fluorescence of A1 in 1–butanol are shown in Figure 3.The time scale per channel is different between two panels in the figure and both time scales are indicated for the decay curve measured at 246K for the comparison. The obtained values of β and τ are summarized in Table 2. The correlation time τ increases and β decreases with decreasing temperature, indicating that the inhomogeneity of solvent orientational relaxation enhanced at low temperature.

The obtained values of β and τ depend strongly on the sample. For example, the correlation times of the observed fluorescence decay curve in the case of A1 and ATMA in 1–butanol at room temperature are rather close to the fastest longitudinal dielectric relaxation time but close to the bulk relaxation time in the case of P1 and AHMA. It is well known that the dielectric response of alcoholic solvents do not show a simple Debye dispersion. In this respect, it seems very important that the non–exponential behavior in the systems of A1 in 2–mehtyltetrahydrofuran and P1 in hexanenitrile, where the solvents are considered to show simple dielectric relaxation. The decay curve of the LE fluorescence of P1 in hexanenitrile was well analyzed by eq.(1) with β = 0.8 and τ = 5.8ps as indicated in Figure 4(A). The rise curve of the ion–pair state obtained by the transient absorption measurement shown in Figure 2(A) was reproduced very well even in the picosecond time regime by the same values of β and τ obtained by the fluorescence measurement as shown in Figure 4(B), although the time resolution of our single photon counting system was about 10ps. Other systems such as ATMA in 1–butanol and A1 in 2–methyltetrahydrofuran were also analyzed successfully by the same manner.

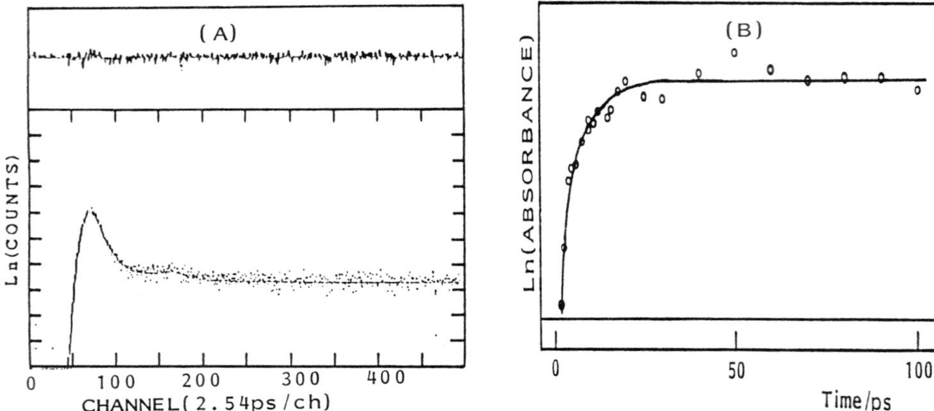

Figure 4. (A) : Time dependence of LE fluorescence of P1 in hexanenitrile at room temperature(dotted) and the calculated curve with β = 0.8, τ = 5.8ps in eq.(1). (B) : Rise curve of IP state of P1 in hexanenitrile, same to the lower panel of Figure 2A(open circle) and the simulated rise curve using β and τ values obtained by the analysis of LE fluorescence(solid line).

One of the possible mechanisms responsible for the extended exponential behavior of the intramolecular electron transfer reaction in polar solvents may be saturation effects of solvation dynamics in nearby solvents strongly coupled to the charge distribution in the excited D-A systems. Such a solute-solvent interaction will give a radial distribution of the orientational relaxation time of the solvent. Another possibility may be related to a relaxation dynamics of the initial distribution of solvation. When the relaxation time between solvation modes is longer than the rapid electron transfer time in a solvation mode favorable for intramolecular electron transfer reaction, the distribution of solvation may be deviated in the course of the reaction from that of a equilibrium one, leading to the observed non-exponential behavior of the fluorescence decay curve.

ACKNOWLEDGEMENT

The author would like to thank Prof. N.Mataga for useful discussions and collaboration and to thank Messrs. S.Nishikawa and K.Kanaji for their collaboration. The present work was partially supported by a Grant-in Aid for Scientific Research on Priority Area of "Non-equilibrium Processes in Solution" (No.02245013) from Ministry of Education, Science, and Culture of Japan.

REFERENCES

1. For example, H.Sumi and R.A.Marcus, J. Chem. Phys. 84,(1986)4894, and references cited therein.
2. V.Nagarajan, A.Brearley, T-J.Kang, and P.F.Barbara, J. Chem. Phys. 86,(1987)3183.
3. W.Jarzeba, G.C.Walker, A.E.Johnson, and P.F.Barbara, Chem. Phys. 152,(1991)57.
4. M.Maroncelli and G.Fleming, J.Chem.Phys.86,(1987)6221.
5. P.F.Barbara and W.Jarzeba, Adv. Photochem.15,(1990)1.
6. T.Okada, N Mataga, W.Baumann, and A.Siemiarczuk, J. Phys. Chem. 91,(1987)4490.
7. N.Mataga, S.Nishikawa, T.Asahi, and T.Okada, J. Chem. Phys. 94,(1990)1443.
8. M.A.Kahlow, T.J.Kang, and P.F.Barbara, J.Chem. Phys.88,(1988)2372.
9. T.Okada, S.Nishikawa, K.Kanaji, and N.Mataga, "Ultrafast Phenomena 7" ed., C.B.Harris, E.P.Ippen, G.A.Mourou, A.H.Zewail, Springer-Verlag, Berlin, 1990,p.397.

DISCUSSION

Tominaga

Why does AHMA have much slower ET (electron transfer) rates than those of ATMA? Does the rotational motion of the amino group play an important role on the electron transfer reaction or is AHMA a totally different molecule from ATMA?

Okada

The rotational motion of the amino group play very important role. In the case of AHMA, the electronic interaction between the amino group and phenyl ring is very small because of the substitution of the two methyl groups into 2 and 6 positions of phenyl ring. Therefore, the electron transfer takes place from the amino group directly to the anthracene in the excited AHMA, where the phenyl ring play as a spacer, while in the case of ATMA, the electron transfer takes place from N,N–dimethylaniline to the anthracene.

De Schryver

As the slow decay is due to product, does its contribution increase upon repeated use of the solution? Repopulation of the exciplex from the CT state can be excluded.

Okada

Yes, the fluorescence intensity due to the photoproducts increases with repeated irradiation of the laser pulses. The lifetime of the photoproduct's fluorescence was completely different from that of the exciplex. We measured the LE fluorescence at the blue edge of the fluorescence band and we could exclude the exciplex fluorescence by this manner.

Rettig

Compound A_1 shows very fast ET rates in butanol, P_1 much slower ones. The nonexponentiality parameter β which presumably reflects the broadness of the distribution function of solvent relaxation times is determined much smaller for A_1 (broader distribution) than P_1. How are these findings related?

Okada

The electronic interaction between donor and acceptor in the excited state of A_1 is stronger than the case of P_1. So, the electron transfer time of the excited A_1 is much faster than that of P_1. Therefore, the fast components of the relaxation time of the solvent may be averaged in the course of the electron transfer reaction of excited P_1. This difference of the electronic interaction between P_1 and A_1 would be the most important factor to explain the obtained β in P_1 and A_1.

Dynamics and Mechanisms of
Photoinduced Transfer and Related Phenomena
N. Mataga, T. Okada and H. Masuhara (Editors)
1992 Elsevier Science Publishers B.V.

Condensed phase studies of radical ions in photoionization and radiolysis

A. D. Trifunac, A.-D. Liu and D. M. Loffredo

Chemistry Division, Argonne National Laboratory, Argonne, IL 60439

Abstract

Studies of radical ions in pulse radiolysis and photoionization reveal that ion-molecule reactions in the condensed phase are widespread. Observations of various radical cations and the products of two-photon ionization of aromatic compounds in hydrocarbon solutions are consistent with the proposed mechanism of photoionization in which proton transfer reactions between excited radical cations and solvent molecules occur. Flash photolysis studies in which radical cation yields are compared with the electron and/or free ion yields as a function of photon energy provide further support for the proposed mechanism.

1. INTRODUCTION

Numerous studies of charge pair creation and dynamics have elucidated several aspects of radiolysis and photoionization . When one uses energetic particles or UV photons, one can cause ionization or photoionization of various compounds in condensed phase, where RH is a solvent, e.g., cyclohexane (c-hexane), and AH is an aromatic molecule, e.g., anthracene:

$$RH \xrightarrow{e^- \text{ beam}} RH^{+\cdot} + e^- \quad (1)$$

$$AH \xrightarrow{2h\nu} AH^{+\cdot} + e^- \quad (2)$$

In the initial pair of charged species created, the geminate radical ion pair, the distance that the electron finds itself after the thermalization process in a given medium reflects the excess energy input. Much of the chemistry that follows is the consequence of charge pair dynamics, and in solvents of low polarity, e.g., hydrocarbons, a majority of geminate ions (~95-97%) recombine, giving rise to excited states:

$$RH^{+\cdot} + e^- \longrightarrow RH^* \tag{3}$$

$$AH^{+\cdot} + e^- \longrightarrow AH^* \tag{4}$$

Such excited states have been examined via fluorescence, and it became apparent that ionic processes are the dominant source of excited states produced in radiolysis.[1]

More recently, we have examined such processes and have focused on the radical cation species produced.[2,3] We have developed tools that allow us to detect magnetic resonance spectra of such radical ions and thus characterize them structurally, and to observe their reactions on the time scale of tens to hundreds of nanoseconds. While most of the geminate ion pairs ($RH^{+\cdot}$,e^-) have very short lifetimes in hydrocarbons (picoseconds) use of appropriate scavengers converts the highly mobile electron into a less mobile anion, $AH^{\cdot -}$, (eq. 5) thus extending the geminate pair lifetime to hundreds of nanoseconds. This allows the study of such processes by applying Fluorescence Detected Magnetic Resonance (FDMR) and other methods such as dc conductivity[4] or flash photolysis.[5] Many studies at these slower time scales have been carried out.

In magnetic resonance studies one observes the excited states produced by geminate ion recombination:

$$e^- + AH \longrightarrow AH^{\cdot -} \tag{5}$$

$$RH^{+\cdot} + AH^{\cdot -} \longrightarrow AH^* \tag{6}$$

In radiolysis we observe fluorescence from excited states produced via radical ion recombination as in eq. 6, while in photoionization under certain conditions we have been able to observe direct electron-radical cation recombination as in eq. 4. We have been able to obtain snapshots, on the tens to hundreds of nanoseconds time scale, of the elusive radical cations which were previously only accessible to study by EPR in special low temperature matrices[6] or by fast optical spectroscopy which gave little information about the identity and structure of such radical cations.[7] The main finding in our work was that all alkane radical cations undergo facile ion molecule reactions in condensed phases so that typically alkane radical cations disappear in room temperature radiolysis in a few nanoseconds.[3]

The ion molecule reactions occurring must be either proton transfer or the symmetric process of H-atom transfer:

$$RH^{+\cdot} + RH \longrightarrow R\cdot + RH_2^+ \tag{7}$$

$$RH^{+\cdot} + RH \longrightarrow RH_2^+ + R\cdot \qquad (8)$$

The occurrence of ion-molecule reactions explain observations of shorter lifetime of geminate radical cation vis a vis its partner electron. The other remarkable observation was the diversity of rates at which such proton transfer reactions occurred. For reasons yet to be determined, some are much more rapid than others.

Picosecond emission studies in our laboratory of some of the systems studied by magnetic resonance have provided a measure of these rates, showing that e.g., c-hexane radical cation (c-$C_6H_{12}^{+\cdot}$) has ~300 picosecond lifetime in c-hexane radiolysis at room temperature, while n-hexane lives ten times longer.[8]

Our findings and many other puzzling observations made by several groups and ourselves, such as the observation of the fast conductivity species "hole" in c-hexane and trans-decalin radiolysis,[9,10] in photoionization[11] and several observations of very low yield of excited states in radiolysis and in photoionization studies,[12-14] need to be explained.

2. RADICAL IONS IN PHOTOIONIZATION

There have been many studies of photoionization of aromatic compounds in alcohols[15] and in hydrocarbons.[16] The photon order of such reactions was established, and the geminate pair dynamics was examined in our group.[4,16]

Recently, we have been able to apply FDMR to photoionization as well, allowing us to directly observe recombining ions, assign their structure, and establish that radical ions are predominantly of singlet spin multiplicity. This implies that photoionization of aromatic compounds in alcohols occurs via the singlet manifold in a two-photon process.[17] We also observed that radical ion yield (as determined by FDMR) decreases with increasing photon energy.

In our group, detailed studies of dc conductivity were also carried out on aromatic compounds in hydrocarbons, and similar conclusions about the photon order and the lack of triplet state involvement were obtained.[18]

Our previous dc conductivity studies have also observed the "fast conductivity" species in photoionization of anthracene and similar solutes in c-hexane and trans-decalin.[11] Clearly, the same species observed in radiolysis must be involved. The species invoked for radiolysis of, e.g., c-hexane was the c-hexane radical cation ("hole"), c-$C_6H_{12}^{+\cdot}$ Thus, a novel mechanism invoking "hole injection" was advanced to explain the photoionization results.[19] It was speculated that, in addition to the usual photoionization process (eq. 2), the "hole injection" process occurs in special cases, i.e.,

$$AH \xrightarrow{2h\nu} AH^{**} + RH \longrightarrow AH^{-\cdot} + RH^{+\cdot} \qquad (9)$$

so that the solvent radical cation is produced. Following this explanation of our results other groups have entertained this idea.[20,21] The existence of a long-lived solvent radical cation, e.g., (c-$C_6H_{12}^{+\bullet}$) as a charge carrier observed in dc conductivity with a lifetime of a few hundred nanoseconds is not compatible with our findings from FDMR and picosecond emission studies, which indicate that all alkane radical cations have a lifetime <10-20 nsec at room temperature, with some living considerably shorter times than that.[3,8]

Photoionization of aromatics in hydrocarbons can yield a variety of products that cannot be explained by the usual photoionization mechanism where only the process shown in eq. 2 is occurring.[21] However, no comprehensive study of products of photoionization of aromatic compounds in hydrocarbons was carried out.

We mention the seminal work of Ausloos and coworkers,[13,22] who have examined radiolysis[22] and photoionization of hydrocarbons.[13] These studies have concluded that excited state fragmentation is the dominant process, e.g., in c-hexane:

$$\text{c-}C_6H_{12}^* \longrightarrow \text{c-}C_6H_{10} + H_2 \qquad (10)$$

and

$$\text{c-}C_6H_{12}^* \longrightarrow \text{c-}C_6H_{11}\cdot + H\cdot \qquad (11)$$

where these excited states were produced by radiolysis and by UV-photoionization.

The process, eq. 10, was considered to be more important at or below ionization, and process eq. 11 becomes more prominent as the energy of excitation increases above the ionization threshold. While Ausloos and coworkers have enumerated several puzzling aspects of hydrocarbon photoionization, the possible role of ionic species was only briefly mentioned.[13] Another outstanding problem of hydrocarbon photoionization is the observation of very low quantum yield of fluorescence ($\sim 10^{-3}$-10^{-4}) in simple hydrocarbons, and the further decrease of this quantum yield as the energy of photons is increased above the photoionization threshold.[12] Thus, any mechanism of photoionization must address these puzzles:

1. The observation of fast conductivity species.

2. The low quantum yield of fluorescence in radiolysis and photoionization of neat hydrocarbons, and its decrease with increasing photon energy.

3. The observation of a lower than expected yield of excited states in radiolysis.[14]

4. Products of photoionization.[13,21,23]

The following section outlines a mechanism that can account for the above enumerated "unresolved issues" in photoionization and radiolysis.

3. MECHANISM OF "HIGH-ENERGY" CHEMISTRY

Consider photoionization of aromatic compounds in hydrocarbons:

$$AH \longrightarrow AH^{+\bullet *} + e^- \tag{12}$$

$$AH^{+\bullet *} \longrightarrow AH^{+\bullet} \tag{13}$$

The modification of the accepted scheme is the introduction of an excited radical cation species, $AH^{+\bullet *}$, which must be rather short lived. The "normal" or "relaxed" radical cations $AH^{+\bullet}$ are produced from this excited species.

It seems reasonable to consider that, as is often observed in neutral excited states, the excited radical cation is more acidic; i.e., it undergoes facile proton transfer to the surrounding solvent:

$$AH^{+\bullet *} + RH \longrightarrow A\bullet + RH_2^+ \tag{14}$$

Our mechanism predicts that with increasing energy of excitation, a greater number of excited radical cations will be produced.

The products expected from the ion-molecule reaction (eq. 14) are various radical products from the solvent, since it is known that phenyl-type radicals can abstract hydrogen from hydrocarbons:

$$A\bullet + RH \longrightarrow AH + R\bullet \tag{15}$$

$$R\bullet + R\bullet \longrightarrow R - R + \text{olefin} \tag{16}$$

Other products of coupling of phenyl radical and alkyl radical are possible as well.

The primary reactive intermediates formed during photolysis of c-hexane are the cyclohexyl radicals which react via radical coupling or disproportionation to give cyclohexene and bicyclohexyl respectively (eq. 16) . Another avenue for cyclohexene formation is the unimolecular decomposition of excited state c-hexane. Thus, we examined the products of c-hexane photolysis in the presence of an aromatic substrate (Table 1). At both wavelengths studied, photolysis in the presence of an aromatic substrate results in an increase in bicyclohexyl. This can be seen by comparing product yield during photolysis of c-hexane alone (background) to the yield of products resulting from photolysis in the presence of *p*-terphenyl. Since radical coupling is considered to be the sole pathway to bicyclohexyl, an increase in bicyclohexyl formation indicates an increased production of cyclohexyl radicals.

The aromatic substrate is the primary absorbing species, yet its presence results in an increase in products derived from the solvent. Also consistent with the proposed mechanism is an energy dependence of product formation, i.e. the observation of more radical products at higher photon energy.

Table 1.
Product formation during photolysis of cyclohexane[a] at 248 nm and 308 nm[b] in the presence of *p*-terphenyl.

	product yields[c] during photolysis of cyclohexane only		product yields during photolysis of cyclohexane in the presence of *p*-terphenyl[d]	
λ(nm)	308	248	308	248
bicyclohexyl	1	3.3	2.8	8.5
cyclohexene	24	65	17	50
$\frac{\text{cyclohexene}}{\text{bicyclohexyl}}$	24	20	6	6

[a]The samples consisted of 1.0 ml solution in a 5 mm path length cell, which were degassed then saturated with SF_6, which served as an electron scavenger. [b]The light source was a Questek series 2000 excimer laser at powers of 88 and 69 mJ per pulse for 308 and 248 nm operation respectively. The beam was focused to an area of 0.1 cm^2. Each sample received a dose of 1000 pulses. [c]Relative product yields based upon the yield of bicyclohexyl at 308 nm in neat solvent (actual yield ~10^{-5} M). Yields were quantified by gas chromatography using an internal standard technique. [d]The *p*-terphenyl concentrations were 6 x 10^{-4} M and 2 x 10^{-4} M for experiments at 308 and 248 nm respectively, in order to achieve a similar optical density (1.2).

The only other conceivable pathway to such products is the fragmentation of the excited state:

$$AH^{**} \longrightarrow A\cdot + H\cdot \tag{17}$$

Such a process would suggest an increase of H_2 yield, which is *not* observed. Thus, qualitative examination of some of the products supports the proposed mechanism.

In the proposed "hole injection" mechanism a threshold energy (>>308 nm) is required for the process to occur. Since radical products are formed at both 248 and 308 nm, this would suggest that the notion of "hole injection" is inconsistent with the observed occurrence of solvent derived radicals at low photon energies. A further limitation of the hole injection idea is that if solvent radical cations were so long-lived, as required by conductivity, they would not react by ion-molecule reaction to give the solvent radical R• to any appreciable extent. Again, as mentioned above, the existence of a long-lived alkane cation is incompatible with our FDMR and picosecond emission studies.

Further studies were carried out using nanosecond photolysis with optical and dc conductivity detection. The proposed mechanism predicts that with increasing energy of excitation, more excited radical cations are produced (eq. 12), which undergo ion-molecule reactions (eq. 14) so that fewer relaxed or "normal" radical cations are produced. On the other hand, we expect that the yield of photoionization will increase with increasing energy. We monitored the yield of photoionization by dc conductivity observation of (a) electron yield, and (b) free ion yield. Since we do not have complete knowledge of all the excited state parameters, we cannot ascertain the efficiency of photoionization at different wavelengths. We do monitor the number of photons absorbed by the sample and try to compare solutions of similar concentration and/or optical density. The important issue is the behavior of the relative yield of the aromatic radical cation and the electron.

We find that, as expected, comparison of electron and/or free ion yield shows an increase with energy (not decrease!), while the yield of $AH^{+\cdot}$ (relaxed) radical cations show a decrease with increasing energy. As an example, Figure 1 shows the results of quantum yields of cation radicals (by optical measurement), electron and free ion signals (by dc conductivity measurement) obtained in 248 nm and 308 nm laser flash photolysis of hydrocarbon solutions containing 2×10^{-4} M acenaphthene. Acenaphthene has the same extinction coefficient at both wavelengths. The same trends were observed in hydrocarbon solutions of anthracene, naphthalene, perylene and *p*-terphenyl. With increasing photon energy, the optically detected concentrations of cation radicals were decreasing or were unchanged while the dc conductivity detected electron and free ion signals all rose with increasing photon energy. In other words, the quantum yield of cation radicals does not parallel the yield of electrons and free ions.

Note that with hole injection mechanisms one would expect a decrease of $AH^{+\cdot}$ yield, but one would expect a decrease of e^- yield and of the free ion yield, which is not observed. Thus, the nanosecond flash photolysis and dc conductivity studies are consistent with our proposed mechanism and are not consistent with the "hole injection" idea.

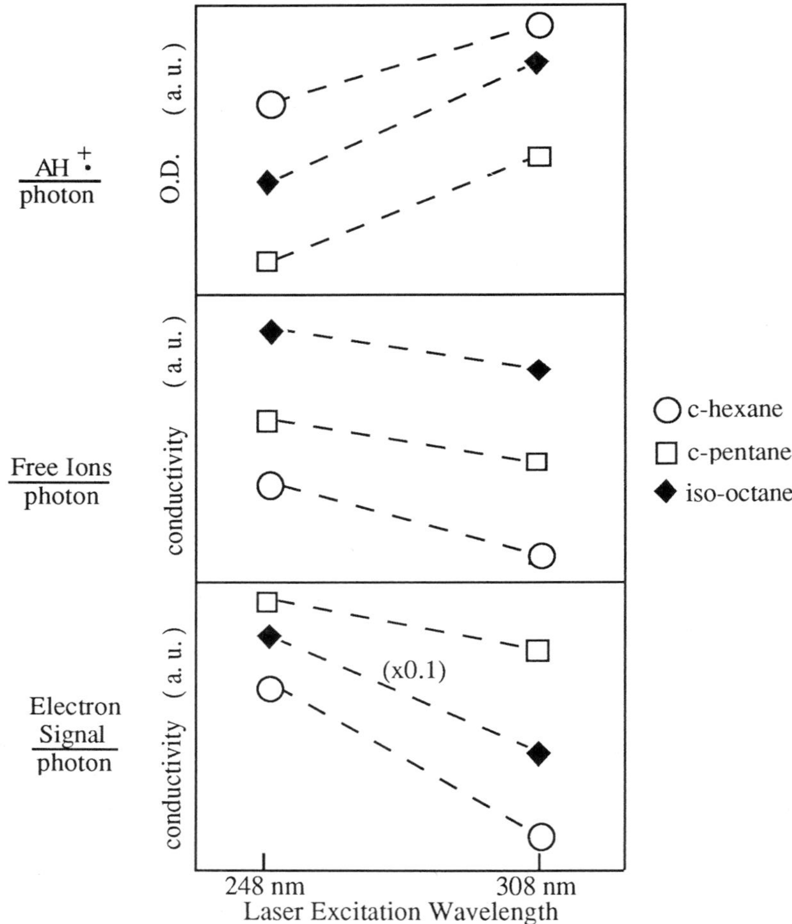

Figure 1. The comparison of quantum yields of cation radicals, free ions and electrons obtained in 248 and 308 nm laser flash photolysis of hydrocarbon solutions containing 2 x 10^{-4} M acenaphthene. The quantum yield of cation radicals is expressed as (O.D. at 670 nm)/(number of photons absorbed). 670 nm is the maximum absorption wavelength of acenaphthene cation radicals in non-polar solvent. The incident light intensity was measured by monitoring the absorption of the anthracene triplet and the ratio of absorbed light was estimated by the O.D. of acenaphthene solution at the excitation wavelength. The electron and free ion signals were measured by kinetic dc conductivity and are in the unit of nanosiemens.

Considering other aspects of the proposed mechanism, we now address evidence for the existence of other transient species which should occur. The protonated species RH_2^+ is the same species proposed in our FDMR studies of alkane radical cations in pulse radiolysis and suggested as the only species that could be considered a fast conductivity species, both in radiolysis and in photoionization.[24] We have observed that two solvents, c-hexane and trans-decalin, exhibit very fast proton transfer from the respective radical cations.[3]

The problem is that such species, RH_2^+, do not have either electron spin or a convenient optical signature in the visible spectrum. However, we could expect that such species, which according to the dc conductivity observations can last several hundred nanoseconds,[11] would undergo further reactions, e.g.,

$$RH_2^+ + AH \longrightarrow RH + AH_2^+ \qquad (18)$$

where the species AH_2^+ are aromatic proton adducts, i.e., arenium ions. Such species were observed by NMR and optical spectroscopy in super acids some years ago.[25] Recently, specialized solvents such as $(CF_3)_2CHOH$ were developed that allowed observations of aromatic carbocations by flash photolysis.[26] Indeed, we have tried to observe such species for several aromatic compounds in this solvent. However, the chemistry in alcohol and hydrocarbon solvents is complicated by the presence of strongly oxidizing products of electron scavenging by SF_6 which may subsequently give rise to aromatic radical cations. These observations can explain several puzzling experimental observations in photoionization and in radiolysis.[5]

As an example of our studies, Figure 2 shows the results of 308 nm laser flash photolysis and pulse radiolysis of anthracene in isopropanol. SF_6 was used as the electron scavenger. Note the same slow formation and the same absorption spectrum of this species at μs scale in both cases. The formation could be fitted by first order kinetics, and the rate constant is proportional to anthracene concentration. The bimolecular rate constant of this formation is obtained as 8 x 10^9 $M^{-1}s^{-1}$ by laser photolysis and 6 x 10^9 $M^{-1}s^{-1}$ by pulse radiolysis. The formation and decay of this species was found to depend on added KOH. The rate constant of KOH + precursor of this species ($k_{(ROH_2^+ + KOH)}$) is measured as 4.4 x 10^9 $M^{-1}s^{-1}$. Adding amines has a similar effect. This species is assigned to the anthracene radical cation. Whether aromatic carbocations can be observed in these solutions is still an open question.

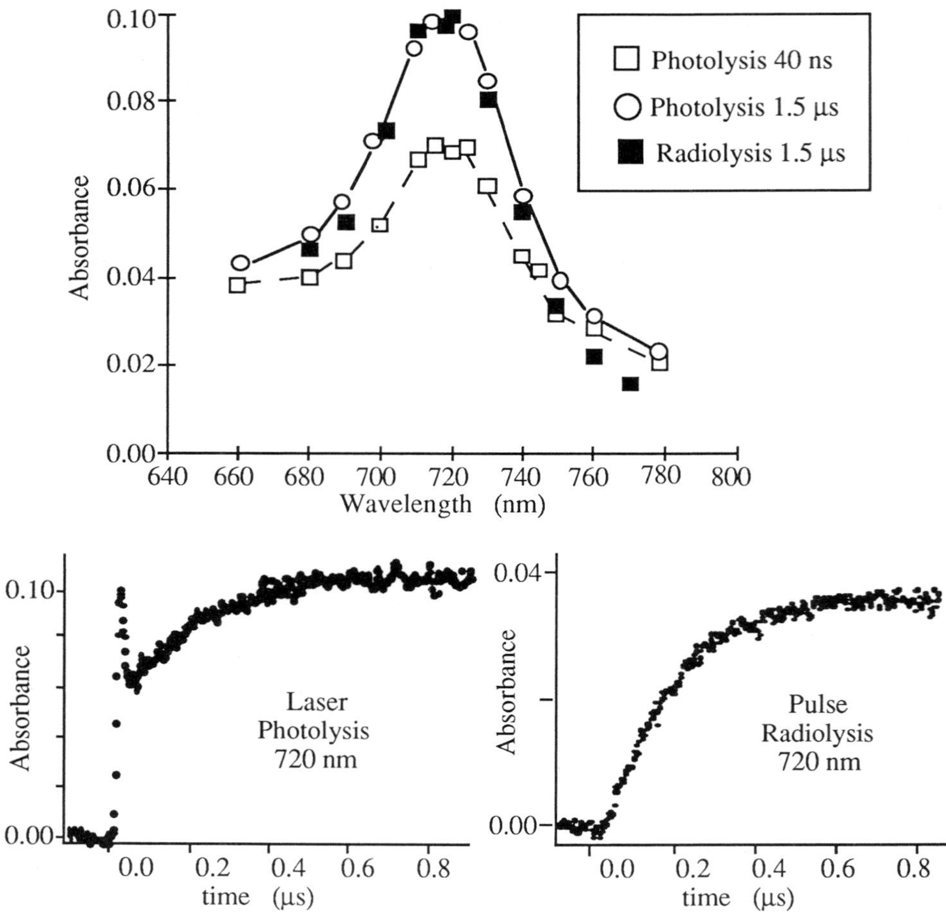

Figure 2. The transient absorption spectrum observed in laser flash photolysis (308 nm laser pulse) and pulse radiolysis of 5 x 10^{-4} M anthracene in isopropanol. The data points from pulse radiolysis were multiplied by 2.2 in order to match the maximum value of laser flash photolysis data for easier comparison. The bottom figures are the time profiles at 720 nm in both laser flash photolysis (left figure) and pulse radiolysis (right figure).

We have outlined a mechanism of photoionization which can qualitatively account for "puzzling" and seemingly conflicting observations in photoionization.

Our observations show that the hypothesis of hole injection fails to account for several findings, both in terms of products observed and the observation of transient species. Carbocations may play a significant role in photoionization and in radiolysis. While we cannot directly observe alkane solvent proton adducts, arenium may be observable.

A mechanism such as the one proposed in eq. 12 and 13 can be easily generalized to neat hydrocarbon radiolysis and high energy photoionization, i.e.,

$$RH \rightsquigarrow RH^{**} \longrightarrow RH^{+\cdot *} + e^- \xrightarrow{RH} R\cdot + RH_2^+ \qquad (19)$$

$$RH^{+\cdot *} \rightsquigarrow RH^{+\cdot} + e^- \longrightarrow RH^*$$

Thus, we would speculate that the low quantum yield of scavenged ions and of fluorescence could be accounted for by the occurrence of $RH^{+\cdot *}$ and its ion-molecule reactions. Ausloos has pointed out that only between 7 and 33% of ions produced recombine to give excited species [13] so the reason for the loss of ions could very well involve the mechanism we propose. Furthermore, our studies of excited state yields using picosecond emission methods have found a smaller yield of excited states from scavenged ions that could not be explained by any reasonable model.[14]

This simple mechanism could be a key needed to explain a considerable number of conflicting observations in radiolysis. Clearly, more work is needed to quantitatively test these ideas.

This work establishes that while charge transfer is the dominant process in condensed phase "high-energy chemistry", ion-molecule reactions play a very significant role in these same systems.

ACKNOWLEDGMENTS

Work performed under the auspices of the Office of Basic Energy Sciences, Division of Chemical Sciences, US-DOE under contract number W-31-109-ENG-38.

REFERENCES

1. Sauer, Jr., M. C.; Jonah, C. D.; Le Motais, B. C.; Chernovitz, A. C. *J. Phys. Chem.* **1988**, *92*, 4099.
2. Werst, D. W. and Trifunac, A. D. *J. Phys. Chem.* **1988**, *92*, 1093.
3. Werst, D. W.; Bakker, M. G.; Trifunac, A. D. *J. Am. Chem. Soc.* **1990**, *112*, 40.
4. Sauer, Jr., M. C.; Schmidt, K. H.; Liu, A. -D. *J. Phys. Chem.* **1987**, *91*, 4836.
5. Liu, A.-D.; Trifunac, A. D. to be published.
6. Radical Ionic Systems. Topics in Molecular Organization and Engineering, Lund A. and Shiotani, M., Eds.; Maruani, J., Ed.; Kluwer Academic:Dordrecht, The Netherlands, 1991.
7. Le Motais, B. C.; Jonah, C. D. *Radiat. Phys. Chem.* **1989**, *33*, 505.
8. Sauer, Jr., M. C.; Jonah, C. D.; Naleway, C. *J. Phys. Chem.* **1991**, *95*, 730.
9. de Haas, M. P.; Warman, J. M.; Infelta, P. P.; Hummel, A. *Chem. Phys. Lett.* **1975**, *31*, 382.
10. de Haas, M. P.; Hummel, A.; Infelta, P. P. *J. Chem. Phys.* **1976**, *65*, 5019.
11. Sauer, Jr., M. C.; Trifunac, A. D.; McDonald, D. B.; Cooper, R. *J. Phys. Chem.* **1984**, *88*, 4096.
12. Walter, L. Lipsky, S. *Int. J. Rad. Phys. Chem.* **1975**, *7*, 175.
13. Schwarz, F. P.; Smith, D.; Lias, S. G.; Ausloos, P. *J. Chem. Phys.* **1981**, *75*, 3800.
14. Jonah, C. D.; Sauer, Jr., M. C.; Cooper, R. *J. Phys. Chem.* **1991**, *95*, 728.
15. Hirata, Y.; Mataga, N. *J. Phys. Chem.* **1985**, *89*, 4031.
16. Brearley, A. M.; Patel, R. C.; McDonald, D. B. *Chem. Phys. Lett.* **1987**, *140*, 270.
17. Bakker, M. G.; Trifunac, A. D. *J. Phys. Chem.* **1991**, *95*, 550.
18. Schmidt, K. H.; Sauer, Jr., M. C.; Lu, Y. -L.; Liu, A. -D. *J. Phys. Chem.* **1990**, *94*, 244.
19. Warman, J. M. *Chem. Phys. Lett.* **1982**, *92*, 181.
20. Konuk, R.; Cornelisse, J.; McGlynn, S. P. *J. Chem. Phys.* **1985**, *82*, 3929.
21. Lamotte, M.; Pereyre, J., Joussot-Dubien, J.; Lapouyade, R. *J. Photochem.* **1987**, *38*, 177.
22. Ausloos, P.; Rebert, R. E.; Schwarz, F. P.; Lias, S. G. *Rad. Phys. Chem.* **1983**, *21*, 27.
23. Loffredo, D. M.; Trifunac, A. D., to be published.
24. Trifunac, A. D.; Sauer, Jr., M. C.; Jonah, C. D. *Chem. Phys. Lett.* **1985**, *113*, 316.
25. Olah, G. A.; Steral, J. S.; Asencio, G.; Liang, G.; Forsyth, D. A.; Mateescu, G. D. *J. Am. Chem. Soc.* **1978**, *100*, 6299.
26. Steenken, S.; McClelland, R. A. *J. Am. Chem. Soc.* **1990**, *112*, 9648.

DISCUSSION

Tagawa

In gas phase, ion molecular reactions occur for methane and ethane, but ion molecular reactions are not popular for alkanes. In liquid phase, ion molecular reactions are popular for alkanes. Do you have good explanations for that?

Trifunac

In the gas phase unimolecular processes dominate. In the matrix isolation studies it was observed that alkane radical cations readily undergo ion–molecule reactions. The condensed phase observations imply that gas phase reactions are of limited use in obtaining insights about reactions of ions in the condensed phase.

Sakaguchi

What is the character of the excited cation which releases proton?
Is it accessible by 715 nm excitation of ground state cation? Or needs more high energy?
Did you try the excitation of ground state cation by multi–pulse technique?

Trifunac

We do not know the character of the excited radical cation ; we only require that it is more acidic so that it undergoes fast proton transfer to the surrounding solvent molecules.
The excitation of the ground state cation should be feasible but we have not yet tried this multipulse experiment. The aromatic radical cations are good candidates for such an experiment while alkene radical cations are less suited since their extinction is quite low. We have tried such experiments with alkene radical cations without success.

Tachiya

What is the reason you use the Monte Carlo treatment instead of using the Smoluchowski equation? Even for geminate recombination which involves the conversion of chemical species to others with different diffusion coefficients the solution based on the Smoluchowski equation is available and simple enough to use for experimental analysis. Actually Prof. Tagawa is already successfully using it for analysis of his data.

Trifunac

You are quite right. We have a very good and convenient Monte Carlo simulation program, so we have not tried to use alternative methods.

Tagawa

What kinds of excited radical cations, do you think, in ion molecular reactions, electronic or vibronic excited state?
Do you estimate the lifetimes of excited radical cations?

Trifunac

We do not know what kind of excitation we are dealing with. All we require that proton transfer is enhanced in such an excited radical cation AH^{+*}. The lifetime of such an excited radical cation is expected to be very short i.e. shorter than the ion-recombination time in a given system. That means that such species can not live longer that few picoseconds. We will attempt to measure lifetimes of such cation in aromatic molecules.

Mobius

Which type of energy transfer processes did you correct for in the evaluation of your photoradiolysis data?

Trifunac

In radiolysis one creates small amounts of excited states of a solvent. Depending on the lifetime of this excited state (cyclohexane excited state lifetime is I nsec) and the concentration of the aromatic scintillator used, one sees some excited state of the scintillator produced via such excited state energy transfer.

Mataga

The relaxed radical cation of anthracene AH^{+} undergoes the usual charge pair dynamics, $AH^{+}+e^{-}$ ——> AH^{*}, but the excited ion AH^{+*} undergoes the ion molecule reactions with solvent (cyclohexane), $AH^{+*}+c\text{–}C_6H_{12}$ ——> A +c–$C_6H_{13}^{+}$. What is the reason for this difference? Why is the recombination reaction with electron not observed in the case of AH^{+*}? And, why is the ion molecule reaction not observed in relaxed AH^{+}?

Trifunac

We have made observations but have not established all the explanations. I have outlined that we find the evidence for occurrence of ion–molecule reactions with presumably excited radical cations AH^{+*}. What is the exact of this excitation we do not now so far. We do know that the "relaxed" cations AH^{+} are quite stable and do not undergo ion–molecule reactions. The excited radical cation presumably reacts in an ion–molecule reaction, faster than the ion–recombination. Excess energy input clearly opens a channel which reduces the yield of "relaxed" radical cations AH^{+}.

Faure

Prof. Haselbach (Basel) performed very interesting chemistry involving excited aromatic radical cations. It would be interesting to link your results with the photochemistry of excited radical ions.

Trifunac

Haselbach has studied photochemistry of various radical cations in a matrix at low temperature. Usually he observes isomerization and transformations of radical cations which have varying degree of unsaturation but are usually not aromatic. He usually works at such levels of dilution in the matrix to exclude reactions between substrate molecules. Prof. Shida, Kyoto University, has also carried out such studies.

Fukuzumi

The excited states of radical cations (AH^{*+}) should be extremely strong oxidants. Why does no electron transfer from a solvent to AH^{*+} take place? In other words why is proton transfer much faster than electron transfer in this case? Both processes may be highly exergonic.

Trifunac

We suggest that excited states of radical cations are more acidic. If your suggestion is correct and electron transfer from solvent takes place, this would be an energy uphill process. Our observations of "relaxed" alkene radical cations suggests that proton transfer (or symmetric H atom transfer) needs very little (other than thermal) excitation. The electron transfer – ion recombination is determined by the distance of the charge pair separation and electron mobility in a given medium. The proton transfer from the excited cation would occur to the surrounding solvent. i.e. there are many solvent molecules so that even a "slower" proton transfer could compete with a faster electron transfer.

Tagawa

1. What is the main factor for controlling the rates of ion molecular reactions of alkane radical cations with alkanes?

2. Do you have good explanation for that the radical cations and the proton aducts have the same absorption spectra?

Trifunac

1. Structural aspects of ion–molecule reactions are presumably involved. Steric and/or relative orientation of radical cation and proton acceptor must be important. Overall, very little is known about the factors influencing carbon center proton transfer.

2. We observe that the condensed phase absorption spectra of aromatic radical cations and aromatic ions are almost identical. We do not have the detailed analysis of the causes for this, but presumably the chromophore has similar electronic structure in both instances. May be more highly resolved spectroscopy would reveal a difference.

Dynamics and Mechanisms of
Photoinduced Transfer and Related Phenomena
N. Mataga, T. Okada and H. Masuhara (Editors)

MAGNETOKINETIC INVESTIGATIONS OF SPIN-FORBIDDEN ELECTRON BACK TRANSFER IN EXCIPLEXES AND RADICAL PAIRS

U.E. Steiner, D. Baumann and W. Haas

Fakultät für Chemie, Universität Konstanz, Germany

Abstract

In this paper we report on spin-orbit coupling induced magnetokinetic effects on the free radical yield of triplet electron transfer reactions. From the magnetic field dependence observed in a variety of mixed solvents the solvent dependent dynamics of short-lived triplet exciplexes is quantitatively deduced.

1. Introduction

One of the most important aspects of photoinduced electron transfer is its superior potential for achieving fast conversion of electronic excitation energy into chemical energy. A separation of the transferred electron from the hole it left behind is a crucial step in natural and artificial light energy conversion systems. In liquid solution such a separation is usually achieved by diffusional motion, whereby a primary pair of redox products separates into free radicals. Since the pioneering work of Weller, Mataga and others [1] intermediates of photoelectron transfer reactions in solution, viz. exciplexes and radical (ion) pairs (cf. Scheme 1) and their dynamics have been studied intensively [2]. Time-resolved fluorescence and laser-flash induced excited state absorption spectroscopy have been the most important experimental techniques in these studies. One important kinetic requirement for a direct, time-resolved study of the dynamics of exciplexes and correlated radical pairs is that formation of these species can be made faster than their decay. This condition is easily met if the electron transfer reactions involve excited singlet states since the time of their production is only limited by the duration of the excited light pulse or the initial photoelectron transfer step. In the case of reacting triplets the intramolecular ISC process following optical excitation usually causes a time lag not much shorter than about a nanosecond before the excited state of interest can react. It is therefore difficult to obtain direct information on subnanosecond dynamics of short-lived electron transfer intermediates in such cases.

This is exactly the situation encountered when studying photoelectron transfer reactions with typical sensitizer dyes as, for example, the phenothiazine dyes thionine or methylene blue in which we have been interested for some time [3-6]. These dyes are cations and acting as photoelectron acceptors they form neutral radicals so that no Coulombic interaction prevents the primary redox pair from dissociation. Furthermore, since rather polar solvents must be used with them, the bonding interaction due to resonant charge delocalization is weak because solvation in polar solvents tends to stabilize localized charge. Thus the primary intermediate of photoelectron transfer reactions with such reactants are very short-lived species with decay times comparable to or shorter than the S_1-T_1 ISC process in the dye.

We will describe here, how the method of magnetokinetics [8,9] in combination with the internal heavy atom effect can be utilized for obtaining quantitative information on the solvent dependent dynamics of the short-lived redox intermediates in such systems. The basis of our method is the kinetic competition between dissociation of the primary redox product (in general a two-step process leading from the triplet exciplex to a solvent-shared radical pair (RP) and furtheron to uncorrelated free radicals) and backward electron transfer (BET) either from the exciplex or the geminate RP. Since, according to the principle of spin conservation, the primary redox product originates with triplet spin and BET leads to the singlet ground state

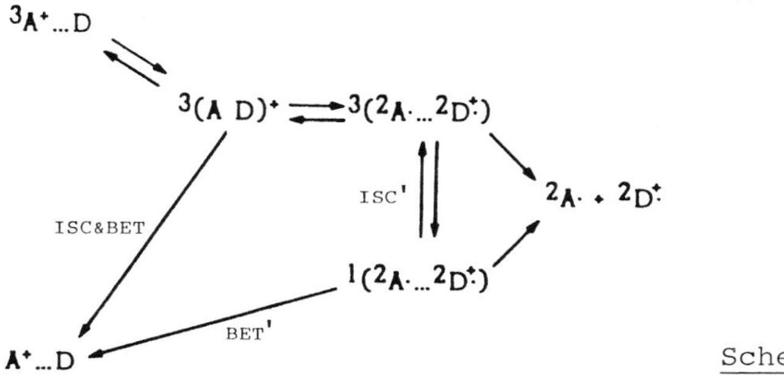

Scheme 1

of the pair a spin conversion has to take place before or synchroneously with the BET process. Control of this spin conversion by the internal heavy atom effect and its modulation by an external magnetic field is reflected (in a time-integrated manner) by a corresponding change in the yield of free radicals that can be conveniently measured on a long time scale. If the details of the magnetic field dependent processes are quantitatively understood it is possible to determine the dynamics of the intermediate redox pair from the magnetic field dependence of the free radical yield.

2. Solvent Dependence of Free Radical Yield

To a good approximation spin-orbit coupling (SOC) as induced by internal heavy atoms in the reacting species is only effective if the primary redox pair is in close contact, i.e. on the stage of the exciplex [7]. Spin conversion (ISC) and BET cannot be separated on this stage and the rate constant of the combined process, to be considered in more detail below, will be denoted k_{isc}. Using k_{er} as the rate constant of exciplex dissociation into radicals and neglecting at this stage of approximation BET contributions due to possible reencounters in the diffusion of the correlated geminate RP, we can represent the yield of free radicals from the exciplex as

$$\Phi_{fr} = k_{er}/(k_{er} + k_{isc}) = 1/(1+k_{isc}/k_{er}) \qquad (1)$$

indicating that this experimental quantity may be used e.g. as an indicator of solvent effects on k_{er} or k_{isc}. Note, however, that relative changes of these rate constants are reflected as changes of Φ_{fr} only if $k_{isc} > k_{er}$. For this reason the application of heavy atom substituents proves to be most useful [6].

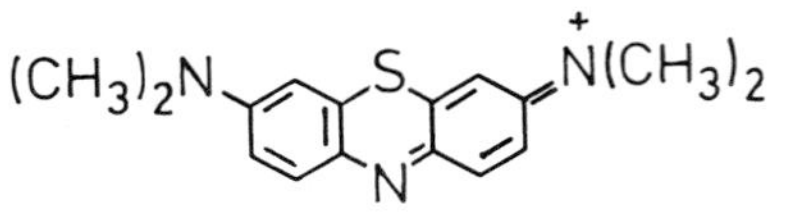

methylene blue (A^+)

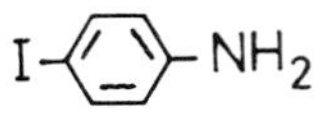

p-I-aniline (D)

The experimental results reported and analyzed in this paper have been obtained by investigating the electron transfer reaction between methylene blue triplet and p-I-aniline, the iodine substituent providing for a strong SOC in the redox pair. Observations of free radical yield were made by ns-time-resolved laser flash spectroscopy (for details cf. [7]). The quantity Φ_{fr} has been measured in a series of solvents including, in particular, binary solvent mixtures allowing for a fairly steady change of solvent properties. A survey of the results obtained is given in Table 1.

The variation of Φ_{fr} extending between 0.017 and 0.255 is considerable. A simple analysis might be based on eq (1) from which one obtains

$$k_{isc}/k_{er} = 1/\Phi_{fr} - 1 \qquad (2)$$

Since from Φ_{fr} only the ratio of k_{isc} and k_{er} can be obtained the individual solvent dependence of each of these parameters remains ambiguous. Nevertheless one might assume that k_{er} mainly reflects solvent viscosity effects, whereas k_{isc} characterizing a process that involves electron transfer (BET) should also depend on solvent polarity. It was on the basis of these assumptions that a reasonable interpretation of the solvent dependence of Φ_{fr} for the reaction of thiopyronine triplet with p-I-aniline and p-Br-aniline could be given previously [6]. As will be shown in the following, however, with the help of the magnetokinetic technique one can determine the absolute values of the rate parameters of the triplet exciplex and, in addition, assess the relative importance of BET within the exciplex and the correlated RP.

Table 1
Characteristics of solvents used and free radical yield Φ_{fr}, observed in zero magnetic field

Solvent v/v (%)	Polarity[a] E_T (kcal)	Viscosity (cP)	free rad. yield Φ_{fr}
MeOH/EGLY			
100/ 0	55.5	0.600	0.100
90/ 10		0.828	0.085
80/ 20		1.155	0.066
60/ 40		2.143	0.042
MeOH/ACN			
50/ 50	55.3	0.337	0.113
30/ 70	54.8	0.332	0.165
25/ 75	54.5	0.330	0.174
10/ 90	52.9	0.330	0.211
2/ 98	50.2	0.332	0.239
0/100	46	0.341	0.255
MeOH/H_2O			
90/ 10	56.0	0.84	0.059
80/ 20	56.5	1.09	0.038
70/ 30	57.0	1.30	0.021
60/ 40	57.5	1.47	0.017
MeOH/Dioxane			
80/ 20	54.4	0.580	0.103
50/ 50	52.2	0.653	0.107
25/ 75	49.4	0.799	0.121
ACN/H_2O			
98/ 2	50.4	0.35	0.191
90/ 10	53.5	0.41	0.110
80/ 20	54.9	0.50	0.090
75/ 25	55.4	0.55	0.070
50/ 50	56.8	0.80	0.040

a) cf. ref [14]

3. Magnetokinetic Effects on Spinforbidden BET

Various types of spin-inversion mechanisms (cf [8,9]) must be considered when analyzing the spinforbidden BET in Scheme 1. In the triplet exciplex the exchange interaction leads to a considerable T_1-S_1 energy splitting and the combined spininversion/ BET process takes place as a SOC induced T_1- S_0 radiationless process governed by the SOC properties of the heayy atom substituent. This process is magnetic field dependent according to the triplet mechanism (TM, cf. Figure 1). In the solvent-separated geminate RP T and S spinstates are in effect degenerate and the independent motions of the two unpaired electron spins will eventually lead to a change of radical spin correlation

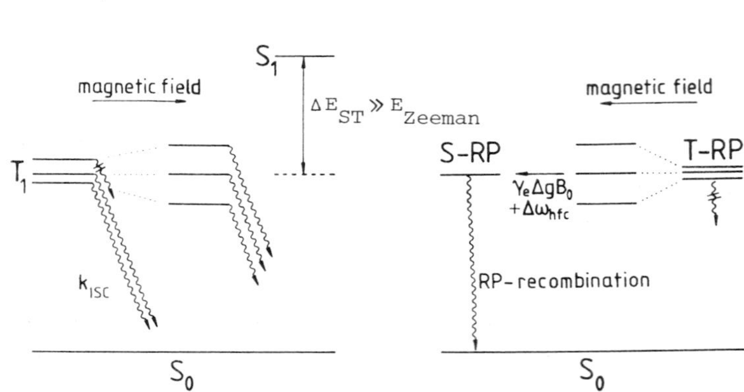

Figure 1. Two cases of SOC induced magnetokinetic effects. In the radical pair mechanism recombination to a singlet product is only possible form the radical pair singlet substate. A magnetic field may support recombination of triplet radical pairs by Δg-dependent mixing of radical pair T_0 and S spin substate. The triplet mechanism, applying e.g. to heavy atom containing triplet exciplexes, requires that fast direct SOC-induced transitions from triplet substates to the singlet ground state can occur. Due to symmetry selection rules SOC does not affect the triplet substates equally and they decay at different rates. In a magnetic field the zero-field triplet substates are mixed. Thereby their kinetic distinction is lost. Increasing uniformity of triplet sublevel decay enhances the overall efficiency of the T - S_0 process. The excited singlet state is not involved in case of this mechanism because the T/S energy splitting is too wide. Note that the splitting of the triplet levels is represented in a strongly enlarged scale as compared to the T/S_0 energy gap.

from T to S, so that a fast (now spin-allowed) BET may take place in the next re-encounter. It may be reasonably assumed [7] that of the various mechanisms of radical spin motion the T_o-S mixing due to the difference in Larmor frequencies (Δg- or Zeeman mechanism, cf. Figure 1), which increases linearly with the magnetic field, is the only one strong enough to become clearly detectable even for rapidly separating RPs. Thus our analysis of the magnetic field effect (MFE) is based on a combination of the TM in the triplet exciplex and the Δg-type RPM in the geminate RP. The mathematical details can be found in ref [7]. Here we will confine ourselves on defining the model parameters entering into the calculations and on a few demonstrations of how these parameters affect the observable magnetic field effect R(B) (eq.(3)) on the yield of free radicals.

$$R = [\Phi_{fr}(B) - \Phi_{fr}(0)]/\Phi_{fr}(0) \qquad (3)$$

The parameters characterizing the TM are k_{isc}, k_{er} and D_r. Here k_{isc} is the rate constant of the T_1-S_o process for two of the exciplex triplet substates, the rate constant for the third one being considered as negligible; k_{er} is the rate constant of exciplex dissociation into a solvent-shared, spin-correlated geminate RP; D_r is the rotational diffusion constant of the exciplex, entering as a parameter to describe spin relaxation among the zero-field triplet substates of the exciplex [10].

The parameters characterizing the Δg-type RPM are Δg, D, a, Λ_S, Σ_{hfc}. Here Δg is the difference of electronic g-factors of the two radicals estimated as 0.015 for the present system [7], and D is the sum of the translational diffusion constants estimated as $3.7\text{x}10^{-5}\eta^{-1}cm^2s^{-1}cP$, with η the dynamic viscosity of the solvent. Parameter 'a' denotes the reaction distance for BET in a singlet spin correlated RP, and Λ_S the probability for a singlet RP at distance 'a' to react in this or any following re-encounters. The quantity Σ_{hfc} (6 Gauss for the present system) denotes the rms deviation of longitudinal hyperfine field differences between the two radicals for all possible nuclear spin configurations.

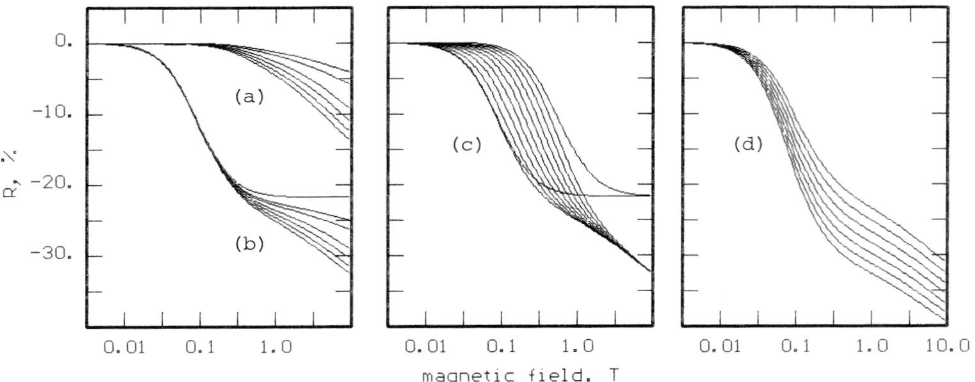

Figure 2. Theoretical parameter dependence of magnetic field effect on free radical yield.

(a) Pure RPM contribution, for curves from below values of (a/Å, Λ_S) are: (7, 1.0), (6, 1.0), (5, 1.0), (4, 1.0) (4, 0.65), (4, 0.5). Other RPM-parameters are $D=2x10^{-5}cm^2s^{-1}$, $\Delta g= 0.015$, $\Sigma_{hfc}= 6$ Gauss.

(b) upper curve: pure TM ($k_{isc}=2x10^{10}s^{-1}$, $k_{er}=10^9s^{-1}$, $D_r= 8x10^9s^{-1}$), other curves: TM in combination with RPM, parameters as in (a).

(c) Variation of TM for Φ_{fr} fixed to 0.1. Parameters for curves from left to right are $\log(k_{er}/s^{-1})$ = 9.0, 9.1, 9.2, 9.3, 9.4, 9.5, 9.6, 9.7, $k_{isc}= 18.5\ k_{er}$, $D_r= 2.6\ k_{er}$. The RPM contribution is constant corresponding to a=0.7nm, Λ_S=1.0. The first and the last case are also shown without contribution of the RPM.

(d) Variation of TM: $k_{er}= 10^9s^{-1}$, other parameters for curves from left to right are k_{isc}= (1.80,1.85, 1.90, 1.95, 2.00, 2.05)$x10^{10}s^{-1}$; D_r= (2.83, 2.61, 2.42, 2.27, 2.14, 2.03, 1.93)$x10^9s^{-1}$ is determined such that Φ_{fr}= 0.1 remains constant. Parameters of the RPM are as in (c).

In order to provide a feeling for how the two mechanisms contribute to the overall MFE, results of some model calculations employing various sets of the characteristic parameters are presented in Figure 2. Here set (a) of curves demonstrates the effect of the RPM only. The lowest curve (strongest MFE) in the set shown corresponds to the most favorable conditions of electron transfer that can be reasonably assumed (unity reaction probability at 0.7 nm separation). It is noteworthy that in the system considered the RPM does not contribute significantly to a MFE below 0.1 Tesla and does not exceed an absolute value of 8% at 3.5 Tesla the highest field investigated. Set (b) in Figure 2 shows the typical behaviour of the MFE due to the TM with a characteristic sigmoid dependence of R on log(B), implying saturation at high fields (cf. top curve in set (b), case of pure TM). Combining the TM with the RPM (cf. other curves of set (b)), the high field saturation behaviour is absent.

When fitting the TM to the experimental R(B) curve, only two parameters of the set k_{isc}, k_{er} and D_r are actually free to choose, if the additional constraint of the absolute experimental value of Φ_{fr} at zero field is taken into account. Thus, choosing k_{isc} and k_{er} the value of D_r is uniquely determined from $\Phi_{fr}(B=0)$ [11]. The effects of a variation of k_{isc} and k_{er} for a fixed value of Φ_{fr} are demonstrated by the sets (c) and (d) of curves in Figure 2. Changing k_{er} and keeping the ratio k_{isc}/k_{er} and D_r/k_{er} fixed (so that $\Phi_{fr}(0)$ remains constant) that part of the magnetic field depencence which is due to the TM undergoes a horizontal shift (if a log(B) scale is applied). The limiting behaviour at high fields (determined by the RPM), however, is independent of k_{er}. A change of k_{isc} at constant k_{er} with D_r adjusted such that Φ_{fr} at zero field is constant, leads to the variation of MFE demonstrated by set (d) in Figure 2. It appears that such a change has a major influence on the amplitude of the MFE.

In the following the experimental results of the MFE on Φ_{fr} in the solvents quoted in Table 1 are reported and analyzed in terms of the magnetokinetic mechanisms described above. As a general remark to the experimental technique for determining the magnetic field dependence of Φ_{fr} (actually defined as the efficiency of free radical formation from every triplet quenched by electron transfer) we note here that the results from the laser flash technique have been corroborated by applying also a more accurate, though more indirect, photostationary technique [7, 12] yielding the MFE at rather precision even for low absolute values of Φ_{fr}. The data points shown are from the latter experiments.

4. Solvent Viscosity Dependence of Magnetic Field Effect

The solvent series of methanol/ethylene glycol (EGLY) mixtures was chosen with the intention to realize a variation of solvent viscosity without a simultaneous change of polarity. This assumption seems to be fairly well justified in view of the chemical structures and the similar values of DK (32.7/37.7) or polarity parameter E_T (55.5/56.3) of the neat solvents. The MFEs observed in this series of solvents together with their theoretical simulations are shown in Figure 3. The parameters of the RPM correspond to the case of a diffusion controlled electron transfer reaction for BET with a reaction distance of 0.7 nm. The change of solvent viscosity is accounted for by the $D \sim 1/\eta$ relation mentioned above. The solvent viscosity dependence of the TM parameters obtained from the theoretical fit of the experimental MFE is displayed in the diagrams of Figure 4. The rate constant k_{er} shows a good proportionality to $1/\eta$ with a slope of about 20 times lower than to be expected for free diffusional separation of a pair of particles at 0.5 nm separation on the basis of the Eigen equation [13]. Therefrom we estimate a binding energy for the triplet exciplex of about 11 kJ mol^{-1}.

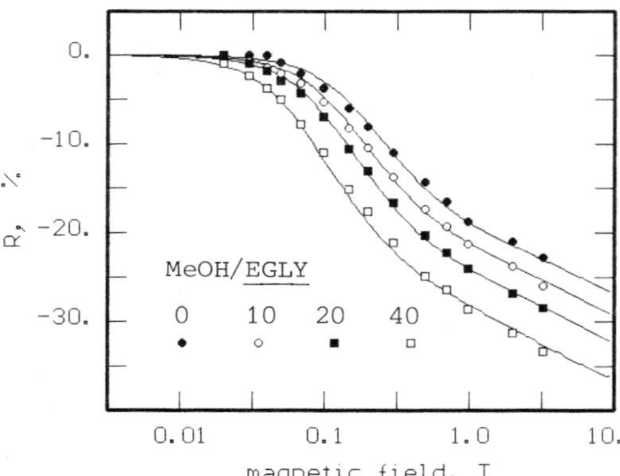

Figure 3. Magnetic field effect in MeOH/EGLY mixtures. Volume percentages of EGLY are assigned. The solid lines represent best fits using the TM parameter values indicated in Figure 4. The parameters of the RPM correspond to those of set (d) in Figure 2, except for D which is adapted to solvent viscosity as given in the text.

The rotational diffusion constant D_r is almost proportional to the inverse of the solvent viscosity. Applying Debye's expres-

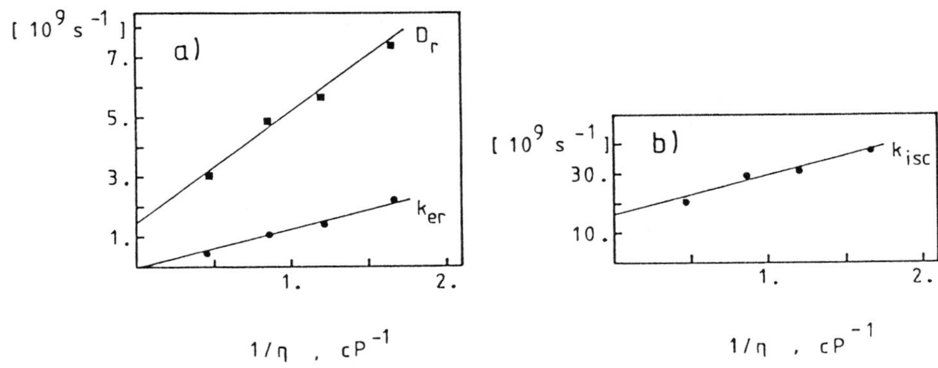

Figure 4. Viscosity dependence of the parameters k_{isc}, k_{er}, D_r determined for the triplet exciplex $^3(MB^{\cdot\cdot}p\text{-}I\text{-}An)^+$.

sion for the rate constant of rotational diffusion one can estimate an effective hydrodynamic radius of the exciplex of about 0.36 nm [7] which fits rather well with what one expects from a molecular model. The rate constant of ISC in the exciplex has the weakest dependence on solvent viscosity, as expected. The value of about 3×10^{10} s^{-1} corresponding to 30 ps lifetime is very fast for an ISC process. It is due to the strong internal heavy atom effect of the iodine substituent.

5. Solvent Polarity Dependence of Magnetic Field Effect

On the basis of the present method of evaluation no preassumptions have to be made when analyzing for the solvent dependence of the various parameters characterizing the dynamics of the triplet exciplexes. This may be demonstrated with the magnetokinetic results for the other solvents listed in Table 1, where the variations of k_{isc}, k_{er} and D_r causing the solvent dependence of Φ_{fr} are less obvious than in the solvent viscosity series described above. The MFE curves observed (cf. Figure 5) span a large range of variations. The kinetic parameters evaluated for the triplet exciplexes are presented in plots versus $1/\eta$ the inverse of solvent viscosity or versus the empirical solvent polarity parameter E_T (Figures 6 and 7).

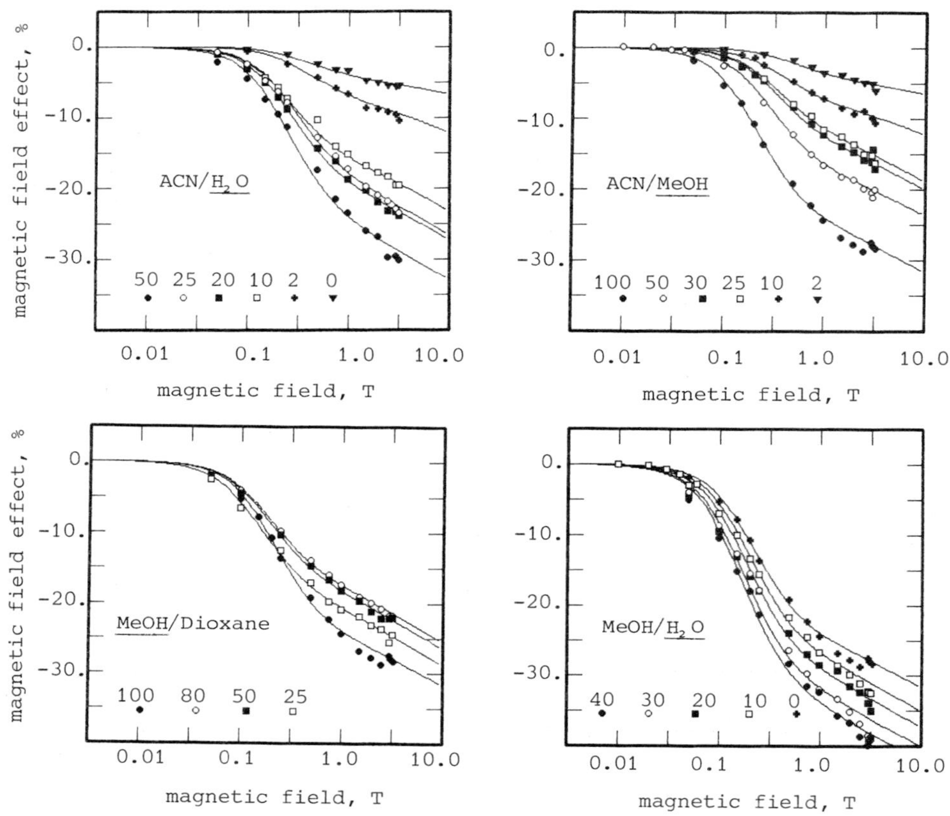

Figure 5. Magnetic field effect on the free radical yield in various solvent mixtures. Volume percentages of the underlined solvent components are indicated for each series of mixtures. The solid lines represent best fits using the TM parameters plotted in Figure 6. The parameter values for the RPM are those of set (d) in Figure 2, except for neat ACN (a= 4Å, Λ_S= 0.4), ACN/2% H_2O (a= 4Å, Λ_S= 0.8), ACN/2% MeOH (a= 4Å, Λ_S= 0.4), ACN/10% MeOH (a= 4Å, Λ_S= 0.8).

From the MFE in the H_2O/MeOH mixtures we obtain that k_{isc} is constant, with a very high value of about $4x10^{10}$ s^{-1}. The change in Φ_{fr} and in its magnetic field dependence is mainly due to changes in k_{er}, varying linearly with $1/\eta$ (Figure 6) which, along with the variation of D_r, is in line with the $1/\eta$ dependence of the MeOH/ EGLY mixtures.

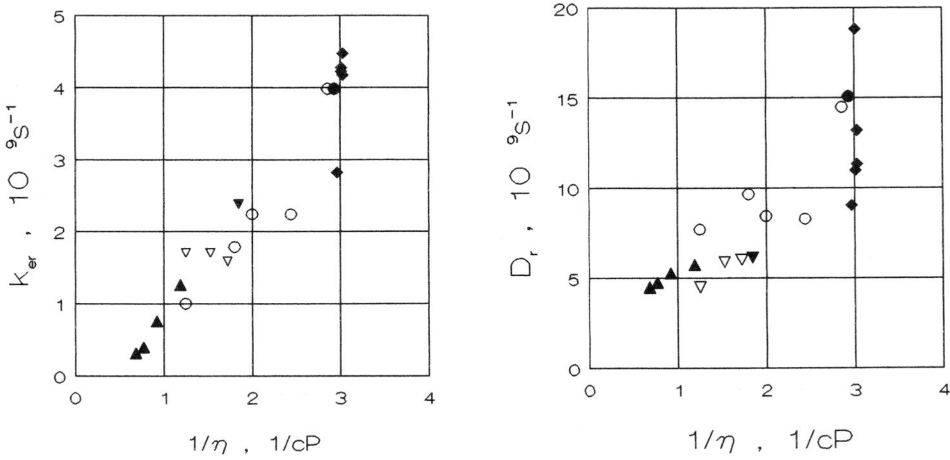

Figure 6. Correlation of exciplex paramters k_{er} and D_r with the inverse of solvent viscosity $1/\eta$.
Symbols are assigned as follows:
● ACN, ○ ACN/H_2O, ◆ ACN/MeOH, ▲ MeOH/H_2O, ▽ MeOH/Dioxane, ▼ MeOH

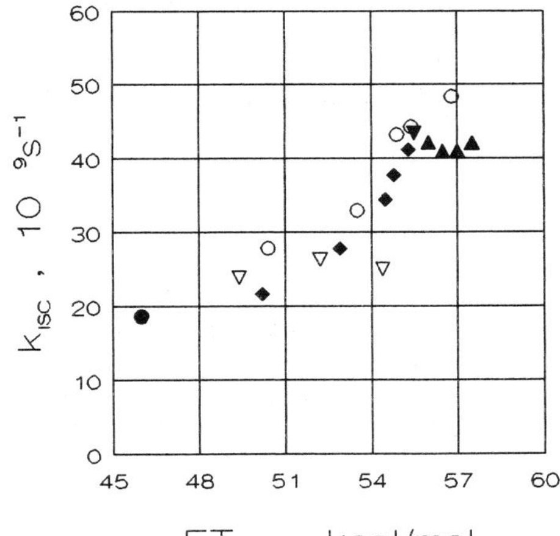

Figure 7. Correlation of exciplex parameter k_{isc} with solvent polarity parameter E_T. Assignment of solvents as in Figure 6.

The MFE curves obtained with ACN and its mixtures with H_2O or MeOH (Figure 5) span sets that are clearly different from those of the H_2O/MeOH mixtures. From the evaluation of the MFE it follows that with increasing amount of ACN k_{isc} is diminished whereas k_{er} and D_r increase. The trends of k_{isc} and k_{er} support each other in their effect on Φ_{fr} adopting its largest value in neat ACN. In the ACN mixtures k_{isc} shows a good correlation with the po-larity parameter E_T, whereas k_{er} and D_r fit the general $1/\eta$ trend although their scatter is considerable. It should be noted that for the small MFE in ACN rich solutions the TM and the RPM are no longer clearly separated on the log(B) axis. So the accuracy at which the triplet exciplex parameters can be determined is only about 30% in these cases, whereas it is better than about 10% in the others.

The mixtures of MeOH/Dioxan, in spite of an appreciable variation in E_T, do not show a great variation either in Φ_{fr} nor in the MFE on it. The rate parameters of the triplet exciplex determined from the MFE curves fit, however, into the general correlation of k_{isc} with E_T and of k_{er} and D_r with $1/\eta$.

In concluding, we point out that the determination of absolute values of individual rate constants characterizing the decay of the short lived triplet exciplexes, as has been demonstrated in this paper, confirms our previous interpretation of the solvent viscosity and polarity dependence of Φ_{fr} [6]. The argument provided for an increase of k_{isc} with solvent polarity was that increasing the solvent polarity should have a decreasing effect on the Franck-Condon energy gap between exciplex and ground state pair of reactants.

Acknowledgement
Financial support of the present investigations by the Deutsche Forschungsgemeinschaft is gratefully acknowledged.

References

[1] For references on the early work cf. for example M. Gordon and W.R. Ware (eds.), The Exciplex, Academic Press, New York 1975.

[2] A recent comprehensive treatise on the field of electron transfer is: M.A. Fox and M. Chanon (eds.), Photoinduced Electron Transfer, Vol. A-D, Elsevier, Amsterdam 1988.

[3] U.E. Steiner, M. Hafner, S. Schreiner and H.E.A. Kramer, Photochem.Photobiol. 19 (1974) 119

[4] U.E. Steiner, G. Winter and H.E.A. Kramer, J.Phys.Chem. 81 (1977) 1104

[5] U.E. Steiner and G. Winter, Chem. Phys. Lett. 55 (1978) 364

[6] G. Winter and U.E. Steiner, Ber. Bunsenges. Physik. Chem. 84 (1980) 1203

[7] U.E. Steiner and W. Haas, J. Phys. Chem. 95 (1991) 1880

[8] U.E. Steiner and T. Ulrich, Chem. Rev. 89 (1989) 51

[9] U.E. Steiner and H.-J. Wolff, in J.J. Rabek and G.W. Scott (eds.), Photochemistry and Photophysics, Vol.IV, CRC Press, Boca Raton, 1991, in press.

[10] U.E. Steiner, Ber. Bunsenges. Physik. Chem. 85 (1981) 228

[11] T. Ulrich, U.E. Steiner and R.E. Föll, J. Phys. Chem. 87 (1983) 1873

[12] W. Schlenker and U.E. Steiner, Ber. Bunsenges. Physik. Chem. 89 (1985) 1041

[13] M. Eigen, Z. Phys. Chem., Frankfurt am Main, 1 (1954) 176

[14] C. Reichardt, Solvent Effects in Organic Chemistry, Verlag Chemie, Weinheim 1979

DISCUSSION

Rettig

You showed strong magnetic field effects in exciplex systems (radical cation and neutral radical) with presumably strong exchange interaction. Would you expect that a similarly strong magnetic field effect might be observable in TICT systems with charge shift (radical cation of neutral radical need together by a single band)? What about TICT systems with charge separation?

Steiner

The essential condition for observing a triplet-mechanism type magnetic field effect is that the triplet under consideration undergoes a rapid, spin-sublevel selective intersystem crossing decay. Spin-sublevel selectivity is the usual property of spin-orbit coupling and not a very restrictive condition. The rate of the ISC process, however, must be faster or at least comparable to the rate of spin-relaxation among the zero-field substates. In liquid solution this process occur at a rate approximately equal to the orientational relaxation rate of the molecule. Comparably fast ISC usually requires enhanced spin-orbit coupling as may be achieved by heavy atom substituents.
The value of the exchange interaction (energy gap to configurationally related singlet) is not of relevance in the triplet mechanism.
In the case of linked donor-acceptor systems or TICT states the triplet mechanism can operate as in triplet exciplexes. The effect could be observed as a magnetic-field dependent shortening of the triplet lifetime or, if another decay channel besides ISC exists, as a magnetic field effect on the yield into this channel (analogous to the free radical yield in our triplet exciplex case). In the case of linked system such a monitoring channel might perhaps be realized by some fast photochemical reaction.

Matsuo

It is a pity that you did not have enough time to discuss $Ru(bpy)_3^{2+}$-viologen system. I would like to hear about the effects of the ligand and acceptor on the magnetokinetic parameters. Could you explain them quickly?

Steiner

Photoelectron transfer between $Ru(bpy)_3^{2+}$ and methylviologen is followed by efficient backward electron transfer (BET), such that only about 25% of free radicals are formed. By a magnetic field of 3.5T the BET is enhanced such that the free radical yield is reduced by 23%. Replacing the bipyridine (bpy) ligand by 4,4'-diethoxycarbonyl-bpy (dce) reduces this magnetic field effect to 6%. When mixing the two types of ligands it appears that the magnetic field effect reflects independent additive contributions of the ligands. Recent results suggest that this behavior might be attributed to the different contributions of the ligands to electron spin relaxation in the Ru(III) complex. The validity of this assumption will be investigated by ^{1}H–NMR of the paramagnetic Ru(III) complexes. Since electron spin relaxation in Ru(III) complexes is due to vibrational modulation of the g–tensor, our magnetokinetic results may provide a novel access to information about the bonding potential of the ligands in such complexes.

Ohno

I am interested in the dependence of k_{isc} on solvent polarities. You mentioned the polarity of solvent shifted ΔE to smaller one. My question is concerned with ΔE–dependence of k_{isc}, which may display the maximum of k_{isc} at higher ΔE. What is the energy gap involved in the spin converted ET process?

Steiner

The free energy change driving the spin–inverted backward electron transfer may be approximated as the negative free energy of the reaction $A^{+}\cdots D$ —> $A\cdots D^{+}$. In methanol this amounts to about 1.2 eV.

Dynamics and Mechanisms of
Photoinduced Transfer and Related Phenomena
N. Mataga, T. Okada and H. Masuhara (Editors)

INTRAMOLECULAR ELECTRON TRANSFERS IN BIMETALATED COMPOUNDS of Ru(II) AND Rh(III)

K. Nozaki[a], T. Ohno[a], and M.-A. Haga[b]

[a]Chemistry Department, College of General Education, Osaka Univeristy, Toyonaka, Osaka 560

[b]Department of Chemistry, Faculty of Education, Mie Univeristy, Tsu, Mie 514

Abstract

Charge transfer excited states and intramolecular electron transfers are examined for three kinds of ligand-bridging binuclear Ru(II) and Rh(III) compounds. The rate of ET were determined by means of kinetic transient absorption spectroscopy in a wide temperature range. Charge transfer excited states of Ru(II) sites are characterized in comparison with absorption spectra of the ligand reduced and the Ru(II) oxidized. Excited electron on the bridging ligand (L-L) are transferred to Rh(III) ion coordinating to the other coordination site of bridging ligand. Protonated moieties of the bridging ligands, oxidized the Ru(II) site in the charge transfer excited state. The rates of ET-to-($L\text{-}LH_2^{2+}$) as well as those of ET-to-Rh(III) are discussed in connection with nuclear rearrangement accompaning ET and vibronic interaction.

1. Introduction.

Rates of nonadiabatic electron transfer (ET) reactions are controlled not only by Franck-Condon factor wighted density but also by vibronic interaction between reactants. The quantum theory of nonadiabatic ET predicted a bell-shaped energy gap-dependence of ET rate [1], whose maximum depends on vibronic interaction between reactants. Vibronic interaction has been taken as being weak for long-range ET [2] and for spin-inverted ET processes [3]. Intermolecular vibronic interaction in a transition state of ET, however, has seldom been estimated in a quantitative sense because both distance and orientation between reactants in the transition state are unknown. It is worthwhile to investigate intramolecular ET processes within bi-chromophore compounds for which the extent of chromophore-chromophore vibronic interaction can be estimated. Actually, dependence of ET rates on the distance between chromophores has been pursued instead of directly measuring vibronic interaction [4], since the vibronic interaction are too small to be directly detected. Mode and path of vibronic interaction enhancing ET rate has been also tried to clarify [5].

Extent of vibronic interaction between metal-chromophores in ligand bridging homo-binuclear compounds has been examined from intensity of metal-to-metal CT transition [6]. ET processes examined here are photoinitiated spin-allowed ones in ligand bridging hetero-binuclear compounds. The electron donor is an excited RuL_2^{2+} (L = 2,2'-bipyridine (bpy) or 4,4'-dimethyl-2,2'-bipyridine (dmbpy)) and the electron acceptor is $RhL'_2{}^{3+}$ (L' = bpy and 1,10-phenanthroline (phen)), which are linked by a tetradentate ligand (L-L),

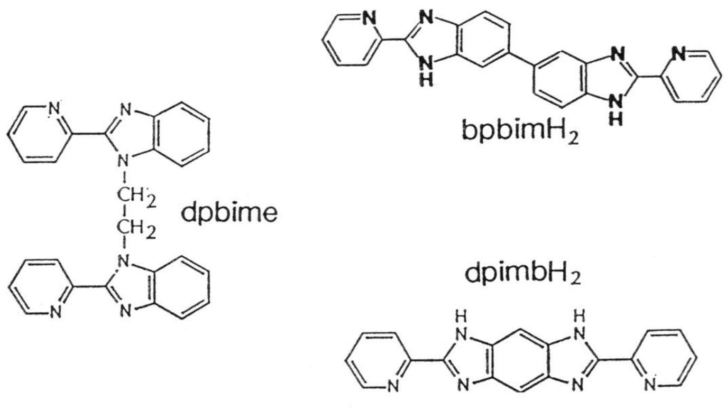

bpbimH_2

dpbime

dpimbH_2

2,2'-bis(2"-pyridyl)-5,5'-bibenzimidazole (bpbimH_2) [7], di-[(2-(2'-pyridyl)benzimidazoyl)]ethane (dpbime) [8], and di-2-(2'-pyridyl)bis(imidazo)benzene ($dpimbH_2$) [9]. $RuL_2(L-L)^{2+}$ in the metal-to-ligand charge transfer (CT) excited state is capable of transferring an excited electron to $Rh(L')_2(L-L)^{3+}$ because of ergonicity (-0.16 eV).

The bridging ligands consist of chromophores of 2-(2'-pyridyl)benzimidazole (pbimH) for $bpbimH_2$ and dpbime, and 2-(2'-pyridyl)imidazol (pimH) for $dpimbH_2$. The chromophores having two protons in place of coordinating to Rh^{3+} are capable of accepting an electron from the excited Ru(II) site. Moreover, the diprotonated forms of the chromophores are expected to interact with the Ru(II) site more than $Rh(L')_2{}^{3+}$ is, because the diprotonated moiety is didrectly linked to the reduced one. ET to $pbimH_3{}^{2+}$, $pimH_3{}^{2+}$ and $Rh(L')_2(L-L)^{3+}$ from the excited Ru(II) site were examined in a wide temperature range. The apparent activation energy of ET quenching was determined from the temperature-dependent rate. They will be discussed in correlation with nuclear rearrangements accompanying ET and vibronic interaction.

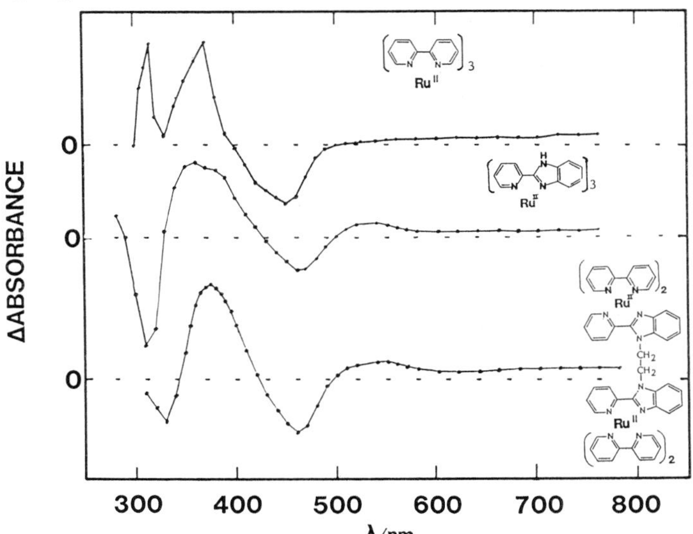

Figure 1. Transient absorption spectra at 100 ns after the laser excitation of $HClO_4$ (1 mM) CH_3CN solution at ambient temperature.
The top : $Ru(bpy)_3{}^{2+}$, the middle : $Ru(pbimH)_3{}^{2+}$, and the bottom : $Ru(bpy)_2(dpbime)^{2+}$.

2. CT excited state of binuclear ruthenium compounds, $[Ru(bpy)_2]_2(bpbimH_2)^{4+}$ and $[Ru(bpy)_2]_2(dpimbH_2)^{4+}$ [8-11]

The lowest excited states of ruthenium(II) polypyridine compounds are regarded as a triplet Ru(II)-to-ligand charge transfer state, in which a π-radical of ligand is bound to Ru(III) ion via Ru-N σ-bonbds. Transient absorption (TA) spectra of laser-excited compounds give rise to the strong evidence for the assignment of a reduced ligand in the CT excited state as is shown in Fig. 1. A TA spectrum of $Ru(bpy)_2(dpbime)^{2+}$ (Fig. 1-c) is a hybrid of TA spectra of laser excited $Ru(bpy)_3^{2+}$ (Fig. 1-a) and $Ru(pbimH)_3^{2+}$ (Fig. 1-b), indicating that the excited electron resides on both bpy and dpbime. As Figs. 2a and 3a show, TA spectra of $Ru(bpy)_2(L\text{-}L)^{2+}$ (L-L = $bpbimH_2$ and $dpimbH_2$) and $Ru(dmbpy)_2(bpbimH_2)^{2+}$ exhibit a π-π* band at 25.6×10^3 cm^{-1} ($\Delta\varepsilon$= 12,300 $M^{-1}cm^{-1}$) due to $bpbimH_2^-$ and at 24.2×10^3 cm^{-1} ($\Delta\varepsilon$ = 23,200 $M^{-1}cm^{-1}$) due to $dpimbH_2^-$, a strong bleaching of

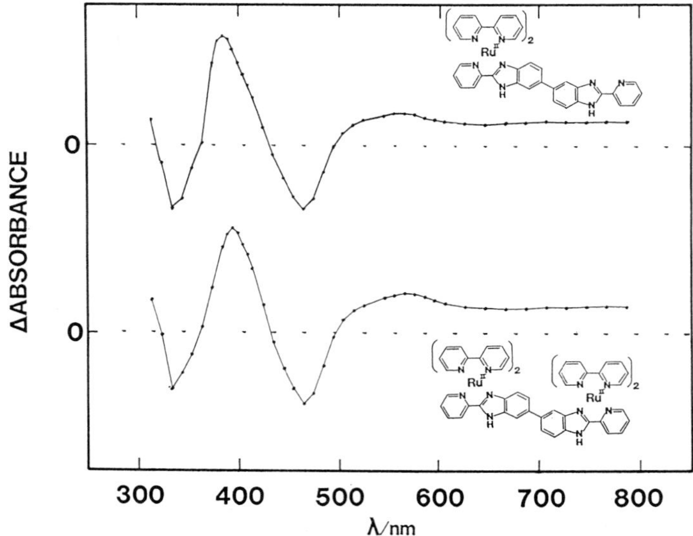

Figure 2. Transient absorption spectra at 100 ns after the laser excitation of $HClO_4$ (1 mM) CH_3CN solution at ambient temperature.

The top : $Ru(bpy)_2(bpbimH_2)^{2+}$, and the bottom : $[Ru(bpy)_2]_2(bpbimH_2)^{4+}$.

the π-π* band at $(26.3\text{-}28.6)\text{x}10^3$ cm^{-1} of $bpbimH_2$ and $dpimbH_2$, and a broad $(L\text{-}L^-)$-to-Ru(III) CT band in a red region ($\Delta\varepsilon$ = 3,000 and 9,200 $M^{-1}cm^{-1}$ at 14.7 $\text{x}10^3$ cm^{-1} for $bpbimH_2^-$ and $dpimbH_2^-$, respectively), which are not ascribed to the formation of bpy^- ($dmbpy^-$). The π-π* at $34.6\text{x}10^3$ cm^{-1} of bpy coordinating to Ru(II) is shifted to $32.2\text{x}10^3$ cm^{-1} in the CT excited state. Since the shifted π-π* band is seen for all the $Ru(bpy)_2(L\text{-}L)^{3+}$ prepared electrochemically, the lowest excited state of $Ru(bpy)_2(dpbime)^{2+}$ is characterized as Ru(II)-to-L CT.

A similar TA of $[Ru(bpy)_2]_2(bpbimH_2)^{4+}$ to that of $Ru(bpy)_2(bpbimH_2)^{2+}$ (Fig. 2b) are indicative of very weak exchange interaction between $d_{\Pi-\Pi}$ electrons of Ru(II) ion and Ru(III) ion. In the case of the binuclear $dpimbH_2$ compound whose phosphorescence is of lower energy by 59 meV than that of the mononuclear counterpart, the TA is slightly changed ;

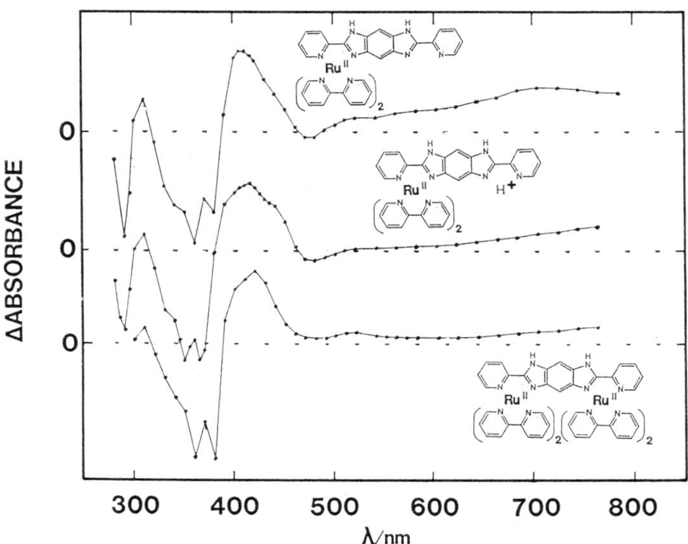

Figure 3. Transient absorption spectra at 100 ns after the laser excitation of CH_3CN solution at ambient temperature.
The top : $Ru(bpy)_2(dpimbH_2)^{2+}$, the middle : $Ru(bpy)_2(dpimbH_3)^{3+}$ in $HClO_4$ (10 mM), and the bottom : $[Ru(bpy)_2]_2(dpimbH_2)^{4+}$ in $HClO_4$ (1 mM).

the band ($\Delta\varepsilon$ = 24,000 $M^{-1}cm^{-1}$) in (18.2-25.0)x10^3 cm^{-1} is strong and shifted to lower energy and the broad band in a red region reduced the intensity to half compared with the mononuclear one (Fig. 3b). Since the mono-protonation of $Ru(bpy)_2(dpimbH_2)^{2+}$ similarly modifies the TA (Fig. 3c), the excited electron in the CT state is delocalized through the whole ligand.

Coordination of $Rh(bpy)_2{}^{3+}$ to $RuL_2(L-L)^{2+}$ (L-L:bpbimH_2 and dpbime) shortened the lifetime of the excited Ru(II) site. The CT excited state localized in the Ru(II) site lengthened the lifetime with decrease in temperature. Diprotonated species of $RuL_2(L-L)^{2+}$ has too short a lifetime of the excited state to be detected at ambient temperature. The TA spectrum of the CT excited state with lifetime of - 10^{-6} s at 140 K was observed to be very similar to that of $[Ru(bpy)_2]_2(L-L)^{2+}$ (L-L : bpbimH_2 and dpbime).

Meanwhile, coordination of $Rh(bpy)_2{}^{3+}$ and two H^+ to $Ru(bpy)_2(dpimbH_2)^{2+}$ changed both the TA spectrum and lifetime of the excited state. TA peak of $(bpy)_2Ru(dpimbH_2)Rh(bpy)_2{}^{5+}$ at 168 K was shifted to 20.8x10^3 cm^{-1} as well as $Ru(bpy)_2(dpimbH_3)^{4+}$ at ambient temperature. Since the $dpimbH_2{}^-$, which was produced in an intermolecular ET with an electron donor (phenothiazine), exhibits a similarly shifted TA with diprotonation, the *-electron of $dpimbH_2{}^-$ may be delocalized throuth dpimbH_2 to be subject to the charge of protons.

3. Intramolecular ET in Photoexcited $RuL_2(L-L)RhL'_2{}^{5+}$ [10]

Quenching of the CT excited state of the Ru(II)-site are ascribed to ET between the excited Ru(II)-site and the Rh(III)-site in $Ru(dmbpy)_2(bpbimH_2)Rh(phen)_2{}^{5+}$ on the basis of redox potentials of the sites (see Table). Rates of the ET to Rh(III) were measured at low temperatures (165-280 K) by monitorring the decay of TA after the laser excitation of Ru(II) site. The rates of the ET are evaluated to be 10^6-5x10^7 s^{-1} from a difference in the decay rates between the Ru(II)-Rh(III) compound and the reference Ru(II) compound

$(Ru(bpy)_2(bpbimH_2)^{2+})$. Temperature dependence of the ET rate gives rise to a pre-exponential term of $5x10^{11}$ s^{-1} by extrapolation and an activation energy (E_a) of 0.20 eV. Similar values of k_{et} at every temperature were measured for the Ru(II)-Rh(III) compound bridged by a dpbime. The same analysis of temperature dependent ET rate obtained for $Ru(bpy)_2(dpbime)Rh(bpy)_2{}^{5+}$ affords a similar pre-exponential term ($2x10^{11}$ s^{-1}) and the same activation energy (E_a=0.20 eV) compared with the $bpbimH_2$ compound.

As for the $dpimbH_2$ brdidging compound, $[Ru(bpy)_2(dpimbH_2)Rh(bpy)_2]^{5+}$, the ET-to-Rh(III) was faster than those of the $bpbimH_2$ and the dpbime compounds. The apparent activation energy is slightly smaller (0.17 eV). The rate of ET-to-Rh(III) in the dpbime bridging compound was similar to that of the $bpbimH_2$ mediated ET-to-Rh(III). E_a of dpbime mediated ET-to-Rh(III) is the same as those of the $bpbimH_2$ mediated ET-to-Rh(III). The ET rates at ambient temperature are estimated to be in the order of 10^8 s^{-1} from the temperature-dependent ones.

The back ET from the Rh(II) to the Ru(III) was too fast to be directly observed. No time lag was found between the TA decay at $25.3x10^3$ cm^{-1} and the recovery of the MLCT band at $21.7x10^3$ cm^{-1} nm. The possible slowest rate of the back ET is 1.3 x 10^8 s^{-1} at 300 K because no ET product was detected by means of n-sec laser photolysis.

4. Intramolecular ET in Excited $RuL_2(L\text{-}LH_2)^{4+}$ [11]

The rapid quenchings of the excited Ru(II)-site by the diprotonated moieties of the tetradentate ligand are ascribed to ET-to-$(L\text{-}LH_2{}^{2+})$ by taking the ergonicity into consideration. The direct measurement of the ET in $RuL_2(bpbimH_4)^{4+}$ by means of p-sec laser photolysis provided the rate constant of 72 x 10^8 s^{-1} for L=bpy and $170x10^8$ s^{-1} for L=dmbpy at ambient temperature. Cooling the sample in a mixture of butyronitrile and propionitrile to 165 K reduced the rate to ca. 0.5 x 10^8 s^{-1}.

The ET rate of the excited Ru(II) moiety depends on the

tetradentate ligands of which the di-protonated site act as an electron acceptor. Diprotonated 2-(2'-pyridyl)benzimidazole of dpbime and $bpbimH_2$, in which the electronic coupling between the chromophores are presumed to be weak, oxidized the excited moiety of $Ru(bpy)_2(L\text{-}L)^{2+}$ with the pre-exponential term of $2.2 \times 10^{12}\ s^{-1}$ and $4.5 \times 10^{12}\ s^{-1}$, respectively. The ET quenching of the excited Ru(II) moiety by diprotonated 2(2'-pyridyl)-imidazo moiety (pimH) of $dpimbH_2$ was the biggest pre-exponential term ($9.5 \times 10^{12}\ s^{-1}$), in which the pimH moiety is expected to moderately interact with each other. This trend indicates the ET processes are not fully adiabatic, though the prefactors are pretty large. The activation energy (E_a) was 0.14 for the $dpimbH_2$ compound and 0.16 eV for the $dpimbH_2$ and the dpbime compounds. The pre-exponential terms (Table) are 10-100 times as large as those of ET to Rh(III).

5. Role of Vibronic Interaction between Electron Acceptor and Electron Donor in ET.

The rate of nonadiabatic ET process is given in Equation 1,

$$k_{ET} = (2\pi/\hbar)V^2(2\hbar\lambda k_BT)^{-1/2}\ (e^{-S}S^w/w!)\exp[-(\Delta E+\lambda+wh\nu)/4\lambda k_BT]$$

$$S = \lambda'/h\nu, \qquad (1)$$

where V, ΔE, ν, λ and λ' are the electronic coupling matrix element, the energy gap involved in ET, the frequency of skeletal vibration of the reactants, the rearrangement energy of the surrounding solvent, and the rearrangement energy of high frequency mode, respectively [1]. The temperature-independent pre-exponential term is proportinal to the square of vibronic interaction. The pre-exponential term obtained here is small ($(2\text{-}5) \times 10^{11}\ s^{-1}$) for ET to Rh(III) compared to those ($(2\text{-}9) \times 10^{12}\ s^{-1}$) of ET-to-($L\text{-}LH_2^{2+}$). The reduction factor is never accounted for by the difference in the ergonicity and the rearrangement energy involved in the ET processes, the latter of which has been estimated from the activation energies to be 0.8 - 1.1 eV by assuming no contributuion of vibrational excitation in the ET process [11].

The electronic interaction between Ru(II) and Ru(III) in the dpbime, $bpbimH_2$, and $dpimbH_2$ bridging compounds is estimated to be <0.1 meV, 0.1 meV, 5 meV, respectively, from the intensities and transition energies (- 1 eV) of Ru(II)-to-Ru(III) CT in a near-infrared region. The broadness of Ru(II)-to-Ru(III) CT does not imply pure electronic but vibronic interaction between Ru(II) and Ru(III) ions [6]. Since the vibronic interaction is mediated by the bridging ligand, there must be a more vibronic interaction between the moieties, pbimH of $bpbimH_2$ and dpbime, and pimH of $dpimbH_2$. Meanwhile, Ru(II)-Rh(III) interaction mediated by the bridging ligand is much weaker than Ru(II)-Ru(III) interaction, because there is a holl in d_λ orbitals allowing Ru(II)-to-Ru(III) CT interaction in the latter. This difference in the electronic interaction is responsible for the difference in the pre-exponential term between the bridging ligands..ls1

Table. ΔG, Pre-factor (k_0) and Activation Energies (E_a) of Electron Transfer to Rh(III) Ion in $RuL_2(L\text{-}L)Rh(L')_2{}^{5+}$ and Diprotonated Chromophore of Ligand ($L\text{-}LH_2{}^{2+}$) in $RuL_2(L\text{-}LH_2{}^{2+})^{4+}$.

	L-L	L	L'	ΔG/eV	k_0 /$10^{11}s^{-1}$	E_a/eV
ET to Rh(III)	$dpimbH_2$	bpy	bpy	-	3.5	0.17
	$bpbimH_2$	bpy	bpy	0.13	2.5	0.20
	$bpbimH_2$	dmb	phen	0.19	4.5	0.20
	dmbime	bpy	bpy	-	2.5	0.20
ET to $L\text{-}LH_2{}^{2+}$	$dpimbH_2$	bpy		-	95	0.14
	$bpbimH_2$	bpy		0.26	45	0.16
	$bpbimH_2$	dmb		0.31	125	0.17
	dpbime	bpy		-	22	0.17

Smaller activation energies obtained for ET-to-($L\text{-}LH_2{}^{2+}$) can be accounted for in terms of ergonicity involved in the ET process. The larger ergonicity of ET-to-($L\text{-}LH_2{}^{2+}$) by 0.1 eV at most gives rise to a smaller E_a. Another possible explanation for the E_a of ligand mediated ET is vibronic interaction between the electron donor, a reduced ligand in the CT excited state, and the electron acceptor, Rh(III), $pbimH_3^{2+}$, or $pimH_3^{2+}$, because the vibrational mode mixing different electronic states, gets active with temperature. Rotation along the flexible C-C bond between pbimH allows a coplanar structure of $bpbimH_2$ in which electronic coupling is enhanced. Bending vibration of the methylene chain enhances the electronic coupling between pbim moieties of dpbime. However, absence of flexible C-C bond in $dpimbH_2$ is reponsible for the smaller E_a for the ET processes.

The rates of ET-to-Rh(III) in the binuclear compounds are much slower than those (500×10^8 s^{-1}) of ET in the encounter complex consisting of excited $Ru(bpy)_3^{2+}$ and $Rh(phen)_3^{3+}$ [12]. The slower rates of ET in the ligand bridging binuclear compounds than those in the encounter complex does not imply the weaker Ru(II)-Rh(III) coupling in the former system. The mutual orientation and distance between the metal sites in the binuclear compounds are less disturbed by various kinds of thermal motion than those in the encounter complex. Such a disturbance of the electronic interaction with the thermal motion must be needed for a mixing betweeen the initial and the final electronic states of ET process.

ACKNOWLEDGEMENT

We thank Prof. N. Mataga and Mr. T. Asahi of Osaka University for their help in performing the p-sec laser experiment.

REFERENCES

1 J. Ulstrup and J. Jortner, J. Chem. Phys. 63 (1975) 4358.
2 S.S.Isied, A.Vassilian, R.H.Magnuson, and H.A.Schwarz, J. Am. Chem. Soc. 107 (1986) 7432.
3 T.Ohno, A.Yoshimura, H.Shioyama, and N. Mataga, J. Phys. Chem. 94 (1990) 4871.
4 J.M.warman, M.P.De Haas, H.Oevering, J.W.Verhoewen, M.N.Paddon-Row, A.M.Oliver, and N.S.Hush, Chem. Phys. Lett. 128 (1986) 95. H.Oevering, M.N.Paddon-Row, M.Heppener, A.M. Oliver, E. Cotsaris, J.W. Verhoeven, and N.S.Hush, J. Am. Chem. Soc. 109 (1987) 3258.
5 W.A.Glauser, D.J.Raber, and B.Stevens, J. Phys. Chem. 95 (1991) 1976.
6 N.S.Hush, Coord. Chem. Rev. 64 (1985) 135.
7 M.Haga, T.Ano, K.Kano, and S.Yamabe, submitted to Inorg.

Chem.
8 T.Ohno, K.Nozaki, N.Ikeda, and M. Haga, Advance in Chemistry Series No. 228, 1991. T. Ohno, K. Nozaki, and M.Haga, submitted to Inorg. Chem.
9 T. Ohno, K. Nozaki, and M. Haga, submitted to J. Phys. Chem.
10 K. Nozaki, T.Ohno, and M.Haga, to be submitted.
11 K. Nozaki, T.Ohno, M, Haga, T. Asahi, and N. Mataga, to be submitted.
12 T. Ohno, A. Yoshimura, D. R. Prasad, and M. Z. Hoffman, J. Phys. Chem. in press.

DISCUSSION

Steiner

Could you give some information on the method used to measure the electron transfer rates in the binuclear RuII/RhIII complexes? Were the results of stationary luminescence measurements in accord with these of time-resolved ones?

Ohno

The rates of ET to Rh^{3+} were obtained by monitoring the decay of the transient absorption. Since some of the binuclear Ru–Rh compounds contain small amount of the binuclear Ru–Ru compounds, which is strongly luminescent, the steady state measurement of phosphorescence was not utilized for the determination of rate.

Kakitani

You obtained different activation energies ∿0.2 eV and ∿0.16 eV between two groups of compounds. Which makes dominant contribution to this between the reorganization energy difference and the free energy gap difference?

Ohno

The Gibbs energy change (ΔG^o) is nearly constant for three ET processes. The reorganization energy (λ_s) involved in the solvation may not change with the bridging ligands. Presumably, $E_a \equiv (\Delta G+\lambda)^2/4\lambda$ of the benzene bridging compound is not distinguished from those of the ethane and biphenyl bridging compounds because of the less negative ΔG^o and the less positive λ_s. After all, the larger value of E_a obtained for ET in the biphenyl bridging compound can not be understood in terms of Franck-Condon factor.

A possible explanation for the larger E_a is that thermally activated rotation in the biphenyl bridging compound enhances the vibronic coupling between Rh^{3+} and the reduced moiety of the bridging ligand.

Yoshihara

You suggested some specific vibrational modes of ligand molecules as a promoting vibrational mode for electron transfer. What are the criteria of selecting these vibrational modes?

Ohno

I have no other appropriate explanation for the smaller value of E_a which was observed for ET within the benzene bringing compound.

Dynamics and Mechanisms of
Photoinduced Transfer and Related Phenomena
N. Mataga, T. Okada and H. Masuhara (Editors)

Excited-state relaxation of ruthenium (II) complexes at low temperature

N. Kitamura, H.-B. Kim and S. Tazuke

Research Laboratory of Resources Utilization, Tokyo Institute of Technology, 4259 Nagatsuta, Midori-ku, Yokohama 227, Japan

Abstract

Excited-state relaxation of ruthenium (II) complexes was studied by temperature- and pressure-controlled, time-resolved emission spectroscopy. A time-dependent (TD) lower energy shift of the emission at low temperature or at high pressure was ascribed to solvent relaxation in the metal-to-ligand charge transfer excited state of the complex. TD nonradiative decay, accompanied by the TD emission shift, was discussed based on the energy gap law and the emission spectral-fitting for the transient emission.

1. INTRODUCTION

The metal-to-ligand charge transfer (MLCT) excited state of tris-chelate ruthenium (II) complexes, Ru(II), is characterized by its strong redox ability and relatively long lifetime. Ru(II), represented by $Ru(bpy)_3{}^{2+}$ where bpy is 2,2'-bipyridine, have been therefore extensively studied to elucidate the photoredox reactions and excited-state properties [1-3]. Synthetic control of the redox, spectroscopic, and excited state properties of Ru(II) has also been an active field of investigation [4,5].

One of the unique properties of Ru(II) is symmetry reduction of the complex upon photoexcitation [6]. In the ground state, three ligands of tris-chelate Ru(II) are equivalent while an electron is localized on a single ligand in the MLCT excited state. Therefore, the symmetry of the molecule is reduced from D_3 to C_{2v} upon photoexcitation, leading to generation of a large dipole moment in the excited state. The excited-state properties of Ru(II) are thus strongly dependent on temperature [7-9], solvent [10-12], and the nature of the ligand [4,5,13,14]. In particular, the spectroscopic and excited-state properties are highly sensitive to temperature variation. At low temperature, the emission spectrum is very sharp and structured with its lifetime of several μs, while the spectrum becomes broad and structureless with increasing temperature. The lifetime of the complex also dramatically decreases upon heating [7-9]. Such characteristic features of Ru(II) complexes have to be studied to elucidate the photophysical primary processes of the complexes as well as to apply Ru(II) as photosensitizers.

We have already reported the factors governing the redox, spectroscopic, and excited-state properties of Ru(II) complexes at ambient temperatures [4,12, 15-19]. In this study, we focused our interests to investigate excited-state relaxation processes of Ru(II) at low temperature. In particular, the study was intended to analyze time-resolved emission spectra of the complex and to show a time dependence of the nonradiative decay process.

2. TEMPERATURE AND PRESSURE EFFECTS ON EMISSION SPECTRA

All the Ru(II) complexes examined in this study ($Ru(bpy)_3X_2$ where X is Cl^-, ClO_4^-, or PF_6^-, $Ru(dp\text{-}bpy)_3(PF_6)_2$, $Ru(bpyraz)_3Cl_2$, $Ru(bpy)_2(dceb)(PF_6)_2$, and $Ru(phen)_2(CN)_2$) showed a large temperature dependence of the emission

bpy dp-bpy bpyraz dceb phen

spectrum in EtOH-MeOH (4/1, v/v) as typically seen in Figure 1a [20]. Upon heating from 80 to 145 K, the spectrum shifts to lower energy and the band shape becomes broader. Variation of temperature, however, is inevitably accompanied by changes in solvent properties such as viscosity (η), dielectric constant (D_s), and refractive index (n) and, therefore, the effects of temperature and solvent properties cannot be discussed independently. Application of hydrostatic pressure can change the solvent properties at a given temperature, so that we performed temperature- and pressure-controlled experiments to obtain complementary information on the photophysical processes [21,22]. Using a temperature-controlled diamond anvil cell under a microscope, we measured the emission spectrum of $Ru(bpy)_3^{2+}$ at various pressures [23]. Figure 1b shows pressure effects on the emission spectrum. At 160 K in EtOH-MeOH, $Ru(bpy)_3^{2+}$ showed a large high energy shift and sharpening of the spectrum upon application of pressure. Similar trends were observed at various pressures (P = 0.06 ~ 1.0 GPa) and temperatures (T = 100 - 200 K) [23].

Under an appropriate (P-T) combination, a time-dependent (TD) lower energy shift of the emission can be observed [20,23-25]. At atmospheric pressure in EtOH-MeOH, the emission spectrum shifts to lower energy with increasing a delay time after excitation (t) around a glass-to-fluid transition temperature of the medium (T_g ~ 125 K, Figure 2a). Analogous TD emission shift was observed at (160 K - 0.06 GPa) as seen in Figure 2b. It is noteworthy that, at atmospheric pressure, the TD emission shift at 160 K was too fast to be followed by nanosecond spectroscopy.

The important finding is that the lower energy emission shift of the complex can be similarly observed with increasing delay time (Figures 2) or

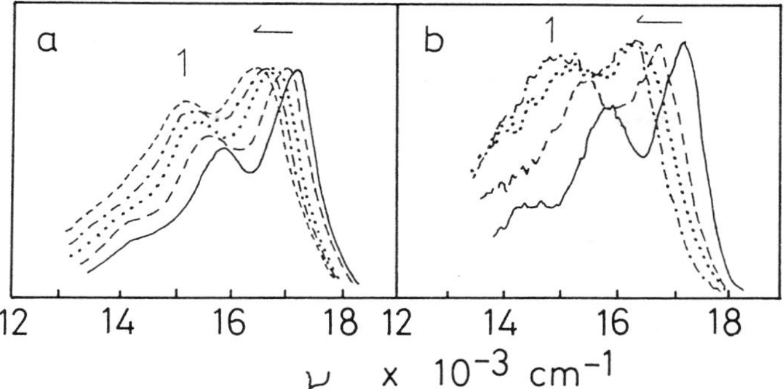

Figure 1. Temperature (a) and pressure (b) effects on the emission spectrum of $Ru(bpy)_3^{2+}$ in EtOH-MeOH (4/1, v/v).
(a) At atmospheric pressure; T = 80 (————), 120 (– – – – –), 130 (• • • • • • • •), 135 (—•—•—•—), and 145K (- - - - - - -).
(b) At 160K; P = 0.06 (————), 0.19 (– – – – –), 0.41 (. . . . . . .) 0.41, and 0.91 GPa (—•—•—•—•).

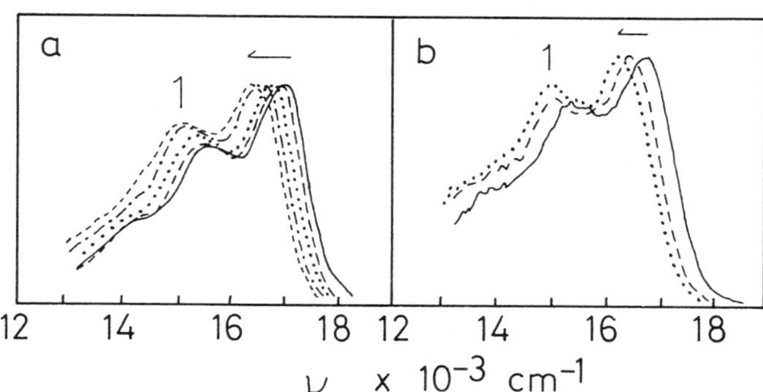

Figure 2. Time-resolved (gate width ~5 ns) emission spectra of Ru $(bpy)_3^{2+}$ in EtOH-MeOH at low temperature (a) and at high pressure (b).
(a) At 125 K - 0 GPa; t = 0 (————), 50 (– – – – –), 200 (• • • • • • •), 600 (—•—•—•—), and 2000 ns (- - - - - - - -).
(b) At 160 K - 0.06 GPa; t = 20 (————), 200 (– – – – –), and 1000 ns (• • • • • • • •).

temperature (Figure 1a), or with decreasing pressure (Figure 1b). A change in D_s or n with pressure cannot explain the present results [23]. The common effect is a change in the viscosity of the medium with these external perturbation. It is easily concluded, therefore, that the degree of solvation or the rigidity of the medium around the MLCT excited state of the complex determines the emission maximum energy (υ^{em}) at a given delay time, temperature, or pressure.

3. SOLVENT INTERACTION IN THE MLCT EXCITED STATE

Interactions between the MLCT excited state of Ru(II) and solvents are very important to explain the present spectroscopic characteristics at low temperature or at high pressure. For quantitative discussion, we define the amount of the TD emission shift as $\Delta\upsilon^{em} = \upsilon_0 - \upsilon_\infty$, where υ_0 and υ_∞ are υ^{em} at t = 0 and ∞ (typically t = 3 ~ 4 μs), respectively. $\Delta\upsilon^{em}$ were comparable at 530 ~ 650 cm^{-1} for $Ru(bpy)_3X_2$, $Ru(dp\text{-}bpy)_3^{2+}$, and $Ru(bpy)_2(dceb)^{2+}$ while other complexes, $Ru(bpyraz)_3^{2+}$ and $Ru(phen)_2(CN)_2$, showed much larger $\Delta\upsilon^{em}$ (850 ~ 1030 cm^{-1}) [20]. The results indicate that the nature of the counteranions (X^-) and the bulky substituents on the dp-bpy ligand have no effect on $\Delta\upsilon^{em}$. Relatively large $\Delta\upsilon^{em}$ for $Ru(bpyraz)_3^{2+}$ and $Ru(phen)_2(CN)_2$ will be explained by specific interactions between the complex and solvent molecules. For $Ru(phen)_2(CN)_2$, both ground and excited states have been known to be subjected to electron donor-acceptor interaction with electron accepting solvents [12]. The 4 and 4' nitrogen atoms of bpyraz will be also susceptible to solvent interactions [26]. Indeed, it has been demonstrated that the vibrational structures of the emission of these complexes vary appreciably with solvent properties as well as with delay time in a given solvent [20], suggesting that solvation modes around the complex influence the excited and spectroscopic properties.

The present TD emission shift was further analyzed by the following Stokes shift correlation function.

$$C(t) = \frac{\upsilon_t - \upsilon_\infty}{\upsilon_0 - \upsilon_\infty} = \sum_{i=1}^{n} A^i \exp(-t/\tau_s{}^i) \qquad (1)$$

where υ_t is υ^{em} at a given delay time (t = t'). In Figure 3, C(t) of $Ru(bpy)_3^{2+}$ in EtOH-MeOH at 125 K is shown as a function of t. As clearly seen in the figure, C(t) does not decay single exponentially but can be fitted by using a double exponential function. Nonlinear least-squares analysis of C(t) vs t plots for five Ru(II) complexes at a given temperature yielded sets of the relaxation times, $\tau_s{}^1$ and $\tau_s{}^2$. Linear Arrhenius plots for both $\tau_s{}^1$ and $\tau_s{}^2$ gave the corresponding activation parameters ($E_a{}^1$ and $E_a{}^2$) as summarized in Table 1.

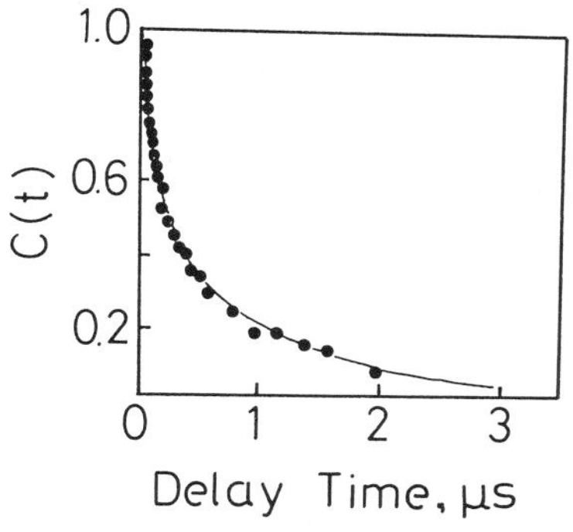

Figure 3. Time dependence of C(t) obtained for $Ru(bpy)_3^{2+}$ in EtOH-MeOH at 125 K.
•; observed, solid line; best-fit curve

Table 1. Relaxation Times and Their Activation Parameters of Ru(II) Complexes in EtOH-MetOH (4/1, v/v) at 125 K[a]

	τ_s^1 (int %[b]), ns	τ_s^2 ns	E_a^1 (log A^1), cm^{-1}	E_a^2 (log A^2), cm^{-1}
$Ru(bpy)_3^{2+}$	150 (10.3)	1000	960 (11.7)	1200 (11.8)
$Ru(dp\text{-}bpy)_3^{2+}$	97 (7.0)	1200	790 (11.1)	820 (10.1)
$Ru(bpyraz)_3^{2+}$	140 (2.1)	1900	2200 (16.9)	1400 (13.9)
$Ru(bpy)_2(dceb)^{2+}$	150 (7.8)	1200	1000 (11.9)	1200 (12.0)
$Ru(phen)_2(CN)_2$	130 (20.5)	940	2500 (19.4)	1600 (14.4)

[a]Uncertainties for the values are τ_s^1, ± 15%; τ_s^2, ± 10%.
[b]Fraction of the component in percent.

The observed τ_s^i and E_a^i for $Ru(bpy)_3^{2+}$, $Ru(dp\text{-}bpy)_3^{2+}$, and $Ru(bpy)_2(dceb)^{2+}$ gave similar values. These values are almost comparable with Debye and longitudinal relaxation times of pure EtOH at 125 K and their activation energies: 1400 ns (1100 cm^{-1}) and 40 ns (1400 cm^{-1}), respectively [20]. The TD emission shift of Ru(II) is, however, not observed in non-alcoholic solvents such as dimethyl sulfoxide, propylene carbonate and acetonitrile, so that the present TD emission shift should be closely related to the specific interaction between the excited complex and alcoholic solvents. Indeed, E_a for $Ru(phen)_2(CN)_2$ and $Ru(bpyraz)_3^{2+}$, interacting with solvent molecules, are much larger than those of other complexes. Furthermore, the TD emission shift of $Ru(bpy)_3^{2+}$ in EtOD-MeOD was slower than that in EtOH-MeOH, manifesting that the motion of -OH (or -OD) or hydrogen bonding strongly influences the solvent relaxation process.

4. EXCITED-STATE RELAXATION OF Ru(II) COMPLEXES

The spectroscopic and excited-state characteristics of Ru(II) are governed by interactions between the excited complex and solvent molecules, depending on delay time, temperature, and pressure, as discussed in the preceding sections. The solute-solvent system initially produced upon photoexcitation possesses a nonequilibrium configuration and, thus, must relax toward the equilibrium one as clearly demonstrated by the TD lower energy shift of the emission (Figure 3). υ^{em} or the emission energy of the 0-0 transition band (E_{00}) at a given delay time, temperature, or pressure will be explained based on the degree of solvation around the excited complex. Such an argument suggests that the TD emission shift (i.e., dynamic solvent relaxation) should accompany TD nonradiative decay in the excited state if the energy gap law is held for the transient emission. For Ru(II), in particular, the nonradiative decay rate constant (k_{nr}) mainly determines the excited-state lifetime, so that detailed analysis on the TD emission shift and k_{nr} is crucial for elucidation of the relaxation processes in the MLCT excited state.

According to the energy gap law, k_{nr} is given as [5,10]

$$\ln k_{nr} = \ln \beta - S_M - \frac{\gamma E_{00}}{\hbar\omega_M} + S_L \frac{\omega_L}{\omega_M}(\gamma+1) + \frac{b\chi}{\hbar\omega_M} \tag{2}$$

$$\beta = (C_k{}^2\omega_k)(\pi/2\hbar\omega_M E_{00})^{1/2} \tag{2a}$$

$$\gamma = \ln(E_{00}/\hbar\omega_M S_M) - 1 \tag{2b}$$

$$\pi = (\Delta\upsilon_{1/2})^2/(16 k_B T \ln 2) \tag{2c}$$

$$b = (k_B T/\hbar\omega_M)(\gamma + 1)^2 \tag{2d}$$

$\Delta\upsilon_{1/2}$ is a full width at half-maximum for the 0-0 and other vibrational components of the emission. ω_M and ω_L are the vibrational frequencies related to bpy ring and Ru-N stretching modes, respectively. S_M and S_L are the vibrational coupling parameters related to the distortion in the molecular coordinates corresponding to ω_M and ω_L, respectively. The unknown factors of E_{00}, $S_{M,L}$, $\omega_{M,L}$, and $\Delta\upsilon_{1/2}$ involved in eq. (2) can be estimated from the emission spectral-fitting based on the Franck-Condon analysis; eq. (3) [27]

$$I(\upsilon) = \sum_{n=0}^{5}\sum_{m=0}^{5}\left(\frac{E_{00}-n\hbar\omega_M-m\hbar\omega_L}{E_{00}}\right)^4\left(\frac{S_M{}^n}{n!}\right)X$$

$$\left(\frac{S_L{}^m}{m!}\right)\exp\left[(-4\log 2)\left(\frac{\upsilon - E_{00} + n\hbar\omega_M + m\hbar\omega_L}{\Delta\upsilon_{1/2}}\right)^2\right] \quad (3)$$

where $I(\upsilon)$ is an emission intensity at a frequency υ and, n and m are the vibrational quantum numbers. The observed time-resolved emission spectra were satisfactorily fitted with E_{00}, S_M, $\hbar\omega_M$, and $\Delta\upsilon_{1/2}$ as parameters and with S_L and $\hbar\omega_L$ as constants fixed at 1.1 and 400 cm^{-1}, respectively (Figure 4). Since the contribution of β (related to the electronic coupling integral, C_k) to k_{nr} is small in the present case, k_{nr} was calculated for a given delay time (or temperature) on the basis of eq. (2) and the parameters estimated from the spectral fitting [28].

Energy gap analysis of ln k_{nr} was performed for both time and temperature dependent emission of $Ru(bpy)_3{}^{2+}$. ln k_{nr} increases linearly with decreasing E_{00} (with increasing delay time or temperature) as clearly seen in Figure 5. It should be emphasized that the ln k_{nr} vs E_{00} data of the time-resolved spectra and the temperature-controlled spectra at t = 0 lie on the same line. Since E_{00} (or υ^{em}) is determined by the degree of solvation around the excited complex as discussed in the preceding sections, k_{nr} is concluded to be governed by solvation around the complex, depending on delay time or

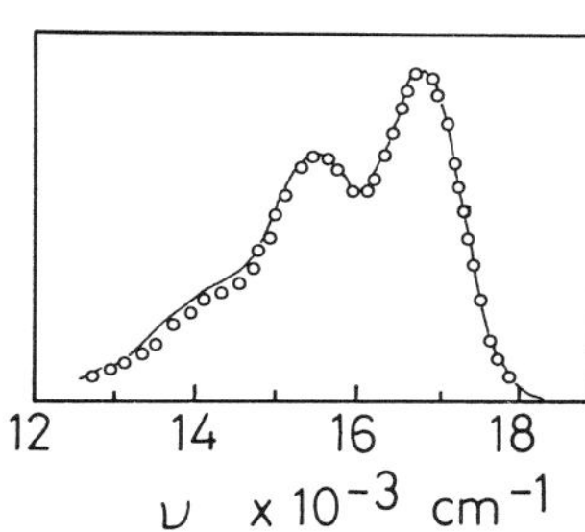

Figure 4. Calculated (○) and observed emission spectra of $Ru(bpy)_3{}^{2+}$ in EtOH-MeOH at 125 K; t = 10 ns.

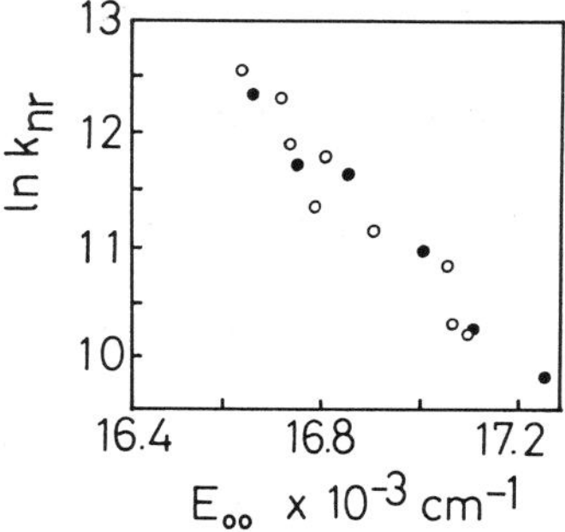

Figure 5. Energy gap plot for temperature (●, t = 0) and time (○) dependent emission spectra of $Ru(bpy)_3{}^{2+}$ in EtOH-MeOH.

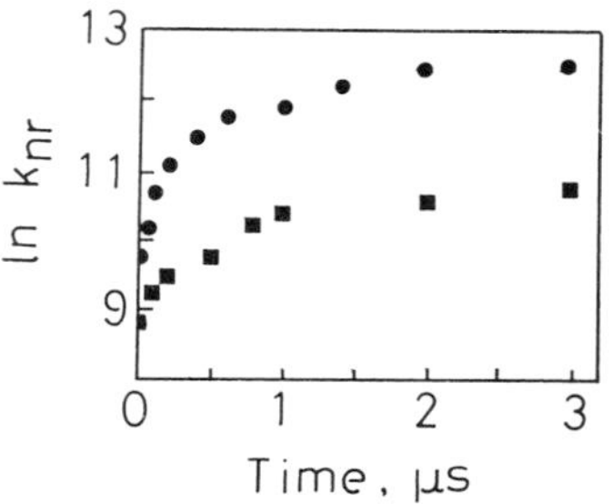

Figure 6. Time dependent k_{nr} for $Ru(bpy)_3^{2+}$ in EtOH-MeOH (●) and EtOD-MeOD (■) at 125 K.

temperature. As expected from TD υ^{em} and the energy gap law, furthermore, k_{nr} was shown to be time dependent in the initial stage of excitation ($t < 1$ μs) as demonstrated in Figure 6. It is noteworthy that the time domain showing the TD k_{nr} corresponds to that undergoing solvent relaxation in the excited state (Figure 4).

The most important factor determining k_{nr} is S_M, which represents a dimensionless displacement of the bpy ring modes related to the difference in the equilibrium coordinates between the ground and excited states. Such structural displacement of the bpy ring modes can be phenomenologically observed as the changes in the vibrational structure of the spectrum (Figures 1 and 2). Indeed, an increase in the intensity of the 0-1 vibrational band relative to that of the 0-0 band with increasing in delay time or temperature was reasonably reproduced by that in S_M (0.90 ~ 1.10 for t = 0 ~ 3 μs at 125 K and 0.85 ~ 1.08 for T = 80 ~ 145 at t = 0) in the emission spectral-fitting procedures. The change in S_M does not affect k_{nr} directly but contributes to k_{nr} through γ (eq. (2)). Although k_{nr} is a complex function of γ, ln k_{nr} was shown to almost linearly increase with decreasing ln $(1/S_M)$ (increasing S_M); eq. (2b). It is clear that the increase in S_M with time or temperature plays a decisive role for time- or temperature-dependent k_{nr} through γ.

ln k_{nr} is also influenced by solvent deuterization. A time dependence of k_{nr} in EtOD-MeOD is very different from that in EtOH-MeOH, as demonstrated in Figure 6. Values k_{nr} in EtOD-MeOD are always smaller than those in EtOH-MeOH, and the increment of k_{nr} during the first 1 μs is much smaller in EtOD-MeOD. It is now apparent that the explanation of the TD k_{nr} should include hydrogen-bonding interactions between the excited-state complex and alcoholic solvents. It has been reported that solvent and/or ligand deuterization influence the nonradiative decay of $Ru(bpy)_3^{2+}$ through the charge-transfer-to-solvent (CTTS) interaction in the excited state [29,30]. In the initial stage of excitation ($t < 1$ μs, Figure 6), the interactions between the excited state of $Ru(bpy)_3^{2+}$ and solvent molecules change with time during the relaxation processes. In the solvent relaxation processes, the surrounding solvent molecules translate and rotate toward an equilibrium configuration, which induces changes in the hydrogen-bonding network. Since the interaction of alcoholic solvents with the excited Ru(II) complex induces the nonradiative decay path through CTTS interaction, the relaxation in these solvents brings about a dramatic increase in k_{nr} with time.

5. CONCLUSION

The spectroscopic and excited-state properties of Ru(II) complexes were shown to be governed as follows: Temperature, pressure, and delay time after excitation determine the degree of solvation around the excited complex, which is reflected on the emission energy (E_{00}) and the distortion in the molecular coordinate (S_M) of the complex. The change in the degree of solvation with time (solvent relaxation), including specific interactions between the complex and alcoholic solvents, renders a time dependence of k_{nr}. Besides Ru(II) complexes, analogous temperature and/or pressure effects on photophysical primary processes are expected for various polar excited molecules. Ultrafast time-resolved spectroscopy, combined with temperature and/or pressure variations as well as with the spectral-fitting methods, will be promising to elucidate further the dynamics of excited-state relaxation.

NOTE AND ACKNOWLEDGEMENT

Prof. Shigeo Tazuke was deceased on July 11, 1989 at the age of 54.

Any correspondence on this article should be addressed to the present address of N.K; Microphotoconversion Project, ERATO, Research Development Corporation of Japan, 15 Morimoto-cho, Sakyo-ku, Kyoto 606, Japan.

The authors are indebted to Dr. T. Hiraga for his collaboration in high pressure experiments. Special thanks are also due to Miss S. Hitomi, Microphotoconversion Project, for her kind help to prepare this manuscript.

REFERENCES

1 K. Kalyanasundaram, Coord. Chem. Rev., 46 (1982) 159.
2 A. Juris, V. Balzani, F. Barigelletti, S. Campagna, P. Belser and A. von Zelewsky, Coord. Chem. Rev., 84 (1988) 85.
3 T.J. Meyer, Pure Appl. Chem., 58 (1986) 1193.
4 Y. Kawanishi, N. Kitamura and S. Tazuke, Inorg. Chem., 28 (1989) 2968.
5 G.H. Allen, R.P. White, D.P. Rillema and T.J. Meyer, J. Am. Chem. Soc., 106 (1984) 2613.
6 M.K. DeArmond and M.L. Myrick, Acc. Chem. Res., 22 (1989) 364.
7 F. Barigelletti, A. Juris, V. Balzani, P. Belser and A. von Zelewsky, Inorg. Chem., 22 (1983) 3335.

8 F. Barigelletti, A. Juris, V. Balzani, P. Belser and A. von Zelewsky, J. Phys. Chem., 91 (1987) 1095.
9 A. Juris, F. Barigelletti, V. Balzani, P. Belser and A. von Zelewsky, Inorg. Chem., 24 (1985) 202.
10 J.V. Caspar and T.J. Meyer, J. Am. Chem. Soc., 105 (1983) 5583.
11 P. Belser and A. von Zelewsky, Gazz. Chim. Ital., 115 (1985) 723.
12 N. Kitamura, M. Sato, H.-B. Kim, R. Obata and S. Tazuke, Inorg. Chem., 27 (1988) 651.
13 D.P. Rillema, G. Allen, T.J. Meyer and D. Conrad, Inorg. Chem., 22 (1983) 1617.
14 A. Juris, S. Campagna, V. Balzani, G. Gremaud and A. von Zelewsky, Inorg. Chem., 27 (1988) 3652.
15 N. Kitamura, S. Rajagopal and S. Tazuke, J. Phys. Chem., 91 (1987) 3767.
16 H.-B. Kim, N. Kitamura, Y. Kawanishi and S. Tazuke, J. Am. Chem. Soc., 109 (1987) 2506.
17 N. Kitamura, H.-B. Kim, S. Okano and N. Kitamura, J. Phys. Chem., 93 (1989) 5750.
18 H.-B. Kim, N. Kitamura, Y. Kawanishi and S. Tazuke, J. Phys. Chem., 93 (1989) 5757.
19 N. Kitamura, R. Obata, H.-B. Kim and S. Tazuke, J. Phys. Chem., 93 (1989) 5764.
20 H.-B. Kim, N. Kitamura and S. Tazuke, J. Phys. Chem., 94 (1990) 1414.
21 T. Hiraga, T. Uchida, N. Kitamura, H.-B. Kim and S. Tazuke, Rev. Sci. Instrum., 60 (1989) 1008.
22 H.-B. Kim, T. Hiraga, T. Uchida, N. Kitamura and S. Tazuke, Coord. Chem. Rev., 111 (1989) 8265.
23 T. Hiraga, N. Kitamura, H.-B. Kim, S. Tazuke and N. Mori, J. Phys. Chem., 93 (1989) 2940.
24 N. Kitamura, H.-B. Kim, Y. Kawanishi, R. Obata and S. Tazuke, J. Phys. Chem., 90 (1986) 1488.
25 H.-B. Kim, N. Kitamura and S. Tazuke, Chem. Phys. Lett., 143 (1988) 77.
26 R.J. Crutchley, N. Kress and A.B.P. Lever, J. Am. Chem. Soc., 105 (1983) 1170.
27 J.V. Caspar, T.D. Westmoreland, G.H. Allen, P.G. Bradley, T.J. Meyer and W.H. Woodruff, J. Am. Chem. Soc., 106 (1984) 3492.
28 H.-B. Kim, N. Kitamura and S. Tazuke, J. Phys. Chem., 94 (1990) 7401.
29 J. van Houten and R.J. Watts, J. Am. Chem. Soc., 97 (1975) 3843.
30 S.F. McLanahan and J.R. Kincaid, J. Am. Chem. Soc., 108 (1986) 3840.

Dynamics and Mechanisms of
Photoinduced Transfer and Related Phenomena
N. Mataga, T. Okada and H. Masuhara (Editors)

Transient Hole-Burning Spectra of Organic Dyes in Solution

H. Murakami[a], S. Kinoshita[a,*], Y. Hirata[b], T. Okada[b] and N. Mataga[b]

[a]Department of Physics, Faculty of Science, Osaka University, Toyonaka, Osaka 560, Japan

[b]Department of Chemistry, Faculty of Engineering Science, Osaka University, Toyonaka, Osaka 560, Japan

Abstract

A picosecond transient hole-burning (THB) spectroscopy has been performed for organic dyes in solution. The THB spectra obtained have been found to show a time-dependent spectral change. This phenomenon corresponds to the solvent relaxation effect observed frequently in the time-resolved fluorescence (TRF) spectrum. Although TRF spectrum is related only to the excited-state relaxation, THB spectrum is affected by both ground- and excited-state relaxations. Comparing with the TRF spectrum measured under the same exciting energy, we have clarified the presence of the ground-state relaxation. The observed results are well understood using a configuration coordinate model.

1. Introduction

When a molecule dissolved in solution is excited by a light irradiation, its electronic configuration changes abruptly and various relaxation phenomena will occur in order to minimize the total energy through interaction with surrounding solvent molecules. Among them, the relaxation process due to the rearrangement of solvent molecules through their translational and rotational motions is known to play an important role. By means of time-resolved fluorescence (TRF) spectroscopy, the above relaxation phenomena can be clearly observed as a time-dependent peak shift [1]. This is particularly prominent at low temperatures or in viscous solvents, where the time constant of the relaxation is comparable with the fluorescence lifetime.

On the other hand, with the light excitation, only solute molecules whose transition energies are nearly equal to the excitation energy are selectively excited. Then, a hole is created in the ground-state population after the pulsed excitation. This means that even the ground-state population becomes in a non-equilibrium state, and consequently the relaxation phenomenon occurs not only within the excited state but also within the ground state. However, by means of the TRF spectroscopy, the relaxation phenomenon only in the excited state can be observed.

The transient hole-burning (THB) spectroscopy is suitable for observing this ground-state relaxation. The THB spectrum is obtained by subtracting a stationary absorption spectrum from a transient spectrum obtained after the optical excitation. Since the magnitude of light absorption is proportional to the difference between the ground- and excited-state populations, the relaxation processes in both states contribute to the THB spectrum.

The purpose of the present paper is to clarify the solvent relaxation effect on the THB spectrum and is to obtain the energy relaxation processes in the ground and excited states separately. To this aim, we have performed a picosecond THB spectroscopy for several organic dyes in solution. Results are then compared with the TRF spectra obtained under the same excitation wavelength.

2. Configuration Coordinate Model

The principles of the THB and TRF spectroscopies are easily understood by using a configuration coordinate (CC) model, in which the solute-solvent interaction is replaced by a harmonic adiabatic potential curve. This model has been successfully applied to explaining the optical response of organic dyes in solution, e.g., the lineshape of the absorption spectrum, the relationship between the Stokes shift and the linewidth, and also the transient behavior of the fluorescence [1,2].

In Fig. 1, we show a schematic diagram of two harmonic potential curves; each corresponds to the excited and ground states, respectively. The minimum positions for the two potential curves are assumed to differ by an amount of Q_0, but the curvatures are to be the same. Light irradiation to this system causes the transient population changes for ground and excited states. Then it is easy to notice that the time-dependent change in the shape of the spectrum gives rise to by the relaxation within each state and the nonadiabatic relaxation between the states. We have employed a stochastic approach to describe the relaxation processes in the potential curves and the detailed calculation on the time-dependent line shape of the THB and TRF spectra gives the following result [3]:

$$I_{THB}(\omega_1,\omega_2,t) \propto -\{[a]+[b]+[c]\},$$

$$I_{TRF}(\omega_1,\omega_2,t) \propto [b],$$

where

$$[a] = \int d\omega g(\omega)(\pi\sigma^2)^{-1}(1-\rho(t)^2)^{-\frac{1}{2}}\exp[-\Delta\omega^2/\sigma^2] \times \exp[-\{(\Delta\omega-\Delta\omega_2)-\Delta\omega(1-\rho(t))\}^2/\sigma^2(1-\rho(t)^2)],$$

$$[b] = e^{-\gamma t}\int d\omega g(\omega)(\pi\sigma^2)^{-1}(1-\rho(t)^2)^{-\frac{1}{2}}\exp[-\Delta\omega^2/\sigma^2] \times \exp[-\{(\Delta\omega-\Delta\omega_2)-(\Delta\omega+2aQ_0^2)(1-\rho(t))\}^2/\sigma^2(1-\rho(t)^2)],$$

$$[c] = -\gamma\int_0^t d\tau\int d\omega g(\omega)(\pi\sigma^2)^{-1}(1-\rho(\tau)^2)^{-\frac{1}{2}} \times \exp[-\{\Delta\omega_2+2aQ_0^2(\rho(t-\tau)-\rho(\tau))-\rho(\tau)\Delta\omega\}^2/\sigma^2(1-\rho(\tau)^2)]e^{-\gamma\tau}.$$

Here, $\Delta\omega = \omega-\varepsilon$ and $\Delta\omega_2 = \omega_2-\varepsilon$ with the center energy of the stationary absorption spectrum ε and that of the probe beam or emitted photon ω_2. $g(\omega)$ is the spectrum of the pump beam and σ expresses the broadening of

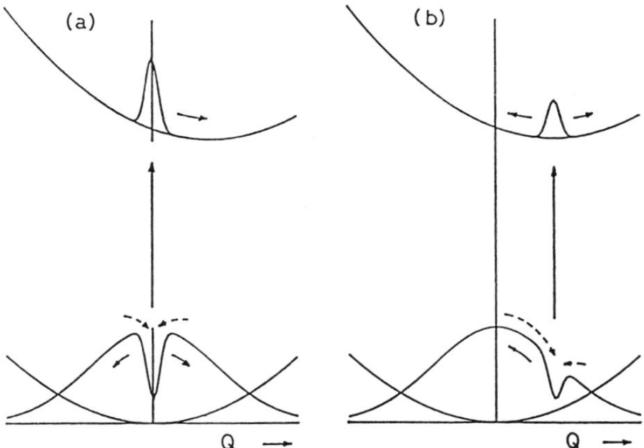

Figure 1. Schematic diagram of the configuration coordinate model for the light excitation (a) around the center and (b) at the low-energy tail of the absorption band.

the stationary absorption spectrum. $\rho(t)$ is the normalized correlation function of the fluctuation of the transition energy and if the fluctuation obeys the Gaussian-Markovian process, $\rho(t) = \exp[-\gamma_m t]$.

[a] and [b] express the contributions by the ground- and excited-state populations to the transient spectra, while [c] is ascribed to the relaxation from the excited state to the ground state. The term [c] is originated from the additional population formed on the ground-state potential curve through the decay from the excited state and is only important when the solvent relaxation rate is comparable to the excited-state decay rate. It is noticeable that the term [b] appears both in TRF and THB spectra.

When the solvent relaxation is fast enough as compared with the pump-probe process and the excited-state lifetime, by putting $\rho(t)=0$, the spectral shapes given by the terms [a] and [b] are reduced to those for the stationary absorption and fluorescence spectra. Then the THB spectrum is expected to be expressed by the sum of the stationary absorption and fluorescence spectra. With decreasing the solvent relaxation rate, but still larger than the excited-state decay rate, time-dependent changes in the THB and TRF spectra are expected to be observed. In the case of the TRF spectrum, this is what we call a solvent relaxation effect. Then, it is easy to obtain the contribution from the ground-state population by subtracting the TRF spectrum from the minus of the THB spectrum. When the solvent relaxation rate is decreased much more and becomes comparable to the excited-state population decay rate, then the term [c] becomes important and the spectral shape deforms. The effect by this term appears in the opposite direction to the hole, i.e., hole-filling. In the following sections, we will show experimentally these effects on the lineshape of the THB spectrum.

3. Experimental Procedures

Samples employed in this experiment were rhodamine dyes and styryl 8, and were dissolved into ethanol or a mixture of ethanol and methanol. The concentration of the dyes were 5 x 10^{-5} M for the THB measurement and 10^{-5} M for the TRF measurement. The measurements were performed at room temperature or under cooling down to 170 K in a flow-type cryostat.

The THB spectroscopy were mostly performed using a mode-locked Nd:YAG laser with the repetition rate of 10 pps and a dye laser excited by it. The dye laser or second-harmonics of YAG laser were used as a pump pulse, while a probe pulse was obtained by the continuum generation by the YAG fundamental [4]. The temporal resolution of this system was about 15 ps. To avoid any saturation effect and the damage of the sample, the spectral change with changing the pump-beam intensity was monitored before every measurement.

The TRF spectroscopy was performed using the combination of a CW mode-locked dye laser and a time-correlated single-photon counting [5]. Typical time resolution of this system is 90 ps. To obtain the TRF spectra, the fluorescence time responses were measured at wavelengths of every 2.5 nm, deconvoluted by the instrumental function and then synthesized by a computer. The obtained TRF spectra were corrected for the monochoro-mator/detector sensitivities and then divided by the square of the emitted-light frequency in order to compare the TRF spectra directly with the THB spectra.

4. Experimental Results and Discussion

First, we show the experimental result on rhodamine 6G in ethanol at room temperature (Fig. 2a) [7]. This sample corresponds to the case of a very fast solvent relaxation process, i.e., much faster than the probing process and excited-state lifetime. Negative ΔOD (difference of the optical density) around 18700 cm^{-1} and positive ΔOD around 23000 cm^{-1} are originated from the transient population change in the ground and excited states, and from the excited-state absorption to the higher excited state, respectively. Both of their magnitudes are found to decrease with time and the time constant of this change is found coincident with the excited-state lifetime. Owing to the above expectation, we have plotted the sum of the stationary absorption and fluorescence spectra in Fig. 2b (curve c). Here, the areas of the absorption and fluorescence spectra are made equal to each other. The obtained result has perfectly reproduced the whole region of the spectrum except for the higher-energy tail, where the excited-state absorption band is superimposed. We subtract the calculated spectrum from the observed spectrum and obtain the excited-state absorption band shown as a curve d in Fig. 2b.

Second, we show the experimental result on rhodamine 640 in ethanol/methanol at 170 K. At this temperature, the solvent relaxation becomes slow enough to directly observe the relaxation phenomena with our apparatus but still faster than the excited-state population decay. The THB spectra are shown in Fig. 3a. The spectra consist of negative ΔOD around $\sim$17000 cm^{-1} and positive ΔOD around $\sim$22000 cm^{-1}, and resemble to the above result for rhodamine 6G. The former is ascribed to the population change by the light absorption, while the latter to the

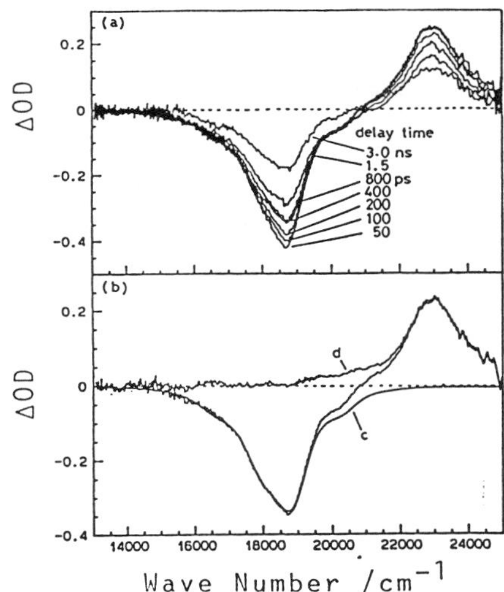

Figure 2. (a) THB spectrum of rhodamine 6G in ethanol excited at 18797 cm^{-1}. (b) The THB spectrum is compared with the sum of the absorption and fluorescence spectra (curve c). A solid curve d is the estimated excited-state absorption [7].

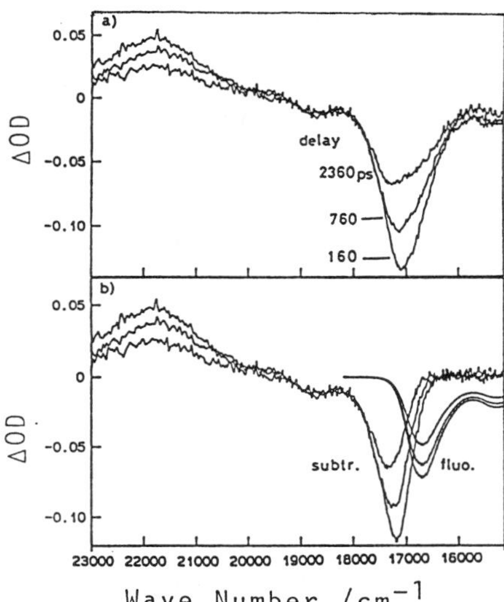

Figure 3. (a) THB spectrum of rhodamine 640 in ethanol/methanol at 170 K. (b) The THB spectrum is subtracted by the TRF spectrum obtained under the same conditions.

excited-state absorption. It is noticeable that the position of the hole around ~17000 cm^{-1} shifts to the higher energy with increasing the delay time, while the excited-state absorption does not change its shape. Accompanied by this spectral shift, the width of the THB spectrum broadens with time. At earlier times after the excitation, the width of the THB spectrum is narrower than the sum of the stationary absorption and fluorescence spectra, which is in contrast to the above case. These spectral changes are considered to be originated from the solvent relaxation effect.

In order to confirm this phenomenon more quantitatively, we have measured the TRF spectrum employing the same temperature and excitation energy. The intensity of the obtained spectrum is so adjusted that the lower-energy region of the THB spectrum is well reproduced by the minus of the TRF spectrum. Then the TRF contribution is eliminated from the THB spectrum. The result is shown in Fig. 3b. Clearly, the spectrum shifts to the high-energy side with time and finally it seems to almost coincide with the stationary absorption spectrum. Since the subtracted spectrum comes from the contribution of the ground-state population, we have clarified the presence of the ground-state relaxation. It is noteworthy that only the subtracted spectrum shows the peak shift, while the TRF spectrum does not. This is because the excitation energy we have employed is 16949 cm^{-1}, which corresponds to the low-energy tail of the absorption band of this dye. At low-energy excitation, the population created on the excited-state adiabatic potential curve is located around the bottom of the curve and does not contribute to the peak shift of the TRF spectrum. On the other hand, the hole in the ground state is far from the peak of the population and then shifts with broadening toward the center of the equilibrium distribution.

Finally, we show the result on styryl 8 in ethanol/methanol at room temperature. In this sample, the excited-state lifetime is so short, ~140 ps at room temperature, that the hole-filling effect is not considered to be neglected. In Fig. 4a, we show the absorption and fluorescence spectra obtained under the stationary condition. Obviously, this dye shows a extraordinarily large Stokes-shift and the widths of both spectra do not coincide with each other. Hence the ordinary CC model is not applicable, possibly because the deformation of the molecular structure takes place in the excited state. In Fig. 4b, we show the THB spectra excited at 16000

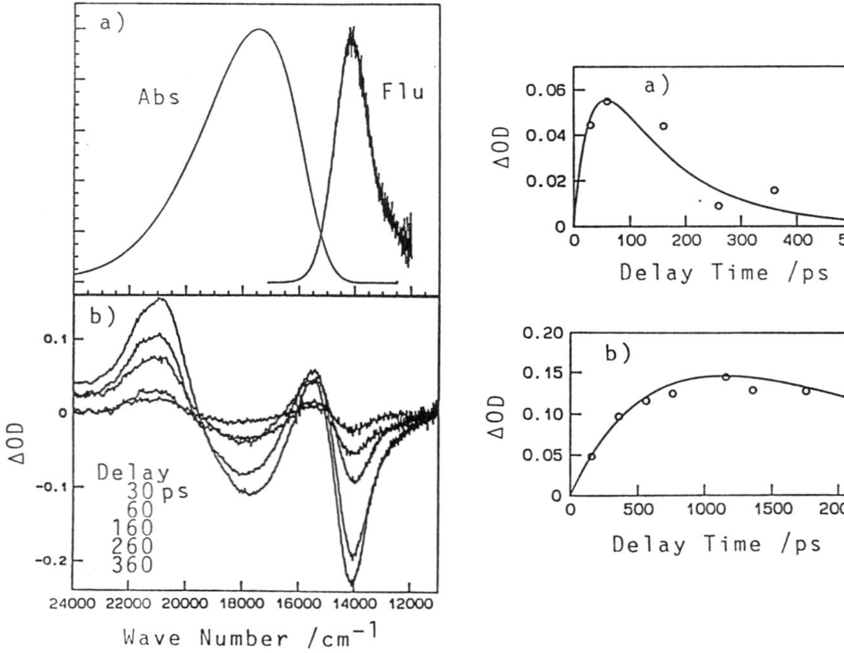

Figure 4. (a) Absorption, fluorescence and (b) THB spectra of styryl 8 in ethanol/methanol at room temperature.

Figure 5. Time dependence of the intensity of THB spectrum at 15500 cm^{-1} obtained (a) at room temperature and (b) at 170 K. Solid curves are simulated results.

cm^{-1}. It may be reasonable to assign the low-energy hole located around 14000 cm^{-1} and the high-energy hole around 18000 cm^{-1} as the contributions by the ground- and excited-state populations, since their energy positions are almost coincident with those of the absorption and fluorescence spectra. The positive ΔOD around 21000 cm^{-1} is ascribable to the excited-state absorption.

The most remarkable is an existence of positive ΔOD between the two holes, ∿15000 cm^{-1}. This region is possibly ascribed to the hole-filling effect described in Section 2. The difference between the hole-filling effect and the ordinary excited-state absorption, although both can give positive ΔOD's, is their time dependences. Since the former is caused by the additional population formed on the ground-state potential curve, its time dependence is essentially expressed by the two relaxation parameters, solvent relaxation rate γ_m and population decay rate γ, while the latter is determined only by the population decay rate γ. Then the time dependence expected by the hole-filling effect is roughly expressed by the balance of two exponential curves, i.e., $(\gamma-\gamma_m)^{-1}\{\exp(-\gamma_m t)-\exp(-\gamma t)\}$. We have simulated the time dependence at two different temperatures by using the observed relaxation rates, $\gamma_m{}^{-1}$ = 30 ps and γ^{-1} = 140 ps at room temperature and $\gamma_m{}^{-1}$ = 700 ps and γ^{-1} = 2000 ps at 170 K. As shown in Fig. 5, fairly good agreement supports our conjecture. More detailed discussion on this point will be presented elsewhere.

In summary, we have performed a picosecond THB and TRF spectroscopies under the same temperature and excitation wavelength, and have clarified the various relaxation effects affecting on the THB spectrum for organic dyes in solution. Further we have obtained the relaxation processes within the ground and excited states separately.

References

* Present address: Research Institute of Applied Electricity, Hokkaido University, Sapporo 060, Japan.

1 S. Kinoshita and N. Nishi, J. Chem. Phys. 89 (1988) 6612, and the references therein.
2 S. Kinoshita, N. Nishi, A. Saitoh and T. Kushida, J. Phys. Soc. Jpn. 56 (1987) 4162.
3 S. Kinoshita, J. Chem. Phys. 91 (1989) 5175.
4 Y. Hirata and N. Mataga, J. Phys. Chem. 95 (1991) 1640.
5 S. Kinoshita and T. Kushida, Anal. Instrum. 14 (1985) 503.
6 S. Kinoshita, H. Itoh, H. Murakami, H. Miyasaka, T. Okada and N. Mataga, Chem. Phys. Lett. 166 (1990)123.

Dynamics and Mechanisms of
Photoinduced Transfer and Related Phenomena
N. Mataga, T. Okada and H. Masuhara (Editors)

Theoretical studies of excited state intramolecular electron transfer in polar solvents

Shigeki Kato, Koji Ando and Yoshiaki Amatatsu

Department of Chemistry, Faculty of Science, Kyoto University, Kitashirakawa, Sakyo-ku, Kyoto 606, Japan

1. INTRODUCTION

Photoinduced electron transfer in polar solvent is one of the most fundamental processes in organic photochemistry and has been received much attention from experimental and theoretical points of view. Thanks to the development of pico and femto-second laser spectroscopy, the attention has been paid to the dynamics of electron transfer in recent years [1-4]. In particular the dielectric relaxation process of solvent associated with the electron transfer has been extensively studied as an important factor to understand the mechanism of reactions.

Among many subjects for the experimental studies of photoinduced electron transfer, 4-(N,N-dimethylamino)benzonitrile (DMABN) is known as a prototype to cause the intramolecular electron transfer in polar solvents [5,6]. The fluorescence emerges at a longer wavelength region after the photoabsorption has been attributed to the formation of a twisted intramolecular charge transfer (TICT) state. Although many experimental works have been performed to elucidate the machanism of CT state formation, the details of mechanism are still unclear because of the lack of theoretical models at molecular level based on the potential energy surfaces of DMABN and the solute-solvent interaction.

In describing the chemical reactions in solution, it is convenient to use the reaction free energy surfaces instead of the potential energy surfaces themselves, because the reactions are very complicated processes involving solute molecules and a large number of solvent molecules [7,8]. The concept of reaction free energy surface has also been utilized to describe the dynamics of nonadiabatic transition involved in many reactions in solutions such as the electron transfer. It would be, therefore, the first step to examine the free energy surfaces in theoretical understanding of the dynamics of electron transfer.

In this report, we will present the results of our theoretical studies on the mechanism of intramolecular electron transfer of excited state DMABN in polar solvents.

2. FREE ENERGY SURFACES FOR TICT STATE FORMATION OF DMABN

We carried out ab initio configuration interaction (CI) calculations for the ground and excited state DMABN to obtain the information for the potential energy surfaces in the gas phase [9]. The potential energy surfaces were represented as the functions of two coordinates; the torsional angle of dimethylamino group with respect to the aromatic plane, τ, and the wagging angle, θ. The cross section of potential surfaces at $\theta = 0°$ are displayed in Fig. 1. The S_2 state correlates to the TICT state at $\tau = 90°$. As seen in Fig. 1, the TICT state is not likely to be formed in the gas phase because the S_2 state energy increases along the torsional coordinate and does not cross with the S_1 surface.

The intermolecular potential functions between DMABN and H_2O were developed with the aid of electron distributions obtained by the ab initio calculations. The atom-atom pair functions were represented by the sum of electrostatic and exchange-exclusion term. It was found that the wave function of S_2 state is expressed by a superposition of three electronic configurations and the weight of each configuration strongly depends on the torsional angle τ whereas the S_0 and S_1 state wave functions do not change their character for the change of geometry. Since the S_2 state wave function contains an ion pair component, the potential surface is expected to be modified by the solute-solvent interaction in polar solvents. We therefore defined the diabatic state functions, one is of the ion pair type and the other two of the locally excited ones in the aromatic moiety, which are stable to the electric field from solvents. With the use of potential functions determined here, the stable geometries of DMABN-H_2O complex were calculated. The calculated results were consistent with the available spectroscopic informations obtained from the supersonic jet experiments [10,11].

Monte Carlo calculations were carried out to see the potential surface features of DMABN in the aqueous solution. The potentials of mean force for the torsional angle were first examined. It was found that the S_2 state potential profile is remarkably altered due to the solvation and the TICT state becomes a stable point on the surface while this point corresponds to the top of potential barrier in the gas phase. On the other hand, the potentials of mean force for the S_0 and S_1 state were very simailar to the gas phase potential energy curves. Figure 2 shows the simulated emission profile of DMABN in water which was calculated as the distributions of energy differences

$$P_{0\leftarrow I}(\Delta E) = \langle \delta(W_I - W_o - \Delta E) \rangle_I, \tag{1}$$

where $\mathbf{W}_I$ is the potential energy for the S_I state. The broad peak at a lower energy region is attributed to the emission from the S_2 state which is stabilized by the solvation. The emission band from the locally excited S_1 state would be assigned to the narrow peak at a shorter wavelength region. The results in Fig. 2 are consistent with the experimental findings.

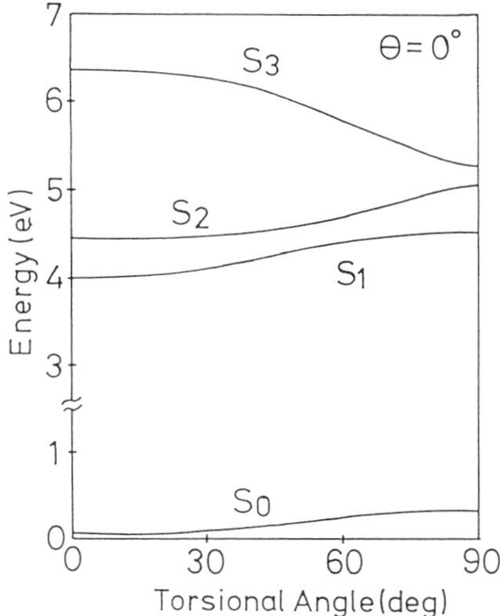

Figure 1. Gas phase potential energy curves of DMABN.

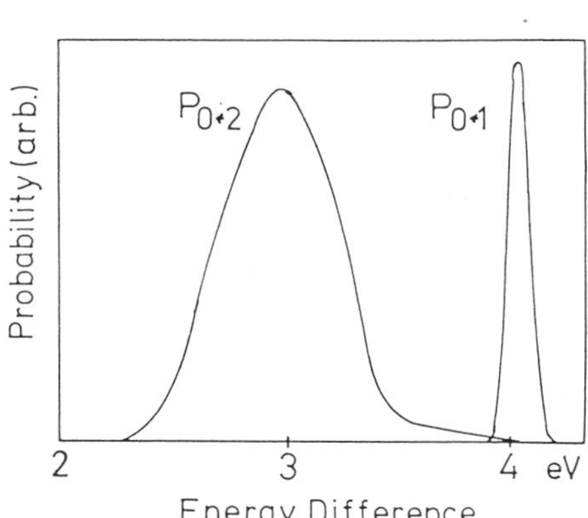

Figure 2. Distributions of S_1-S_0 and S_2-S_0 energy differences.

The reaction free energy surface is one of the basic concepts in describing the dynamics of electron transfer reactions. The energy difference of two state is conveniently taken as the solvation coordinate, s,to treat the nonadiabatic transition between them [12]. In this work we calculated the reaction free energy surfaces as the functions of two coordinates, τ and s, because the torsional motion plays an essential role in the CT state formation of DMABN. As the coordinate s, we took the quantity $s = \beta(W_1 - W_2)$. The calculated free energy surfaces are displayed in Fig. 3. Monte Carlo technique was applied for this purpose. Although the free energy curves are assumed to be parabolic in the previous treatments of electron transfer reactions, the shape of free energy curve of S_2 state in Fig. 3 is far from the parabolic form along the minimum energy path. This is because the charge distributions of DMABN in the S_2 state strongly depends on the torsional angle τ. It is noteworthy, however, that the S2 state curve along the coordinate s at a fixed value of τ is nearly parabolic. As seen in Fig. 3, the torsional coordinate is required to undergo a deformation to reach the transition state region.

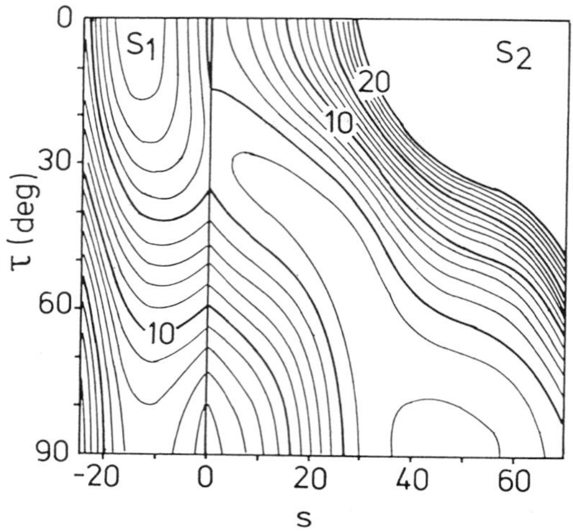

Figure 3. Reaction free energy surfaces for CT state formation of DMABN.

3. DIELECTRIC RELAXATION DYNAMICS INDUCED BY THE IONIZATION OF DMA

The dielectric relaxation of solvents associated with the change of charge distribution of solute has been recognized to be one of the essential factors in determining the rates of electron transfer reactions. This can be represented by the stochastic process on the reaction free energy surface under the influence of friction and random forces. Molecular dynamics (MD) method has been applied to simulate the relaxation processes and examine the assumptions such as the fluctuation-dissipation relation inherent to the stochastic models [13-18]. Apart from DMABN, we studied the solvation dielectric relaxation processes induced by the ionization of dimethylaniline (DMA), a closely related molecule to DMABN, in water and methanol solvents. DMA is known as a typical donor in many experimental studies of photoinduced electron transfer. We intended to obtain a clear-cut picture for the solvation dynamics of DMA, which would provide the key information to understand the dynamics of more complicated DMABN system.

The potential energy surfaces of ground and cation state of DMA were calculated as the functions of torsional and wagging angles , τ and θ, of dimethylamino group by the ab initio SCF method. The intermolecular potential functions between DMA and solvent molecules, H_2O and CH_3OH, both in the neutral and the cation state were also determined. We carried out the equilibrium and nonequilibrium MD trajectory calculations using the potential functions obtained here. The two important internal degrees of freedom, τ and θ, were explicitly taken into account in the MD trajectory calculations.

Figure 4 shows the potentials of mean force along the coordinate s, which is defined by

$$s = W_1(\mathrm{r}) - W_0(\mathrm{r}) - h\nu \tag{2}$$

where W_0 and W_1 are the potential energies of neutral and cation state, respectively, and $h\nu$ the photon energy required to cause the ionization of DMA in solutions. The values for $h\nu$ were taken to be 6.8 and 7.2 eV for H_2O and CH_3OH solutions. As seen in the figures, the free energy curves can be well approximated by the parabolic functions and the force constants were 0.25 and 0.25 eV^{-1} for H_2O and CH_3OH solution, respectively. One of the important points in Fig. 4 would be that the curves for methanol is very similar to those for water. This is because the dipole moments of H_2O and CH_3OH are close to each other, 1.94 and 1.69 Debye, respectively.

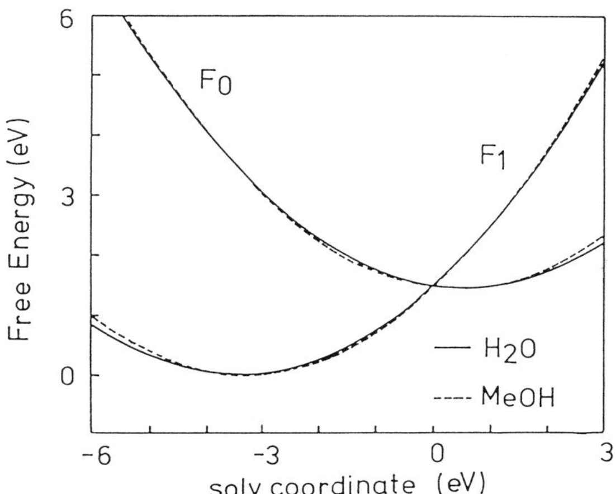

Figure 4. Free energy curves along the solvation coordinates. Solid and dashed lines are for H_2O and CH_3OH solution.

In the stochastic models for chemical reaction dynamics in solutions, the fluctuation dissipation relation, the key assumption of linear response theory, is usually utilized to derive the equation of motion. It is therefore worthwhile to examine whether the linear response assumption holds or not for the results of MD calculations. We compared in Fig. 5, the response function

$$S(t) = \frac{< s(t) > - < s(\infty) >}{< s(0) > - < s(\infty) >} \tag{3}$$

obtained from the nonequilibrium trajectory calculations with the time correlation function

$$C(t) = \frac{< \delta s(0)\delta s(t) >_I}{< \delta s(0)^2 >_I} \tag{4}$$

calculated from the equilibrium simulations. In the linear response theory, these two quantities are equated. As seen in Fig. 5(a), the linear response relation seems to be satisfied quite well for the water solvent. It is not easy to judge the applicability of linear response theory for methanol solution. The response function in Fig. 5(b) accords with the time correlation function obtained from the calculations for the cation state while there is an apparent deviation between them in the case of neutral state. We may say, however, that the linear response relation is approximately satisfied for the methanol.

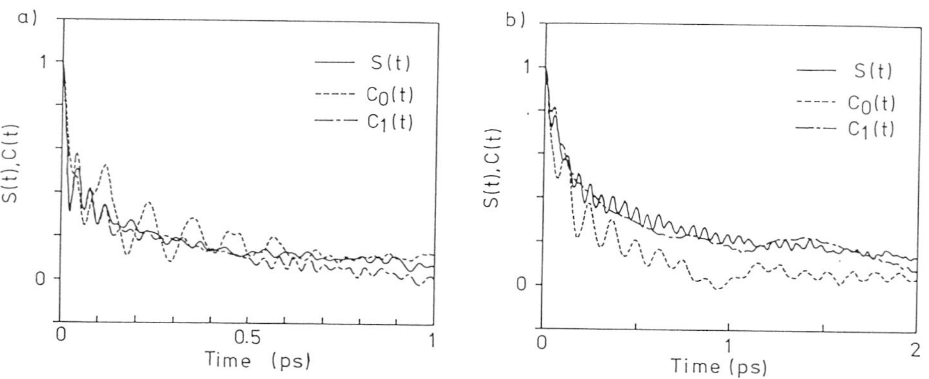

Figure 5. Dielectric relaxation dynamics.
(a) H_2O and (b) CH_3OH solution.

4. GENERALIZED LANGEVIN DESCRIPTION OF SURFACE HOPPING PROCESSES

We can represent the dynamics of electron transfer reactions by the propagation of systems along the solvation coordinate s which is defined by the difference of potential energies of two states. Such the motion may be described by the stochastic equation of motion because it is regarded as the projection of complicated motions in solution onto the one dimentional motion. It would be meaningful to derive the stochastic equation of s for the processes including the surface hopping as in the electron transfer reactions. For the surface hopping, the crossing of potential surfaces plays a critical role. It is natural to choose the transition state point to be the minimum energy point on the crossing surface

$$f(\mathrm{x}) = W_1(\mathrm{x}) - W_0(\mathrm{x}) = 0, \tag{5}$$

where x is the coordinates, and the reaction coordinate to be the steepest descent path in the mass-weighted Cartesian coordinate space, which passes through this point [19,20]. With this definition, we obtain two branches of reaction coordinates, one on the surface W_0 and the other on W_1. It is easily proved that the directions of these two branches of reaction coordinates coincides each other and are normal to the crossing surface at the minimum energy crossing point. We can therefore use the reaction coordinate and the normal coordinates perpendicular to it to describe the dynamics near the surface crossing region. As seen in the case of DMA, the free energy curves are almost parabolic along the solvation coordinate s and the linear response relation seems to be satisfied at least qualitatively. It would be therefore worthwhile to examine the harmonic model in describing the dynamics along the solvation coordinate. The potential energy functions are then expressed as

$$W_I(\mathrm{x}) = \sum_i \frac{1}{2}\omega_i^2 x_i^2 + \sum_i g_i^I x_i + W_I(0) \qquad (I = 0, 1), \tag{6}$$

where the photon energy $h\nu$ is included in $\mathrm{W}_0(0)$. For these potential functions, the direction vector of the reaction coordinate, ${}^t\tilde{\mathrm{s}} = (\tilde{s}_1, \tilde{s}_2, ...)$, at the minimum energy crossing point is given by

$$\tilde{s}_i = \frac{dx_i}{d\tilde{s}} = \frac{\Delta g_i}{(\Sigma_j \Delta g_j^2)^{\frac{1}{2}}} \qquad (i = 1, 2, ...), \tag{7}$$

where

$$\Delta g_i = g_i^1 - g_i^0. \tag{8}$$

If we take the coordinate s as the linear line with the direction of $\tilde{\mathrm{s}}$,

$$\tilde{s} = {}^t \mathrm{s} \cdot \tilde{\mathrm{x}}, \tag{9}$$

the potential energy difference of two states is easily shown to be propotional to the coordinate s.

$$s = W_1(\mathbf{x}) - W_0(\mathbf{x}) = (\sum_j \Delta g_j^2)^{1/2}\tilde{s}. \tag{10}$$

Using the matrix projection operator **P** with the element

$$\hat{P}_{ij} = \frac{\Delta g_i \Delta g_j}{\Sigma_k \Delta g_k^2}, \tag{11}$$

it is straightforward to derive the generalized Langevin equation (GLE)

$$\ddot{s} + \Omega^2 s + \mu^{-1/2}\bar{g} + \int_0^t \zeta(\tau)\dot{s}(t-\tau)d\tau + \zeta(t)s(0) = \mu^{-1/2}R(t), \tag{12}$$

for the motion along the energy difference coordinate s. In Eq. (12), μ may be regarded as the effective mass, $\mu = (\Sigma_k \Delta g_k^2)$. The friction kernel ζ(t) is given by

$$\zeta(t) = \tilde{\omega}_{PQ}^2 \tilde{\omega}_{QQ}^{-2} cos(\tilde{\omega}_{QQ} t)\tilde{\omega}_{QP}^2, \tag{13}$$

and is related to the random force R(t) by the fluctuation-dissipation theorem

$$< R(0)R(t) >= \zeta(t)/k_B T. \tag{14}$$

The elements of $\tilde{\omega}_{QP}^2$ and $\tilde{\omega}_{QQ}^2$ are

$$\tilde{\omega}_{QP,i}^2 = (\omega_i^2 - \bar{\omega}^2)\tilde{\omega}_i, \tag{15}$$

$$\tilde{\omega}_{QQ,ij}^2 = \omega_i^2 + (\bar{\omega}^2 - \omega_i^2 - \omega_j^2)\tilde{s}_i\tilde{s}_j, \tag{16}$$

and the vector $\tilde{\omega}_{PQ}^2$ is the transpose of $\tilde{\omega}_{QP}^2$. The average quantities in eq.(12) are

$$\bar{\omega}^2 = \sum_k \omega_k^2 \tilde{s}_k^2, \tag{17}$$

and

$$\bar{g} = \sum_k (g_k^I + \omega_k^2 x_k^\dagger), \tag{18}$$

respectively.

Figure 6 shows the components of $\tilde{s}$ obtained from the Fourier transform of the correlation functions, $C(t)$, for the cation state. For DMA-H_2O system, a broad band coming from the librational mode of H_2O is observed at the region of 750 - 900 cm^{-1}, while the main band for DMA-CH_3OH syatem is centered at about 650 cm^{-1}. As seen in Fig.6, the main band for water solution is composed by several separated narrow peaks.

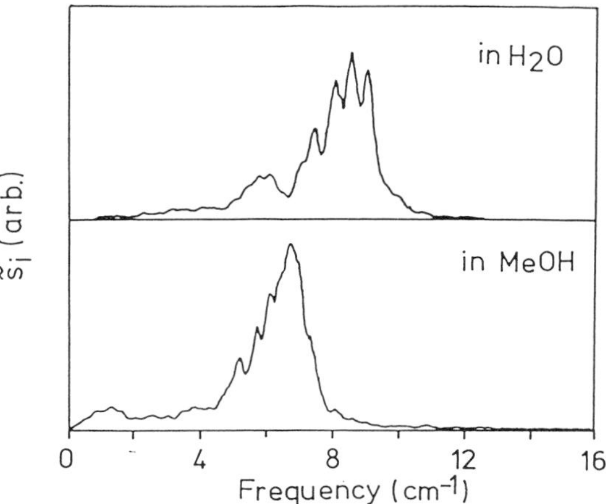

Figure 6. Components of the solvation coordinate vector s.

We could obtain all the quantities in the GLE, Eq. (12), from the present MD calculations. The effective frequency Ω for the harmonic free energy curve is calculated by

$$\Omega^2 = \bar{\omega}^2 - \zeta(0) = \bar{\omega}^2 - \frac{\sum_i (\omega_{PQ,i})^2}{\omega_{QQ,i}^2}, \tag{19}$$

and the results are 417 and 196 cm^{-1} for H_2O and CH_3OH solutions, respectively. The effective masses μ were also calculated to be 0.42$\times$ 10^{-4} and 1.99 $\times$ 10^{-4} eV^{-1}, respectively. It is noted that these effective frequencies are in good agreement with the values directly calculated from the free energy curves given in Fig. 4, 440 and 190 cm^{-1}. Using the components of $\tilde{s}$ in Fig.6, it is straightforward to calculate the time-dependent friction, $\zeta(t)$, and the results are shown in Fig.7 both for H_2O and CH_3OH solutions. The time-dependent friction for CH_3OH solution decays very rapidly and is follwed by an oscillation with a very low frequency. Since the initial decay of $\zeta(t)$ is much faster than the time scale of dielectric relaxation in Fig. 5, the relaxation dynamics can be described by an overdamped motion in this case. On the other hand, the time scale of initial decay is comparable with that of dielectric relaxation in the case of aqueous solution. This result accounts for the oscillatory behavior of dielectric relaxation for H_2O solution.

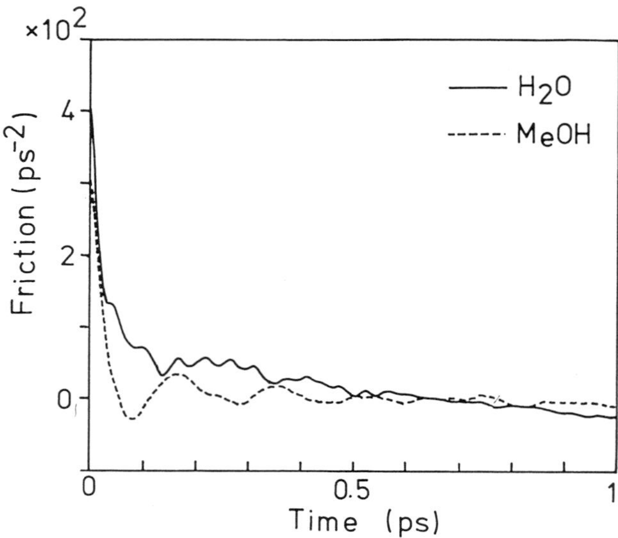

Figure 7. Time dependent friction. Solid and dashed lines are for H_2O and CH_3OH solution.

5. CONCLUSION

We have presented the results of Monte Carlo calculations for the reaction free energy surfaces of excited state DMABN in the aqueous solution. The free energy surfaces were given as the functions of two coordinates; one is the torsional angle of dimethylamino group and the other is the difference of potential energies between the S_1 and S_2 state. Although the free energy curves along the solvation coordinate defined by the energy difference of two states are usually assumed to be parabolic, they are far from the parabolic form in the present case. This is because the S_2 state is represented by the superposition of neutral and ion pair states.

In order to obtain the information of the dynamics of electron trasfer reactions, we carried out the MD calculations for the ionization processes of DMA in H_2O and CH_3OH solutions. We also derived the GLE equation for the surface hopping process in solutions by utilizing the reaction path model for the potochemical processes originally developed to describe the gas phase processes and found that the present model well reproduces the characteristic feature of MD simulation results. Considering that the free energy curves along the solvation coordinate is regarded as a parabolic form at each value of torsional angle even for the case of DMABN, it is possible to extend the present simple dynamic model in describing the dynamics of intramolecular electron transfer process of DMABN in polar solvents. We are now working along this line.

6. REFERENCES

1 P.Madden and D.Kivelson, Adv. Chem. Phys. 66 (1986) 467.
2 E.M.Kosower and D.Huppert, Ann. Rev. Phys. Chem. 37 (1986) 127.
3 P.F.Barbara and W. Jarzeba, Acc. Chem. Res. 21 (1988) 195.
4 G.E.McMains and M.J.Weaver, Acc. Chem. Res. 23 (1990) 294.
5 E.Lipper, W.Rettig, V. Bonacic-Koutecky, F. Heisel and J. A. Mieha, Adv. Chem. Phys. 68 (1987) 1.
6 D.Huppert, S.D.Rand, P.M.Rentzepis, P.F.Barbara, W.S.Sturev, and Z.R.Grabowski, J. Chem. Phys. 75 (1987) 413.
7 R.A.Marcus, J. Chem. Phys. 24 (1956) 966, 979.
8 J.T.Hynes, in The Theory of Chemical Reaction Dynamics, edited by M.Baer (CRC,Boca Raton,FL,1985), Vol.4, p.171.
9 S.Kato and Y.Amatatsu, J. Chem. Phys. 92 (1990) 7241.
10 T.Kobayashi, K.Futakami,and O.Kajimoto, Chem. Phys. Lett. 130 (1986) 63.
11 J.A.Warren, E.R.Bernstein, and J.I.Seeman, J. Chem. Phys. 88 (1988) 871.
12 J.K.Hwang and A.Warshel, J. Am. Chem. Soc. 109 (1987) 715.
13 M.Rao and B.J.Berne, J. Phy. Chem. 85 (1981) 1498.
14 M.Maroncelli and G.R.Fleming, J. Chem. Phys. 89 (1988) 5044.
15 J.S.Bader and D. Chandler, Chem. Phys. Lett. 157 (1989) 501.
16 M.Maroncelli, J. Chem. Phys. 94 (1991) 2084.
17 T.Fonseca and B.M.Ladanyi, J. Phys. Chem. 95 (1991) 2116.
18 E.A.Carter and J.T.Hynes, J. Chem. Phys. 90 (1991) 5961.
19 S.Kato, R.L.Jaffe, A.Komornicki, and K.Morokuma, J. Chem. Phys. 78 (1983) 4567.
20 S.Kato, J. Chem. Phys. 88 (1988) 3045.

Dynamics and Mechanisms of
Photoinduced Transfer and Related Phenomena
N. Mataga, T. Okada and H. Masuhara (Editors)

Microscopic solvation of 9,9'-bianthryl studied in its clusters with polar molecules

Kenji Honma[a] and Okitsugu Kajimoto[b]

[a]Department of Material Science, Himeji Institute of Technology, Kamigori, Hyogo, Japan

[b]Department of Chemistry, Kyoto University, Sakyo-ku, Kyoto, Japan

Abstract

Dynamics of the electronically excited state of 9,9'-bianthryl(BA) in clusters with some polar molecules was studied in a free jet condition. The BA-X_n clusters were formed by supersonic expansion. The number of "solvent" molecules, X, were determined by the TOF mass selected resonance enhanced multiphoton ionization (REMPI) spectrum. The dispersed fluorescence and fluorescence lifetime were measured for each BA-X_n cluster. BA in the clusters with nonpolar molecules, Ar and cyclohexane, shows fluorescence from photo-excited state. In the clusters with polar molecules, red-shifted fluorescence has been observed. Using acetone as the "solvent", the fluorescence is shifted toward longer wavelength and the lifetime becomes longer with increase the number of "solvent". These results were explained by the occurrence of polar excited state of BA in cluster with polar molecules.

1. Introduction

Recent developement of the supersonic jet technique combined with the use of laser spectroscopy provides an oppotunity to study "solvated" molecules in isolated conditions. A plenty of reseaches are devoted to study the role of solvent by using gas phase clusters. In favorable cases, the study of cluster with specific number of "solvent" molecules provides versatile information bridging the isolated molecule to molecule in solution. By applying this technique, we have been studied the dynamics of the electronically excited state of 9,9'-bianthryl(BA) in some polar "solvents"[1].

BA has been a favorite object to study the dynamics of its excited state in polar solution[2][3]. Especially, the red-shifted emission in polar solution has been studied by many researchers and the origin of that emission has been ascribed to the twisted intramolecular charge-transfer(TICT) state[4] which is stabilized in polar solvent. That is, the torsional angle between two anthracene moieties, being 90 deg in the ground state, is relaxed from 90 deg in the excited state so that the electron can transfer. However, a mechanism of "charge-transfer" still has some controversial points. Recently, the red-shifted emission was observed under the conditions of which the torsional motion of BA is highly restricted[1][5], and those results imply that the "charge-transfer" occurs without orbital overlap between two antracene moieties. In this paper, we wish to report the behavior of the electronic excited state of BA in the clusters with nonpolar and polar molecules. The number of "solvent" molecules in the clusters are specified and the properties of electronic state of BA has been studied as a function of those number as well as molecular properties of "solvent".

2. Experimental

The experiments were carried out by using supersonic jet apparatue. The clusters were formed by coexpansion of BA and solvent molecules with He through a pulsed nozzle heated at 300°C. A tunable dye laser pumped by XeCl excimer laser was irradiated to the free jet, and the laser induced fluorescence was collected by lens system onto a photomultiplier (HAMAMATSU R-928). A monochromator (NIKON P-250) was used for the measurement of dispersed fluorescence. The fluorescence lifetimes were also measured by using digital storage oscilloscope (PHILLIPS PM3320) with time resolution of 4ns. The time-of-flight (TOF) mass spectrometer was used to determine the number of "solvent" molecules. The tunable dye laser was tuned for bands observed by the laser-induced fluorescence and the second laser (308nm) was irradiated to ionize that cluster. By measuring the time-of-flight of that cluster ion, the number of the "solvent" molecules were determined for each band. Studied polar "solvents" were acetone, diethyl ketone, methylethyl ketone, acetonitrile, water, ethanol, 2-propanol and CF_3H. The clusters with Ar and cyclohexane were also studied as the nonpolar "solvents".

3. Results and discussion

3.1. BA-acetone

Figure 1 shows the laser-induced fluorescence spectra of BA-acetone and BA-cyclohexane clusters with the spectrum of bare BA in the "ground" vibronic band region[6].The LIF spectrum of BA has sharp structure assigned to be the transition to the vibrational excited state of torsional motion. The spectrum of BA-cyclohexane has broad feature which is shifted by -190cm^{-1}. The spectrum of BA-acetone has some bands which have different dependence on the concentration of acetone. By using TOF mass

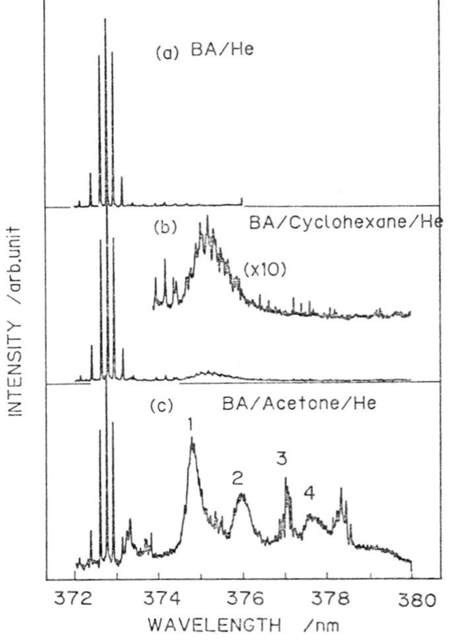

Fig.1 LIF spectra of (a) BA (b) BA-cyclohexane (c) BA-acetone

selected REMPI spectra, they have been assigned as the clusters with different number of acetone molecules and those numbers are also shown in the figure. The dispersed fluorescence spectra measured by excitation of bands indicated by arrows in Fig.1 are summarized in Fig.2. In the fluorescence spectrum of bare-BA, there are structured peaks and they are assigned as the transitions to ground state vibrations. Although each peak shows broad feature, the fluorescence spectrum of BA-cyclohexane cluster consists of peaks which have similar spacing to those of bare-BA. Broad bands are ascribed to the congestion of transitions to the cluster vibration, and it is concluded that the fluorescent state of BA in cyclohexane cluster has same character as the isolated BA molecule.

On the other hand, BA-acetone shows largely red-shifted and much more broad fluorescence. The amount of this Stokes shift increases with increase of the number of acetone molecules as summarized in Table 1. These results are consistent with the fluorescence of BA in solution, i.e. BA in acetone shows largely red-shifted fluorescence compared with BA in cyclohexane. The fluorescence lifetimes of BA clusters are also consistent with those of BA in solution. As summarized in Table 1, the lifetime of BA-cyclohexane is similar to that of bare-BA. Those of BA-acetone is longer and increases with increase of the number of acetone molecules.

These results of BA-acetone clusters can be explained by the occurrence of polarity change in the electronic excited state of BA in BA-acetone clusters. The polar electronic state is stabilized by the interaction with acetone molecules and its energy is lowered by increase of the number of acetone molecules. That state has forbidden character for the transition to the nonpolar ground state and that character causes the longer lifetime. This polar electronic state can be same as so called intramolecular charge transfer state which

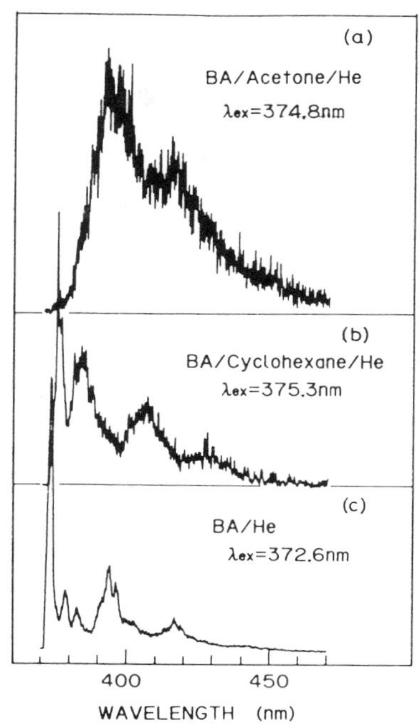

Fig. 2 Dispersed fluorescence (a) BA-acetone (b)BA-cyclohexane (c)BA

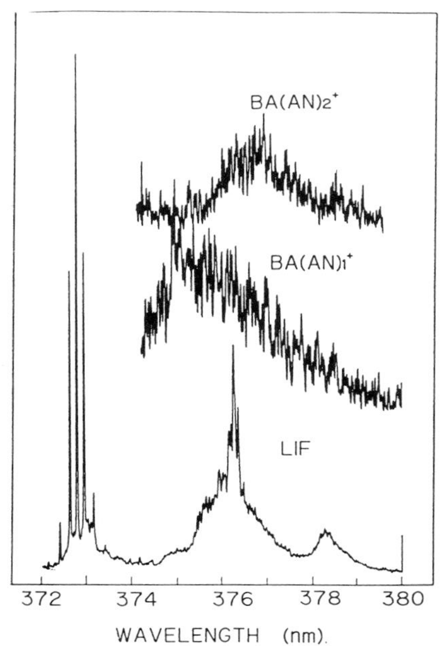

Fig. 3 LIF and REMPI spectra of BA-acetonitrile clusters

has been observed in the polar solvents. The increase of acetone in cluster may exhibit the gradual change from isolated BA to "solvated" BA.

3.2. BA-acetonitrile,CF_3H

The behavior of the electronic excited state of BA-acetone cluster implys that the cluster bridges between isolated BA and BA in solution, i.e. the increase of number of "solvent" molecules changes the electronic state of BA from isolated BA to BA in solution. However, it may not be true in the clusters of BA with acetonitrile and CF_3H. The laser-induced fluorescence spectra of BA-acetonitrile are summarized in Fig. 3. The TOF mass selected REMPI spectra are also shown in this figure. In spite of the poor signal to noize ratio, the first and second bands can be assigned to be BA-(acetonitrile)$_1$ and BA-

Table 1 Stokes shifts and fluorescence lifetimes of BA-X_n clusters.

species	λ_{abs} (nm)	Stokes shift (cm^{-1})	lifetime (ns)
bare-BA	372.6		10.8
BA-cyclohexane	375.3		9.9
BA-(acetone)$_1$	374.8	1300	24.5
BA-(acetone)$_2$	375.9	2331	36.5
BA-(acetone)$_3$	377.6	2387	38.0
BA-(acetone)$_3$	378.3	2625	40.0
BA in acetone solution		3500[a]	32[b]

[a] N. Nakashima and D. Phillips, Chem. Phys. Lett. 97, 337 (1983).
[b] N. Nakashima, M. Murakawa, and N. Mataga, Bull. Chem. Soc. Japan 49, 854 (1976).

(acetonitrile)$_2$, respectively. The amount of Stokes shift and fluorescence lifetime are summarized in Table 2. According to these results, the first acetonitrile make the Stokes shift of 1365cm^{-1} and elongate the lifetime, however, second and more acetonitrile molecule make no further change. Acetonitrile has more polar character like dipole moment than acetone, and these results imply that some solvent character other than polarity plays important role in the dynamics of electronic excited state of BA.

The results of BA-$(CF_3H)_n$ clusters also show another discrepancy in the behavior of electronic state between clusters and solution. The Stokes shift and fluorescence lifetime are summarized in Table 2. In spite of the less polar character of CF_3H molecule, the increase of the number of CF_3H molecule enhances the polar character of BA more drastically than acetonitrile. In our laboratory the dispersed fluorescence of BA was

Table 2 Stokes shifts and fluorescence lifetimes of BA-(acetonitrile)$_n$ and BA-$(CF_3H)_n$ clusters.

species	λ_{abs} (nm)	Stokes shift (cm^{-1})	lifetime (ns)
BA-(acetonitrile)$_1$	374.8	1365	20.8
BA-(acetonitrile)$_2$	376.3	1259	26.0
BA-(acetonitrile)$_n$[a]	378.3	1118	
BA in acetonitrile solution		4040[b]	35[c]
BA-$(CF_3H)_1$	374.4	903	16.9
BA-$(CF_3H)_2$	375.4	1794	27.0
BA-$(CF_3H)_n$[d]	376.5	3304.6	41.7

[a,d] The number of solvent molecules were not specified.
[b] N. Nakashima and D. Phillips, Chem. Phys. Lett. 97, 337 (1983).
[c] N. Nakashima, M. Murakawa, and N. Mataga, Bull. Chem. Soc. Japan 49, 854 (1976).

measured in supercritical CF_3H fluid and no red-shifted fluorescence was observed. These results indicate that the polar electronic state of BA appears only in clusters with CF_3H.

3.3. Structured bands in BA-ketone clusters

As discussed in the section 3.1, the most of laser induced fluorescence bands of BA-acetone clusters are broad and structureless. However, bands with sharp peaks are also observed between broad bands. The fluorescence spectra of these bands are very similar to those of BA-cyclohexane and bare-BA. One of the dispersed fluorescence is shown in Fig. 4. The lifetime of these bands are also similar to those of BA-cyclohexane. Although the resolution of our TOF-REMPI spectrum is not good enough to assign those origin unambiguously, their dependence on the concentration of acetone imply that the number of acetone in these bands is

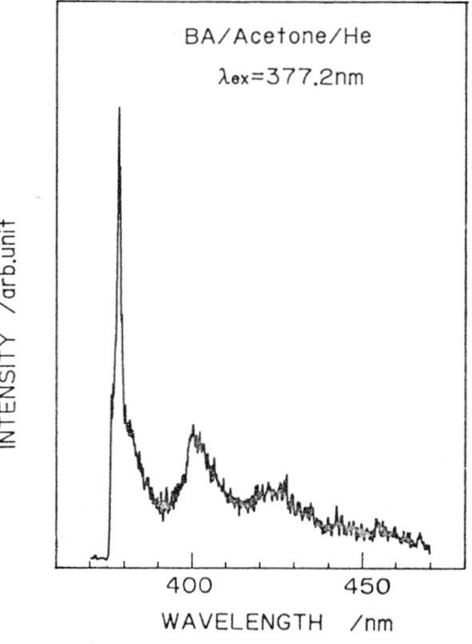

Fig. 4 Dispersed fluorescence of BA-acetone excited at structured band

same as that in the adjacent broad band. These bands are possibly structural isomers. Among the clusters of BA with polar molecules studied, these kinds of structured bands can be observed only in the cluster of BA with symmetric ketone, i.e. acetone and diethyl ketone. The laser induced fluorescene of BA-diethyl ketone is shown in Fig. 5 with that of BA-acetone. In these two clusters, broad band and structured one appeared alternatively. The dispersed fluorescence spectra are shown in Fig. 6 for both excitation of broad band and that of structured one. Structured band shows structured spectrum which is similar to the fluorescene of BA-cyclohexane. The lifetime was measured for both bands. The structured band has short lifetime which is similar to the lifetime of BA-cyclohexane, i.e. 9.0ns, and broad band has long lifetime, i.e. 24.7ns. These results indicate that there are two kinds of BA-ketone cluster. In one of those clusters, polar state of BA can be observed, however, in the other kind of clusters, the electronic state of BA does not make any change even within the cluster with polar molecules.

For comparison, we have measured fluorescence of BA-methylethyl ketone and the result is shown in Fig. 5. In the cluster of this asymmetric ketone, no structured band was observed in the laser induced fluorescence. The dispersed fluorescence is broad and lifetime is long, i.e. 27.7ns. These results implies that the appearance of the polar state of BA in the BA-ketone clusters is controlled by the structure of the solvents, i.e. nonsymmetric structure of solvent is important.

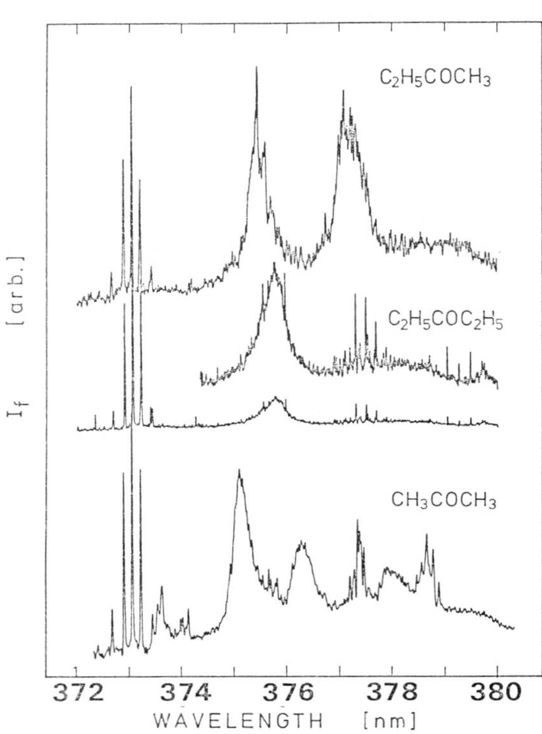

Fig. 5 LIF spectra of BA-ketone clusters

3.4. Mechanism of polar state formation BA

The results are summarized as the following.

(1) Only fluorescence from the excited state was observed for BA-cyclohexane cluster. Same results were observed for other BA-nonpolar molecule cluster.

(2) Red-shifted fluorescence were observed in BA-polar molecule clusters, and the lifetime of these fluorescene were longer than that of bare-BA fluorescence.

(3) In case of clusters with acetone and CF_3H, increase of the number of "solvent" enhance the Stokes shift and elongate the fluorescence lifetime. On the other hand, two

and more "solvent" show no further change in the Stokes shift and lifetime in case of acetonitrile.

(4) Only in case of clusters with symmetric ketones, strutured bands were observed and at those bands, the fluorescence spectra were similar to those of bare-BA and BA-cyclohexane cluster.

These results imply that not only the polar character of "solvent" but also the structure of cluster is important in the appearance of polar state of BA. Since the red-shifted fluorescence was observed only in the cluster with polar molecules, the origin of this emission is polar excited state of BA. This polar state is lowered its energy by the interaction with polar "solvent" molecules. The extent of the stabilization of this state depends on both the polarity of "solvent" and the structure of cluster since the interaction of BA with "solvent" is local and the relative configuration of BA and "solvent" is rather ligid in cluster. Therefore one possible explanation for the result of BA-(acetonitrile)$_2$ is that two acetonitrile molecules attach to BA such that dipole moments of two are partially canceled each other. This could be possible since the clusters are formed with BA in nonpolar gound state within the jet.

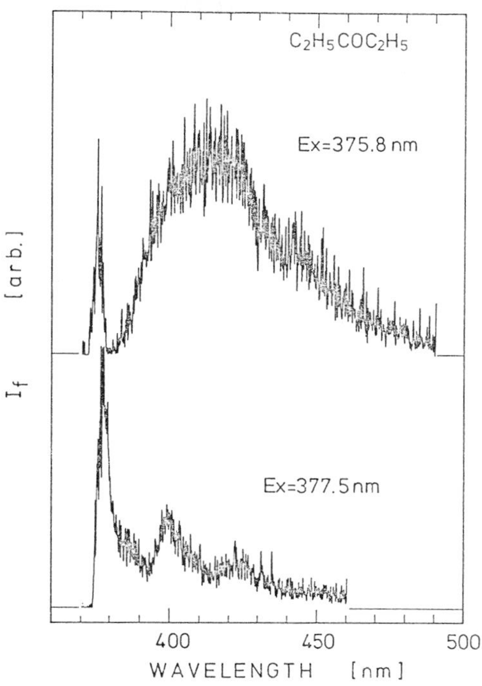

Fig. 6 Dispersed fluorescence of BA-diethylketone clusters.

The importance of cluster structure must be emphasized since only clusters with symmetric ketone show both structured and broad bands. Symmetry of BA system has been demonstrated to be important for "charge-transfer"[7]. In most clusters, "solvent" probably attaches to BA nonsymmetrically, and that provides broken symmetry to BA. Under these conditions, degeneracy of the charge-resonance state of BA is removed and lower state in which electron is localized on one anthracene moiety is stabilized by the interaction with polar "solvent". However, in case of symmetric ketone, "solvent" may possibly form symmetric clusters in which two methyl or ethyl groups face with two anthracene moieties. In this kind of cluster, charge-resonance state of BA is still degenerated and the fluorescence appears to be same as that of bare-BA.

References

1. K. Honma, K. Arita, K. Yamasaki, and O. Kajimoto, J. Chem. Phys. 94, 3496 (1991); O. Kajimoto, K. Yamasaki, K. Arita, and K. Hara, Chem. Phys. Lett. 125, 184 (1986).

2. F. Schneider and E. Lippert, Ber. Bunsenges. Phys. Chem. 72, 1155 (1968); F. Schneider and E. Lippert, Ber. Bunsenges. Phys. Chem. 74, 624 (1970).

3. N. Nakashima and D. Phillips, Chem. Phys. Lett. 97, 337 (1983); N. Nakashima, S. Nagakura, and N. Mataga, Bull. Chem. Soc. Japan 49, 854 (1976).

4. Z. R. Grabowski, K. Rotokiewicz, A. Seimiarczuk, D. J. Cowley, and W. Bowman, Nouv. J. Chim. 3, 443 (1979).

5. M. A. Kahlow, T. J. Kang, and P. F. Barbara, J. Phys. Chem. 91, 6452 (1987);T. J. Kang, M. A. Kahlow, D. Giser, S. Swallen, V. Nagarajan, W. Jarzeba, and P. F. Barbara, J. Phys. Chem. 92, 6800 (1988).

6. They are assigned as "ground" vibronic bands since all vibration except for the torsion are in ground level.

7. N. Mataga, H. Yao, T. Okada, and W. Rettig, J. Phys. Chem. 93, 3383 (1989).

Part 3:
Electron and Energy Transfer in Molecular Aggregates and Polymers

Dynamics and Mechanisms of
Photoinduced Transfer and Related Phenomena
N. Mataga, T. Okada and H. Masuhara (Editors)

Relaxation in Spatially-Restricted Structures

J. Klafter[a] and J.M. Drake[b]

[a]School of Chemistry, Tel Aviv University, Tel Aviv 69978, Israel

[b]Exxon Research and Engineering, Annandale, New Jersey 08801, U.S.A.

Abstract

In this paper we review different approaches to analyzing nonexponential decay functions, with emphasis on the stretched exponential form. We concentrate on a few solvable models for which the stretched exponential law follows in a natural fashion. Although each model describes a different mechanism, it is shown that they have the same underlying reason for the stretched exponential pattern: the existence of scale invariant relaxation times. We bring examples for applications in some real systems.

1. INTRODUCTION

There has been accumulating evidence that nonexponential relaxation patterns in time are common to many disordered condensed matter systems. They appear to be the rule rather than the exception. A few examples for such decays include: fluorescence decay in porous solids [1-3], in polymers [4,5], in self-organized systems such as micelles and vesicles [6], solvation dynamics in polar solvents [7], dielectric relaxation in glasses and in polymers [8], electron scavanging in glasses [9] and birefringence studied in polymers [10].

A number of fitting methods have been proposed in order to analyze nonexponential relaxations and to obtain information about the underlying systems from them. Most of these methods focus on calculating distributions of lifetimes or rate constants in multiexponential expressions [5,11,12].

An interesting empirical observation has been made recently suggesting that the decay of correlation functions for many diverse systems follows the same stretched exponential law,

$$\Phi(t) = \exp\left[-(t/\tau)^{\alpha}\right], \quad o<\alpha<1 \tag{1}$$

The parameters α and τ depend on the material and can be a function of external variables such as temperature. Equation (1) was first proposed by Kohlrausch to describe mechanical creep [13], and was later used by Williams and Watts to describe dielectric relaxation in polymers [14]. The stretched exponential has since been employed to fit data in other systems, some of which are mentioned as examples above. Here we show that this ubiquitous decay law can be derived from several models which describe different physical mechanisms. Each of the models generates scale invariant relaxation rates which are responsible for the stretched exponential behavior [15].

Another decay form which has been used to fit relaxation results, although less extensively than the stretched exponential, is the enhanced power law

$$\Phi(t) \sim (t/\tau)^{-B\ln^{\beta-1}(t/\tau)};\ \beta\geq 1,\ t>\tau \tag{2}$$

which is essentially the exponential-logarathmic form of Inokuti and Hirayama [16] and reduces to a simple power law for $\beta=1$. Equation (2) has been introduced to explain electron scavanging in glasses and transport properties in amorphous semiconductors [9,17].

In this paper we briefly review three models which lead to stretched exponentials. A more detailed discussion is presented in ref.[15]. We will analyze the Forster direct transfer mechanism [18] which is an example of relaxation via *parallel* channels and relate its mathematical structure to the *serial* hierarchichally constrained dynamics model [19]. The third related theory is that of the target model [20]. In all three cases the relaxation is given by

$$\Phi(t) = \int dR f(R)\ e^{-t/\tau(R)} \tag{3}$$

where f(R) is a weight factor and $\tau(R)$ is a microscopic relaxation time which varies with the relevant variable in the problem (distance or level, vide infra). We show that:

(a) for those cases where $\tau(R)$ scales with R, $\tau(R)\sim R^{\delta}$, a stretched exponential is obtained

(b) for $\tau(R)\sim \exp[\gamma R]$ an enhanced power law is obtained.

Our conclusions concerning the stretched exponential behavior are then extended to spatially restricted structures and applied to direct energy transfer in porous systems and to birefringence experiments in polymer systems.

2. MODELS FOR STRETCHED EXPONENTIALS [15]

The first model we discuss, which esults in a stretched exponential decay, is the direct energy transfer model (DET). This model has been extensively ocvered in the literature [18,21,22] and is schematically presented in Fig.1. It describes the decay law of an initially prepared static donor due to direct transfer to *randomly* distributed static acceptors. Although DET offers a relatively simple example of spatial disorder, it is still rich enough to obtain the complex relaxations of Eqs. (1) and (2). The donor in Fig.1 can simultaneously relax through many competing channels, a process that leads to the non-exponential decay. To each donor there corresponds a *hierarchy* of donor-acceptor distances. The relaxation function in each channel is given by

$$\phi_i = \exp[-tW(R_i)] \tag{4}$$

where $W(R_i)$ is the relaxation rate that depends on the relative locations of a donor-acceptor pair. This defines a relaxation time $\tau(R_i)=[W(R_i)]^{-1}$. One has to include now all the relaxation channels and to configurationally averge (see ref.[21] for details). When a substitutional occupancy of site by acceptors with probability p is assumed,

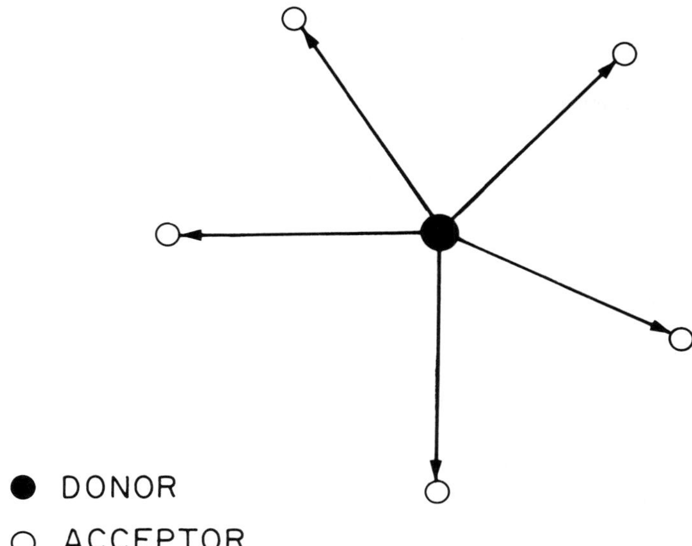

Figure 1. A schematic representation of DET

$$\Phi(t) = \exp[-p \sum_i \{ 1- \exp [-tW(R_i)]\}] \quad ; \qquad p << 1 \tag{5}$$

By introducing a site density function [22]

$$\rho(R) = \sum_i{}' \delta (R-R_i)$$

we obtain

$$\Phi(t) = \exp [-p\int dR\rho(R) \{1- \exp[-tW(R)]\}]. \tag{6}$$

Two types of isotopic rates are usually considered [15,22]:

(a) $W(R)=AR^{-s}$, $s \geq 6$, which implies scale invariance $\tau(R)\sim R^{s}$ (7)

(b)) $W(R)=B\exp[-\gamma R]$, which implies $\tau(R)\sim\exp[\gamma R]$

On regular underlying spatial structures $\rho(R)$=const. and we find for case (a) a stretched exponential

$$\Phi(t) = \exp[-(t/\tau)^{d/s}] \qquad \text{(in d dimensions)} \qquad (8)$$

For case (b)

$$\Phi(t) \sim t^{-B} \ln^{d-1}(Bt) \qquad \text{(in d dimensions)} \qquad (9)$$

This exhibits an enhanced power law which takes the form of an algebric decay in one dimension.

The nonexponential decays of Eqs. (8) and (9) are both a reuslt of a *parallel* relaxation scheme and a *hierarchy of distances.*

The same types of relaxation patterns can be derived when we restrict the transfer to include only the nearest neighbor acceptors. Then

$$\Phi_{NN} = \int_0^\infty f(R) \exp[-tW(R)]\, dR \qquad (10)$$

where f(R) is the probability of having a nearest neighbor at distance R. For randomly placed acceptors in one-dimension

$$f(R) = p\exp(-pR) \qquad (11)$$

For W(R) in Eq. (7), case (a),

$$\Phi_{NN} = p\int e^{-pR} \exp(-tAR^{-s})\, dR \qquad (12)$$

which by the method of steepest descent gives

$$\Phi_{NN} = p \exp[-Ct^{\frac{1}{1+s}}] \qquad (13)$$

again a stretched exponential but with a smaller exponent than for d=1 in Eq. (8), because the influence of more distant defects has been truncated.

The direct transfer decay laws follow a stretched exponential behavior both for the many parallel channels case and for the fastest decay channel when $W(R) \sim R^{-s}$, namely when the position dependent relaxation time $\tau(R)$ is scale invariant.

Recently Palmer et. al. [19] introduced a model of relaxation which is *serial* rather than parallel. This hierarchical model supposes that relaxation occurs in stages, and the constraint imposed by a faster degree of freedom must relax before a slower degree of freedom can relax. This implies that the time scale of relaxation on one level is subordinated to the relaxation below.

In one possible realization they considered a system with a discrete series of levels n=0,1,2,..., with the degrees of freedom on level n represented by N_n spins which point either up or down as in Fig.2. The spins in level n+1 are only free to change their state when μ_n spins in level n attain one of their 2^{μ_n} possible states. The relaxation time τ_{n+1} of level n+1 is then

$$\tau_{n+1} = 2^{\mu_n} \tau_n \tag{14a}$$

$$= \tau_o \exp \left(\sum_{k=o}^{n} \tilde{\mu}_k \right) \tag{14b}$$

when $\tilde{\mu}_k = \mu_k \ln 2$. The relaxation function $\Phi(t)$ is given by

$$\Phi(t) = \sum_{n=o}^{\infty} \omega_n \exp(-t/\tau_n) \tag{15}$$

where $\omega_n = N_n / \sum_{n=o}^{\infty} N_n$ is a weight factor for level n. Note that the hierarchy of relaxation times generated by Eq. (14b) is similar to the hierarchy of transition rates discussed in the direct transfer model.

One can now choose specific forms for μ_n and ω_n and calculate the corresponding relaxation function. The choice $\mu_n = \mu_o$ implies

$$\tau_n = \tau_o \exp (\tilde{\mu}_o n) \tag{16}$$

which is essentially case (b) of Eq. (7). Choosing

$$\omega_n = \omega_o \lambda^{-n} = \omega_o e^{-n\ln\lambda} \tag{17}$$

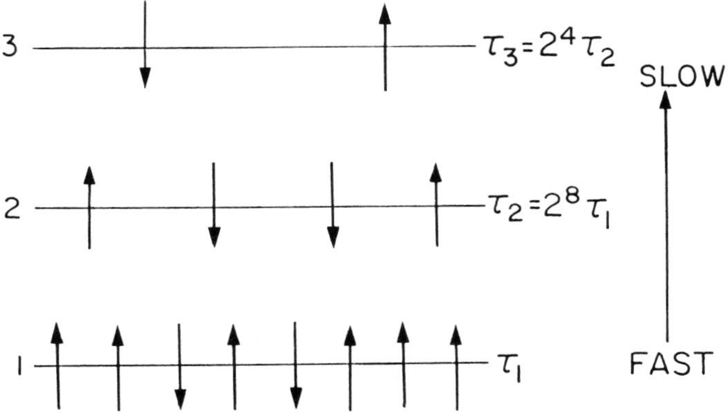

Figure 2. The hierarchically constraint dynamics model

which corresponds to Eq. (11), and converting the sum in Eq. (15) to an integral yields

$$\Phi(t) = \omega_o \int_0^\infty e^{-n\ln\lambda} \exp[-t \exp(-\tilde{\mu}_o n)/\tau_o]dn \tag{18}$$

which we recognize to be the integral that leads to the algebraic relaxation law that corresponds to d=1 in Eq. (9), namely $\Phi(t) \sim t^{-(\ln\lambda)/\tilde{\mu}_o}$.

The same choice for ω_n, but now coupled with the

$$\mu_n = \mu_o \cdot n^{-1}$$

implies

$$\tau_n = \tau_o \exp\left(\tilde{\mu}_o \sum_{l=1}^{n} l^{-k}\right)$$

$$\simeq \tau_o\, n^{\tilde{\mu}_o} \text{ (for k=1)} \tag{19}$$

which corresponds to case (a) of Eq. (7). For k=1 this leads to the relaxation integral

$$\phi(t) = \omega_o \int_0^\infty e^{-n\ln\lambda} \exp[-tn^{-\tilde{\mu}_o}/\tau_o]dn \tag{20}$$

which, as in Eq. (12), produces the stretched exponential law with exponent $\beta = \dfrac{1}{(1+\tilde{\mu}_o)}$.

Although the physical pictures of relaxing through a serial arrangement of hierarchically constrained levels and relaxing through a direct transfer to a nearest neighbor defect are quite different they both lead to the same relaxation integrals. The weight factor ω_n and the relaxation time τ_n for each level in the hierarchical model correspond to the weight factor f(R) for the defect position, and the transition rate W(R) respectively in the direct transfer model.

The target model, which is schematically shown in Fig.3, offers yet another mechanism for a stretched exponential decay. Unlike the DET case where the donor and acceptors are static, here the acceptors (quenchers) diffuse towards the initially prepared donor (the target) and quench it upon encounter [20]. The target problem can be recast into the same mathematical framework as the two previous models. We assume a one dimensional example where a target is located at the origin and diffusing acceptors are initially randomly distributed around it. Let $f(R_1)$ be the probability of having no acceptors at distance R_1 from the target. As before, for randomly placed, acceptors $f(R_1) = \exp(-pR_1)$. The probability that an acceptor at R_i has not reached the origin by time t is given by $\exp[-t/4R_i^2]$, where the diffusion constant of the acceptors has been set equal to unity. The relaxation law for the target is then

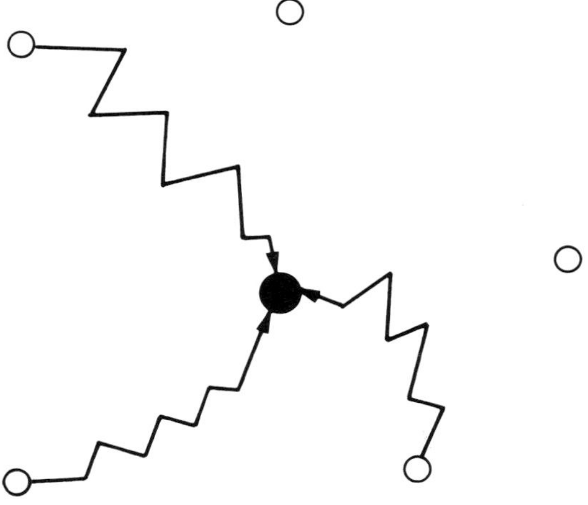

Figure 3. A schematic representation of the target model (diffusion towards donor)

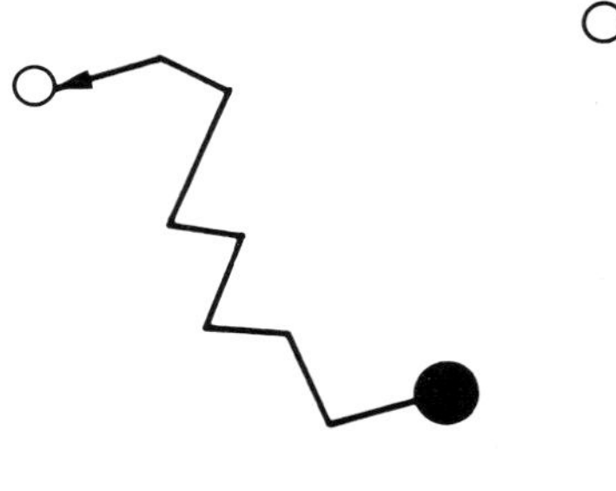

Figure 4. The trapping scheme (diffusion towards acceptors)

$$\phi(R_1,t) \simeq \exp(-pR_1) \prod_{i=1} \exp[-t/4R_i^2]$$

$$\simeq \exp(-pR_1) \exp\left[-tp\int_{R_1}^{\infty} \frac{dR}{4R^2}\right]$$

$$= \exp[-pR_1 + p\frac{t}{4R_1}] \tag{21}$$

Averaging over R_1 we arrive at

$$\Phi(t) = \int_0^{\infty} \exp(-pR_1)\exp(-pt/4R_1)dR_1 \tag{22}$$

which is again of the form of Eq. (12) in the Förster case and of Eq. (20) in the Palmer et. al. case, now with $\tau(R)=(4/p)R$, and yields the stretched exponential law with $\beta = 1/2$. This result can be generalized to other dimensions (see ref.[23]).

Another relaxation scheme which should be commented on here is the trapping mechanism. This relaxation scheme is complementary to the target model and is schematically presented in Fig.4. In the trapping case the donor diffuses towards *static* acceptors (or an excitation diffuses towards randomly distributed traps). This problem has been discussed in detail and has been shown to differ from the target case (the idea of relative diffusion does not apply here!). The trapping relaxation does not result in a pure stretched experimental. It is characterized by decay forms which continuously slow down due to spatial fluctuations of the acceptors [20,24].

We have demonstrated that a common mathematical framework underlies different physical models which lead in a natural way to the common emperically found stretched exponential relaxation law. The unifying feature of the theories is the generation of a scale invariant distribution of relaxation times. One should be able to differentiate among physical mechanisms underlying the relaxation via the variation behaviour of α and τ in Eq. (1) as a function of external variables such as temperautre and pressure.

3. EXTENSIONS OF DET TO RESTRICTED GEOMETRIES [25]

The different models discussed in the previous section have been applied to explain stretched exponential decays in a variety of systems. Here we extend the DET model to restricted geometries. We show that the scale invariant distribution of relaxation times $\tau(R)$ dominates and that even when restrictions are imposed we obtain stretched exponential decays. In order to be able to relate the DET to experimental systems the coefficient A in Eq.(7) case (a) has to be specified:

$$A = \frac{1}{\tau_D} R_0^S \tag{23}$$

where τ_D is the donor lifetime and R_0 is the Förster critical radius, which defines a length in the system. In homogeneous systems the role of R_0 is obvious [18]. However, in the case of DET from a donor to randomly distributed acceptors in restricted geometries a second length enters, which characterizes the spatial restrictions. The model-restricted geometries we discuss here are fractal structures and cylinders. Although fractals introduce the concept of self-similarity, regular shapes such as cylinders and spheres mimic the geometrical properties of pores and molecular assemblies.

In order to apply DET to these model systems we use Eq.(6) which expresses the decay in terms of the site density function $\rho(R)$, which is essentially the two-point correlation function on the structure.

Fractal structures are examples of restricted geometries. They are usually disordered, tenuous but self-similar (such as percolation clusters) [20,22,26]. The self-similar nature of these structures means that there is no typical length that characterizes them and so, when DET is considered, R_o remains the dominating length. The site density function for fractals is [3]

$$\rho(R) = \frac{F}{V_d} \rho_o \left[\frac{\bar{d}}{d}\right] R^{\bar{d}-d} \tag{24}$$

where $\bar{d}$ is the fractal dimension ($1 \leq \bar{d} \leq 3$), d is the Euclidean embedding space, ρ_o is the density of acceptor sites, and F is an unknown shape factor. The survival of a donor on a fractal is therefore [3]

$$\Phi(t) = \exp[-t/\tau - p\rho_o F R_o^{\bar{d}}\ \Gamma(1 - \bar{d}/S) t^{\bar{d}/S}] \tag{25}$$

again a stretched-exponential but with an exponent $\bar{d}/S$. The prefactor $p\rho_o F R_o^{\bar{d}}$ is equal to the number of acceptors with a radius R_o in $\bar{d}$ dimensions. It has been used to interpret DET experiments on porous Vycor glass [27] on silica gels [3], on zeolites [28] and on Langmuir-Blodgett films [29]. It should be noted, however, that fitting decay curves to the stretched exponential in Eq.(25) may not confirm or even imply that the underlying structure is really a fractal. Behavior similar to that described by Eq.(25) with nonintegral values of $\bar{d}$ can also occur as a result of crossover processes between dimensions. Nevertheless, Eq.(25) is useful in interpreting DET in restricted geometries especially when corroborated by other characterization techniques [2].

More conventional shapes that serve as useful models for restricted geometries are cylinders and spheres. These geometric systems are characterized by their radius R_p. When DET experiments are performed on these geometries, it is the relation between R_o and R_p that determines the decay patterns of $\Phi(t)$. Cylindrical pores have been used to model local pore characteristics. Many properties of porous materials are being studied in the framework of a cylindrical pore, for example, adsorption, wetting, and diffusion. Spheres are usually chosen to approximate micelle or microemulsion shapes. In order to be able to utilize DET as a tool for structural studies of these geometries, it is essential that one understands the dependence of the energy transfer reaction on the donor-acceptor distributions on or within these geometries. Fig. 5 shows schematically donor and acceptor distributions in cylindrical shapes.

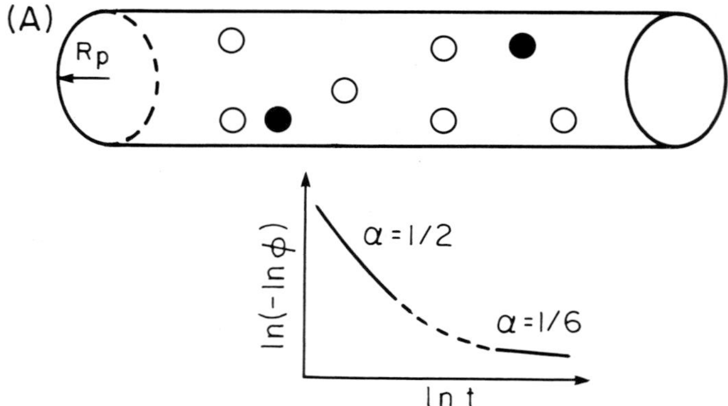

Figure 5. Example of a spatial confinement: the cylinder with a scheme of the corresponding DET decay

We now address the case of a cylinder of radius R_p with a donor and randomly distributed acceptors on its surface. Calculating the corresponding site density function and using Eq.(6), one finds that the survival probability can be numerically evaluated. Analytical expressions are easily derived for the short and long time behavior of $\Phi(t)$. The crossover time is estimated from the ratio of the two relevant length scales R_o and R_p:

$$\Phi(t) = \exp\left[-t/\tau - p\rho_o \pi R_o^2 \, \Gamma(1 - \tfrac{2}{S})(t/\tau)^{2/S}\right] \tag{26}$$

for $t<\tau(2R_p/R_o)^S$. The short time decay displays a two-dimensional behavior according to both the exponent 2/S and the prefactor $p\rho_o\pi R_o^2$. The donor senses only acceptors on the surface in its immediate vicinity. For long times $t>\tau(2R_p/R_o)^S$,

$$\Phi(t) = \exp[-t/\tau - p\rho_o 4\pi R_p R_o \Gamma(1-1/S)(t/\tau)^{1/S}] \tag{27}$$

This corresponds to a one-dimensional relaxation according to the exponent 1/S. The prefactor $4\pi p\rho_o R_p R_o$ equals, for $R_p< R_o$, the number of acceptors on the cylinder surface within radius R_o. The donor's decay crosses over from a two-dimensional to a one-dimensional form at a crossover time $t_{cross} \sim (2R_p/R_o)^S$. This is a direct result of the finiteness of the systems and a finger print of a length, R_p, that competes with R_o. Both the short-time and the long-time decays are stretched exponentials for which the exponent and prefactor reflect the dimension and molecular arrangement in the system.

If the acceptors are now distributed within the cylinder with the donor still on the surface, then for short times $t < \tau(R_p/R_o)^S$,

$$\Phi(t) = \exp\left[-t/\tau - p\rho_o \frac{2\pi}{3} R_o^3 \Gamma(1-3/S)(t/\tau)^{3/S}\right] \tag{28}$$

a three-dimensional decay to acceptors in the volume close to the donor, but, as the prefactor indicates, the number of acceptors participating corresponds only to half a sphere of radius R_o. For long times, $t > \tau(R_p/R_o)^S$,

$$\Phi(t) = \exp[-t/\tau - p\rho_o 2\pi R_p^2 R_o \Gamma(1-1/S)(t/\tau)^{1/S}] \tag{29}$$

characteristic of one dimension. The prefactor in Eq.(29), $2\pi p\rho_o R_p^2 R_o$, is the number of acceptors in a cylinder volume of radius R_p and length determined by the critical radius R_o. Here, too, a crossover between stretched-exponential forms of the decay is predicted, but between three-dimensional and one-dimensional.

What emerges from this analysis is that DET is capable of sensing local dimensionalities as well as crossovers between them. This contributes to a more complete picture of the investigated structures. Both the *exponents* and the scaling of the *prefactors* with R_o and R_p contain desirable information that may lead to an understanding of the morphologies of confinements. When local structures are known, the dependence of the prefactor on R_o can provide insight about the distribution of the acceptors. Crossover times, if measurable, can give clues about the typical size of the restriction, R_p.

The proposed procedure for analyzing experimental decays is to fit relaxation data to a stretched exponential, with the two fitting parameters. The exponent α is given the meaning of d/S, with d as an "effective dimension", and the prefactor is proportional to the number of acceptors within a radius R_o on the structure (and therefore should depend also on R_p). A model can then be proposed for the underlying geometry that is consistent with values of both parameters. If the apparent dimension d obtained is not an integer, then the concept of a crossover can be used in the analysis. A modified form can be applied to fit experimental decays:

$$\Phi(t) = \exp[-t/\tau - A_o \Gamma(1-d/S)(t/\tau)^{d/S}] \tag{30}$$

A straightforward connection between the static microstructure of the confinement and DET dynamics is achieved when a site density function $\tau(R)$ can be independently derived. Then, based on Eq.(6), dynamics and statics are directly related. Such an approach is important in establishing the method of using DET to get structural information. In a detailed study on porous Vycor glass, DET measurements and ultrathin transmission-electron-microscopy analysis of $\tau(R)$ were simultaneously carried out . The combined studies confirm the power of DET [30].

DET, using the scheme of a donor transferring energy to randomly distributed acceptors, provides insight into the microstructure of complex restricted geometries. The approach relies on stretched exponential decays and crossover between them and the interplay between two competing lengths: R_p, inherent to the spatial restriction, and R , introduced by the DET method. When DET is applied to study local morphologies, corroboration with other characterization techniques is very useful.

4. APPLICATIONS [2,10]

We now briefly discuss two experimental examples where stretched exponential relaxations occur and can be related to our scaling arguments.: direct energy transfer in porous silica gels and birefringence studies in branched polymers. For more detailed analyses and discussion we refer the reader to ref.[2,3].

Direct energy transfer experiments have been carried out on the well characterized family of porous silicas, Si-40, Si-60 and Si-100 which were studied and describes extensively [2] with rhodamine 6G and malachite green as the donor-acceptor pair. Each of the silicas has a well defined mean pore size R_p=18 Å (Si-40), R_p=35 Å (Si-60) and R_p=60 Å (Si-100). The experiment has been conducted using a time correlated single photon counting system. As stated, R_o provides an estimate of the length scale probed by DET here $R_o \simeq$ 57 Å. A more realistic measure should take into account the experimental limitations of the detection system. This introduces the length scale R_{max} which is approximately 1.5 R_o. The results on the three porous glasses have been fitted by Eq.(30): $\bar{d} \simeq 2$ (Si-100), $\bar{d} \simeq 2$ (Si-60), $\bar{d} \simeq 3$ (Si-40), the regular Euclidean limits of the equation. When corroborated with the characterization studies on these systems, the following picture emerges, in which the relationship between R_p (from structural characterization) and R_{max} (the optical yardstick) determines what features of the morphology are probed by direct energy transfer:

(1) When $2R_p > R_{max}$, (Si-100) the DET probes scale less than R_p. The process senses therefore only a portion of the local surface as shown in Fig.6A. Here $\bar{d}$ is nearly 2, which means a smooth surface. Slight deviations from d=2 occur because of crossovers to higher dimensional configurations.

(2) When $2R_p < R_{max}$, (Si-40) the probing length scale is above R_p. We are then sensitive to the pore network. In other words, R_{max} permits for interpore energy transfer which corresponds to $\bar{d}$=3. This is exemplified in Fig.6B, where the various length involved are presented. A transition to $\bar{d}$ =2, for the locally smooth surface, should be possible at higher acceptor concentrations and short times. The latter case is difficult to achieve experimentally.

(3) When $R_p \sim R_{max}$, (Si-60), $\bar{d} \simeq 2$. Here again the local vicinity of the primary building blocks is observed and crossover effects are more pronounced.

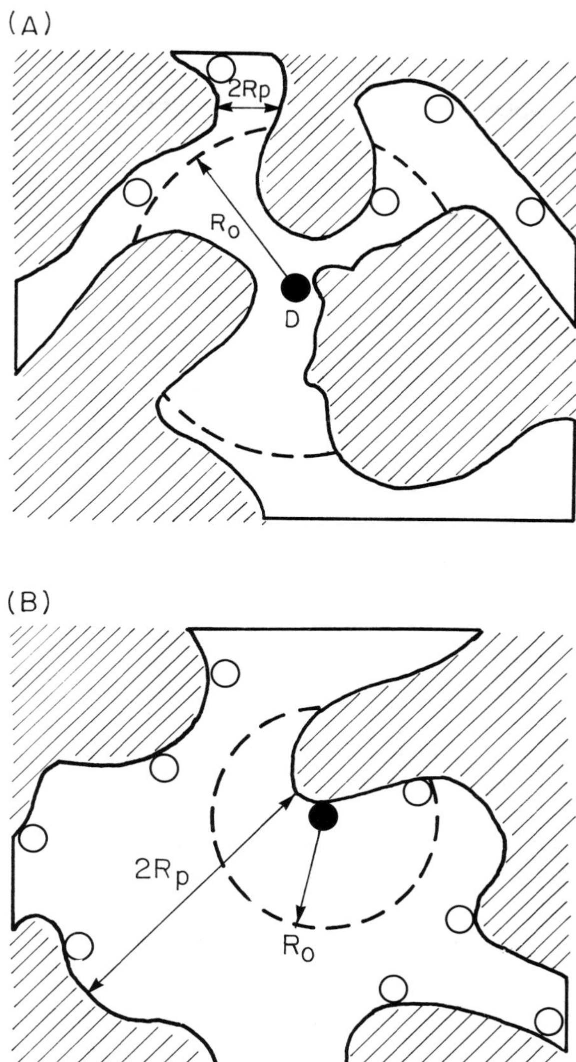

Figure 6. Schematic representation of the two lengths R_o and R_p. (A) $R_o > R_p$. (B) $R_o < R_p$.

These conclusions agree with other characterization studies described in previous work [2,3]. Direct energy transfer provides a spectroscopic method to elucidate the local spatial organization of transparent restricted geometries and can serve as an example for a non-diffusive monomolecular reaction in such systems.

Another example for the appearance of a stretched exponential in a system where $\tau(R)$ displays a scale invariant behavior is the transient electric birefringence experiments on branched polymers [10,31].

In a typical transient electric birefringence experiment [31], one applied a rectangular electric pulse to a polarizable polymer in a dilute solution, and studies the transient evolution of the birefringence towards equilibrium. The advantage of the method is that it tests directly all the possible conformations of a polymer, including the large, improbable, fluctuations included in the distribution f(R). As explained by Degiorgio et al [31], the relaxation of the birefringence B(t) as a function of time in the transient regime is directly related to the distribution of distances:

$$B(t) = \int dR\ S(R)f(R)\ e^{-t/\tau(R)}$$

where S(R) is a signal function behavior, and $\tau(R)$ the relaxation time of a polymer with given distance between surface monomers. The precise form of S(R) is not important for our purpose because we will be interested only in the long time behavior of B(t). Let us note however that its precise form would be interesting for the study of short time behavior.

For monodispensed branched polymers one can show that in the so called Zimm case

$$\tau(R) \sim R^3 \tag{31}$$

and

$$f(R) \sim \exp\left\{-\left[R^{\bar{d}}/N\right]^{1/(\bar{d}-1)}\right\} \tag{32}$$

Using the accepted values $\bar{d}=2$ for d=3 we get

$$B(t) \sim \exp\left\{-\left[t/N^{3/2}\right]^{2/5}\right\} \tag{33}$$

namely a stretched exponential. Different exponents can be obtained if other scaling rules are assumed for $\tau(R)$ [10].

REFERENCES

1 J. Klafter and J.M. Drake, Eds Molecular Dynamics in Restricted Geometries (Wiley, New York, 1989).
2 J.M. Drake and J. Klafter, Phys. Today, 43 (1990) 46.
3 P. Levitz, J.M. Drake and J. Klafter, J. Chem. Phys., 89 (1988) 5224.
4 J.D. Byers, M.S. Friedricks, R.A. Friesner and S.E. Webber, in Ref.1.

5 H.F. Kauffmann, G. Landl and H.W. Engl, in Dynamical Processes in Condensed Molecular System, Ed. A. Blumen, J. Klafter and D. Haarer (World Scientific, Singapore, 1991).
6 N. Mataga, in Ref.1.
7 M. Maroncelli and G.R. Fleming, J. Chem. Phys., 86 (1987) 6221.
8 M.F. Shlesinger, Ann. Rev. Phys. Chem., 39 (1988) 269.
9 J.R. Miller, Chem. Phys. Lett., 22 (1973) 180.
10 M. Daoud and J. Klafter, J. Phys., A23 (1973) 180.
11 D.R. James, Y.S. Liu, P. de Mayo and W.R. Ware, Chem. Phys. Lett., 120 (1985) 460.
12 J.M. Beechem, M. Ameloot and L. Brand, Chem. Phys. Lett., 120 (1985) 466.
13 Some history and references are in E.W. Montroll and J.T. Bendler, J. Stat. Phys., 34 (1984) 129.
14 G. Williams and D.C. Watts, Trans Faraday Soc., 66 (1970) 80.
15 J. Klafter and M.F. Shlesinger, Proc. Natl. Acad. Sci. USA, 83 (1986) 848.
16 Inokuti and Hirayama, J. Chem. Phys., 43 (1978) 1965.
17 H. Scher and E.W. Montroll, Phys. Rev., B12 (1985) 1434.
18 T. Förster, Ann.Phys.(Leipzig) 2 (1948) 55; Naturforscher Teil, 4 (1949) 321.
19 R.G. Palmer, D. Stein, E.S. Abrahms and P.W. Anderson, Phys. Rev. Lett. 53 (1984) 958.
20 A. Blumen, J. Klafter and G. Zumofen, in Optical Spectroscopy of Glasses, Ed. I. Zschokke (Reidel, Dordrecht, 1986).
21 A. Blumen, Nuovo Cimento B63 (1981) 50.
22 J. Klafter and A. Blumen, J. Chem. Phys., 80 (1984) 875.
23 S. Redner and K. Kang, J. Phys., A17 (1984) L451.
24 A. Blumen, G. Zumofen and J. Klafter, J. de Physique, Colloque C7, 46, C7-3 (1985).
25 J.M. Drake, J. Klafter and P. Levitz, Science, 251 (1991) 1574.
26 R. Kopelman, Science, 241 (1988) 1620.
27 U. Even, K. Rademann, J. Jortner, N. Manor and R. Reisfeld, Phys. Rev. Lett. 52 (1984) 2164.
28 C.L. Yang, P. Evesque and M.A. El-Sayed, in ref.1.
29 N. Tamai, T. Yamazaki and I. Yamazaki, Chem. Phys. Lett., 147 (1988) 25.
30 P. Levitz and J.M. Drake, in Dynamics in Small Confining Systems, MRS Extended Abstracts, Eds. J.M. Drake, J. Klafter and R. Kopelman (MRS, Pittsburgh, 1990).
31 V.Degiorgio, T. Bellini, R. Piazza, F. Mantegazza and R.E. Goldstein, Phys. Rev. Lett., 64 (1990) 1043.

DISCUSSION

Masuhara

Your theoretical treatment assumes that the small confining sites are identical with each other. If sites have some distribution with respect to kinetics, is it possible to discriminate two systems with identical sites and inhomogeneous distribution?

Klafter

Adding a distribution of environments to the formalism which I presented is possible. This, however, will complicate the analysis. Applying the various relaxation schemes to real systems is more efficient if one is able to avoid such inhomogeneities. For example make sure that the donor (without acceptor) decays exponentially. Otherwise relating the experimental decay pattern to one of the models is less certain.

Yamazaki

In actual materials, the fluorescence decay curve takes somewhat complicated form not expressed only by the survival function based on the Fractal theory. It seems to be essential to take in actual systems a superposition of different types of decay terms reflecting a multidomain nature of the actual material system. Have you any idea in this respect?

Klafter

As I already mentioned the multidomain nature of actual materials can be accounted for by an additional averaging. In developing our expressions for direct transfer on fractals, for instance, we derived equations which specifically depended on the donor location. This adds another variable to the problem over which we can average.

Mobius

I can easily imagine systems with a distribution of relaxation times in the absence of acceptors. Can you introduce such a distribution in your treatment, and how would that change the form of the decay?

Klafter

Such distributions can be introduced into the theoretical treatments. Generally I believe that if the distribution of the donor relaxation time (usually resulting from a distribution of environments) is not unrealistically broad the stretched exponential nature would not change. In a related problem of birefringence studies of branched polymers we showed (Dr. M. Daoud and myself) that a stretched exponential is obtained for both monodispersed and size distributed polymers.

Tachiya

I quite agree with you that it is not fruitful to try to get the distribution of the rate constant from the observed survival probability by using the inverse Laplace transformation. It does not deepen our understanding of the physics. In fact I pointed out this in 1974 (M. Tachiya and A. Mozumder, Chem. Phys. Lett., 28 (1974) 87). By the way, in the same paper I derived the general equation on energy transfer in the continuum limit for the survival probability in the presence of randomly distributed scavengers.

Klafter

You are correct in stating that a distribution of rates, obtained from the inverse Laplace transform, is not informative if not related to a well defined model or not justified by other means.
Our derivation of the energy transfer in the continuum limit is more general, I believe, as we are able to obtain decays for regular and for fractal structures using the site density function.

Dynamics and Mechanisms of
Photoinduced Transfer and Related Phenomena
N. Mataga, T. Okada and H. Masuhara (Editors)

Photoinduced Charge Transfer Dynamics in Poly(N-vinylcarbazole) Films

Hiroshi Masuhara* and Akira Itaya**

*Department of Applied Physics, Osaka University, Suita 565, Japan

**Department of Polymer Science and Engineering, Kyoto Institute of Technology, Matsugasaki, Kyoto 606, Japan

Abstract

Using time-correlated single photon counting and laser photolysis methods, fluorescence and absorption spectra of poly(N-vinylcarbazole) thin films were measured in ns and μs time domains. The exterplex fluorescence peak in the time-resolved spectra shifted to the long wavelength with time. The fluorescence decays were nonexponential with a slow tail up to a few μs and enhanced by applying an electric field. Transient absorption spectra were composed of bands of polymer cation and doped acceptor anion which are ascribed to migrating hole and trapped electron, respectively. By analyzing decay curves of the polymer cation, diffusion coefficient of the hole was estimated. The photoinduced charge transfer mechanism, involving multiple exterplexes and ionpairs, are proposed and discussed.

1. Introduction

Poly(N-vinylcarbazole) (abbreviated as PVCz) is the most representative polymer whose photophysical and photoconductive properties have been elucidated in detail [1]. Particularly, a clear relation between photophysical properties and tacticity is a very interesting characteristic of this polymer and has received a lot of attention [2, 3]. As model compounds for isotactic and syndiotactic sequences of PVCz, meso- and rac-2,4-di(N-carbazolyl)pentanes, giving sandwich and partial overlap excimer fluorescence, respectively, were studied in solution [3, 4]. By adding electron acceptors, both excimers are quenched effectively , giving some exciplexes. Hoyle and Guillet reported a contribution of an exterplex in addition to the exciplex, namely, two carbazolyl chromophores with plus charge and one acceptor anion form an excited complex [5]. Examining the time-resolved fluorescence spectra of the model compounds in the presence of 1, 3-dicyanobenzene, we revealed that the special geometrical structure of the two carbazolyl groups leads to two kinds of exterplex where the

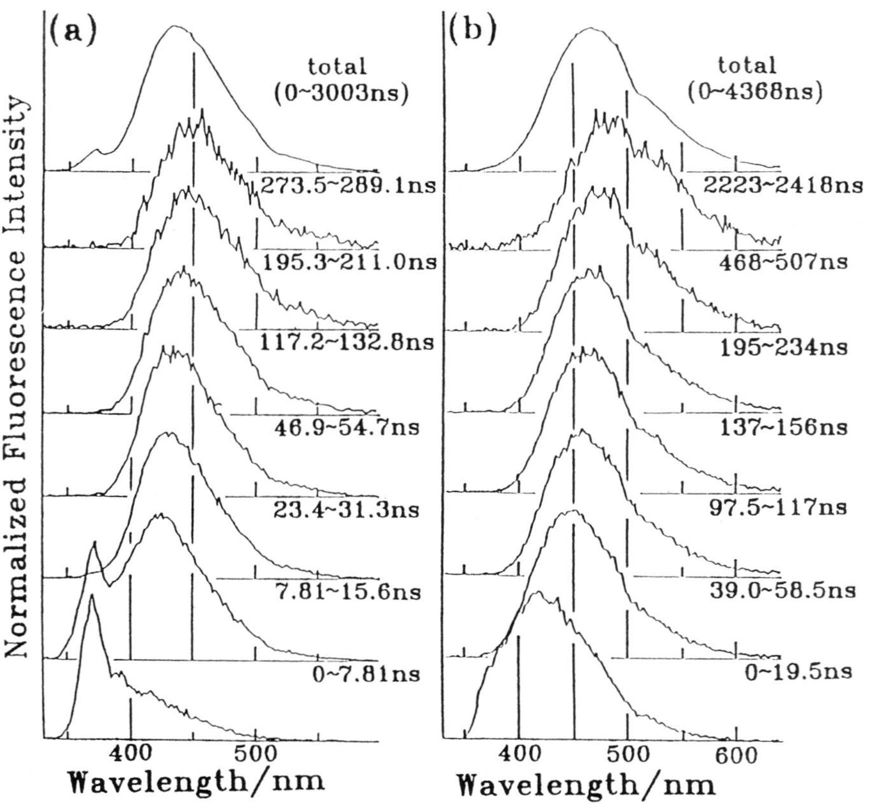

Figure 1. Normalized time-resolved fluorescence spectra of (a) polycarbonate film doped with 40 wt% N-isopropylcarbazole and 0.08 wt% 1, 4-dicyanobenzene and of (b) poly(N-vinylcarbazole) film doped with 1.8 mol% 1, 4-dicyanobenzene. Excitation wavelength is 295 nm. The gate time is given in the figure. All the spectra are not corrected for detector sensitivity.

hole is delocalized in the corresponding sandwich and partial overlap structures [6].

Doping of some electron acceptors such as aromatic anhydrides and nitriles in PVCz films also gives exciplex fluorescence and causes the chemical sensitization of the photoconductivity [7, 8]. Electric and magnetic field effects on steady-state exciplex fluorescence were investigated for elucidating the carrier-photogeneration mechanism [8], and systematic studies on relations between photoconductivity and charge transfer interaction were reported [9]. Nevertheless, time-resolved fluorescence and absorption spectra have been rarely published, and molecular aspects of photoconductive mechanism are still beyond our knowledge. A simple application of time-correlated single photon counting for fluorescence measurement is easy, while absorption spectral measurement is very difficult. Transient absorbance of the films is usually very small due to its short path length, and a very fast decay process of mutual interactions of excited states is induced when an excitation intensity is increased for getting an appreciable absorbance [10]. Recently, we developed a laser photolysis method, monitoring absorption by multi-reflection condition, and have used it for revealing photoprocesses of PVCz films [11]. In the present review, we summarize the time-resolved fluorescence and absorption spectral data, and discuss dynamics and mechanism of the photoinduced charge transfer processes.

2. Fluorescence Spectroscopic Study

Fluorescence spectra of PVCz films with and without electron acceptors are well known to be exciplex and excimer emissions, respectively. As a model system of the doped film, polycarbonate film containing 40 wt% N-isopropylcarbazole (i-PrCz) and 0.08 wt% 1,4-dicyanobenzene (DCNB) was prepared and measured under the same experimental condition. Total and time-resolved fluorescence spectra are shown in Figure 1. The exciplex fluorescence in the model system has a peak around 430nm which is in the shorter wavelength region compared to that of doped PVCz. We consider that the main component in the total fluorescence spectra of the model and PVCz systems are due to exciplex and exterplex, respectively. In the latter case the plus change is stabilized in the plural carbazolyl chromophores. In solution this idea was directly confirmed by transient absorption spectroscopy of some monomer and dimer model compounds of meso- and rac-2,4-di(N-carbazolyl)pentanes [4].

The time-resolved fluorescence spectra of both systems showed very interesting behavior. In the early-gated spectrum, the excimer fluorescence was mainly observed and the exciplex fluorescence in the long wavelength region was weak [12]. The fluorescence peak continuously shifted to the long wavelength with time, 430 to 450 nm and 450 to 480nm for i-PrCz-DCNB and PVCz-DCNB systems, respectively. Their peak wavelength after the shift is longer than that of the total fluorescence spectrum for both systems as in Figure 1. Since the decay time of the sandwich excimer fluorescence is 35 ns and that of

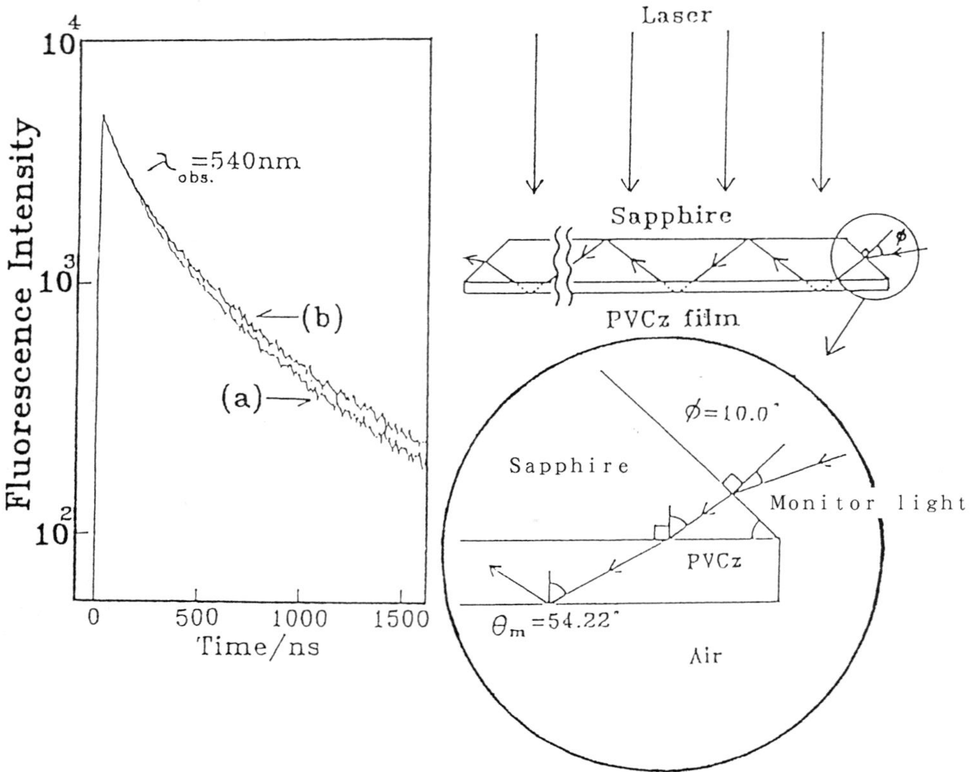

Figure 2. Exterplex fluorescence decay curve of poly(N-vinylcarbazole) film doped with 1.8 mol% 1,4-dicyanobenzene with (a) and without (b) electric field of 600 kV/cm.

Figure 3. Optical arrangement monitoring transient absorption under a multi-reflection condition. The critical angle at polymer/air interface is calculated to be 36.4°.

exciplex as well as exterplex is around 100ns, the present behavior involving the continuous shift cannot be explained by a superposition of excimer and exciplex as well as exterplex fluorescence. We consider that the latters shifted gradually with time.

As expected from Figure 1, the exterplex fluorescence of the PVCz films shows a slow decay and an appreciable fluorescence intensity is observed even in the μs time regions. Actually fluorescence decay has a tail up to a few μs, and obeys a nonexponential function. One of the examples is shown in Figure 2. This novel decay behavior is quite different from the PVCz dynamics in solution and suggests a complicated mechanism. As this polymer is photoconductive, an electric field effect upon

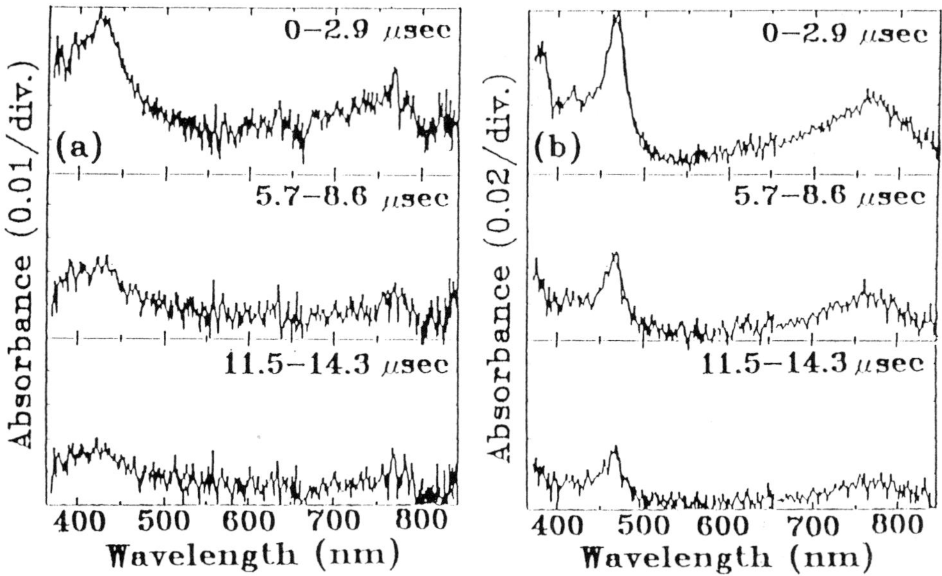

Figure 4. Transient absorption spectra of poly(N-vinylcarbazole) thin films doped with (a) 2 mol% 1, 4-dicyanobenzene and (b) 2 mol% 1, 2, 4, 5-tetracyanobenzene.

the exterplex fluorescence was examined [12]. It is worth noting that the decay was accelerated under high electric field, and the total fluorescence intensity is reduced.

In the case of PVCz film doped with 1, 2, 4, 5-tetracyanobenzene (TCNB) the ground state CT complex is formed, so that the fluorescence behavior is further complex compared to the PVCz-DCNB system. CT fluorescence due to the ground state complex was overlapped with the exterplex emission of the same donor-acceptor pair. The CT emission decays fast in a few hundred ns and shows a blue spectral shift.

3. Absorption Spectroscopic Study

Time-resolved absorption spectra of PVCz thin films with thickness of 800 nm were measured by our new laser photolysis system whose optics is schematically shown in Figure 3. The spectra of PVCz films doped with DCNB and TCNB are shown in Figure 4. In the case of the PVCz-DCNB system, a peak was observed at 430 nm, which can be assigned to DCNB anion. The bands at 375 and 470 nm of the PVCz-TCNB system are due to TCNB anion. The spectrum above 650 nm in both systems are ascribed to the PVCz cation by referring a series of cationic

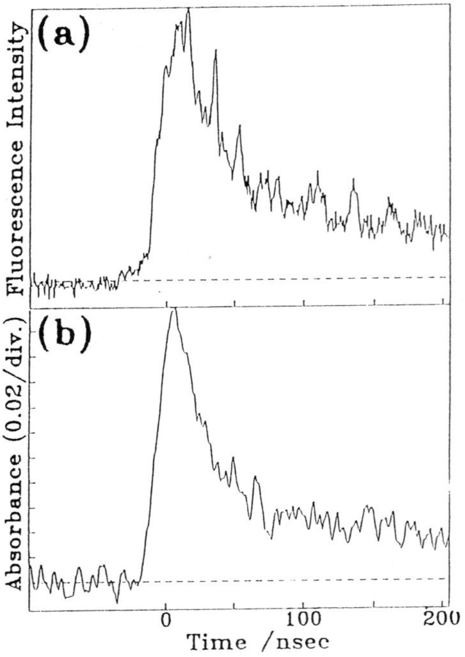

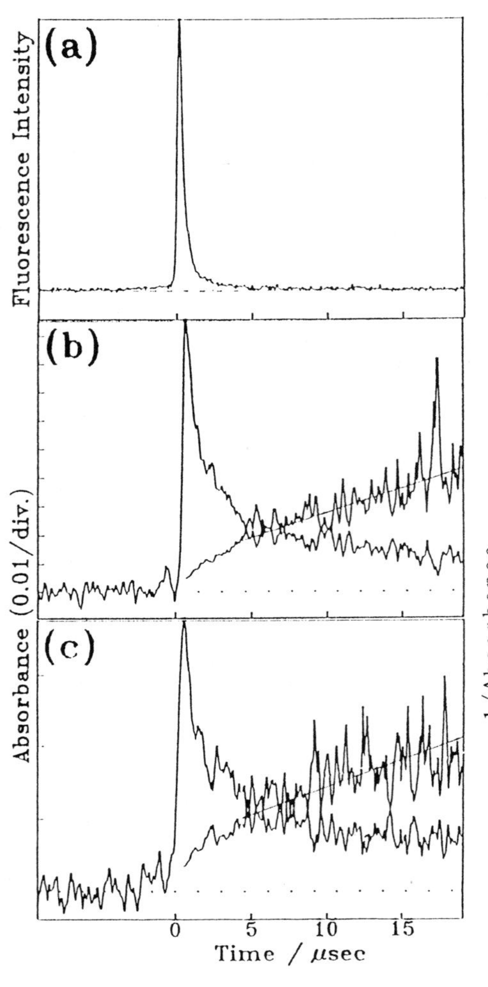

Figure 5. Decay curves of (a) exterplex fluorescence (664-676 nm) and (b) transient absorbance (740-780 nm) of poly(N-vinylcarbazole) film doped with 2 mol% 1, 4-dicyanobenzene.

Figure 6. Decay curves of (a) exterplex fluorescence (553 nm), (b) transient absorbance (419 nm), and (c) transient absorbance (740-780 nm) of poly(N-vinyl-carbazole) film doped with 2 mol% 1, 4-dicyanobenzene.

absorption spectra of carbazole monomer, dimer, and polymers [4, 13]. Absorption spectral shape of carbazolyl dimers is very sensitive to their relative geometry, nevertheless, the peak is always located between 700 and 800 nm and is broad as well as structureless. One explanation we proposed is that the hole is always stabilized as a dimer and the PVCz cation band is a superposition of the bands of sandwich, partial overlap, and open dimer cations. The absorption spectrum of the cation in the PVCz-DCNB system is broader than that in the PVCz-TCNB system, suggesting that partial overlap as well as open dimer cations are more involved in the former system compared to the latter one.

All the absorption spectra decayed monotonously without spectral change. The decay curves of the ionic absorption in the PVCz-DCNB system are shown in Figures 5 and 6. The rise was determined by the response function of the present system and was quite rapid. In the time region of a few hundred ns, the rise and decay profile of absorption is similar to that of exterplex fluorescence, indicating that the present absorption is due to the exterplex. On the other hand, the common behavior to exterplex fluorescence and absorption was not observed at the late stage. Namely, the fluorescence has a tail in 2～3 μs time region and completed around 5 μs, while the absorption decay shows a slow process. This absorption is surely ascribed to nonemissive ionic species, namely, ionpair and/or free carriers. These species are alive over 10 μs.

The decay is not a simple exponential, and the second order decay kinetics roughly holds in spite of the rather low S/N

SCHEME I

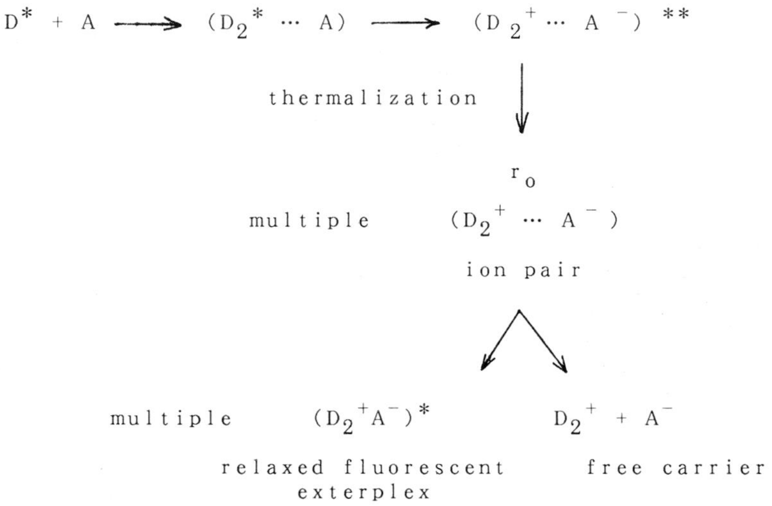

$D^{*} + A \longrightarrow (D_2^{*} \cdots A) \longrightarrow (D_2^{+} \cdots A^{-})^{**}$

thermalization ↓

r_o

multiple $(D_2^{+} \cdots A^{-})$

ion pair

↙ ↘

multiple $(D_2^{+}A^{-})^{*}$ $D_2^{+} + A^{-}$

relaxed fluorescent exterplex free carrier

value in the late stage. This means that a homogeneous charge recombination is the rate-determining step. Now, we conclude that the nonemissive, ionic species are free carriers. In the initial part of the decay process is complicated because of the overlap of absorptions of exterplexes. The decay behavior of absorption spectra of the PVCz-TCNB system is similar to that of the PVCz-DCNB system. However, the contribution of fast component, absorption due to exterplex, is small and the dynamic range of absorption decay measurement is wide in the PVCz-TCNB system. The absorption decay of ionic species in the PVCz-TCNB system was confirmed up to 40 μs.

4. Photoinduced Charge Transfer Mechanism

Before the present time-resolved spectroscope measurement, photophysical and photoconductive processes of PVCz films were elucidated on the basis of total fluorescence and time-of-flight photoconductive measurements [7-9, 14]. Only one species has been assumed for exciplex and ionpair and the nature of the carriers has been beyond our knowledge. The present results reveal molecular aspects of photoinduced processes, and a more direct consideration is made possible. We propose here that exterplex and dimer cation are responsible to photoinduced charge transfer mechanism as shown in the Scheme 1. Singlet excited state D in PVCz films migrates over carbazolyl chromophores, converts to excimer D_2, and encounters an acceptor A, forming a loose encounter complex $(D_2^{*}...A)$. Electron transfer occurs in this complex and a nonrelaxed exterplex $(D_2^{+}...A^{-})^{**}$ is formed. This state dissipates excess energy and undergoes thermalization, giving a loose ionpair $(D_2^{+}...A^{-})$. The electrostatic force between D_2^{+} and A^{-} in the ionpair is just equal to thermal energy, and the interionic distance r_0 is defined as Onsager radius. Diffusion motion starts from $(D_2^{+}...A^{-})$, and this ions recombine with each other or dissociate to free carriers $D^{+} + A^{-}$. Formation yield of the ionpair and r_0 were determined experimentally to be 93-96 % and 23-28 A, respectively [8, 9, 14].

It is worth noting that exterplex fluorescence shows time-dependent shift. Since molecular motion in PVCz films is frozen at room temperature, this behavior cannot be interpreted by stabilization due to molecular reorientation as in solution. We consider that the geometry of D_2^{+} and the relative geometrical structure of D_2^{+} to A^{-} in $(D_2^{+} A^{-})^{*}$ take diverse distributions, and each exterplex emits and decays independently. The continuous shift of the peak of time-resolved fluorescence spectra can be attributed to a superposition of these multiple exterplexes.

A nonexponential and slow fluorescence decay curve is consistent with this idea. The lifetime of the exterplex is usually less than 100 ns in solution. If we consider that the geminate recombination in $(D_2^{+}...A^{-})$ gives a relaxed exterplex $(D_2^{+}A^{-})^{*}$ and that the process is one of rate-determining steps, it is possible to explain the present novel fluorescence

Table 1.

Absorption and kinetic parameters of holes in poly(N-vinylcarbazole) films.

Dopant	1,4-Dicyanobenzene	1,2,4,5-Tetracyanobenzene
Anion band		
in film	430 nm	470 nm
in solution	427 nm	462 nm
ε in solution (M^{-1} cm^{-1})	9.4×10^{3}	4.7×10^{3}
Rate constant ($M^{-1}s^{-1}$)	1.8×10^{8}	1.8×10^{7}
Diffusion coefficient ($cm^{2}s^{-1}$)	2.4×10^{-4}	2.4×10^{-5}

behavior as follows. The geminate recombination in the ionpair ($D_2^+ ... A^-$) and the latter dissociation to free carriers occurs in the μs time region, and these processes are competing with each other and determine the temporal behavior of exterplex fluorescence. The dissociation to free carriers is assisted by an applied electric field, so that the electric field accelerates the exterplex decay and reduces the total fluorescence intensity. Interionic distance and local geometrical structure in ($D_2^+ ... A^-$) may take a wide distribution, so that the geminate recombination gives various rates, giving a complex, nonexponential decay of the exterplex. Multiple ionpairs are thus involved in the mechanism. The important role of multiple exciplexes and ionpairs in photoinduced charge separation dynamics has already been pointed out in solution by Mataga et al. [15]

At the late stage absorption decay curves obey the second order kinetics, indicating that those are due to free carriers. In the present film, dopant concentration is low compared to that of carbazolyl groups, so that only the hole can migrates, giving a homogeneous recombination process with the acceptor anion. In solution, hole migration along the PVCz chain was directly confirmed by laser photolysis method [16].

The bimolecular reaction rate constant can be related to diffusion coefficient by the following Smoluchowski equation.

$$k=4\pi R_0 N_0 D \times 10^{-3}$$

where k and D are bimolecular rate constant and diffusion coefficient, respectively, N_0 is Avogadro's number, and R_0 is the reaction radius of carrier (hole) and anion (electron). The obtained diffusion coefficients of PVCz films doped with DCNB and TCNB and the relevant parameters are summarized in Table 1. These are the first values for the diffusion coefficient of the hole in polymer films which are directly determined by transient absorption spectroscopy.

It is noticeable that the diffusion coefficient of the hole depends upon the doped electron acceptor, although the latter is minor component and the hole hops over carbazolyl chromophores. This result is consistent with the difference of absorption spectrum of the hole between PVCz-DCNB and PVCz-TCNB systems. Since the hole is separated from the anion, spectral difference should be ascribed to a difference of electronic and geometrical structures of holes. As pointed out already above, absorption spectrum of the polymer cation in the PVCz-TCNB system is relatively sharper than that in the PVCz-DCNB system.

We explained that the absorption band of PVCz cation is a superposition of the bands of some dimer cations. The contribution of loose structure like partial overlap and open dimer cations should be larger in the PVCz-DCNB system than in the PVCz-TCNB system. The dimer with loose structure is a shallow trap for the hole, so that migration should be fast. Another interpretation is that the plus charge delocalization occurs in PVCz, which was considered [4, 13] and experimentally proposed [17]. The broader spectrum in the PVCz-DCNB system than in the PVCz-TCNB system may suggest wider delocalization. Namely, both interpretations for absorption spectral shape are consistent with the difference of diffusion coefficient between exterplex and CT complex systems.

Since absorption spectrum is determined by geometrical as well as electronic structures, relative geometry of carbazolyl chromophores in PVCz film should be very much affected by the dopant molecule, although the concentration of the latter is less than a few mol%. The ground state complex is formed in the case of the PVCz-TCNB system, which may result in the change of orientation and association of polymer chains. On the other hand the PVCz-DCNB system is the exterplex system, so that DCNB is incorporated without any particular attractive interaction with the polymer. The difference of morphology is responsible to that of diffusion coefficient, which is consistent with our current result on photoconductivity of the PVCz films [9]. A more detailed study on the films with different acceptors are being performed, and the molecular mechanism will be summarized in the near future.

5. References

1 J. Mort and G. Pfister (eds.), Electronic properties of polymers, Wiley-Interscience, New York, 1982, Chaps. 5 and 6.

2 A. Itaya, K. Okamoto, and S. Kusabayashi, Bull. Chem. Soc. Jpn. 49 (1976) 2082.
3 J. Vandendriessche, P. Palmans, S. Toppet, N. Boens, F. C. De Schryver, and H. Masuhara, J. Am. Chem. Soc. 106 (1984) 8057.
4 H. Masuhara, J. Mol. Struct. 126 (1985) 145; Makromol. Chem. Suppl. 13 (1985) 75.
5 C. E. Hoyle and J. E. Guillet, Macromolecules 12 (1979) 956.
6 H. Masuhara, J. Vandendriessche, K. Demeyer, N. Boens, and F. C. De Schryver, Macromolecules 15 (1982) 471.
7 K. Okamoto, A. Yano, S. Kusabayashi, and H. Mikawa, Bull. Chem. Soc. Jpn. 47 (1974) 749. A. Itaya, K. Okamoto, and S. Kusabayashi, Bull. Chem. Soc. Jpn. 50 (1977) 22. K. Okamoto, S. Kusabayashi, and H. Mikawa, Bull. Chem. Soc. Jpn. 46 (1973) 2613.
8 N. Yokoyama, Y. Endo, and H. Mikawa, Bull. Chem. Soc. Jpn. 49 (1976) 1538. K. Okamoto, A. Itaya, and S. Kusabayashi, Chem. Lett. (1976) 99. M. Yokoyama, Y. Endo, A. Matsubara, and H. Mikawa, J. Chem. Phys. 75 (1981) 3006.
9 H. Yoshida, Mc thesis, Kyoto Institute of Technology (1989).
10 H. Masuhara, in: Photophysical and photochemical tools in polymer science, M. A. Winnik (ed.), Reidel, Dordrecht, (1986) 43.
11 N. Ikeda, T. Kuroda, and H. Masuhara, Chem. Phys. Lett. 156 (1989) 204. H. Masuhara, in: Photochemistry on solid surface, M. Anpo and T. Matsuura (eds), Elsevier, Amsterdam, 1989, 15. A. Itaya, T. Yamada, and H. Masuhara, Chem. Phys. Lett. 174 (1990) 145.
12 H. Sakai, A. Itaya, and H. Masuhara, J. Phys. Chem. 93 (1989) 5351.
13 H. Masuhara, K. Yamamoto, N. Tamai, K. Inoue, and N. Mataga, J. Phys. Chem. 88 (1984) 3971.
14 M. Yokoyama, S. Shimokihara, A. Matsubara, and H. Mikawa, J. Chem. Phys. 76 (1982) 724.
15 Y. Hirata, Y. Kanda, and N. Mataga, J. Phys. Chem. 87 (1983) 1659.
16 H. Masuhara and A. Itaya, in: Macromolecular complexes dynamic interactions and electronic processes, E. Tsuchida (ed.), VCH, New York, 1991, 61.
17 Y. Tujii, A. Tuchida, M. Yamamoto, and Y. Nishijima, Macromolecules 21 (1988) 665. Y. Tsuji, K. Takami, A. Tsuchida, S. Ito, Y. Onogi, and M. Yamamoto, Polym. J. 22 (1990) 319.

DISCUSSION

Isoda

In your talk, you referred to the time–of–flight photoconductivity method. Have you ever compared your results on the photoinduced charge transfer dynamics with the data obtained by the time–of–flight method using Metal–Insulator–Metal structures?

Masuhara

High electric field is applied to the thin film in the case of photoconductivity measurement, which gives a directional flow of carriers. This is quite different from the situation of homogeneous bimolecular recombination of carriers which we observed by multireflection laser photolysis.

Mataga

According to your investigations, the origin of the nonexponential decay of the exciplex emission is due to the distribution of the carbazole dimer cation states. Do you have any idea about various geometrical structures which correspond to such distribution of the dimer cation states?

Masuhara

It was already confirmed by me and you that the dimer cations with geometries such as sandwich dimer, partial overlap dimer, and open dimer are stable in solution. These dimer cations are involved in the present exterplex and furthermore, the acceptor anion takes various orientations to these cations. This gives non exponential decay.

Tagawa

Why the mobilities of free dimer cation radicals are different between exciplex systems and complex systems? Are free dimer cation radicals free from the acceptor anion radicals?

Masuhara

There are two explanations; one is that the donor carbazole of CT complex is partially charged to plus, which is repulsive to migrating hole. The second is to consider that hole migration comprises trapping, detrapping, rapid hole transfer, and again trapping. This trap site in the CT complex system is deeper than that in the exciplex systems. Detrapping is rate–determining step and slower for the CT complex system than for the exciplex system. I suppose that both are due to the difference of film morphology between CT complex and exciplex systems.

Yamamoto

It is very interesting results.

1. As for the carrier, you showed that the dimer cation is free carrier, but then the traps which are usually supposed to be must be another species. What do you think about this point?

2. As for the transient absorption spectra, have you measured it under high electric field?

Masuhara

1. Your discussion is very important, but experimentally it is difficult to discriminate absorptions due to the migrating hole and the trap, because the former and the latter have shorter and longer lifetimes, respectively, and both are involved.

2. We are very much interested in transient absorption spectral measurement under a high electric field, however, we have not tried it yet.

Tagawa

Decay kinetics of free dimer cations follow the second order, you said. But do the decays completely follow the second order? Is there no deep trap or permanent trap?

Masuhara

Transient absorbance is very small because this is done only by multi–reflection laser photolysis of the their poly(N–vinylcarbazole) film, so that we could not exclude a possibility that other kinetics is involved. At the present stage of investigation, bimolecular kinetics is the most probable in the time domain of 1∿10 μs.

Dynamics and Mechanisms of
Photoinduced Transfer and Related Phenomena
N. Mataga, T. Okada and H. Masuhara (Editors)

PHOTOINDUCED ELECTRON TRANSFER: FUNDAMENTAL DIFFERENCES BETWEEN HOMOGENEOUS PHASE AND ORGANIZED MONOLAYERS

Dietmar Möbius, Ramesh C. Ahuja, Gabriella Caminati[a], Li Feng Chi, Wolfgang Cordroch, Zhi-min Li, and Mutsuyoshi Matsumoto[b]

Max-Planck-Institut für biophysikalische Chemie, Postfach 2841, D-3400 Göttingen;

[a]Dipartimento di Chimica, Via G. Capponi, 9, I-50121 Firenze, Italy;

[b]National Chemical Laboratory for Industry, Tsukuba, Ibaraki 305, Japan.

Abstract

Fundamental differences between homogeneous environment and organized monolayer systems are observed in photoinduced electron transfer, in particular with respect to the distance dependence. The role of prominent physical properties of monolayer systems in long range photoinduced electron transfer will be discussed like the effect of thermal fluctuations, the influence of geometrical correlation of donor and acceptor, the effect of energy delocalization and of the remote environment of excited molecules.

1. INTRODUCTION

Photoinduced electron transfer processes have been investigated in homogeneous solution [1-3] and adequate theories were developed to account for the experimental results [4-8]. When attempts were made to imitate various steps in the course of photosynthesis in artificial systems in order to understand the structural requirements, photoinduced electron transfer processes were studied in monolayer organizates [9-11]. According to a model involving electron tunnelling [12] the distance dependence was a particularly prominent issue [13-17]. This was approached alternatively by synthesizing extremely elaborate molecules containing donor, acceptor and connector moieties [18-22]. The behaviour of such complex molecules was investigated in homogeneous solution. The decisive step in understanding the primary processes in photosynthesis was the determination of the structure of the photosynthetic center of *Rhodopseudomonas viridis* [23]. Now energetics and spatial correlation of the essential components were known in a particular case, and a complete theory consistent with the kinetic data was developed [24].

Large discrepancies seem to exist between the results obtained in different artificial systems, in particular with respect to distance dependence of photoinduced electron transfer. Some fundamental differences between photoinduced electron transfer in homogeneous solution and monolayer systems are discussed in the following sections. These are concerned mainly with the role of thermal fluctuations, geometrical correlation of donor and acceptor, energy and charge delocalization, interfacial phenomena and the influence of remote environment.

2. PHOTOINDUCED ELECTRON TRANSFER IN HOMOGENEOUS SOLUTION

The investigation of the relation between rate constant and free energy ΔG in photoinduced electron transfer yielded a particularly intriguing result: the rate constant vs. free energy plot did not show the inverted region [3], i.e. the theoretically expected decrease of the rate constant with decreasing ΔG after having reached the diffusion limited rate constant.

First evidence for an inverted region in the rate constant vs. free energy relationship was obtained with large molecules containing a donor and an acceptor moiety separated by a rigid steroid spacer [25]. The donors in those experiments were quinones reduced by solvated electrons, and the acceptors were aromatic hydrocarbons. The separation of donor and acceptor was identical in this series, however, the orbitals involved at least on the acceptor side and their orientation relative to the donor orbital, are different, and therefore, the electronic factor is not constant. Nevertheless, the results are qualitatively convincing and mark the starting point for a large number of investigations with such type of systems.

A different, extremely interesting issue in addition to that of the inverted region is the distance dependence of photoinduced electron transfer. The systems with rigidly linked donor D and acceptor A are particularly suited for studying the distance dependence, provided that the relative orientation of donor and acceptor is kept constant with variation of the spacer. Keeping all other parameters constant, the distance dependence of the rate constant should be an exponential function of the type $k \sim \exp(\alpha r)$, where r is the distance between D and A and α the damping constant. Observed damping constants are of the order of $\alpha \approx 10\ \mathrm{nm}^{-1}$. Frequently, such damping constants are compared for qualitatively different processes. In the photooxidation type of photoinduced electron transfer the electron comes from the excited state of the donor and ends up in the LUMO (lowest unoccupied molecular orbital) of the acceptor. In the case of photoreduction, the electron comes from the HOMO (highest molecular orbital) of the donor and is transferred to the half-filled HOMO of the excited acceptor. In the first case, the barrier set up by the spacer for the electron to overcome or to tunnel across is much lower than in the second case. According to the simple tunneling theory the probability p of tunneling on the barrier height φ is given by $p = \exp[(r/\hbar)\sqrt{(2m\varphi)}]$, where $\hbar$ is $h/(2\pi)$, h is Planck´s constant, and m is the mass of the electron.

Consequently, in the photoinduced electron transfer processes according to the photoreduction type the damping constant α should be much larger than in photooxidation type processes, provided that there is no hole transfer for which there is only little evidence in the case of organic molecules [26]. Further, an appropriate modification of the barrier, e.g. by introduction of π-electron systems, should lower the barrier. Indeed, the damping constant observed in the photoinduced electron transfer between porphyrins linked by a series of spacers including π-electron systems in the spacer, has a rather small value of $\alpha = 4\ nm^{-1}$ [27].

The complex molecules have been developed successfully into true models of the biological photosynthetic center by attachment of a secondary donor and a secondary acceptor in order to mimick the series of electron transfer steps required for a long-lived charge separation [20,28,29]. This fascinating type of molecules represent examples of a new type of chemistry suggested earlier [30] involving planned molecular interactions.

3. LONG-RANGE PHOTOINDUCED ELECTRON TRANSFER IN MONOLAYER SYSTEMS

The investigations of photoinduced electron transfer in monolayer systems were based on initial studies of electron tunneling across single monolayer sandwiched between metal electrodes [31-33]. Cyanine dyes were used as donors as well as acceptors and an amphiphilic viologen (SV) was the mainly used acceptor in the investigations. When monolayers with these molecules incorporated in appropriate matrices are formed, the active components are located at the hydrophilic side. Only in the case of an amphiphilic pyrene derivative the donor was at the hydrophobic interface, and monolayer systems were investigated that had already the essential functional components of donor, catalyst and acceptor of photosynthetic systems [10]. Donor and acceptor were separated by a fatty acid monolayer of well defined thickness.

3.1 Super long range photoinduced electron transfer

A typical set of results of excited donor steady state fluorescence quenching experiments is shown in Figure 1 (taken from ref. 16). From the fluorescence intensities of the donor in the presence (I) and in the absence (I_o) of the acceptor the quantity $[(I/I_o) - 1]$ which is proportional to the rate constant of the quenching process is determined and plotted on a logarithmic scale vs. the interlayer spacing. Two different acceptors were used in combination with the same donor D, and the experimental points coincide with straight lines having the same slope. This is expected since the barrier for electron tunneling is independent of the acceptor but given by the excited donor state energetic level. The vertical offset reflects the different ΔG of the reaction with the different acceptors. The most remarkable result, however, is the extraordinarily long range of photoinduced electron transfer in the monolayer systems which can even be observed for a spacing of nearly 3 nm. The slope of the two lines is α = 3.2 nm^{-1}, a value much smaller than usually observed in solution or disordered systems. The validity of the results of electron transfer across single monolayers has been frequently questioned. A detailed discussion of the various arguments and of experiments designed to evaluate the possible role of defects in monolayer systems is given in ref. 34.

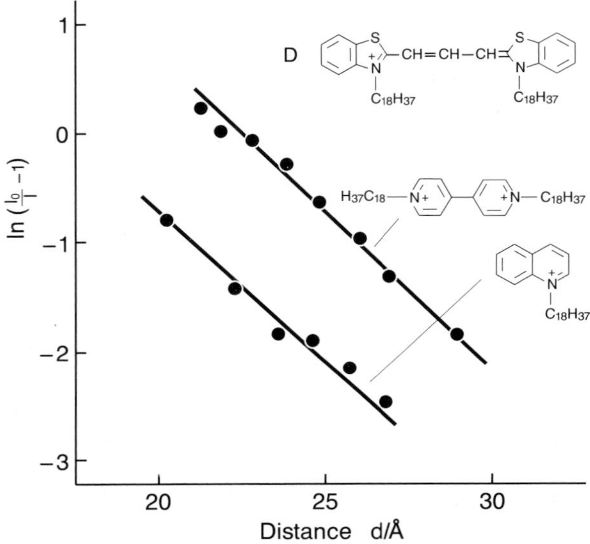

Figure 1. Steady state donor fluorescence quenching due to electron transfer to acceptor monolayer vs. spacing between donor and acceptor layer (see text); structures of the donor D and two acceptors incorporated in matrix monolayers (from ref. 16).

3.2 Role of thermal fluctations

The electron transfer requires energetic match of donor and acceptor level within the limits the uncertainty principle. This energetic match is created and also destroyed by thermal fluctuations of the system. The time between two collisions is t_c. The probability of electron transfer in case of energetic match depends on the perturbation energy 2ε and may be approximated by $p_e = [(\varepsilon/\hbar)t_c]^2$. The rate constant for electron transfer $k = (p_e p_m)/t_c$, where p_m is the probability of energetic match. Two limiting cases can be envisaged: a) t_c is short compared to the time $(\hbar/\varepsilon)$ of electron transfer and b) t_c is long compared to that time. In case a, the rate constant of electron transfer $k \sim \varepsilon^2$, whereas in case b the result $k \sim \varepsilon$ is obtained [35,15,16].

The relatively small value of the slope in Figure 1 is reasonable if the situation of case b is applied with a barrier height estimated from the barrier in tunneling across a single monolayer between metal electrodes and a energetic position of the excited state cyanine obtained from the oxidation potential of the chromophore and the excitation energy.

4. PARTICULAR PROPERTIES OF MONOLAYER SYSTEMS

In addition to the explanation given above of super long range photoinduced electron transfer in monolayer systems in contrast to the situation in homogeneous solution, particular properties of monolayer systems have to be taken into account for understanding the observed phenomena. These properties are mainly based on a collective response of the molecules to an external stimulus.

4.1 Role of symmetry and geometry limitations

The amazing molecules with donor and acceptor connected by a rigid spacer have a built-in 1:1 relationship of donor and acceptor. The rate of electron transfer in such rigid systems may be limited by the symmetry of the interacting molecular orbitals on donor and acceptor (electronic factor). This is not seen in the very simple theoretical approaches where hydrogen functions have been used to approximate the interaction, however, in more detailed calculations the relative orientation has been taken into account and reasonable MO´s have been used [24,16,36]. Under particular conditions (symmetry and geometrical relation of donor and acceptor), electron transfer may be extremely slow. A lateral shift of either molecule would then greatly enhance the rate of electron transfer.

This type of restriction is much less efficient in monolayer systems. Here, an excited donor may interact with several acceptor molecules in various geometric relations with the donor. In essence, this enhances the overall efficiency of photoinduced electron transfer as compared to the systems in solution and is expressed in a larger range of electron transfer.

Another aspect is dynamics of both donor and acceptor. In monolayer systems these components are normally rigidly fixed. In fact, the π-electron systems of the cyanine dyes and of SV that are frequently used are oriented parallel to the layer planes (this is different in the case of the stilbene derivatives used in ref. 17). We have recently studied electron transfer processes in monolayers at the air-water interface involving methyl viologen (MV) adsorbed from the aqueous subphase to the monolayer matrix of DPPA (dimyristoyl-phosphatidic acid) and a pyrene derivative PyPPC (structure see Fig. 3). The adsorbed MV (dynamic equilibrium) can be detected by reflection spectroscopy which discriminates the adsorbed from the dissolved molecules [37-39]. A reflection spectrum of the adsorbed MV is shown in Figure 2, where the reflection is the difference in reflectivity of the monolayer-covered solution and the bare solution surface. The band due to MV at 265 nm is clearly seen.

The pyrene moiety of PyPPC in mixed monolayers of PyPPC:DPPA, molar ratio 1:50, is located at the hydrophobic side of the monolayer and therefore separated from the plane of the head groups by the long hydrocarbon chain. The excited pyrene may interact with adsorbed MV by long range electron transfer.

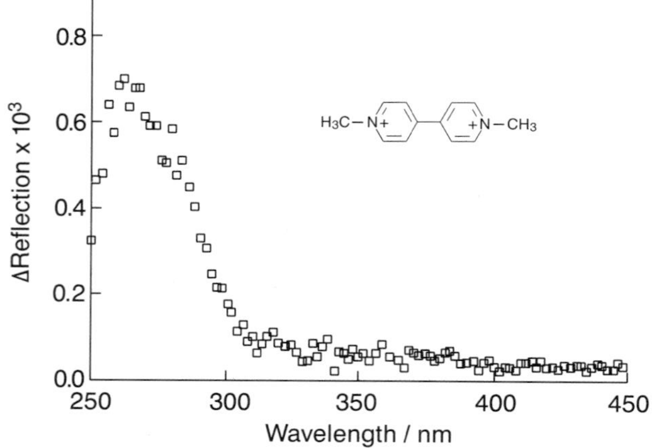

Figure 2. Reflection spectrum of methyl viologen (MV) adsorbed from the aqueous subphase to a monolayer of DPPA; surface pressure: 23 mN/m; [MV]: 10^{-6} M.

Quenching of the monomer pyrene emission at 380 nm is indeed observed as shown in Figure 3, where the relative fluorescence intensity I/I_o (circles, left ordinate) is plotted vs. the bulk MV concentration. From the maximum value of the reflection at 280 nm (squares, right ordinate) it is seen, that saturation of the monolayer with MV is reached at a bulk concentration of 10^{-6} M. At that concentration the fluorescence of PyPPC is practically entirely quenched. This shows that long range photoinduced electron transfer occurs not only in rigid monolayer systems on solid substrates but also in floating monolayers with adsorbed acceptor at the air-solution interface.

4.2 Effect of energy delocalization

Another phenomenon in monolayer systems that is practically not encountered in the investigations in solution is energy delocalization via incoherent or coherent exciton motion. This may have a marked effect on the efficiency of photoinduced electron transfer if donor and acceptor are located at the same interface. In a situation where no acceptor molecule is within a critical radius for electron transfer of the excited donor molecule no electron transfer will occur. In the presence of other donor molecules, that are randomly distributed in the matrix monolayer the excitation energy may hop to a neighbour which may have an acceptor within reach. Then, energy delocalization due to incoherent exciton motion enhances the efficiency of the primary step in photoinduced electron transfer. Such an increase of the apparent rate constant of electron transfer upon increase of the donor density with constant acceptor density has been observed in steady state fluorescence quenching [40,41] and transient [42] measurements with monolayer systems containing donor and acceptor in adjacent monolayers at the same interface.

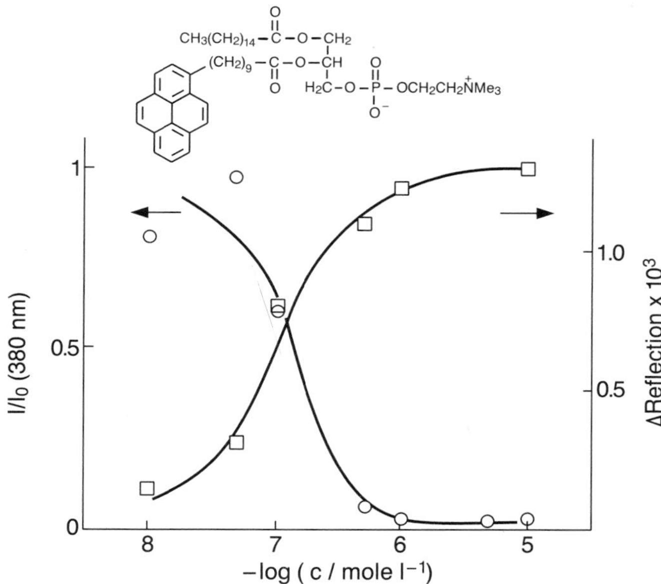

Figure 3. Photoinduced electron transfer across mixed monolayer of PyPPC and DPPA, molar ratio 1:50, to MV adsorbed from the aqueous subphase. Relative pyrene monomer fluorescence intensity at 380 nm (circles, left scale) and reflection signal due to adsorbed MV (squares, right scale) vs. bulk concentration of MV.

This effect is most pronounced in monolayer systems with donor organized in J-aggregates [10]. Energy delocalization in this case is due to coherent exciton motion [35]. The enhancement of the apparent rate of electron transfer, although much stronger than in the case of incoherent exciton hopping, is limited by the rate of fluorescence emission which is proportional to the number of oscillators in the coherent domain (according to the classical picture) and therefore higher than in the case of dye monomers [35]. Taking the increased radiative rate into account, an enhancement of the apparent rate of electron transfer due to energy delocalization by a factor of at least 10^3 was estimated in a particular case [40].

Such an enhancement due to energy delocalization should not be observed in arrangements where all donor molecules are practically in the same situation with respect to the nearest acceptor molecule. This was indeed found in the systems used for the evaluation of the distance dependence where donor and acceptor are separated by one monolayer, and this fact supports the interpretation of donor fluorescence quenching as being due to long range electron transfer [34].

4.3 Interfacial states

The interfaces in monolayer systems are constituted by the contacts between the individual monolayers, i.e. the contacts between head groups of adjacent layers and the hydrophobic end groups in systems formed by the regular Y-type Langmuir-Blodgett transfer process [43,44]. From investigations of photoconduction in multilayer assemblies between metal electrodes the existence of interfacial electronic states with a density of 10^{15} $cm^{-2}eV^{-1}$ was deduced [45,11,46].

These states might serve as electron reservoir and cause photoreduction of appropriate dye molecules. Evidence for such a photoreduction of cyanine dyes in monolayer assemblies has been found upon evacuation of a chamber with the sample. In the absence of oxygen the fluorescence intensity of the dye is smaller than in room air. An example of this effect is shown in Figure 4, where the relative steady state fluorescence intensity of the the cyanine dye is plotted vs. the pressure in the chamber. Upon admittance of air the initial intensity is recovered, whereas atmospheric pressure of Ar or N_2 leaves the reduced intensity unchanged. The phenomenon is rationalized by the formation of a small population of the reduced cyanine dye radical which quenches the fluorescence

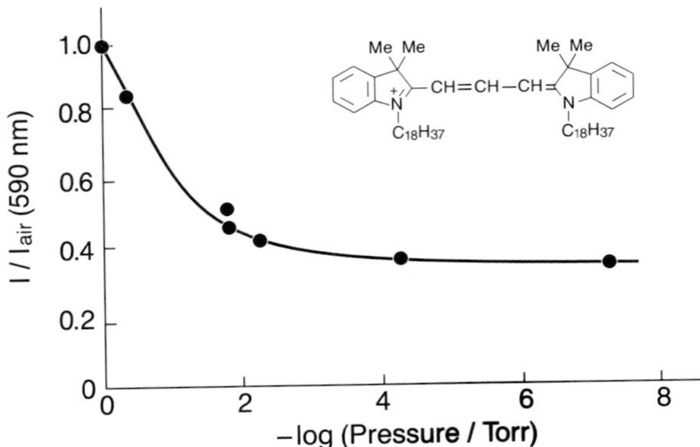

Figure 4. Pressure dependence of the fluorescence of the cyanine dye in monolayer systems. The decrease is due to the decrease in oxygen partial pressure. The effect is ascribed to the formation of reduced cyanine dye acting as energy acceptor.

of the cyanine dye by acting as acceptor in Förster energy transfer [47]. Fluorescence quenching depends on the intensity of the exciting light and on the dye density in the expected way and is strongly enhanced when the monolayer is formed on high pH aqueous subphase before transfer to the glass plate.

In monolayer systems with donor and acceptor (stearylviologen) arranged at the same interface, the formation of the reduced acceptor radical was observed spectroscopically [48,49] and by EPR [50]. In general, the electron recombines with the oxidized donor, however, this path may be blocked when the oxidized donor obtains an electron from the pool of interfacial states or from an additional donor [49]. In fact, a photoreduction of the excited donor by an electron from the interfacial states with subsequent electron transfer in a slow process to the acceptor could also occur.

The formation of the reduced acceptor radical after low quantum yield photoinduced electron transfer can be useful for the detection of such processes if no significant fluorescence quenching of the excited donor can be observed. This is the case when donor and acceptor mixed in matrix monolayers are separated by a bilayer of about 5 nm thickness. With the amphiphilic viologen (structure see Figure 1) as acceptor the formation of the radical can be followed by measuring the absorption change at 400 nm. In monolayer systems with a bilayer of arachidic acid separating donor (the cyanine dye with the structure shown in Figure 4) and acceptor monolayer, no radical formation can be observed. However, a modification of the insulating bilayer by incorporation of quinthiophene as molecular conductor leads to formation of the radical in vacuum [16, 50] as shown in Figure 5 (taken from ref. 16) where the absorption change of a single monolayer system is plotted vs. time of excitation of the cyanine dye at 545 nm.

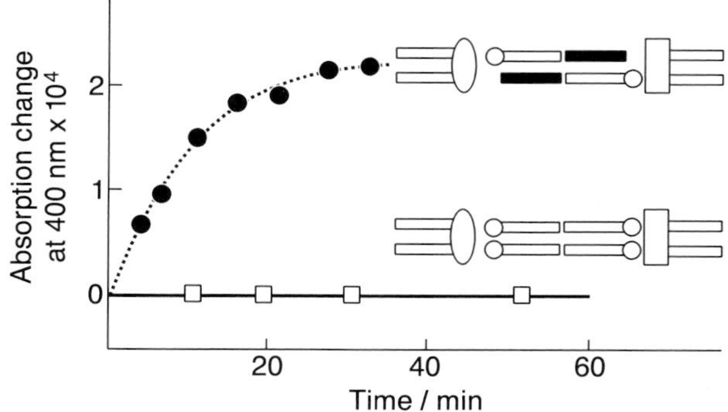

Figure 5. Formation of reduced acceptor radical after photoinduced electron transfer across a bilayer system (see structures). Absorption change at 400 nm (radical band) vs. time of excitation of the cyanine dye. Formation of the radical only in the presence of quinthiophene as molecular conductor in the insulating bilayer.

4.4 Effects of remote environment

Monolayer systems are built up on solid substrates by stepwise transfer of single monolayers. Therefore, the remote environment (2 to 5 nm) of a dye monolayer is not isotropic as in solution but structured and therefore, the influence of dipole layers constituted by the heads and tails of the ordered molecules should manifest itself in modified spectroscopic properties of dyes and of chemical equilibria at interfaces. This is in fact the basis of the response of potential probes [52] used for the estimation of interfacial potentials of biomembranes, vesicles or other microheterogeneous systems. Monolayers provide the possibility to modify systematically the potential at the interface where the probe is located without changing its near environment by partly modifying the tails of the matrix molecules. A particularly elegant technique is the use of a two-component matrix monolayer of a fatty acid and its analog with $-CF_3$ tail [53]. The fluorescence of an amphiphilic cyanine dye in such a matrix monolayer at the air-water interface increases strongly with decreasing surface potential due to an increased fraction of the fluorinated fatty acid [53].

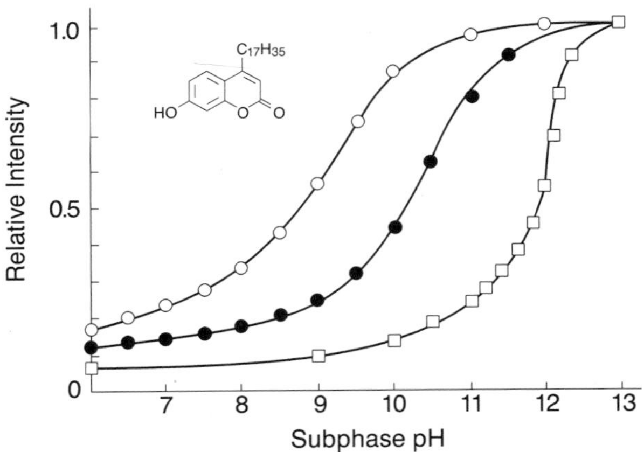

Figure 6. Titration curves of the pH-probe located at the system-water interface of a bilayer system on glass plates. Interface between the two monolayers constituted by methyl tails (full circles); partial replacement of the methyl groups by $-CF_3$ groups in the outer layer (open circles) and the inner layer (open squares), respectively, causes a strong shift of the protonation equilibrium of the probe without change of the near environment.

Protonation equilibria of an amphiphilic pH-probe (structure shown in Figure 6) in monolayers at the air-water interface have been investigated systematically [54 and references therein]. If an appropriate reference monolayer is used for the determination of the pK of the probe, the charge density of a monolayer can be evaluated from the apparent pK of the probe in mixed monolayers according to the Gouy-Chapman theory [53]. It is possible, however, to shift protonation equilibria drastically without variation of the surface charge density simply by changing dipole layers at a distance of 2.5 nm from the interface where the probe is located. This is demonstrated in Figure 6 [52] with a bilayer system transferred to a glass plate in which the probe is at the system-water interface. The fluorescence intensity is proportional to the density of the probe anion and depends on the pH of the aqueous subphase. The titration curve represented by the full circles was obtained with methyl groups at the tails of both monolayers at the interface 2.5 nm from the probe. The curve to the left (open circles) is found after partial replacement of the methyl groups of the outer layer by $-CF_3$ groups, and the curve to the right (open squares) represents the data for a system where the same fraction of methyl groups of the inner layer was replaced by $-CF_3$ groups. This simple variation of the dipole layers causes a shift of the equilibrium at the bilayer system-water interface by nearly three pK units without change of the interfacial charge density.

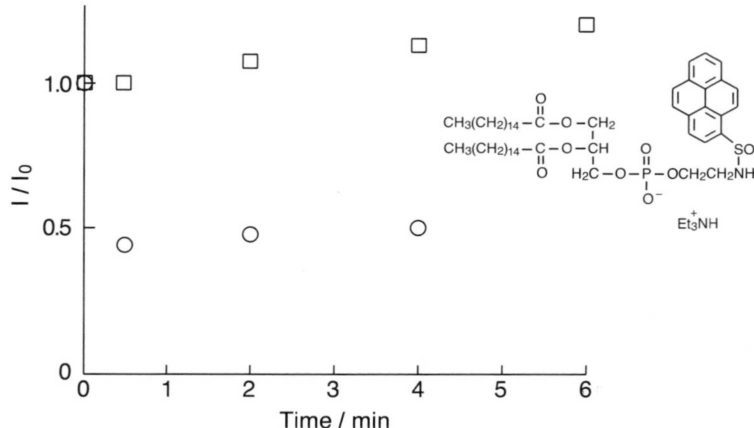

Figure 7. Monomer fluorescence intensity of the pyrene located at the bilayer-water interface with time after addition of MV to a concentration of 10^{-6} M. No quenching due to electron transfer with eicosylamine as first layer (squares) but considerable quenching with cadmium arachidate as first layer (circles) due to different density of adsorbed MV.

Such effects must play an important role in photoinduced electron transfer processes involving interfaces like micelles, vesicles, biomembranes as well as single monolayers and monolayer systems. In particular, the effect has been studied with bilayers on glass plates having an amphiphilic pyrene (structure shown in Figure 7) located at the bilayer-water interface incorporated in a monolayer of DPPA. Methylviologen (MV, structure see Figure 2) dissolved in the adjacent aqueous phase may be adsorbed and then quenches the pyrene fluorescence via photoinduced electron transfer, and this can be observed by monitoring the fluorescence intensity with time after addition of the MV. Figure 7 shows two such traces for two bilayer systems differing in the monolayer only that is in contact with the glass, i.e. rendering the glass surface hydrophobic for the deposition of the second monolayer. No photoinduced electron transfer (squares) is found if this monolayer is eicosylamine, considerable quenching of the pyrene monomer fluorescence is obtained for cadmium arachidate as first monolayer (circles).

The difference in the apparent efficiency of photoinduced electron transfer in these two systems is attributed to a different surface density of adsorbed MV due to the influence of the dipole layers constitued by the head groups and charges on the glass surface on the MV concentration near the bilayer-water interface. The observed effect is qualitatively in agreement with strong shifts of protonation equilibrium of the pH-probe caused by the type of hydrophobization of the glass surface [55].

5. CONCLUSION

Monolayers at the air-water interface and monolayer systems differ from homogeneous solution in many ways. Some prominent properties of monolayer systems and their possible effect on electron transfer have been discussed. In particular, the extremely long range of electron transfer may be due to such structure-dependent physical conditions like geometrical correlation of donor and acceptor and the accessibility of several acceptor molecules per donor, as well as collective phenomena like energy delocalization via incoherent or coherent exciton motion, existence of interfacial electronic states and the presence of internal fields due to dipole layers. As no quantitative treatment of such effects is given, the discussion of the relevant experimental results here is intended to demonstrate the uniqueness of the organized monolayer systems and to raise the interest for photoinduced electron transfer processes in these entities.

Ackowledgments. Financial support of this work by the Bundesministerium für Forschung und Technologie (No. 03M4008D), the Fonds der Chemischen Industrie, and the Consilio Nazionale della Ricerca for one of us (G.C.) is gratefully acknowleged.

6. REFERENCES

1 A. Weller, Z. phys. Chem. N. F. 18 (1958) 163.
2 N. Mataga, T. Okada, and N. Yamamoto, Chem. Phys. Lett. 1 (1967) 119.
3 D. Rehm and A. Weller, Ber. Bunsenges. Phys. Chem. 73 (1969) 834.
4 V.G. Levich and R.R. Dogonadze, Dokl. Acad. Nauk SSSR 124 (1959) 123.
5 R.A. Marcus, Ann. Rev. Phys. Chem. 15 (1964) 155.
6 V.G. Levich, Adv. Electrochem. Electrochem. Eng. 4 (1966) 249.
7 R.A. Marcus and N. Sutin, Biochim. Biophys. Acta 811 (1985) 265.
8 J. Jortner, J. Chem. Phys. 64 (1976) 4860.
9 K.–P. Seefeld, D. Möbius, and H. Kuhn, Helv. Chim. Acta 60 (1977) 2608.
10 D. Möbius, Ber. Bunsenges. Phys. Chem. 82 (1978) 848.
11 E.E. Polymeropoulos, D. Möbius, and H. Kuhn, J. Chem. Phys. 68 (1978) 3918.
12 H. Kuhn, Chem. Phys. Lipids 8 (1972) 401.
13 H. Kuhn in H. Gerischer and J.J. Katz (eds.), Light-Induced Charge Separation in Biology and Chemistry, Berlin: Dahlem Konferenzen 1979, p. 151.
14 D. Möbius, Acc. Chem Res. 14 (1981) 63.
15 H. Kuhn, Mol. Cryst. Liq. Cryst. 125 (1985) 233.
16 H. Kuhn, in Proceedings of The Robert A. Welch Foundation on Chemical Research XXX. Advances in Electrochemistry, Houston, Texas, 1986, p. 339.
17 W.F. Mooney and D.G. Whitten, J. Am. Chem. Soc. 108 (1986) 5712.
18 J.W. Verhoeven, Pure & Appl. Chem. 58 (1986) 1285.
19 G.L. Closs and J.R. Miller, Science 240 (1988) 440.
20 D. Gust and T.A. Moore, Science 244 (1989) 35.
21 M. Fujihira and H. Yamada, Thin Solid Films 160 (1988) 125.
22 J.-M. Lehn, Angew. Chem. Int. Ed. Engl. 27 (1988) 89.
23 J. Deisenhofer, O. Epp, K. Miki, R. Huber, and H. Michel, J. Mol. Biol. 180 (1984) 385.
24 H. Kuhn, Phys. Rev. A34 (1986) 3409.
25 J.R. Miller, L.T. Calcaterra, and G.L. Closs, J. Am. Chem. Soc. 106 (1984) 3047.
26 F. Willig, R. Eichberger, K. Bitterling, W.S. Durfee, and W. Storck, Ber. Bunsenges. Phys. Chem. 91 (1987) 869.
27 A. Osuka, K. Maruyama, N. Mataga, T. Asahi, I. Yamazaki, N. Tamai, J. Am. Chem. Soc. 112 (1990) 4958.
28 D. Gust, T.A. Moore, A.L. Moore, S.-J. Lee, E. Bittersmann, D.K. Luttrulli, A.A. Rehms, J.M. DeGraziano, X.C. Ma, F. Gao, R.E. Belford, and T.T. Trier, Science 248 (1990) 199.
29 M.R. Wasielewski, G.L. Gaines III, M.P. O´Neil, W.A. Svec, and M.P. Niemczyk, J. Am. Chem. Soc. 112 (1990) 4559.
30 H. Kuhn, Pure & Appl. Chem. 11 (1965) 345.
31 B. Mann and H. Kuhn, J. Appl. Phys. 42 (1971) 4398.
32 E.E. Polymeropoulos, J. Appl. Phys. 48 (1977) 2404.
33 E.E. Polymeropoulos, Solid State Commun. 28 (1978) 883.
34 H. Kuhn and D. Möbius in B.W. Rossiter, J.F. Hamilton, and R.C. Baetzold

(eds.), Physical Methods of Chemistry, 2nd Edition, in print.
35 D. Möbius and H. Kuhn, Isr. J. Chem. 18 (1979) 375.
36 R.J. Cave, P. Siders, and R.A. Marcus, J. Phys. Chem. 90 (1986) 1436.
37 H. Grüniger, D. Möbius, and H. Meyer, J. Chem. Phys. 79 (1983) 3701.
38 D. Möbius and H. Grüniger, Bioelectrochem. & Bioenerg. 12 (1984) 375.
39 M. Orrit, D. Möbius, U. Lehmann, and H. Meyer, J. Chem. Phys. 85 (1986) 4966.
40 D. Möbius, Mol. Cryst. Liq. Cryst. 96 (1983) 319.
41 R.C. Ahuja and D. Möbius, Thin Solid Films 179 (1989) 457.
42 N. Tamai, T. Yamazaki, and I. Yamazaki, Thin Solid Films 179 (1989) 451.
43 K.B. Blodgett, J. Am. Chem. Soc. 57 (1935) 1007.
44 I. Langmuir, Proc. Roy. Soc. (London) 170A (1939) 15.
45 M. Sugi, K. Nembach, and D. Möbius, Thin Solid Films 27 (1975) 205.
46 E.E. Polymeropoulos, D. Möbius, and H. Kuhn, Thin Solid Films 68 (1980) 173.
47 Z. Li, Kollektives Verhalten von Farbstoffmolekülen in Monoschichten: Sauerstoffeffekt und molekulare Assoziation, PhD-thesis, Georg-August-Universität Göttingen, F.R.G. 1991.
48 D. Möbius and G.S. Ballard, unpublished, see e.g. D. Möbius in M.L. Hair and M.D. Croucher (eds.), Colloids and Surfaces in Reprographic Technologies, ACS Symposium Ser. No. 200, 1982, p.93.
49 T.L. Penner and D. Möbius, J. Am. Chem. Soc. 104 (1982) 7407.
50 J. Cunningham, E.E. Poymeropoulos, D. Möbius, and F. Baer in J.P. Fraissard and H.A. Resing (eds.) Magnetic Resonance iin Colloid and Interface Science, Reidel Publ. Comp., Dordrecht 1980, p.603.
51 L.F. Chi, Organisation von Farbstoffen in Monofilmen und Schichtsystemen für nichtlineare optische Effekte und Ladungstransport, PhD-thesis, Georg-August-Universität Göttingen, F.R.G. 1989.
52 A.S. Waggoner and L. Stryer, Proc. Natl. Acad. Sci. U.S.A. 67 (1970) 579.
53 D. Möbius, W. Cordroch, R. Loschek, L.F. Chi, A. Dhathathreyan, and V. Vogel, Thin Solid Films 178 (1989) 53.
54 J.G. Petrov and D. Möbius, Langmuir 6 (1990) 746.
55 J.G. Petrov and D. Möbius, Langmuir 7 (1991) in print.

Abbreviations. SV: N,N´-dioctadecyl-4,4´-bipyridinium perchlorate; MV: N,N´-dimethyl-4,4´-bipyridinium perchlorate; DPPA: dimyristoyl-phosphatidic acid; PyPPC: 3-palmitoyl-2-(1-pyrenedecanoyl)-L-α-phosphatidylcholine.

DISCUSSION

Wasielewski

In comparing hole transfer to electron transfer, some work by Miller and Closs suggests that there is little difference in electron transfer rate through a hydrocarbon. What evidence do you have that photooxidative electron transfer processes in monolayer films spaced by hydrocarbons have different rates than photoreductive electron transfer? Also, shouldn't these rates depend on the HOMO & LUMO energies of the spacer?

Mobius

So far we have not been able to see hole transfer across a single monolayer in our systems. Some evidence for such hole transfer was found by Willig (see Proceedings of the 3rd International Conference on LB Films, Gottingen 1986). Therefore, the photoreduction process seems to have a shorter range than the photooxidative process. The rates should be related to the energetic position of the HOMO (photoreductive) and LUMO (photooxidative) of the insulating monolayer as characterized by electron affinity and ionization potential, respectively.

Fujihira

In connection with Dr. Wasielewski's question concerning the role of HOMO and LUMO for the reductive and the oxidative quenching by donor and acceptor, respectively, we recently obtained the experimental results of S–D layered monolayer assembly in which the hole transfer through the HOMO of the alkyl chain of the spacer seems to be responsible for the reductive quenching by D. For estimation of hight of the energy barriers for HOMO and LUMO can be done by the data for the electron affinity and the ionization potential of polyethylene and higher alkane. (see for example for LUMO J. Electroanal. Chem. *119*, 379 (1981)).

Mobius

Comment of Fujihira needs no answer.

Steiner

As I understand your distinction of two cases of electron transfer rate (tunneling rate), where k_{et} is either proportional to the coupling energy linear, or squared this is equivalent to the distriction of coherent or incoherent processes like it is done in the description of electronic energy transfer, e.g. in molecular systems. But what particular fact could favor the 'coherent' case in moonlighters as opposed to the case of selection where it is usually not .

Mobius

The formalisms for electron transfer and energy transfer are equivalent. In your terminology the case $k_{DA} \sim \varepsilon$ corresponds to coherent electron transfer. Several reasons may create a situation favorable for coherent electron transfer in monolayer assemblies, including particularly large ε (strong coupling between D* and A) partly due to a low tunneling barrier (compare eg. A. Osuka, K, Maruyama, N. Mataga, T. Asahi, I Yamazaki, N. Tamai, J. Am. Chem. Soc., 1990, **112**, *4958*), and different τ_c (dephasing time) in the organized systems as compared to a homogeneous medium. Further, I indicated the effect of collective interactions and of interfacial electrostatic potentials. However, I do not claim to have a proof for the case $k_{DA} \sim \varepsilon$, it is more a proposal to take such an interpretation into account.

Fujihira

As to the effect of the density of the acceptor on the slope of log k_q vs. distance, we have done the intramolecular quenching rate experiment of the mixed monolayer of A–S amphiphilic compounds with different intervening alkyl chain lengths diluted with a fatty acid. The slope decreases with the increase in dilution. This indicates that quenching of S* by A proceeds by A directly connected with S (through bond) and also by the A moieties connected to the different S moieties (via through space).

Mobius

Your observation seems to contradict my conclusion that the presence of many acceptor molecules in the plane of the acceptor monolayer causes a decrease of the slope in the log k_q vs. distance plot. However, one has to be very careful in comparing the results in a dilution series, since association phenomena and energy delocalization may change excited state lifetime and thereby influence steady state fluorescence intensities used for the determination of k_q.

Kakitani

In the ET between donor and acceptor molecules separated by a monolayer of hydrophobic chains, the distance between donor and acceptor is more than 20A , while the distance is usually less than 20 A in the homogeneous ET. So, the interaction between the reactants in the monolayer case must be considerably smaller than in the homogeneous case. It is usually accepted that the incoherent mechanism applies in the homogeneous case. Then, it seems to be impossible for the ET in the monolayer case to be coherent. How do you think about this point?

Mobius

In my presentation I have tried to show some phenomena that may cause differences in electron transfer efficiency between monolayer organizates and isolated systems in a homogeneous phase. One possible explanation is a much stronger coupling between excited donor and acceptor (molecular orbital symmetry, geometry) and different relaxation time t_c. Within your terminology I have suggested to consider coherent ET in monolayer organizates. This, however, is not generalized but depends on the system, since we have also observed cases which we tend to rationalize by incoherent ET.

Sumi

You proposed a classifications of electron transfer in which the rate constant becomes proportional to ε in one limit or to ε^2 in another limit. The limit where the rate constant becomes proportional to ε was ascribed to the situation that electron transfer takes place coherently at the transition state, although the whole process of electron transfer takes place incoherently, since it includes thermal activation to the transition state. In terminologies of traditional theories, this situation is called the adiabatic limit. In this limit, however, it has been believed that the rate constant becomes independent of ε. How do you compromise this discrepancy? In other words, how do you interpret your classification in terms of traditional theories of the nonadiabatic vs. the adiabatic limits?

Mobius

In the traditional terms I should classify the case $k_{DA} \sim \varepsilon$ (k_{DA}=rate constant of electron transfer, ε perturbation energy) as adiabatic. However, in my talk I have considered very qualitatively a microscopic situation and given only the result without it (this may be seem in the paper D. Mobius, H. Kuhn, Israel J. Chem. 1979, **18**, *375*). In my opinion, the rate constant k_{DA} depends on ε also in the adiabatic case. This could best be shown by a graph. The barrier height in the double well potential surface depends on ε. Therefore, when you change the distance between D^* and A you change the perturbation energy and consequently the barrier height.

Dynamics and Mechanisms of
Photoinduced Transfer and Related Phenomena
N. Mataga, T. Okada and H. Masuhara (Editors)

FRACTALS AND EXCITATION TRANSFER IN MOLECULAR ASSEMBLIES

I. Yamazaki, N. Tamai and T. Yamazaki

Department of Chemical Engineering, Faculty of Engineering, Hokkaido University, Sapporo 060, Japan

Abstract

Dynamics of electronic excitation transfer under restricted molecular geometries was investigated by means of a picosecond time-resolved fluorescence spectrophotometer. The fluorescence decay function depends on (1) dimensionality of systems (2D, 3D or stacking multilayers) and (2) distribution of functional chromophores in assembly frameworks. The fractal or fractal-like structure is found to be essential in molecular assemblies.

1. INTRODUCTION

Transport and trapping of electronic excitation energy have been the subject of extensive theoretical and experimental works [1,2]. Special attention has recently been paid to dynamics of the Förster-type excitation energy transfer in molecular assemblies of restricted molecular geometries, for examples, vesicles Langmuir-Blodgett (LB) films and photosynthetic light-harvesting antenna [3]. Several examples with which we are concerned are shown schematically in Figure 1. One can expect to observe new aspects of the excitation transfer quite different from those where acceptor molecules are randomly and uniformly distributed in the three-dimensional (3D) rigid or fluid medium. The present paper will review our recent experimental results on excitation transfer in vesicles and LB films. Several types of fluorescence decay kinetics for the excitation transfer will be presented in relation to their assembly geometries and fractal configuration.

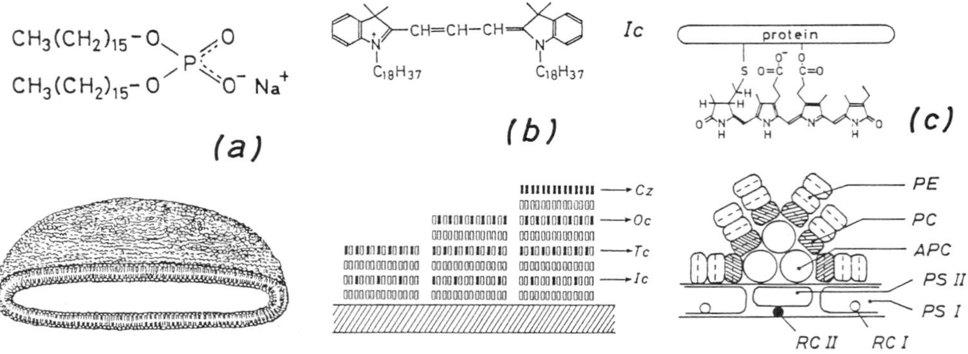

Figure 1. Schematic illustration of organized molecular assemblies: (a) vesicles, (b) LB multilayer films, and (c) photosynthetic antenna of algae.

2. EXCITATION RELAXATION PROCESSES IN MOLECULAR ASSEMBLIES

Electronically excited molecules incorporated in molecular assemblies may undergo various energy relaxation pathways. Let us consider LB monolayer film, as example. Different species, isolated monomer, dimer and higher aggregates, are formed, which are surrounded by a number of fatty acid as diluent molecules. Figure 2 shows a simulation pattern of chromophore distribution. This may suggest that photoexcitation at a site is followed by site-to-site energy migration and quenching at trap sites.

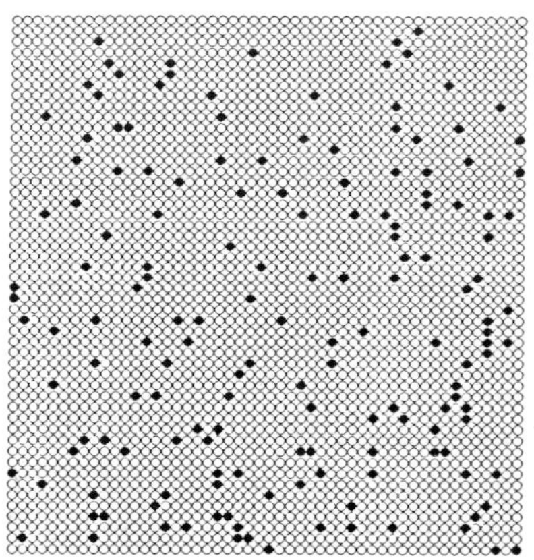

Figure 2. Computer-generated plots of the distribution of functional guest molecules in LB monolayer under the assumption of random dispersion and square lattice (50 x 50).

According to our recent study [4-9], the electronic energy relaxation processes in molecular assemblies can be summarized as follows:

$D^* + Q \longrightarrow D + Q^*$	energy trapping	(1)
$D^* + D \longrightarrow D + D^*$	energy migration between donors	(2)
$D^* \longrightarrow (D)_n^* \longrightarrow (D)_m^*$ (with emission $h\nu_n$ from $(D)_n^*$ and $h\nu_m$ from $(D)_m^*$)	energy migration to aggregates	(3)
$D^* + D \longrightarrow (D \cdots D)^*$	excimer formation	(4)
$D^* + A \longrightarrow D + A^*$	energy transfer to adjacent layer	(5)

Process 3 is observed in LB films; the time-resolved fluorescence spectra shows band shifts to the red associated with the energy migration towards lower energy sites, D_m, D_n and so forth [6,8]. Process 4 is observed in liquid crystals [5] and LB films [4]. Process 5 represents interlayer energy transfer from donor to acceptor layers, which is observed in artificial and biological stacking multilayers [9]. Interlayer energy transfer should compete with the energy dissipation within monolayer (processes 2-4). In photosynthetic antenna, the interlayer transfer is dominant, and the overall efficiency of energy transfer to an inner core of the reaction center is near unity.

3. EXCITATION TRANSFER IN 2D SYSTEM

3.1 Dyes Adsorbed on Vesicle Surfaces [3,7,10]

Vesicles are static colloidal particles consisting of surfactant molecules with two long hydrocarbon chains connected to the polar head group (Figure 1a). Single compartment vesicles contain 2000-10000 monomers per vesicle, and their surfaces are usually as large as 10^5 - 10^6 Å^2. When donor and acceptor molecules are adsorbed on the surface, one can examine kinetics of the 2D energy transfer. We have investigated the excitation transfer between rhodamine 6G (donor) and malachite green (acceptor) and between rhodamine B (donor) and malachite green (ac-

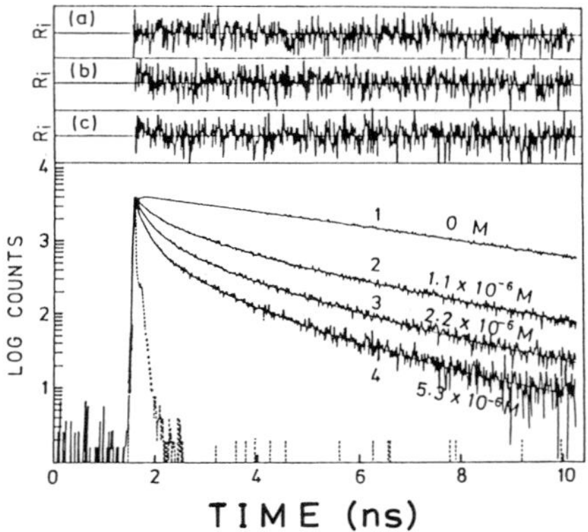

Figure 3. Fluorescence decay curves of rhodamine 6G (donor) in the presence of malachite green (acceptor) adsorbed on vesicle surfaces.

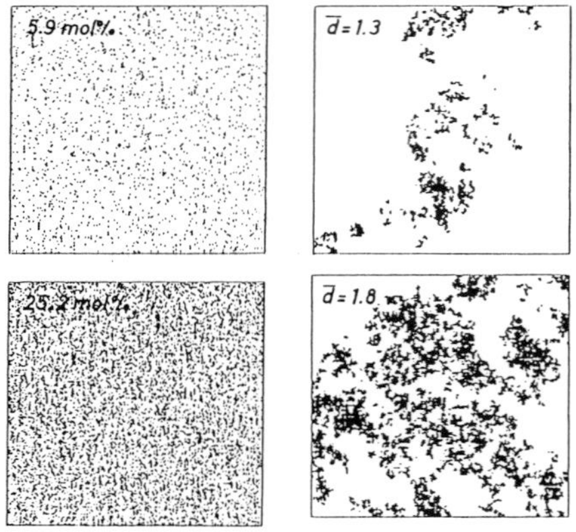

Figure 4. Computer simulation for distribution of dyes on vesicle surfaces. The fractal distribution (right) are compared with an random distribution (left).

ceptor) adsorbed on anionic vesicles consisting of dihexadecylphosphate (Figure 1a). Figure 3 shows the donor fluorescence decay curves together with the results of analyses based on the fractal theory.

The fractal denotes a self-similar structure of dilatational symmetry which will have great potential to describe a multitude of irregular structures. Klafter and Blumen [11] proposed a fluorescence decay function of the donor in a 2D plane where the acceptors are distributed in a space of the fractal dimension:

$$\rho(t) = \exp[\, -t/\tau_D - \gamma_A (t/\tau_D)^{\beta}\,] \tag{6}$$

$$\text{where} \quad \beta = \bar{d}/s, \quad \text{and} \quad \gamma_A = x_A (d/\bar{d}) V_d R_0^{\bar{d}} (1 - \beta) \tag{7}$$

τ_D is the lifetime of donor, and s is the order of the multipolar interaction; x_A is the fraction of fractal sites occupied by acceptors; d and $\bar{d}$ are the Euclidian and fractal dimensions, respectively; and V_d is the volume of unit sphere in d dimension.

The fluorescence decay curves can be analyzed with eq 1 by a conventional curve-fitting method. By varying γ_A and β values with τ_D being fixed, the best-fitting curves were obtained for different surface densities of acceptors (Figure 3). On increasing acceptor concentration, γ_A increases linearly with the concentration, whereas β is almost constant. The fractal dimension $\bar{d}$ can be derived directly from β values through eq 2; $\bar{d} = 1.31 \pm 0.078$ for rhodamine 6G and $\bar{d} = 1.32 \pm 0.049$ for rhodamine B. For the fractal distribution of dyes, a model picture was obtained, as shown in Figure 4, through the computer graphics under an assumption of Levy's dust model. In this model, we assume that, in each step of molecular adsorption, an adsorbed site is grown at distance r apart from the previous site with the probability P(r) expressed as $P(r) \propto r^{-\bar{d}}$, where $\bar{d}$ is the fractal dimension associated with a random-walk of step distance r. The right column of Figure 4 shows the fractal distribution in 200 x 200 square lattice, and the left shows the random and uniform distribution with the same average densities as the right column.

3.2 LB Monolayer Films [8,12]

The LB film is typical of the 2D molecular assembly. The 2D excitation energy relaxation was studied with rhodamine B incorporated in LB monolayers with concentrations of 0.01-30 mol% at 80 K and 295 k. The stacking structure of LB films is shown

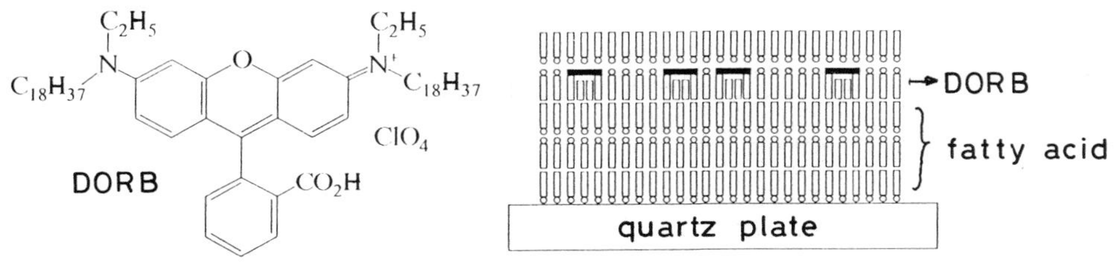

Figure 5. Molecular structure of N,N'-dioctadecylrhodamine B (DORB) and stacking structure of the LB film.

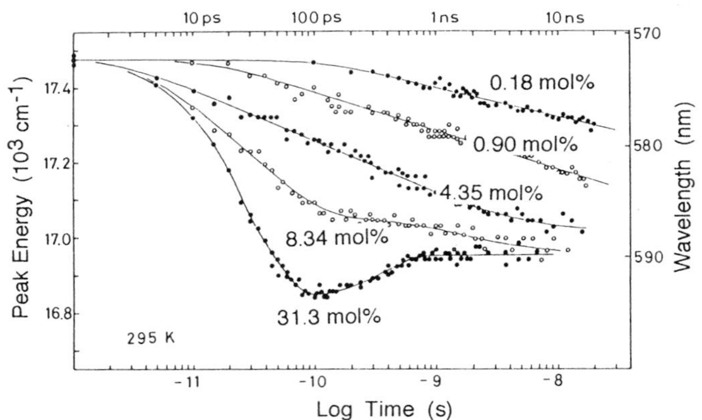

Figure 6. Plots of the dynamic Stokes shift against time in fluorescence of DORB LB monolayer films.

in Figure 5. The time-resolved fluorescence spectra exhibit a dynamic Stokes shift. The peak energy of the monomer fluorescence band, E(t,T), is plotted in Figure 6 as functions of the time and concentration of DORB. The monomer fluorescence band, which locates at 573 nm in the initial time region, shifts to the red with time. The shift occurs very slowly in lower concentration of DORB, but it becomes faster as the concentration is increased. The E(t,T) value reaches a constant value at time considerably shorter at 80 K than at 295 K. Such spectral shift reflects an energy migration among energetically disordered monomer sites which can be represented by the ultrametric space

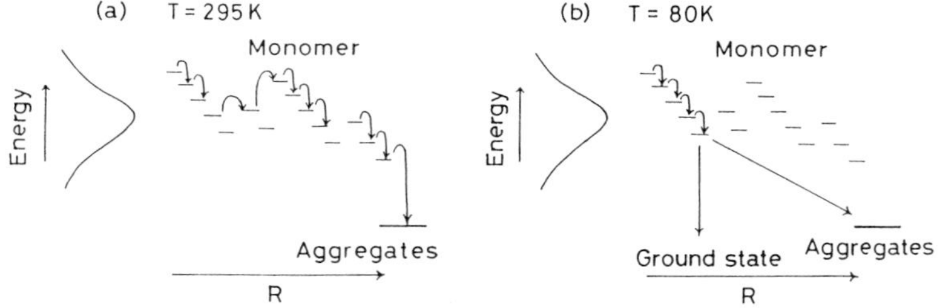

Figure 7. Schematic illustration of excitation energy migration and trapping in DORB LB monolayer films.

of hierarchical energy-level distribution [11], as is shown in Figure 7.

Figure 8 shows the fluorescence decay curves of LB monolayers excited at 545 nm. At concentration less than 0.1 mol%, the fluorescence decay is almost single exponential with a lifetime of 3.6 ns. As the concentration increases, the decay curve deviates largely from exponential form. The decay time of the short decay component coincides with the relaxation time of the dynamic Stokes shift. Thus the fastest decay can be ascribed to fluorescence quenching at trap sites in the course of energy migration between monomer sites of slightly different energy levels. The trap sites are considered to be dimers and/or aggregates of rhodamine B. The fluorescence decays can be fitted with an equation including a stretched exponential function:

$$\rho(t) = A_1 \exp[\,-t/\tau_D - \gamma(t/\tau_D)^{\beta}] + A_2 \exp(\,-t/\tau_D) \qquad (8)$$

The amplitude A_2 is dominant at low concentration, but as the concentration is raised A_1 becomes dominant. This means that the LB monolayer film is a mosaic assembly consisting of the two

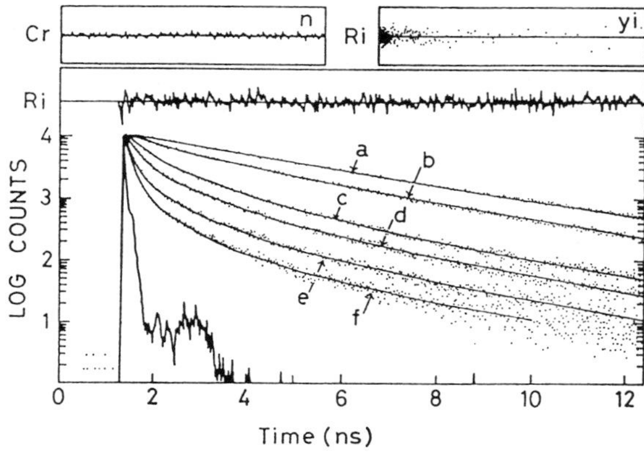

Figure 8. Fluorescence decay curves (······) and calculated best-fit curves (——) of DORB LB monolayer films. Concentrations of DORB are (a) 0.182, (b) 0.90, (c) 4.35, (d) 8.34, (e) 15.4 and (f) 31.3 mol%. The weighted residuals are plotted for curve f.

types of domains; a monomeric domain and a domain containing aggregates with high density. Regarding to the first term, the β value depends on the dye concentration. By using eq 7, the fractal dimension d was evaluated to be 1.7 in higher concentration and 1.9 in the lowest concentration. Rhodamine B molecules in LB film are distributed in a fractal structure illustrated in Figure 4, except for the case of very low concentration. In many cases of LB monolayer films containing different chromophores, the d value falls in between 1.3 and 1.7.

4. INTERLAYER EXCITATION TRANSPORT IN LB MULTILAYERS [9]

With stacking multilayered LB films, the interlayer excitation transfer from outer to inner layers was studied. We have prepared three types of LB films consisting of sequential stack of donor (D) and acceptors (A1, A2 and A3) layers; namely, D-A1 (hereafter referred to as 2L), D-A1-A2 (3L), D-A1-A2-A3 (4L). The stacking structures of these multilayers are illustrated in Figure 9. Six to eight layers were deposited on a quartz plate in the following order; (1) five layers of palmitic acid-cadmium salt, (2) monolayer(s) consisting of palmitic acid and small amounts of dye (A1, A2 and A3), (3) monolayer of carbazole (D), and (4) a monolayer of palmitic acid. An outer layer of palmitic acid prevents the multilayer structure from being destroyed. The concentration of pigment molecules was 10 mol% in each layer.

Figure 9 shows the time-resolved fluorescence spectra of the LB multilayers of 2L and 3L. Each spectrum is normalized to the maximum intensity. It is seen that the spectrum changes significantly with time in the picosecond time range. In 2L, following excitation of the D layer at 295 nm, a fluorescence band of D appears weakly at 350 nm and A1 band at 420 nm. In 3L, the fluorescence bands of D and A1 appear in the initial time region, and then A2 band appears at 470 nm after 40 ps. Similar spectral change can be seen in 4L. It was found that the fluorescence from the inner layer appears more slowly than those of the outer layers, and that the transfer rate depends on the Förster radius (R_0) distance. These sequential time behaviors can be fitted

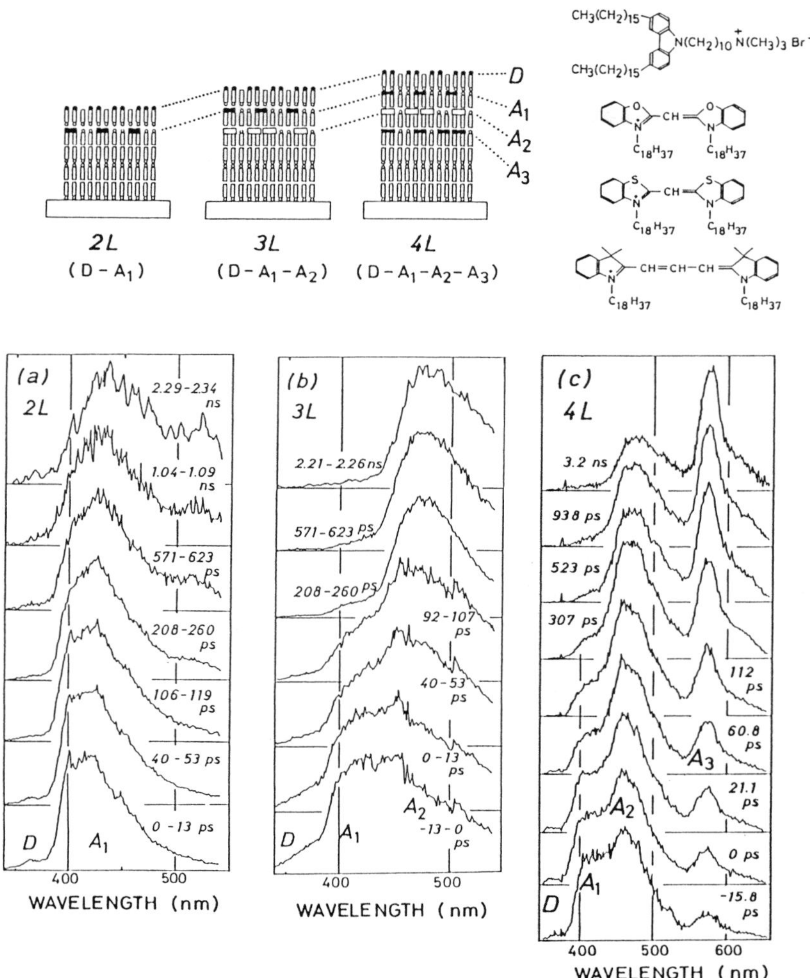

Figure 9. Schematic illustration of the LB films and their time-resolved fluorescence spectra obtained with excitation at 295 nm.

approximately with a decay kinetics of exp $(-2kt^{1/2})$ type. The values of the rate constants fall in between 0.02 to 0.2 $ps^{-1/2}$. These results show that the energy transfer takes place sequentially from the outer to the inner layers, similarly to a photosynthetic light-harvesting antenna pigment system.

5. COMPARISON BETWEEN BIOLOGICAL ANTENNA AND ARTIFICIAL MULTILAYER FILMS [3,9]

The sequential energy transfer kinetics in LB multilayers are on the whole similar to those in the photosynthetic antenna, phycobilisomes. However, the transfer efficiency is different; the efficiency in each step is 0.9 in phycobilisomes whereas it is 0.5-0.8 in LB films. Note that the transfer efficiency is determined by a branching ratio between the energy transfer to adjacent layer (process 7) and the excitation trapping within a layer (process 3 and 6). The low efficiencies in LB multilayers are a consequence of relatively high density of traps due to dimer and/or higher aggregates of chromophores. Aggregation results in irregular distribution or island structure of guest molecules in artificial molecular assemblies. On the other hand, the biological antenna has an uniform distribution of functional chromophores; tetraphyrroles in phycobilisomes are distributed with a regular array in polypeptide networks. As a consequence, the excitation transfer takes place with efficiencies near unity even in fairly long interchromophore distances.

6. REFERENCES

1 Th. Förster, Z. Naturf., 4a (1949) 321.
2 N. Mataga and T. Kubota, Molecular Interactions and Electronic Spectra, Marcel Dekker, New York, 1970.
3 I. Yamazaki, N. Tamai and T. Yamazaki, J. Phys. Chem., 94 (1990) 516.
4 I. Yamazaki, N. Tamai and T. Yamazaki, J. Phys. Chem., 91 (1987) 3572.
5 N. Tamai, I. Yamazaki, H. Masuhara and N. Mataga, Chem. Phys. Lett., 104 (1984).
6 N. Tamai, T. Yamazaki and I. Yamazaki, J. Phys. Chem., 91 (1987) 841.
7 N. Tamai, T. Yamazaki, I. Yamazaki, A. Mizuma and N. Mataga, J. Phys. Chem., 91 (1987) 3503.
8 N. Tamai, T. Yamazaki and I. Yamazaki, Chem. Phys. Lett., 147 (1988) 25.
9 I. Yamazaki, N. Tamai, T. Yamazaki, A. Murakami and Y. Fujita, J. Phys. Chem., 92 (1988) 5035.
10 N. Tamai, T. Yamazaki, I. Yamazaki and N. Mataga, Ultrafast Phenomena V, G. R. Fleming and A. E. Siegman (eds.), Springer-Verlag, Berlin, 1986.
11 J. Klafter and A. Blumen, J. Chem. Phys., 80 (1984) 875.
12 N. Tamai, T. Yamazaki and I. Yamazaki, Can. J. Phys., 68 (1990) 1013.

DISCUSSION

Mobius

Your island formation of the acceptor in monolayers should manifest itself in quenching of the donor fluorescence. I should except a relation between quenching of donor and area fraction of acceptor in the acceptor monolayer.

Yamazaki

In our experiment, the LB monolayer contains donor (rhodamine B) and acceptor (crystal violet) with concentrations of the donor as small as 0.1 mol% and of the acceptor 0.2∿20 mol%. We might expect that substantial number of acceptors surround a donor and that no energy migration takes place among donors.

Sundstrom

Have you tried other fitting functions for your fluorescence decays of the one-component LB-films, and what was the result?

Yamazaki

Fluorescence decay curves in LB films generally take non-exponential form except for cases of extremely low concentration of functional guest molecules. Curve fitting analysis based on superposition of exponential decays requires 4- or 5- components, but makes no physical meanings.

Fujihira

We also observed the fluorescence microscopic picture of the mixed monolayer of pyrene tailed fatty acid and fatty acid. From the various experimental evidence, now I could conclude that the bright parts of the fluorescence micrograph are ascribed to crystal formation in the monolayer. By using the other type of pyrene fatty acids, in which the pyrene moiety is located in the middle of alkyl chains of fatty acid, we could form more homogeneous mixed monolayer. Difficulty in forming crystalline in the latter monolayer is ascribed to the lowering the melting point.

Matsuo

Formation of dye aggregate which is harmful to light-harvesing system, may be prevented by putting bulky substituent around the chromophore unit. As a consequence, the fractal structure may disappear and homogeneous distribution of the chromophores will be achieved. Why do not you try this possibility as a means to establish light-harvesting system by the use of multilayered LB films.

Yamazaki

We have tested this effect by using cyclodextrin LB monolayer in which functional guest molecules are inside cyclodextrin pores. We have found that the fluorescence decay is single exponential as a result of the uniform distribution of monomeric guest molecule in LB monolayer film.

Sundstrom

You assigned the time–dependent spectral shift of the fluorescence of the one–component films due to energy migration between molecules of different sites having different energies. Do you have results such as calculations of energy levels that support your suggestion that different conformations have different energy?

Yamazaki

From the magnitude of spectral shift, we estimated the site–energy distribution to be 300 cm^{-1} as the halfwidth of the Gaussian distribution. Similar site–energy distribution was reported for benzophenone by K. M. Weitzel and H. Bassler, J. Chem. Phys. **84**, *1590* (1986).

Yamamoto

Does the dimensionality depend on the magnification of observation? I ask this, because you showed both the microscope picture of pyrene and molecular fractal picture. These look quite different. Even in pyrene system that you showed, if we see it by STM, the picture may be very simple.

Yamazaki

By means of STM, we might expect to see directly surface structure of LB films for substantially large area in near future. Now we are able to speculate indirectly from the optical microscope images. We pointed out that the changes of distribution pattern with changing the concentration of guest molecules are parallel among the fluorescence decay analyses and the optical microscopic image.

Dynamics and Mechanisms of
Photoinduced Transfer and Related Phenomena
N. Mataga, T. Okada and H. Masuhara (Editors)

Charge Separation and Energy Transfer in Microstructured Co-polymer systems.

J. A. Delaire, J. Faure, R. Pansu, M. Sanquer-Barrié, S. Salhi

Ecole Normale Supérieure de Cachan and Université de Paris-Sud, Centre d' Orsay, Bat. 350, 91405 Orsay France.

Abstract.

Charge separation as a consequence of fluorescence quenching of the singlet state of a donor bound to poly(acrylic/or/ methacrylic acid) and poly(styrene sulfonate) has been investigated using charged and neutral zwitterionic acceptors, with a view to determining the relative contribution of hydrophobic and electrostatic effects in the yield of electon transfer and charge escape. The results are suitable to probe a copolymer of polyacrylic acid and $(Ru(bpy)^2(MeVbpy))^{2+}$, PAH-Rubpy, synthetized by Kaneko *et al*[14] as a model for water splitting by solar radiation. An approach of time-resolved laser studies of charge transfer and charge separation in block-copolymer microstructured films is introduced.

1. INTRODUCTION

Electron transfer as a consequence of excited state quenching is one of the more intensively investigated subjects of the last decades. Taking advantage of theoretical progress as well as the improvement of fast kinetics laser techniques, charge pair separation and ion pair stabilization have focussed the attention in photochemistry. Twenty years ago, Weller[1] and Rehm et al.[2] attempted to separate the various physical parameters that govern the photoelectron transfer and proposed their theory on the effect of exothermicity. Since then, with Mataga[3], Moore *et al* [4] and many other authors[5], distance and mutual orientation between reaction centers have been the subject of fruitful investigations, involving exciting developments of research programs starting from molecular design of donor-acceptor assemblies either chemically bound or solvated in micelles, colloids, microemulsions, polyelectrolytes and various other polymers.

We are interested in polymeric systems, one of the materials with which one can expect the goal of molecular electronics development. Two main types of polymers have been chosen: hydrosoluble polymers, in which the specific polymeric factors capable of influencing the electron transfer and charge separation can be explored, and hydrophilic-hydrophobic block copolymer films, which their ability of structuration into a solid micellar-like system predisposes for assisting charge separation. In both cases, it was interesting to investigate the influence of microstructurations that can appear under specified procedure. Several studies have focussed on the effect of the microenvironment in such polymers, and recently polyelectrolytes-bound chromophores systems has been especially investigated[6] .

2. ROLE OF ELECTROSTATIC INTERACTIONS IN PHOTO-INDUCED CHARGE SEPARATION IN POLYELECTROLYTE-BOUND VINYLDIPHENYL-ANTHRACENE.

The reaction we will discuss in most detail in this section is the electron transfer and charge separation between 9-10 diphenylanthracene DPA covalently bound to three polymers and Methyl viologen MV^{2+} and a zwitterionic viologen, 4,4'-bipyridinium-1,1'-bis(trimethylene sulfonate) SPV. The three polymers have been chosen for exhibiting or not hydrophobic effects : poly(methacrylic acid) PMAH-DPA, polyacrylic acid PAH-DPA, and poly(styrenesulfonate) PSS^-Na^+-DPA. These polymers are depicted in Fig. 1.

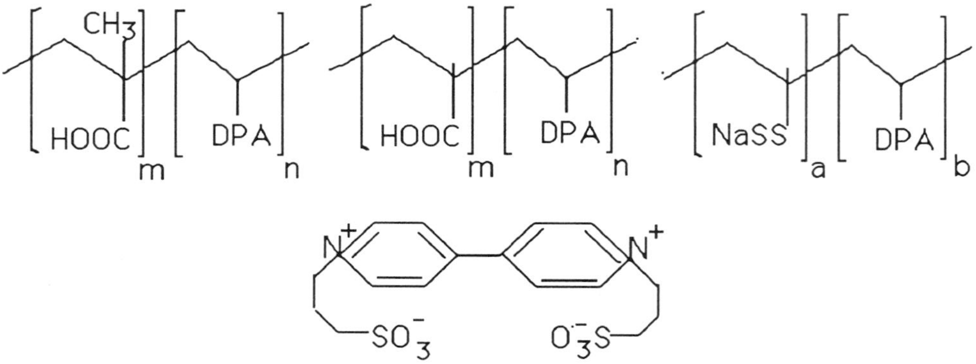

Figure 1. PMAH, PAH, PSS-Na+ /DPA copolymers (m=0.999, n=0.001 a=0.98, b=0.02 ; NaSS= Styrene sodium sulfonate), and SPV quencher.

In an earlier work[7] , one of us reported the effect of pH on the MV^{2+} quenching of the excited singlet state of DPA in PMAH-DPA. At high pH value, a strong quenching is observed, but charge separation does not follow the quenching process ; in contrast, at low pH the fluorescence quenching rate is less than diffusion controlled but the quantum efficiency for charge escape, Φ_{ce} , is ca. 0.7. It was concluded that at low pH, the collapsed PMA hydrophobic coil protects the chromophore and the back reaction between the cations $DPA^{\cdot+}$ and $MV^{\cdot+}$ is less preponderant than when the negative charges of the ionized polyelectrolyte govern the environment of DPA^+.

The fluorescence quenching of $PSS-Na^+$-DPA by MV^{2+} has been reported by Webber[8]. We found that this system is analogous to deprotonated PMAH-DPA/MV^{2+} in that fluorescence quenching is very efficient and occurs in basic solution via a mixed dynamic-static mechanism instead of a single dynamic quenching observed in acidic medium. Typical fluorescence decays in PMAH-DPA/MV^{2+} are reported in Figure 2.

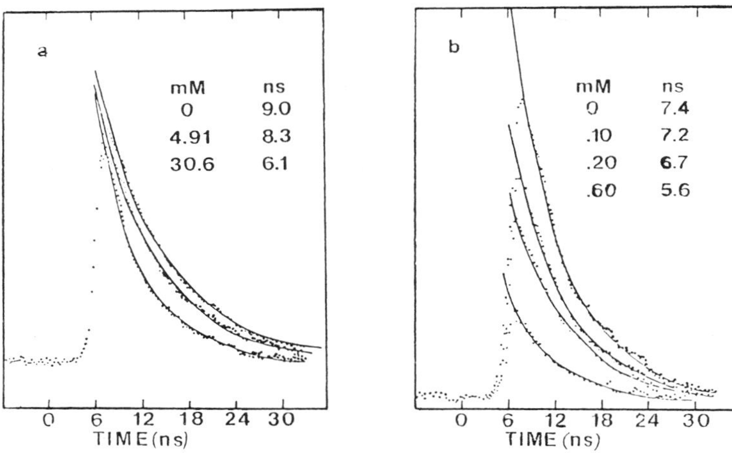

Fig 2- Fluorescence quenching of PMAH-DPA as a function of MV^{2+} concentration. (a) acidic solution, (b) basic solution ; λ=470 nm. From ref.[9]

Hydrophobic and hydrophilic effects, attractive or repulsive electrostatic interactions are therefore some of the major factors governing the charge separation in such polymers. Using several combinations of various polyelectrolytes and fluorescent donors **D** and acceptors **A**, we report experiments from which it has been possible to differentiate the main factors, pH and ionic strength, influencing the reactions we will discuss in detail, which are the following, (the charge separation reaction being of course the result of several processes following the excitation of the donor D),

$$D + A \underset{k_{rec}}{\overset{\Phi_{ce}}{\rightleftarrows}} D^{+} + A^{-}$$

D and A being the species described above. Φ_{ce} and k_{rec} have been measured by nanosecond and picosecond laser spectroscopy.

A- pH effect.

-**PMAH-DPA and PAH-DPA polymers**. From quasi-elastic light scattering experiments[10] , it has been shown that the diffusion coefficient D of the PMAH-DPA polymer decreases abruptly around pH 5 (see fig. 3) typical of a change of conformation of the polymer. As the pH increases, the variation of the Stern-Volmer constant of the PMAH-DPA quenching by SPV is correlated with this change of conformation and shows a sharp growth while the polymer coil is split up and stretches in the basic solution. Chu *et al*[11] reported a similar observation in the quenching of pyrene bound to polymethacrylic acid by Tl^+ ions.

Charge separation efficiencies are compared in Table 1.

Table 1- Φ_{ce} values measured for PMAH-DPA

Quencher	Φ_{ce} (low pH)	Φ_{ce} (high pH)
MV^{2+}---> MV^+	0.73	0
SPV ---> SPV^-	0.33	0.07

Φ_{ce} is obtained from the relation:

$\Phi_i = \Phi_q \cdot \Phi_{ce}$

where Φ_i is the ionization quantum yield given by laser photolysis measurements of the intermediate concentrations:

$$\Phi_i = \frac{[D^+]}{[D^*]} = \frac{[A^-]}{[D^*]}$$

and Φ_q is obtained from fluorescence quenching experiments.

In contrast to the behaviour of MV^{2+} described above, SPV^- is observed whatever the pH may be, but at lower pH, electrostatic attraction between $DPA^{\cdot+}$ and $SPV^{\cdot-}$ plays against hydrophobic protection. At high pH values, one observes with SPV a less efficient, although weak, charge separation yield. The pH effect on the charge separation yield and the recombination rate constant for the quenching of PMAH-DPA by SPV is shown in figure 3. As the pH increases the charge separation yield and the recombination rate have a sigmoid form but correlated opposite variations.

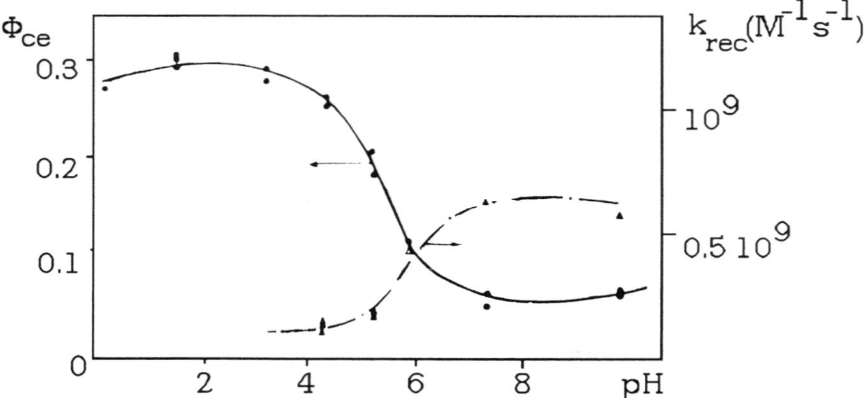

Figure 3. Charge separation yield Φ_{ce} (full line) and recombination rate constant k_{rec} (dashed line) for the quenching of PMAH by SVP as a function of pH.

It should be observed that below pH 5 the decay of $SPV^{\cdot -}$ as well as $MV^{\cdot +}$ no longer obeys second-order kinetics; it has been speculated that this was the consequence of a protonation reaction of the reduced species $A^{-\cdot}$

At the opposite of PMAH-DPA, PAH-DPA does not exhibit hydrophobic microdomains. Quasi-elastic light scattering (QELS) experiments have shown that in acidic medium its conformation is that of a statistical coil, and in basic medium these coils are unfolded when giving a polyanionic conformation. The diffusion coefficient D measured by QELS as a function of pH is shown in fig 4 for both polymers.

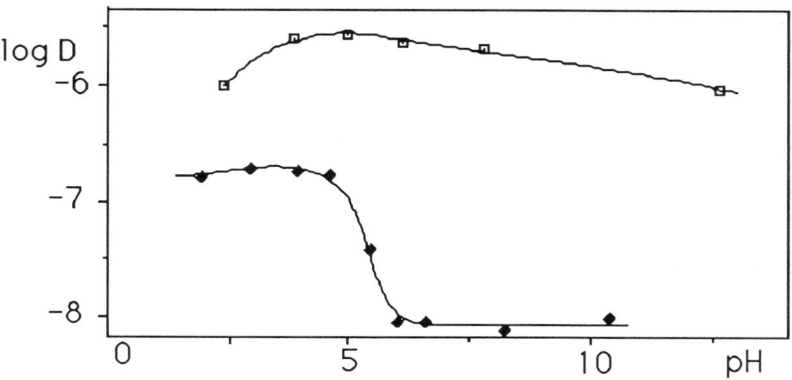

Fig 4- PMAH-DPA (lower curve) and PAH-DPA (upper curve) diffusion coefficients measured by QELS plotted as log D. (absolute values of D are not significant since molecular weights and polydispersities of polymers are different).

As a matter of fact, D values show a smooth variation for PAH-DPA as compared with the sharp variation of PMAH-DPA mentioned above. An interesting feature is the maximum observed at pH 5 for PAH-DPA : when a significant part of the carboxylic groups of the chain are ionized, there is an increase of the hydrogen bonds which causes a shrinking of the statistical coil. The fluorescence lifetimes τ_f of PMAH-DPA and PAH-DPA and the fluorescence quantum yields Φ_f as a function of pH are compared in table 2. There is no significant difference in the results obtained with different polymers and with the change of acidity of the solution. The

Table 2
Fluorescence lifetimes in nanoseconds, and quantum yields for PMAH-DPA and PAH-DPA at low and basic pH. Accuracy about 1%.

	τ_f (pH=2)	Φ_f (pH=8)	Φ_f (pH=8)/Φ_f (pH=2)
PMAH-DPA	8.8	7.1	0.93
PAH-DPA	9.0	7.0	0.89

lifetime variation with pH was correlated with a variation of the local refractive index of the solutions.[12]

Time resolved fluorescence polarization experiments performed by Tan *et al*[13] on 9-methylphenanthrene bound at the end of a PMAH chain led to two rotational relaxation times that have been interpreted as two distinct populations of the fluorescent probe. We have obtained a similar result with PMAH-DPA. Although only a one component decay of 30 ns lifetime has been observed in acidic solutions, very close to the mean value of the two components decay obtained by Tan *et al* (32ns), in basic solution we have detected a two exponential decay of 20 ns and 2 ns respectively, with a mean rotational lifetime much shorter (10 ns) ; both components have the same population. Time resolved polarization experiments are shown in Figure 5 for both polymers.

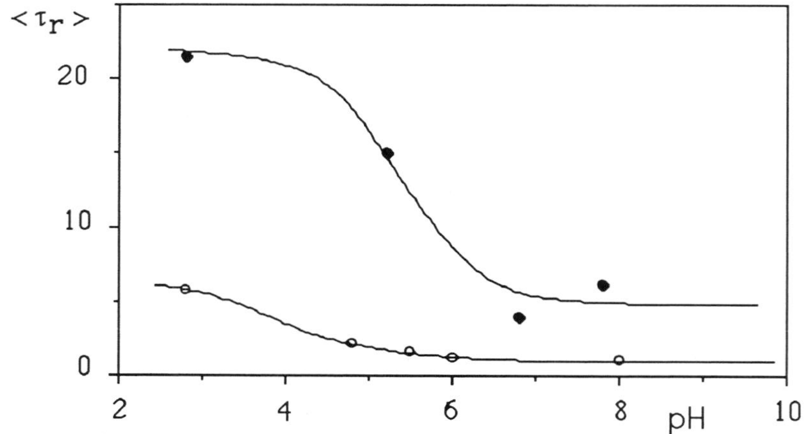

Fig 5. Mean rotational lifetimes $<\tau_R>$ in ns as a function of pH for PMAH-DPA (upper curve) and PAH-DPA (lower curve).

We can conclude that for DPA, even in basic medium, two populations of DPA compete, one protected within the coil and the other one being free. For PAH-DPA, $<\tau_R>$ values are much lower and show a sigmoid form, but with a very smooth variation with pH. This result is consistent with the conclusions drawn with PMAH-DPA, one can expect that charge separation quantum yields may be much lower with PAH than with PMAH in acidic medium, since an additional hydrophobic protection provided by the methyl groups occurs for the latter. Although the Φ_{ce} values in basic solution are comparable, at low pH one obtains 0.16 for MV^{2+} and 0.20 for SPV. As can be shown in figure 6, a significant result is obtained with the PAH-DPA/SPV couple concerning both Φ_{ce} and k_{rec} variation with pH, and confirms the conclusions about hydrophobic effects and charge repulsion.

As a matter of fact, this system does not exhibit the correlated and *reverse* sigmoid form of Φ_{ce} and k_{rec} . This can be interpreted by a subtle

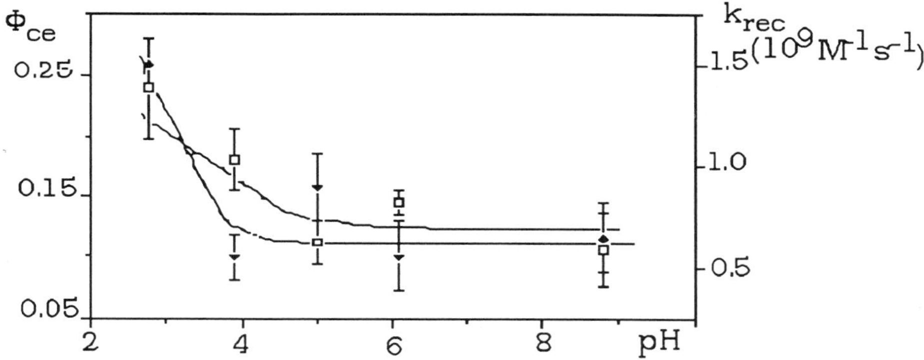

Fig 6- PAH-DPA/SPV system: Charge separation yield (o) and charge recombination rate constant (•) as a function of pH.

balance between electrostatic and conformational effects. Pure electrostatic effect should lead to an increase of Φ_{ce} with increasing pH (as observed in PAH-Rubpy/SPV system, *vide infra*) and a correlated decrease in k_{rec}. But the unfolding of statistical coils is responsible for a non-protection of DPA and this effect seems dominant in this case, leading to the observed higher values of Φ_{ce} in acidic medium. On the other hand, the recombination reaction seems dominated by the electrostatic effect (repulsion between the polyanion and $SPV^{\cdot -}$ in basic solution).

B- Ionic strength effect in PSS^--Na^+-DPA polymer.

Ionic strength effects on fluorescence quenching. As we mentioned earlier, PSS^--Na^+-DPA was found to exhibit a fluorescence quenching behaviour analogous to deprotonated PMAH-DPA, but there is no transition analogous to the one found for PMAH upon deprotonation. However, Drifford *et al* [14] established the diminution of coil size for polyelectrolytes when the ionic strength increases, as a result of screening electrostatic repulsion. Fluorescence quenching of PSS^-Na^+-DPA by SPV leads to the conclusion of static and dynamic quenching similar to that obtained with MV^{2+} (Figure 2). The fluorescence decay, measured on a

Table 3-
K_{SV} data obtained from time resolved quenching experiments can be obtained from the sum of time resolved K_{SV} (second and third columns) and continuous Stern-Volmer measurements (fourth column). In the last column are reported the values obtained with MV^{2+} (from ref [9]).

[NaCl] , M	K_{SV} from I_Q (SPV)	K_{SV} from τ_Q (SPV)	K_{SV} (SPV)	K_{SV} (MV^{2+})
0	348	290	724	$1.4\ 10^5$
2.0	29	25	47	$4\ 10^2$

classical picosecond apparatus, could fit reasonably well to a single exponential : $I(t)= I_Q(0) \exp(-t/\tau_Q)$. The static quenching was measured by the decrease in the initial intensity $I_Q(0)$ while the dynamic quenching is reflected in the shortening of τ_Q. The results given in Table 3 show that the sum of the initial slopes decreases with increasing ionic strength.

It can be concluded that SPV, although concentrated into the vicinity of the polymer coil, is less attracted than MV^{2+}. This effect is probably due to the dipolar interaction between SPV and the electrostatic field of the polyanion. At high ionic strength the screening effect leads to a rate quenching expected for a diffusion-controlled reaction, in agreement with Sasson *et al* [15] and Itoh *et al*[16]. While the neutral quencher is less efficient at high ionic strength, a clear choice cannot be made, from simple ionic strength effects, on the fluorescence quenching, between steric hindrance and concentrating effect due to interaction between the quencher and the polymer anions. Ionic strength effect studies on charge pair separation allowed us to conclude in favor of the latter explanation.

Ionic strength effect on charge separation. The charge escape yield in the PSS^--Na^+-DPA/SPV system is low, about 0.05. Transient measurements required high laser power; so it was necessary to add e_{aq} scavengers in order to study the interference of a two-photon ionization effect. The rate constant of the reaction : e_{aq} + SPV --> SPV^- has been measured using pulse radiolysis (2.3 10^{10} M^{-1} s^{-1}) and compared to that of the reaction of the solvated electron scavenging by acetone (5.9 10^9 M^{-1} s^{-1}) and by H^+ (2. 10^{10} M^{-1} s^{-1})[1]. The observed transient absorption at 605 nm is essentially independent of pH and of added acetone even in strong concentrations compared to SPV. Under such conditions, the variation of the transient concentration and decay at 605 nm as a function of NaCl concentration, represented in Figure 7, is characteristic of charge sepa-

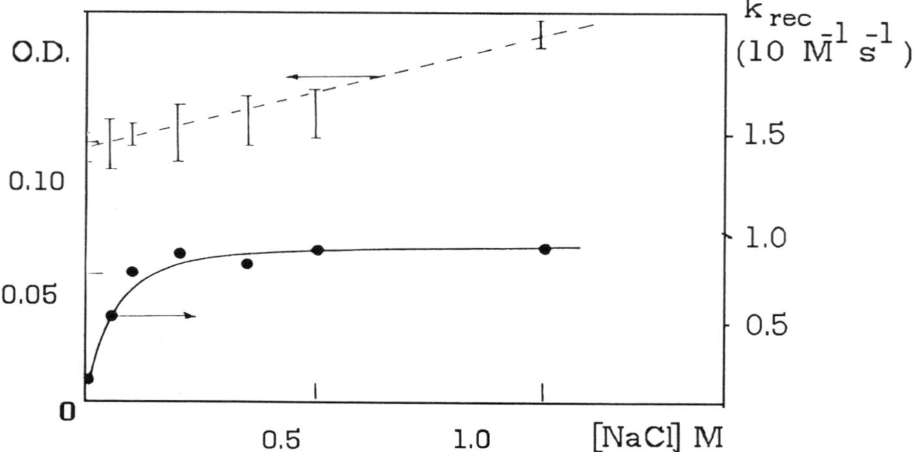

Figure 7. Transient absorption (---) and k_{rec}.(◦) vs NaCl concentration for PSS^--Na^+-DPA/SPV.

ration yield and of charge recombination, respectively. One observes an increase of Φ_{ce} with ionic strength from Φ_{ce}=0.05 at zero ionic strength to 0.07 in NaCl 1M. The ionic strength effect is much less important than the protonation effect for SPV/ PMAH-DPA, but is more dramatic on the back reaction : DPA^+ + SVP^- --> DPA + SPV, although from the charges of the reacting species one would expect a decrease in the reaction rate. This means that the repulsion of SPV^- by the negative charges of the polyanion must dominate the recombination process at low ionic strength, this latter process being decreased by the addition of NaCl which induces a coiling effect of the polymer and therefore a screening effect. This has been demonstrated by one of us[18] , using the Manning theory. For practical applications, low ionic strengths are preferable since the lifetime of the ion pairs is decreased by ca 10 while the yield Φ_{ce} is increased by a factor of 1.4 at high ionic strength.

3. APPLICATION TO A COPOLYMER AS A MODEL FOR WATER SPLITTING AND TO A BLOCK-COPOLYMER FILM.

A- Luminescence quenching of PAH-Rubpy* by viologens. Comparison with the model system in water : Ruthenium tris bipyridyl- Viologen.

Photoinduced electron transfer from polymer-bound ruthenium complexes to viologens was first demonstrated by Kaneko et al[18-19] Since then they synthetized and studied several copolymers with pendant bipyridyl groups complexed with ruthenium di-bipyridyl ions[21-23]. In collaboration with M. Kaneko, some of us[24] published the influence of the

HOOC | bpy | bpy(Rubpy)$_2$

0.939 | 0.037 | 0.024

microstructure of a copolymer of polyacrylic acid and PAH-Rubpy, $(Ru(bpy)^2(MeVbpy))^{2+}$, prepared and purified as indicated in ref.[19].

The variation of the luminescence lifetime of PAH-Rubpy with pH has been described by Kaneko *et al* [18,19]. The lifetime slowly decreases with increasing pH, and is correlated with the uncoiling of the polymer. As in the case of PMAH (see Fig 3) the curves exhibit a characteristic sigmoid form. It has been observed that the lifetime at low pH is slightly longer than in the model compound, and shorter at high pH. At high pH the effect has been attributed to the electric field of the polyanion which modifies the energy levels in the chromophore by a change on the metal to ligand distances; at low pH a protection by the coil is evoked, but the effect is far from those observed in the case of PMAH-DPA, where the hydrophobic effects are prevailing.

Luminescence quenching by MV^{2+}. Kaneko and some of us[18] have studied the quenching of PAH-Rubpy by MV^{2+} and found that in acidic medium the rate constant k_q is very low, only about 10^7 M^{-1} s^{-1}, which is much slower than the model system[23]. Although there is neither an evidence for a coil conformation in acidic medium nor the possibility of hydrophobic domains like those postulated in PMAH, there are some experimental facts[24-25] which could be an indication of the existence of intra- and inter-molecular hydrogen bonding between the coils. The chromophore quenching is very efficient (k_q = 1.9 10^{10} $M^{-1}s^{-1}$)in basic solution. The application of the Bjerrum method has shown that MV^{2+} is complexed in the proportion of one MV^{2+} for two acrylate ions; the proximity of the chromophore and the quencher insures a great efficiency in the quenching.

Luminescence quenching by SPV. The quenching of PAH-Rubpy by SPV was studied with and without added salt between pH 2 and 11. The results are shown in Figure 8.

The presence of added salt smoothes the pH effect, which is qualitatively analogous to that observed for other systems and can also be attributed to electrostatic and conformational effects.

As for MV^{2+}, we measured a complexation effect which can explain the high values of k_q. In basic medium, the rate constants in Fig.8 are close to the diffusion limit. The salt effect at low pH can be attributed to a

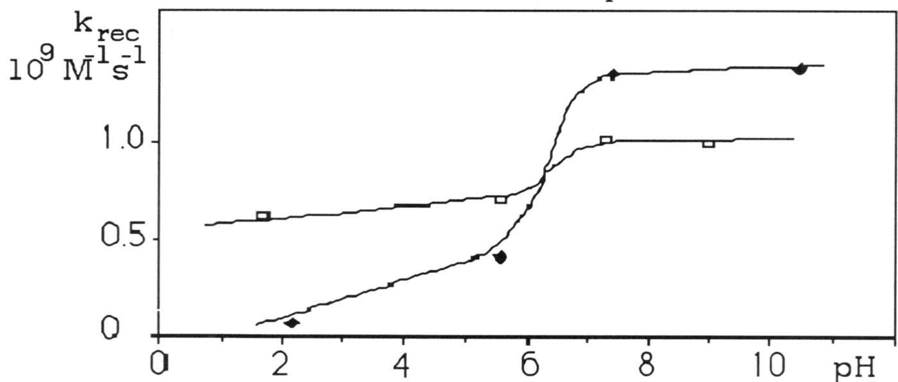

Fig 8- Quenching of PAH-Rubpy by SPV as a function of pH. (•) no salt added ; (°) with NaCl 0.1M.

competition between two opposite effects, one of them being predominant. The electrostatic shielding of the few polymer charges in solutions with added salt would lead to a reverse variation because the dipolar molecules of SPV would be less attracted by the chain which ejects the chromophore out of the coils and makes it accessible to SPV.

In the case of PAH-Rubpy, time resolved fluorescence measurements did not lead to any polarization, probably due to a randomization of the excitation energy on orbitally equivalent states but with different orientations[27-28] But if we extend the results of fluorescence depolarization obtained in the same conditions with PAH-DPA[22], which exhibits an increase in the residual polarization of DPA when a neutral salt is added, we can assume a decrease in the mobility of the fluorescent probe, as a consequence of the shrinking of the coil.

In presence of a complexation of the quencher with the polyanion, in basic solution, the question arises whether the quenching is static or dynamic, or both. In the Figure 9 we have represented the luminescence decay measured with a picosecond apparatus of aerated solutions of PAH-Rubpy with and without SPV in aerated solutions. As one can see, both emissions start from the same intensity; the quenching is purely dynamic.

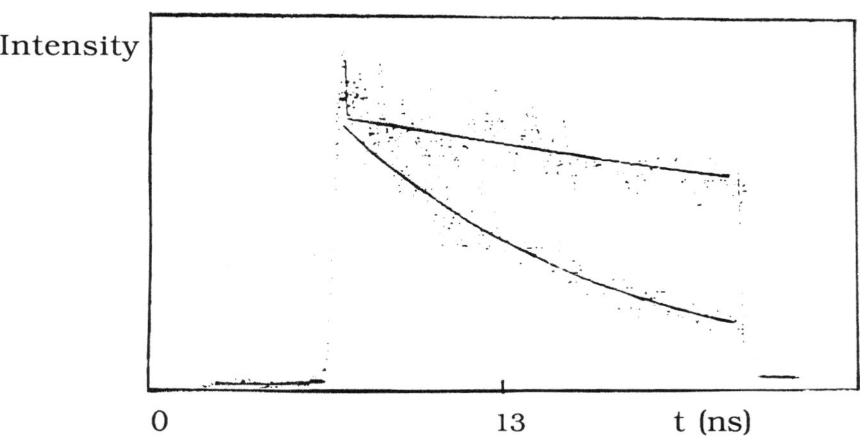

Figure 9- Luminescence decay of PAH-Rubpy (10^{-4} M chromophore unit) in aerated solutions, with and without quencher.

Charge escape quantum yields have been determined under low laser energy to avoid multiphotonic ionization of the chromophore. All laser studies have been performed in comparison with the model system $Ru(bpy)_3^{2+}/MV^{2+}$ in an acetate buffer and at ionic strength of 0.1M of NaCl, conditions of the maximum yield of charge separation[29] . The results are given in table 4.

Furue *et al*[30] recently reported on a similar high charge separation yield with a polysulfonate, in the presence of SPV as acceptor. They also reported the same enhancement with a symmetrical situation, for the same chromophore linked to a polycation, and MV^{2+} as quencher. However the yield has been found lower than for the model system at relatively low ionic strength($\Phi_{ce}/\Phi°_{ce}$ = 0.65), and this lower value has been attributed to hydrophobic binding of $MV^{\cdot+}$ to the polymer. In our case, the electrostatic repulsion of the polyelectrolyte causes an enhancement effect, and the hydrophobic interaction of $SPV^{\cdot -}$ and polyacrylate ions are probably negligible.

Table 4.Relative cage escape quantum yield $\Phi_{ce}/\Phi°_{ce}$ and back recombination rate constants k_{rec} for $Ru(bpy)_3^{2+}$ and for PAH-Rubpy, in presence of MV^{2+} and SPV as quenchers.

System	pH	added NaCl	relative Φ_{ce}	k_{rec} 10^9 M^{-1} s^{-1}
$Ru(bpy)_3^{2+}/MV^{2+}$ (model system)	5.5	0.1	1.0	4.0
$Ru(bpy)_3^{2+}$/SPV	6.2	0	0.4	14.0
$Ru(bpy)_3^{2+}$/SPV	8.2	0	0.5	4.0
PAH-Rubpy/MV^{2+}	2.0	0	0.95	2.0
PAH-Rubpy/MV^{2+}	8.0	0	< 0.02	___
PAH-Rubpy)/SPV	2.1	0	0.39	2.06
PAH-Rubpy)/SPV	10.2	0	1.21	0.11

Concerning water photolysis, since the basic pH gives the highest yield for charge escape, considerations on redox potentials are not favourable for water reduction : $E°(H_2O/H_2)$ = - 0.36 V at pH6 and $E°(SPV/SPV^{\cdot -})$ =-0.39 V. An attempt to produce H_2 with the sacrificial sys-tem PAH-Rubpy/SPV/colloidal Pt/EDTA at PH 6 led to an initial rate of H_2 production comparable with the same system in which PAH-Rubpy is replaced by the parent complex, but the evolution stopped after a short time.

B- Luminescence quenching of Zinc Tetraphenyl porphyrin and phthalocyanin in Polystyrene-Poly(2-vinylpyridine) block copolymers.

When they are prepared under adequate conditions, films of block copolymers such as polystyrene-polyvinyl-2-pyridin (SV_2P) are microstructured [31] . Depending on the solvents employed, one can obtain either lamellar multilayers of the hydrophobic moiety polystyrene (PS) alternate with poly(2-vinylpyridine) (P2VP), or PS domains imbedded in the hydrophilic polymeric medium constituted by P2VP. Using NMR spectroscopy, we determined that less than 1% of pyridine aresubstituted in 4 instead of 2 position. It was possible to dissolve preferentially Zinc-tetraphenyl porphyrin (ZnTPP) or Zn-phtalocyanin (ZnPc) in the P2VP phase, taking advantage of the presence of this small percentage of P4VP which allows the complexation of these molecules on the axial position. On the other hand, if ZnTPP is covalently linked to a PS chain, the ZnTPP is preferentially dissolved in the polystyrene phase. We have revealed these situations by electron micrography. In figure 10 we have reported two typical situations: the first case on the left, the second on the right. Both localisations reveal the existence of microdomains and the localisation of the dye is given by the black spots, typical of the interaction of the electron beam with the Zinc atom.

Crouch *et al*[32-33] have studied ZnTPP in polymer films as coatings on

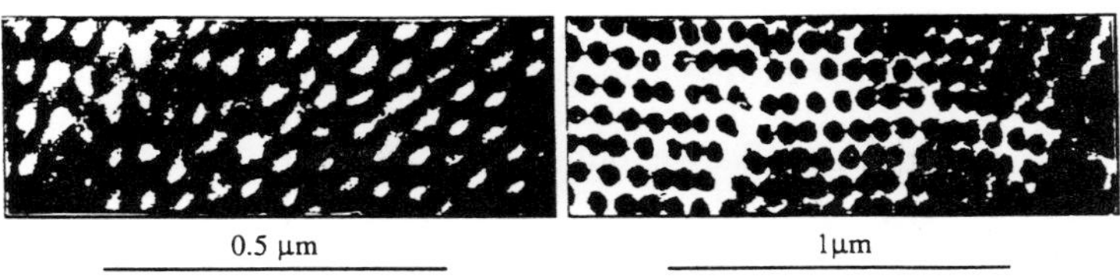

Figure 10. Electron micrograph of Zn-phtalocyanin in the P2VP phase of the block co-polymer (left), and Zn-TPP in the polystyrene domains (right). The dye are inserted in the dark domains.

photoelectrodes. In a photophysical study of this dye dispersed in films of poly(4-vinylpyridine) including a cationic copolymer providing a combination of hydrophobic (polystyrene) and hydrophilic (polystyrene p-substituted by a quaternary ammonium salt) moieties[34], they claim that under excitation there is a charge-transfer between ZnTPP and pyridine. By using picosecond spectroscopy, they observed in a solution of pyridine a transient absorption in the red that has been assigned to the charge transfer product $ZnTPP^{+}$-Py^{-}.

We tried to perform such charge separation in PS-P2VP films. We did not observe any transient such as Crouch *et al* reported for ZnTPP in pyridine. Nevertheless there is a strong effect of the concentration on lifetime, typical of a trapping effect. We have studied the fluorescence lifetime of ZnTPP dissolved in the film either in the polystyrene phase or in the poly(4-vinylpyridine) phase as a function of concentration of ZnTPP. The results are represented in Figure 11.

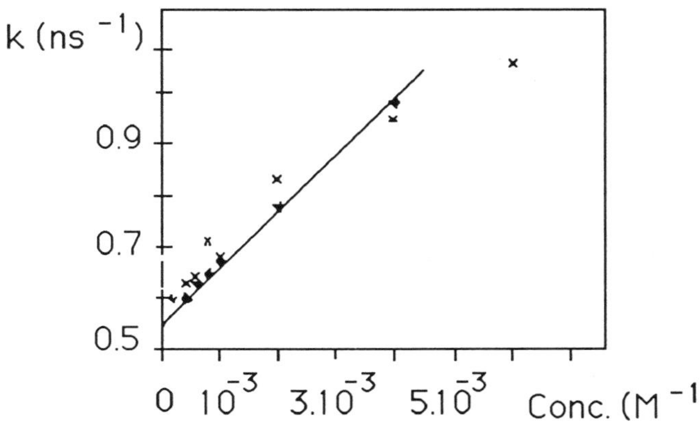

Figure 11. Quenching rate constant of the fluorescence of ZnTPP as a function of the concentration of ZnTPP in films : (bold points: dye in P2VP phase; (+) : dye in the polystyrene phase).

We have evaluated the critical resonant energy transfer distance for ZnTPP in our matrix from the emission spectrum and the fluorescence quantum yield, using the Förster equation.[35]. It has a typical value of

32 nm. This corresponds to a critical transfer concentration of 2 10^{-3}M for the dye in the matrix. Above the critical concentration exciton diffusion occurs. We studied the absorption and the emission spectra as well as the decay kinetics of the fluorescence as a function of ZnTPP concentration from 6 10^{-3} M to 4 10^{-4} M.

The absorption spectra of ZnTPP in polymers do not show the presence of aggregation in the samples. The local concentration of ZnTPP in the polymer is much higher than the saturation concentration of ZnTPP in usual solvents. Nevertheless, most of the ZnTPP molecules are thus solubilized as monomers. Optical densities as high as 0.025 can be obtained at 530nm on 1.2μm transparent films.

In the P2VP phase, the fluorescence decay of the ZnTPP molecule is essentially exponential with a deviation towards multi-exponentiality at high concentration. A linear dependence of the decay rate on the dye concentration is observed. This can be rationalised using a model of exciton diffusion toward traps. The exciton migration in disordered matrices is known to be non-diffusive.[36] Nevertheless, as we use high concentrations, the diffusion formula gives an acceptable approximation of the decay.[37] The decay of a population of excited molecules $n_D(t)$ where resonant exciton transfer and trapping takes place is given by the equation :

$$\rho(t)=\exp(-t/\tau_D - 35.5\, g^{1/2} R_{DD}^3 R_{DA}^{3/2} n_D n_A t - g^{4/3}\pi^{3/2} n_A R_{DA}^3 (t/\tau_D)^{1/2})$$

where τ_D is the excited donor lifetime in the absence of acceptors, n_A is the trap number density, g is a reorientation factor, R_{DD} and R_{DA} are the dye-dye and dye-trap Förster distances. We can neglect the third term in the bracket. If the traps are assumed to be linked to the matrix, n_A is constant and a linear variation is expected. If the traps are impurities mixed with the dye, n_A may be proportional to n_D and a quadratic dependence is expected. The dependence becomes cubic or more if we assume that traps are proper to aggregates. The slope of Figure 11 enables us to define an equivalent trapping volume of the P2VP phase of the polymer matrix : $R_{DD}^{9/2} R_{AD}^{3/2} n_A$.

As indicated in Figure 11, in the case of the dye inserted in P2-VP we are almost in the case of traps linked to the matrix since we obtain a rather straight line. It seems that in the PS phase a non-linear dependence occurs. The difference could be explained by the complexation of the ZnTPP molecule with the P4VP moiety that does not occur in PS. We measured a mean value of 10^{12} molecule.$m^{-3/2}$.

This enables us to evaluate the concentration center to be used in order to get a significant fraction of collected energy reaching the reactive site. As an exemple if a reaction center has a Förster distance R_{DR} of 39 Å (H_2TPP linked to a quinone) the critical concentration at which 50% of the energy flows to the reaction center as a value of $n_R = n_A (R_{DA}/R_{DR})^{3/2} = 6\ 10^{-3}$ M.

The PV2P phase can be swollen by water; there is a strong pH dependance of the amount of absorbed water, with a critical value around pH 3.5. In the more acidic region, the charge separation postulated by

Crouch *et al* may be favored. The excitation of ZnTPP dissolved in the PS phase and the energy transfer to ZnPc selectively dissolved in the P2VP phase has been observed. An interesting feature is the constant lifetime observed for ZnPc in P2VP phase when the film is swollen by addition of acidic water. This is in good agreement with the predominant migration energy transfer along the chain, the change in volume obtained in this case decreasing the bulk concentrations of ZnPC and ZnTPP but of course leaving unchanged the situation of the dyes linked to the chain.

4. CONCLUSION AND OUTLOOK.

Polyacids can be used to improve photoinduced charge separation, depending on several structural factors of the polymers, the charge of the acceptor... The yield of charge escape can be relatively high and is strongly influenced by the pH domain, according to the coiling effect under the ionization of the acidic groups. An additional hydrophobic effect can be obtained with methyl groups present in the chain. Hydrophobic effects seem generally more influent than electrostatic ones. It has been possible to find conditions under which the charge escape in an acrylic acid polymer with pendant $Ru(bpy)_3^{2+}$-like complex is higher than that of free $Ru(bpy)_3^{2+}$ in solution.

Copolymer films of Polystyrene and Poly(Vinyl-2-Pyridine) present microstructurations that are suitable to realize segregation of donors and acceptors as in a micellar system, but with the advantage of the properties of a polymeric material. The energy migration along the chain is a favorable factor but the concentration of the acceptor may be higher than that of an intrinsic trap that has been detected in both moieties of the microstructured film. The PV2P phase giving the possibility of dissolution of a large amount of acidic water, PS-PV2P block copolymers films are promising media for applying the results of the study of charge separation in hydrosoluble polymers.

5. REFERENCES

1 A. Weller, Pure and Appl. Chem. No.16 (1968) 115.

2 D. Rehm and A. Weller, Isr. J. Chem. No 8 (1970) 259.

3 N. Mataga, Pure Appl. Chem. No 56 (1984) 1255.

4 T.A.Moore, D. Giest, P. Mathis, J.C. Mialocq, C. Chachaty, R.V. Bensasson, E.J. Land, D. Doizi, P.A. Liddel, G.A. Lehman, A.L. Moore, Nature (London) No 307 (1984) 630.

5 J. Faure and J.A. Delaire, "Photoinduced electron transfer", part.B Fox and Chanon (eds), Elsevier, (1988) 1.

6 Y. Morishima, T. Furui, S. Nozakura, T. Okada, N. Mataga, J. Phys. Chem. No 93 (1989) 1643.

7 J.A. Delaire, M.A. Rodgers, S.E. Webber, Eur. Polym. J. No 22 (1986) 189.

8 S. E. Webber, Macromolecules, No 19 (1986) 1658.

9 J.A. Delaire, M. Sanquer-Barrié, S.E. Webber,J. Phys. Chem. No 92 (1988) 1252.

10 J.A. Delaire, M.A. Rodgers, S.E. Webber, J. Phys. Chem. No 88 (1984) 6219.
11 D.Y. Chu, J.K.Thomas, Macromolecules, No17 (1984) 2142.
12 M. Sanquer-Barrié, J.A. Delaire, to be published.
13 K.L.Tan, F.E. Treloav, Chem. Phys. Letters No 170 (1990) 509.
14 M. Drifford, J.P. Dalbiez, Biopolymers No 24 (1985) 1501.
15 R.E. Sassoon, Z. Aisenshtat, J. Rabani, J. Phys. Chem. No89 (1985), 1182.
16 Y. Itoh, S.I. Nozakura, Photochem. Photobiol. No 39 (1984) 451.
17 M. Anbar, M. Bambenek, A.B. Ross, "Selected Specific Rates of Reactions of Transients from Water in Aqueous Solution. I. Hydrated electron NSRDS, N.B.S., US Department of Commerce : Washington DC, (1973).
18 M. Sanquer-Barrié, Sc. D. Thesis (1989). To be published.
19 M. Kaneko, S. Nemoto, A. Yamada, Y. Kurimura, Inorg. Chim. Acta, No 44 L (1980) 289.
20 M. Kaneko, A. Yamada, E. Tsuchida, Y. Kurimura, J. Polym. Sci., Polym. Lett. Ed. (1982).
21 M. Kaneko, A. Yamada, E. Tsuchida, Y. Kurimura, J. Phys. Chem., No 88 (1984) 1061.
22 X.H. Hou, M. Kaneko, A. Yamada, J. Polymer Sc. Part A; Polym. Chem Ed., No 24 (1986) 4406.
23 M. Kaneko, A. Yamada, Adv. Polym. Sci., No 55 (1984) 1.
24 M. Sanquer-Barrié, J.A. Delaire, M. Kaneko, New J. Chem. No 15 (1991) 65.
25 B. Valeur, C. Noël, P. Monjol, L. Monnerie, J. Chim. Phys. No 68 (1971) 97.
26 A.R. Mathieson, J.V. MacLaren, J. Polym. Sci. A3 (1965) 2555.
27 M.L. Myrick, R.L. Blakely, M.K. DeArmond, J. Am. Chem. Soc., No 109 (1987) 2841.
28 E. Krausz, Inorganic. Chem., No27 (1988) 2392.
29 E. Amouyal, B. Zidler, Isr. J. Chem., No 22 (1982) 117.
30 M. Furue, K. Sumi, H. Tomita Y. Maeda, S. Nozakura, Books of Abstracts, the Sixth International Conference on Photochemical Conversion And Storage of Solar Energy, Paris, E5 (1986).
31 C. Sadron, Journal de Chimie-Physique, No 72 (4) (1975) 539.
32 A.M. Crouch, C.H. Langford, J. Electroanal. Chem., No 221 (1987) 83.
33 A.M. Crouch ; C.H. Langford, J. Photochem Photobiol.A : Chemistry, No 52 (1990) 55.
34 D.D. Montgomery ; K. Shigehara ; E. Tshuchida ; F. C. Anson, J. Am . Chem. Soc., 1984, 106, 7991.
35 I.B. Belman, in Energy Transfer Parameter of aromatic compounds Academic Press London 1973.
36 C.R.Gochanour H.C. Andersen M.D. Fayer, J.Chem.Phys. No75 (1981) 3649.
37 Fleming G.R. in Chemical Application of Ultrafast Spectroscopy Clarendon Press Oxford. (1986)

DISCUSSION

Steiner

The changes of Φ_{ce} in the polymers with Ru as a function of pH were quite remarkable. Did you see any indication of a fast geminate recombination, particularly at high pH where Φ_{cc} is low?

Faure

As mentioned at the beginning of my talk, the time resolution of our picosecond set up is about 60 ps. Prof. Mataga's work in collaboration with Prof. Morishima's have shown that the geminate recombination occurs in a shorter time in polymer systems.

Harriman

Did you confirm the observation of Crouch and Langford relating to photoionization of ZnTPP pyridine? Do you accept this observation? To me, it seems incredulous!

Faure

We tried to observe a signal in the red range as Crouch and Langford did, and in various conditions. We have never seen any transient.

Yamamoto

As for the microstructure of the block copolymer films, the structure of interface may determine all aspects of the phenomena. Do you know anything about this?

Faure

The energy transfer that I mentioned may occur at the interface, since the dimension of the microdomains are about ten times that of the energy transfer. It should be very interesting to investigate the interface but till now we did not.

Mobius

Did you see an increase in charge separation in the Ru^{++} (bpy)/MV^{2+} system with increasing ionic strength? This seems to increase to me.

Faure

We did not study free $Ru(bpy)_3^{2+}$ by itself. The ionic strength effect that we observed is attributed to the interaction with the polyanion, according to Manning's theory.

Dynamics and Mechanisms of
Photoinduced Transfer and Related Phenomena
N. Mataga, T. Okada and H. Masuhara (Editors)

Photoionization of excitons in aromatic hydrocarbon crystals

Masahiro Kotani and Ryuzi Katoh

Faculty of Science, Gakushuin University
Mejiro, Tokyo 171, Japan

Abstract

Generation of electrons and holes in organic crystals via higher excited states has been studied. Cross sections for photoionization and for photoabsorption of excitons have been determined as a function of the excitation energy. It has been found that higher excited states relax rapidly to a lower-lying state which autoionizes.

1. PHOTOCONDUCTIVITY

Electrical insulators can be classified in two categories. One is a group of materials in which electrons are inherently not very mobile and may be termed as "mobility insulators". Most polymers, due to their disordered structures, belong to this class of materials. The other is a group which may be termed as "band-gap insulators" which, having large band-gaps, lack charge carriers in thermal equilibrium at ambient temperatures. When, by some means, charge carriers are supplied, materials belonging to this group can exhibit a variety of electronic transport processes. Aromatic hydrocarbon crystals belong to this group, together with rare gas solids (also liquids) and some ionic crystals. A distinct feature of aromatic hydrocarbon crystals is the coexistence of charged states, such as electrons and holes, and of excitons which are electrically neutral.

Transient absorption, luminescence such as fluorescence,

phosphorescence and delayed fluorescence, modulations of them by electric and magnetic fields, action spectrum of photoconductivity and quantum yields of charge carriers can all be used as experimental probes for studying the relaxation of photoexcited states in this class of crystals.

2. PHOTOIONIZATION OF EXCITONS

With the aim of clarifying the initial step of charge carrier generation in aromatic hydrocarbon crystals, we have been developing a method for measuring the efficiency of the singlet exciton photoionization [1]. Excitons are generated by weakly absorbed light homogeneously in the crystal and photoionized by light for which the crystal is transparent (Fig.1). The ionization can be measured quantitatively through the photocurrent with Time-of-Flight technique. Ionization "spectrum" of the exciton can be obtained when the wavelength of the ionizing light is varied. Optical absorption due to exciton can also be measured. The cross

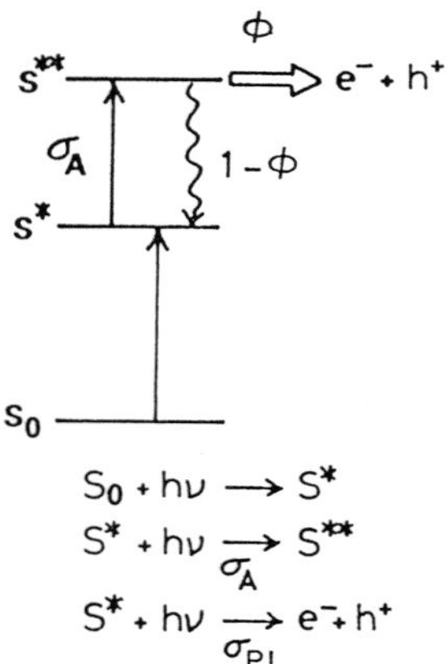

Fig. 1 Photoionization of singlet exciton S*. Of the higher excited states S** a fraction Φ ionizes into electrons and holes.

section for the optical absorption is the total cross section for the transition whereas the cross section for the photoionization is a partial cross section for the process ending up with an electron and a hole. The ratio of the two cross sections gives the efficiency with which a higher excited states, generated with a two-step excitation, ionizes. This procedure eliminates the difficulty encountered with most compounds that the penetration depth of light changes as the photon energy is varied. It is well known that a high density excitation near the crystal surface gives rise to many other processes, such as exciton-exciton collisional ionization and exciton dissociation at surfaces. Figure 2 shows photocurrent pulses observed with a high purity anthracene crystal. The lower trace is observed when the crystal is excited with light which generates excitons. The current in this case is proportional to the square of the exciting intensity when the density of charge carriers is kept low, so that volume recombination of electrons and holes is not important. Obviously the photon energy is not sufficient

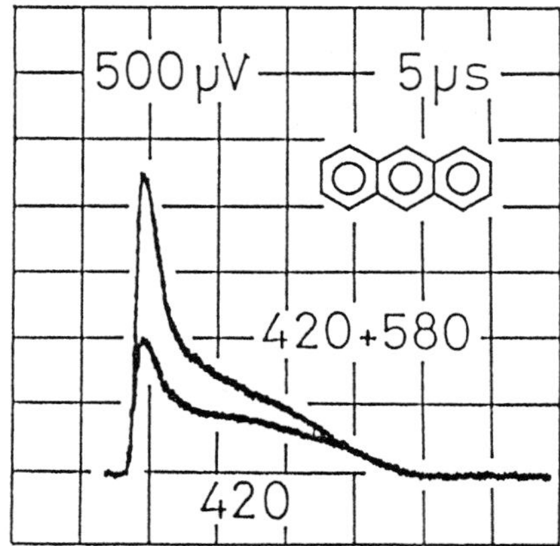

Fig. 2 Photocurrent pulses observed on an oscilloscope. The increase in the photocurrent by the additional light of 580 nm corresponds to the photoionization of excitons [R.Katoh and M.Kotani, Mol. Cryst. Liq. Cryst. 183 (1990) 447].

for the internal ionization and two quanta of photons are necessary for generating an electron and a hole. The pulse shape reflects the spatial distribution of the generated charge and is not caused by any temporal history of the carrier genration. The area under the curve corresponds to the amount of collected charge (The charge carriers may be

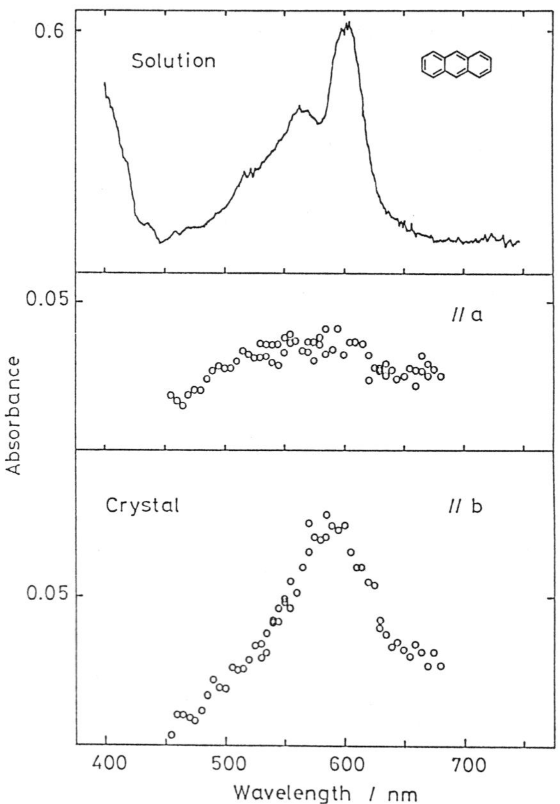

Fig. 3 Transient optical absorption of anthracene. Top: Cyclohexane solution [H.Miyasaka, PhD Thesis, Osaka University]. Middle: Crystal spectrum (// a). Bottom: Crystal spectrum (// b).

generated by the photoioniztion of excitons, but other possibilities such as exciton-exciton collisional ionization can not be excluded from the observation of the photocurrent alone). The upper trace is obtained when another pulse of light, in this case at 580 nm, irradiates the crystal simultaneously. This light is not absorbed by the crystal in the ground state. The increase in the photocurrent is proportional to the intensity of the second-step pulse. This indicates that carriers are generated by photoionization of excitons.

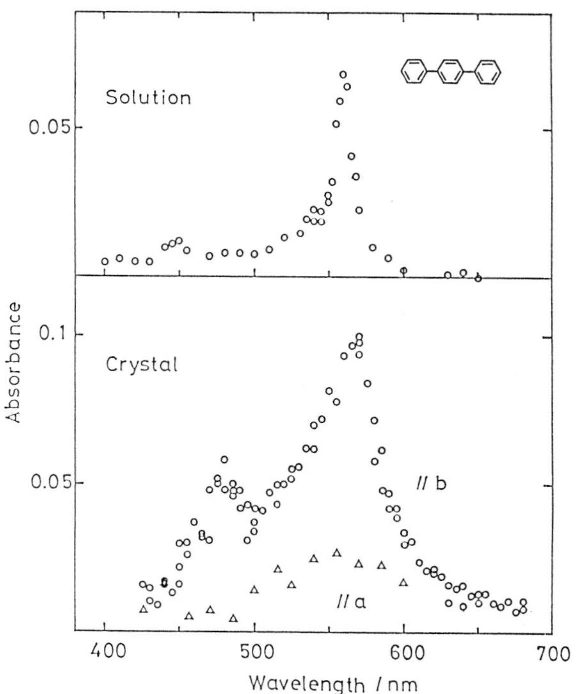

Fig. 4 Transient optical absrption of p-terphenyl. Top: Cyclohexane solution. Bottom: Crystal spectrum. The peak around 480 nm (450 nm in solution) is a triplet-triplet absorption.

3.OPTICAL ABSORPTION OF EXCITONS

Transient absorption spectrum due to excitons has been measured with direct pump-and-probe method. Measurements of transient absorption is generally considered to be difficult with crystals, because of the possible damage by intense laser pulses and also due to the optical quality of organic crystals which is often not sufficient for transmission measurements. We have found that the optical damage can be avoided by selecting a proper excitation wavelength, so that the absorption is sufficiently weak and the energy is deposited over a relatively large penetration depth of light.

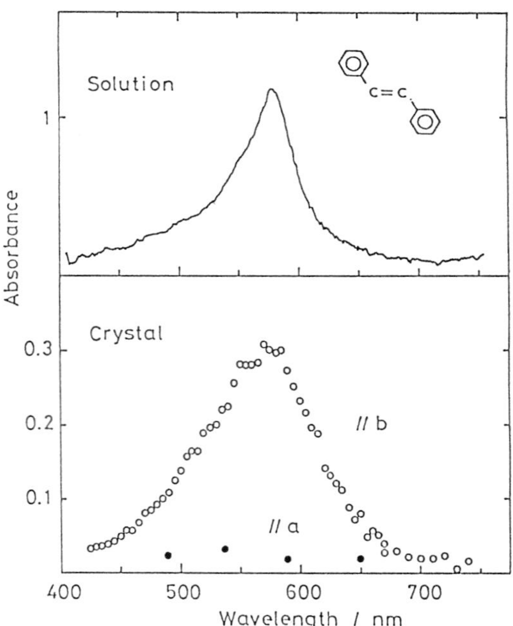

Fig. 5 Transient optical absorption of t-stilbene. Top: Cyclohexane solution [H.Miyasaka, PhD Thesis, Osaka University]. Bottom: Crystal spectrum.

Transient spectrum could be measured with p-terphenyl, anthracene and t-stilbene, using thick (ca. 1 mm) crystals. The spectra are similar to those of solutions, indicating that the final states of the two-step excitation are essentially molecular excited states and not the ionized states (Fig. 3-5).

4.EFFICIENCY OF THE PHOTOIONIZATION OF EXCITONS

In all three compounds studied the photoionization efficiency has been found to be constant over a considerable

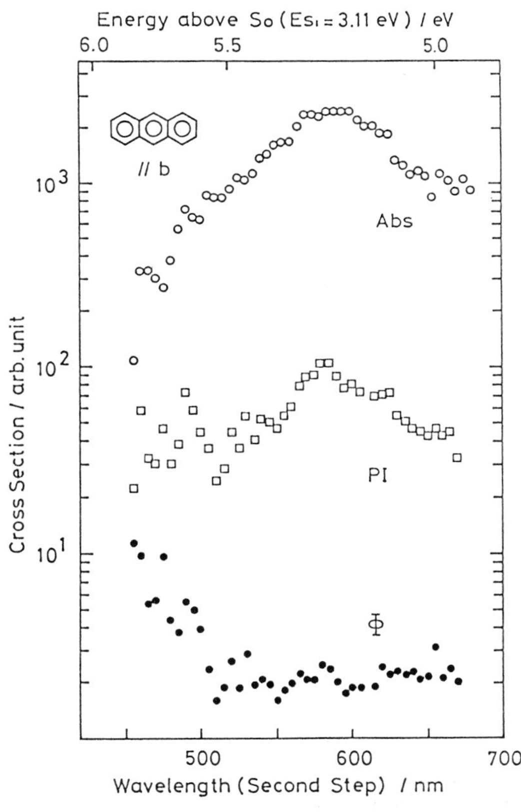

Fig. 6 Absorption (Abs), photoionization (PI), and ionization efficiency (Φ) of singlet exciton in an anthracene crystal.

energy range, as may be seen in Figs. 6-8 [2-4].

This indicates that the excess energy over the (internal) ionization threshold is not converted to the kinetic energy of the photogenerated electrons, but is dissipated as other energies, such as intramolecular vibrations. Charge separation seems to occur from a lower-lying excited state without generating an electron with a kinetic energy. This may be considered as autoionization of a molecule embedded in the crystalline environment, or electron transfer to a

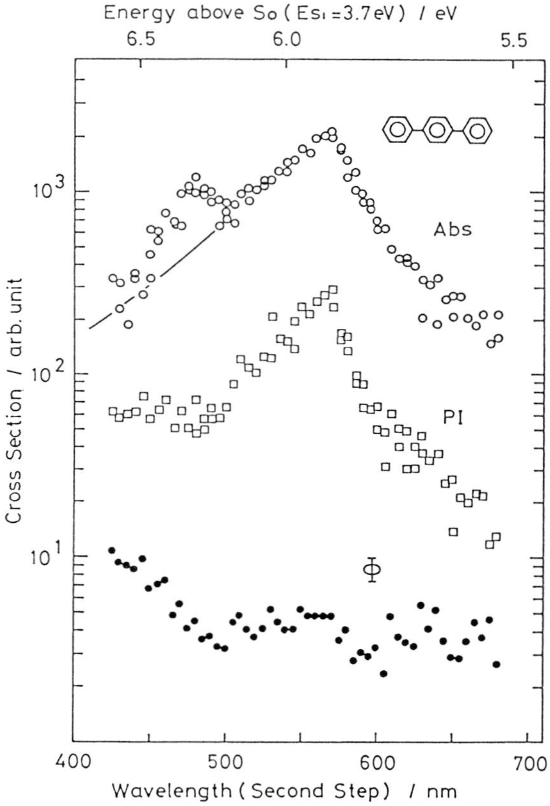

Fig. 7 Absorption (Abs.), photoionization (PI) and ionization efficiency (Φ) of singlet exciton in a p-terphenyl crystal.

neighbouring molecule (charge transfer state in a crystal). This absence of energetic electrons is a distinct feature of aromatic crystals, as compared to most polymers and liquids, where an increase in the excitation energy is very often accompanied by a corresponding increase in the "initial separation" in Onsager terms [5]. With anthracene and p-terphenyl a steep increase is observed in the carrier

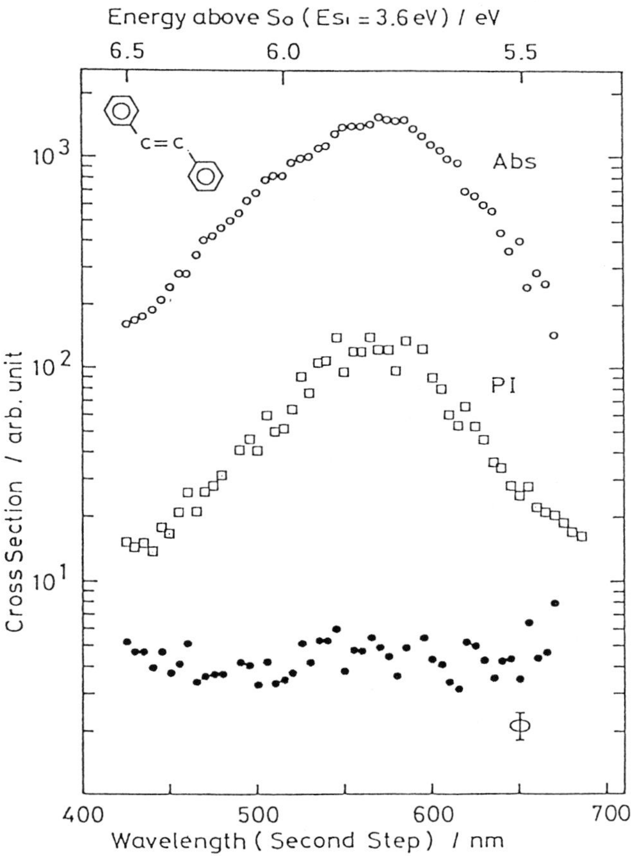

Fig. 8 The optical absorption (Abs), photoionization (PI), and ionization efficiency (Φ) of singlet exciton in a t-stilbene crystal.

generation efficiency above a certain energy (Figs. 6,7). This may indicate the onset of the generation of energetic electrons.

The question whether the ionization of an excited state proceeds by ejection of an electron with a certain kinetic energy, as can be observed in external photoemission of electrons into vacuum, or by electron transfer to a neighbouring molecule (generation of a charge transfer exciton), the central theme of this symposium, is difficult to answer, but our results seem to indicate that the former sets in above a certain energy, while the latter is important with low energy excitations.

Excitonic generation of charge carriers is not limited to photoionization of excitons. In some cases exciton-exciton annihilative ionization can be observed [6].

ACKNOWLEDGEMENTS

The authors are grateful to Dr. H. Miyasaka for supplying them with solution spectra of the compounds studied.

REFERENCES

1. M.Kotani, E.Morikawa, and R.Katoh, Mol. Cryst. Liq. Cryst. 171 (1989) 305 and references cited therein.
2. R.Katoh and M.Kotani, Chem. Phys. Letters 174 (1990) 541.
3. E.Morikawa, Y.Isono, and M.Kotani, J. Chem. Phys. 78 (1983) 2691. Results presented in the present work are greatly improved over those published earlier.
4. R.Katoh and M.Kotani, J. Chem. Phys. 94 (1991) 5954.
5. L.Onsager, Phys. Rev. 54 (1938) 554.
6. A.Kurabayashi and M.Kotani, Mol. Cryst. Liq. Cryst. 183 (1990) 193.

Dynamics and Mechanisms of
Photoinduced Transfer and Related Phenomena
N. Mataga, T. Okada and H. Masuhara (Editors)

Dynamics of geminate ion pairs in liquid alkanes studied by LL-twin picosecond pulse radiolysis

Y. Yoshida[a] and S. Tagawa[b]

[a]Nuclear Engineering Research Laboratory, Faculty of Engineering, University of Tokyo, 22-2 Shirane Shirakata, Naka-gun, Ibaraki, 319-11 Japan

[b]Research Center for Nuclear Science and Technology, University of Tokyo, 22-2 Shirane Shirakata, Naka-gun, Ibaraki, 319-11 Japan

Abstract

The decay of electrons and alkane cation radicals in radiolysis of neat liquid alkanes is observed in picosecond time range by LL twin picosecond pulse radiolysis. Lifetimes of electrons and alkane cation radicals are different because of the proton transfer reactions of alkane cation radical, especially for cycloalkanes and lower n-alkanes. The initial spatial distributions and distances of electrons and parent cation radicals are estimated by analysis of the decays of electrons in neat liquid alkanes and the decay of solute ions in solute-solvent systems.

1. INTRODUCTION

The primary processes of the radiation chemistry in liquid alkanes have been paid much attention for 30 years, especially for cyclohexane. However, the primary processes of neat liquid alkanes had not been made clear for a long time, because of the following reasons.

(1) The lifetimes of geminate pairs of electrons and alkane cation radicals are very short because of the high mobility of electrons.

(2) The identification of alkane cation radicals and excited states is very difficult because of overlapping of the absorption spectra.

(3) The decay kinetics are very complex because of several different decay processes of alkane cation radicals, such as recombination, protonation and decomposition in addition to the overlapping decay and formation of alkane excited states.

Especially the identification of very broad transient spectra observed in picosecond pulse radiolysis[1~4] and multiphoton laser photolysis[5] of neat liquid alkanes was difficult. In neat liquid cyclohexane, there are several different assignments of the broad absorptions as follows.

1. Mainly excited states[1,3,5]
2. Mainly cation radicals[2]
3. Partly excited states[4]

In the geminate ion recombination processes,[6~9] a pair of a positive parent cation radical and an electron produced by the ionization of high energy radiation, diffuse in the Coulomb potential. Most of the pairs recombine geminately in liquid alkanes due to the long range of the Coulomb potential. The ionization and the following geminate ion recombination are very important as the first step of the primary processes of the radiation chemistry.

Although the decomposition processes of alkanes had been discussed on the base of the product analysis, recent picosecond pulse radiolysis studies[5] showed the very rapid formation of the alkyl radicals which play an important role in the decomposition mechanism. This showed that hydrogen atoms are detached from the highly excited alkanes and the excited cation radicals. Thus, the primary decomposition processes are closely related to the geminate recombination process.

It is necessary for the elucidation of the primary processes to study the direct observation of short-lived intermediates, such as radical cations, electrons, and excited states by picosecond pulse radiolysis. The Twin-Linac pulse radiolysis system[10] which was composed of two electron linear accelerators was developed in 1985. One linac was used for irradiation sources, and the other was used for the production of the Cherenkov analyzing light. The system has been advanced by using picosecond diode pulse lasers to measure the optical absorption in the infrared region. The new system, "LL-twin" system is a very powerful method to investigate the primary intermediates in the visible, near infrared, and infrared regions.

The present paper describes the first measurement of the decay of electrons in neat liquid alkanes and the composition of the decay of electrons with that of cation radicals.

2. NEW PICOSECOND PULSE RADIOYSIS SYSTEM

Fig. 1 shows the block diagram of the LL-twin system. A sample is irradiated by 10 ps electron pulses from 28 MeV S-band linac.[10] The picosecond photodiode laser is synchronized with the electron pulse. The time-profiles of the optical absorption can be obtained by changing the delay inserted to the laser trigger line. The available wavelengths of the lasers are presently 660, 720, 750, 806, 850, 1300, and 1500 nm. The laser pulse widths are within several tens picoseconds and the peak power is less than 100 mW. The light is detected by Si or Ge photodiodes. The current signal from the detectors are digitized by A/D and treated by computer. The time-resolution of the system is about 50 ps at each wavelength.

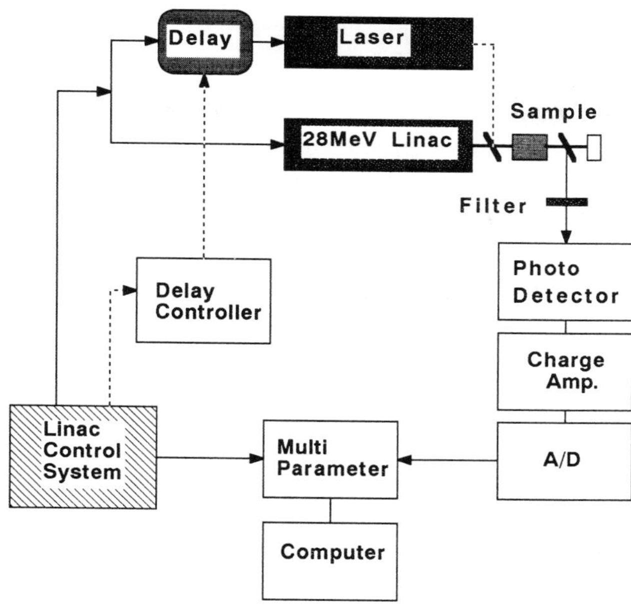

Fig. 1 Block diagram of the LL-Twin picosecond pulse radiolysis system.

3. SOLUTE-SOLVENT SYSTEM

The geminate ion recombination can be described by the diffusion theory (Smoluchowski equation[11]). The rapidness of the recombination depends on the sum of the diffusion coefficients of the cation radicals and the electrons. The observation of the geminate processes was usually done on the solute geminate pairs in solute-solvent system, because the diffusion coefficients of the solute geminate pairs are smaller than those of the electron-radical cation pairs.[6,12]

The processes of the geminate ion recombination in solute-solvent system are analyzed by the modified Smoluchowski eq. considering the charge transfer reactions from the original geminate pairs (alkane cation radicals and electrons) to solute geminate pairs.[13] Fig. 2 shows the theoretical calculation on the time-dependent behavior of geminate pairs in 10 mM solution of biphenyl in cyclohexane.[8]

Table 1 shows the initial spatial distributions[7] and characteristic lifetimes (the time when 43% of geminate pairs are survived) in various neat alkanes obtained by the analysis of the geminate decays of solute ions.

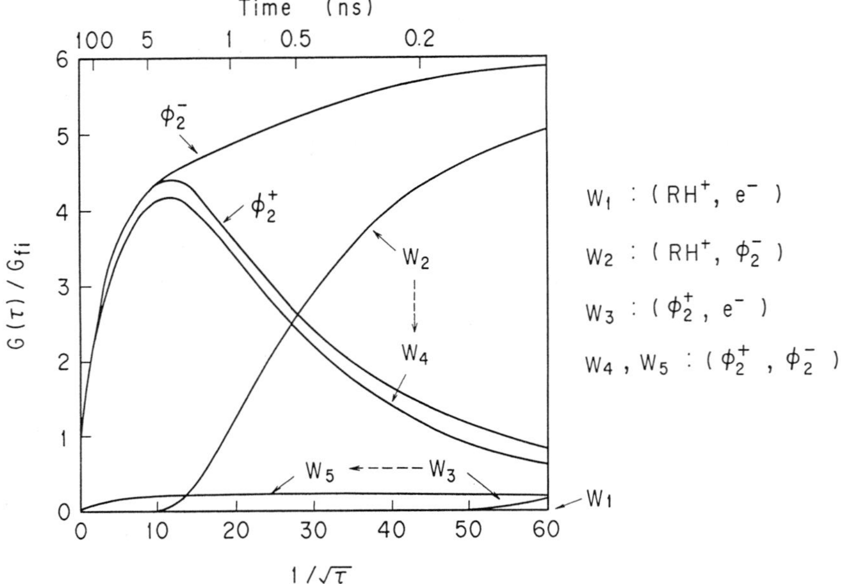

Fig 2. The simulation of the dynamics of the geminate ion recombination in 10 mM solution of biphenyl in cyclohexane, based on the modified Smoluchowski equation. RH^+, $\phi 2^+$, and $\phi 2^-$ represent solvent radical cations, biphenyl cation, and biphenyl anion, respectively. W_1~W_5 represent each geminate pair, as shown in the figure. $G(\tau)$ is the G-values of each biphenyl ionic species or geminate pair at the generalized time, $\tau=Dt/r_c^2$. D is the diffusion coefficient of the solute geminate ion pairs, and r_c is the Onsager length. G_{fi} is the G-value of the free ions. Pair W_1 is converted to W_4 through W_2 or to W_3 through W_5 under the geminate recombination process.

Table I Mobilities of electrons, μ, characteristic lifetimes, τ_g, initial spatial distributions and separations, r_0.

liquids	μ (cm^2/Vs)	τ_g (ps)	distributions	r_0 (nm)
trans-decalin	0.013	35	exponential	6.1
methylcyclohexane	0.044	16	exponential	6.2
n-hexane	0.071	9.4	exponential	6.0
cis-decalin	0.10	5.6	exponential	5.6
cyclohexane	0.23	3.2	exponential	6.1
isooctane	5.3	0.43	exponential	9.3
neopentane	50	0.63	gaussian	13.2
TMS	100	0.31	gaussian	13.2

4. ABSORPTION SPECTRA IN ALKANES

The absorption spectra in alkanes had not been elucidated, because the the intermediates were very short-lived and the spectra are broad. Recent nanosecond pulsed radiolysis studies showed the assignments of the intermediates in n-alkanes.[3] Fig.3 showed the absorption peaks of the cation radicals and excited states of n-dodecane. Both bands overlap in small number of carbon atoms and separate with increasing of the carbon numbers.[3]

The transient absorption spectra due to the electrons in the infrared region were measured by nanosecond pulse radiolysis. Fig. 4 shows the absorption spectrum obtained in neat n-dodecane. Major four bands were observed. The bands at 650 and 850 nm are assigned to the excited states and cation radicals, respectively, by the scavenging experiments and the comparison of the lifetime with the fluorescence lifetimes.[3,14] The band observed above 1000 nm disappeared by electron scavengers, and was assigned to the electrons. The bands around 240 nm is due to alkyl radicals.

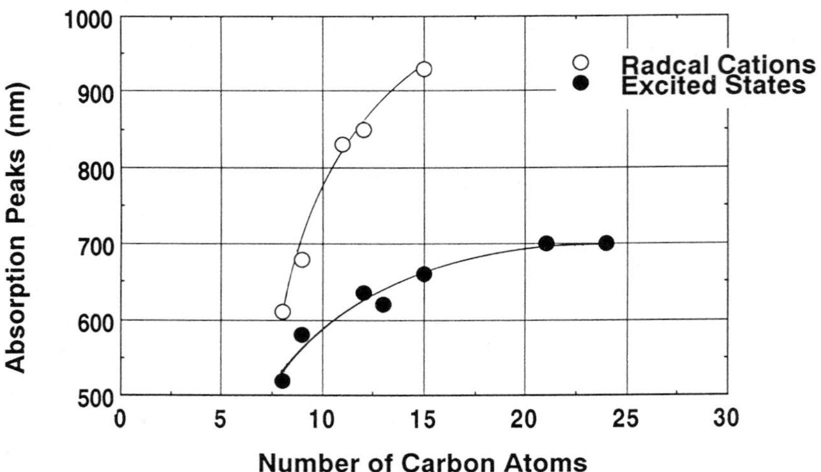

Fig.3 Absorption peaks of the cation radicals and excited states of n-alkanes.

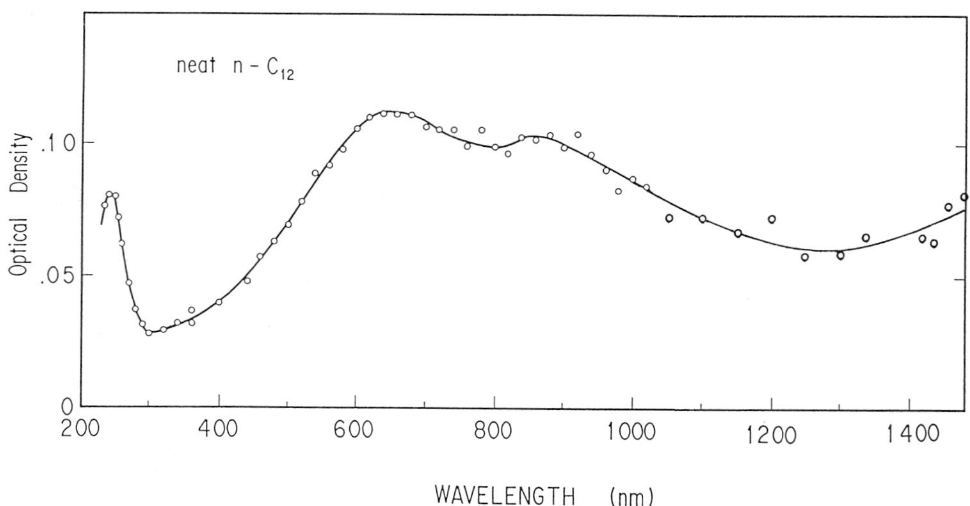

Fig.4 Transient absorption spectrum obtained in the nanosecond pulse radiolysis of neat n-dodecane

5. PRIMARY PROCESSES IN NEAT ALKANES

Fig. 5 shows the time-dependent absorptions obtained in the LL-twin pulse radiolysis in neat n-dodecane monitored at 660, 806, and 1300 nm, respectively.

The decay of the electrons observed at 1300 nm agrees with the theoretical decay on the diffusion theory with the initial exponential type distribution and initial separation of 6.6 nm. The distribution is almost same as those of other alkanes, such as cyclohexane and n-hexane, obtained in the solute-solvent system[7].

It is known that the decay processes of the alkane cation radicals are the geminate ion recombination with the electrons, the ion-molecular reaction of the cation radicals with alkane molecules[14], and the dissociation processes of the cation radicals, such as detachment of an H atom and elimination of a molecule of hydrogen. The decay of the cation radicals of n-dodecane at 806 nm agrees with the decay of the electrons in the time-region from 50 ps to 1 ns. Therefore, the cation radicals decay mainly due to the geminate ion recombination in picosecond time range. From the nanosecond pulse radiolysis results, the ion-molecular reaction occurs in the several tens nanoseconds in n-dodecane. On the other hand, the dissociation occurs from short-lived excited cation radicals which relax within the time-resolution of the experiment system.

The decay at 660 nm is complicated because of the overlap of the decay of cation radicals, the formation of excited states from the geminate ion recombination, and the decay of the excited states with the lifetimes of 4 ns.

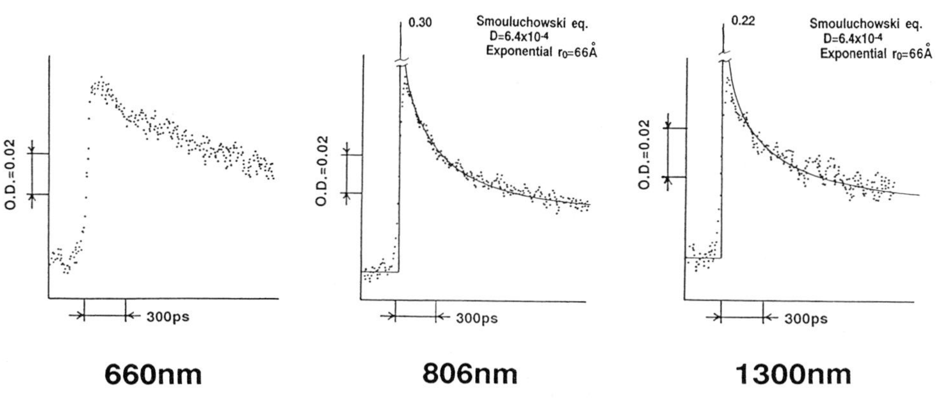

Fig. 5 Time-dependent absorptions obtained in the LL-twin pulse radiolysis in neat n-dodecane monitored at 660, 806, and 1300 nm.

In other alkanes, such as cyclohexane and n-hexane, the decays of the electrons agree with the Smoluchowski eq. by using the initial distributions obtained in the solute-solvent system. On the other hand, the decays of the cation radicals could not be observed clearly, because of both the overlap of the cation radical bands and excited states bands, and very short lifetimes of cation radicals due to the rapid ion-molecular reactions.

6. CONCLUSION

Lifetimes of the alkane cation radicals and electrons are different in neat liquid alkanes, especially for cycloalkanes and lower n-alkanes, because of the ion molecular reactions of the cation radicals. The initial spatial distributions and distances of electrons and parent cation radicals can be estimated from the decay of electrons in neat liquid alkanes and the decay of solute ions in solute-solvent system but they cannot be estimated from the decay of the cation radicals in neat liquid alkanes.

Acknowledgement --- The authors are grateful to Mr. T. Ueda and Mr. T. Kobayashi for their help in these pulse radiolysis experiments.

REFERENCES

1. S. Tagawa, M. Washio, H. Kobayashi, Y. Katsumura and Y. Tabata, Radiat. Phys.. Chem., 21 (1983) 45.
2. C. D. Jonah, Radiat. Phys. Chem., 21 (1983) 53.
3. S.Tagawa, N. Hayashi, Y. Yoshida, M. Washio and Y. Tabata, Radiat. Phys. Chem., 34 (1989) 503.
4. M.C. Sauer, Jr. and C. D. Jonah, Pulse Radiolysis, ed. Y. Tabata, CRC Press, Boston, (1991) 321.
5. H. Miyasaka and N. Mataga, Chem., Phys., Lett., 126 (1986) 219.
6. Y.Yoshida, S. Tagawa and Y. Tabata, Pulse Radiolysis, ed. Y.Tabata, CRC Press, Boston, (1991) 343.
7. Y. Yoshida, S.Tagawa and Y. Tabata, Radiat. Phys. Chem., 28 (1986) 201.
8. Y. Yoshida, S.Tagawa and Y. Tabata, Radiat. Phys. Chem., 30 (1987) 350.
9. Y. Yoshida, S.Tagawa M. Washio, H. Kobayashi and Y. Tabata, Radiat. Phys. Chem., 34 (1989) 493.
10. H. Kobayashi, Y. Tabata, T. Ueda and T. Kobayashi, Nucl. Instrum. Meth., B24/25, (1987) 1073.
11. K. H. Hong and J. Noolandi, J. Chem. Phys., 68 (1978) 5163.
12. C. A. M. van den Ende, L. Nyikos, J. M. Warman and A. Hummel, Radiat. Phys. Chem., 15 (1980) 273.
13. M. Tachiya, Radiat. Phys. Chem., 30 (1987) 75.
14. Y. Katsumura, Y. Yoshida, S. Tagawa and Y. Tabata, Radiat. Phys. Chem., 21 (1983) 103.
15. D. W. Werst and A. D. Trifunac., J. Phys. Chem., 92 (1988) 1093.

Dynamics and Mechanisms of
Photoinduced Transfer and Related Phenomena
N. Mataga, T. Okada and H. Masuhara (Editors)

Layered structure and energy transfer in poly(vinyl alkylal) Langmuir-Blodgett films

Masahide Yamamoto, Shinzaburo Ito, and Satoru Ohmori

Department of Polymer Chemistry, Faculty of Engineering, Kyoto University, Yoshida Sakyo-ku, Kyoto 606

Abstract

Structural relaxation of the polymer LB films by heating was investigated by the interlayer energy transfer technique. The structural change could be sensitively detected by this method. Above a certain critical temperature Tc, the layered structure was irreversively disordered and mixed through interlayer diffusion of polymers.

1. INTRODUCTION

Recently, extensive investigations on Langmuir-Blodgett (LB) films have been made on the design and the construction of functional thin films [1]. In the field of photophysics and photochemistry, Kuhn and his coworkers have studied fundamental photophysical processes, such as energy transfer and electron transfer in LB films, and they have used these processes as tools to reveal the structural properties of the LB films. They built an interlayer energy transfer system with the long chain fatty acids using cyanine dyes as the hydrophilic moiety [2].

As for materials of LB films, some preformed polymers were found to form a stable monolayer at the air-water interface and to be transferable to solid substrates [3-12]. One of these polymers is the poly(vinyl octal) reported by Ogata et al. [12]. Using this polymer as a base polymer, we built a system having a uniform chromophore distribution in a two dimensional plane, in which the chomophores are linked to the polymer chain with covalent bonds. Such preformed polymer LB films have some favorable characteristics for photophysical and photochemical processes [13-15]. One feature is a uniform distribution of chromophores in the monolayer, i.e., polymer LB films have a fairly low concentration of energy trap sites such as excimer-forming sites. The other is thinness of polymer LB films, e. g., the monolayer of poly(vinyl octal) has a thickness of ca. 1 nm which is thinner than the conventional LB films made from long chain fatty acids. The thinness of LB films favors energy transfer and migration between neighbouring layers.

In this study, we investigated the structural relaxation of the polymer LB films by using the interlayer energy transfer method to probe the layered structure. Photostational fluorescence intensities of an energy donor and an energy acceptor and the time-resolved fluorescence decays were measured and analyzed.

2. EXPERIMENTAL

Phenanthrene (P) and anthracence (A) moieties were introduced to a polymer as an energy donor and an energy acceptor chromophore, respectively (Fig. 1). The base polymers were poly(vinyl octal) (PVO) and poly(vinyl butylal) (PVB). The Förster radius of the present donor-acceptor pair P-A is 2.12 nm [16]. The sample preparation was described elsewhere [13-15]. The fractions of P and A in the polymers are 12 and 7 mol% in base unit

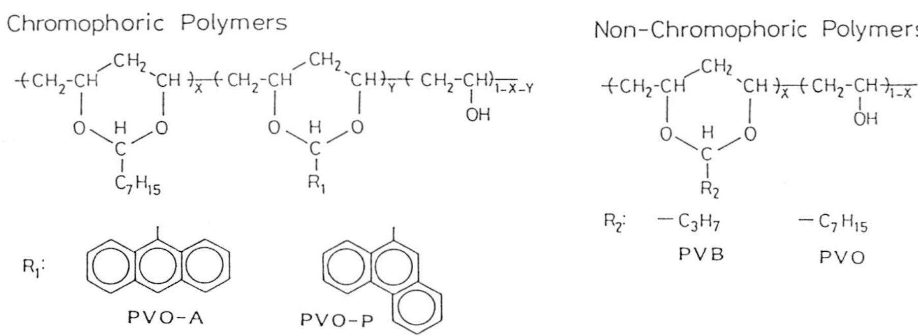

Fig. 1. Sample formula.

Table 1
Composition of polymer, surface pressure at the deposition, and glass transition temperature (Tg) for each polymers

Sample	x %	y %	Surface pressure $mN\ m^{-1}$	Tg ^{o}C
PVB	70	0	20.0	63
PVO	73	0	25.0	25
PVO-P	57	12	22.5	50
PVO-A	55	7	21.0	48

(Table 1). The fraction of hydroxy group is ca. 30 - 40 mol%.

LB films were prepared in the following way. The benzene solution of each polymer (0.01 wt%) was spread on pure water in a trough. The temperature of the water subphase was set to 19 ^{o}C for PVO and 7 ^{o}C for PVB. The surface pressure-area isotherm was recorded using a Wilhelmy type film balance. Fig. 2 shows the surface pressure-area isotherm for PVO, and PVO-P and PVO-A. Solid substrates were a non-fluorescent quartz plate which was made hydrophobic by dipping it in a 10 % solution of

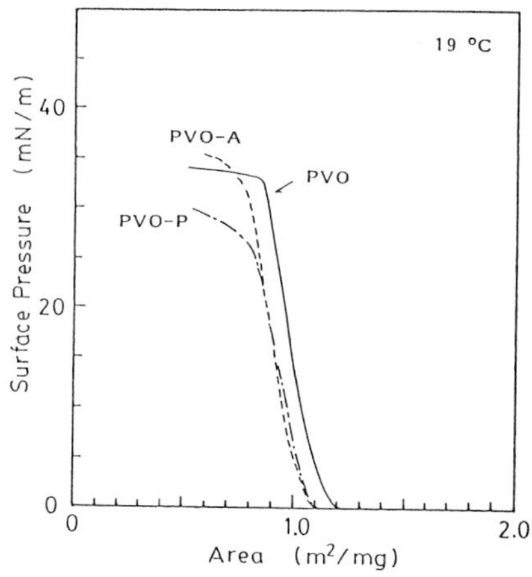

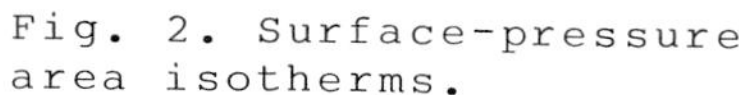
Fig. 2. Surface-pressure area isotherms.

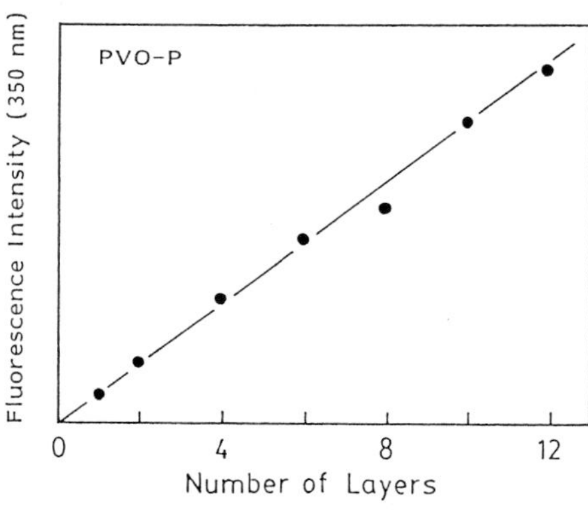

Fig. 3. Fluorescence intensity vs. number of layers (on one side).

trimethylchlorosilane in toluene. At a fixed surface pressure (20 - 25 mN m^{-1}) the LB film was prepared by dipping the substrate vertically at a rate of 15 mm min^{-1}. For the hydrophobic quartz plate, the deposition was possible in both up- and down-modes from the first dip, yielding a Y-type built-up film. Multilayered sample films were prepared as follows: at first, 4 layers of PVO were deposited on a quartz plate and then, 2 layers of PVO-P as energy donating layers, n layers (n = 0 - 8) of PVO as the spacer, and 2 layers of PVO-A as energy accepting layers were transferred, and finally, PVO layers were again deposited so that the total number of LB layers became eighteen. All samples have the same number of layers and the same compositions of PVO-P, PVO-A, and PVO polymers: these are abbreviated as PVO-P(n)A and PVB-P(n)A films for PVB polymer. Fig. 3 shows the fluorescence intensity vs. number of layers for PVO-P. This linear relationship indicates that the multilayers were deposited satisfactorily. Ellipsometric study of the PVO LB films has demonstrated that the film thickness of PVO is ca. 1 nm per layer [12, see also Fig. 9].

3. RESULTS AND DISCUSSION

3.1 Interlayer energy transfer

Fig. 4 shows the fluorescence spectra of PVO-P(n)A films having different number of spacer layers. For reference, the fluorescence spectrum of two layers of PVO-P is shown in the upper left of the figure. As the number of spacer layers is reduced from 8 to 0, the intensity of phenanthrene emission decreases and in place of it anthrancene emission increases at longer wavelengths. This indicates that the excitation energy of phenanthrene chromophore transfers to the anthracence chromophore over the spacer layers and that the efficiency is controlled by the distance between the donor layer and acceptor layer.

3.2 Structural relaxation measured by interlyer energy transfer

Previously [15], we studied the structural relaxation of the polymer LB films at room temperature by the interlayer energy transfer. The ratio of fluorescence intensity at 438 nm (I_A) to that at 350 nm (I_P) approximately represents the energy transfer efficiency. We followed this ratio of I_A/I_P with the time, during which the LB film was left at room temperature after

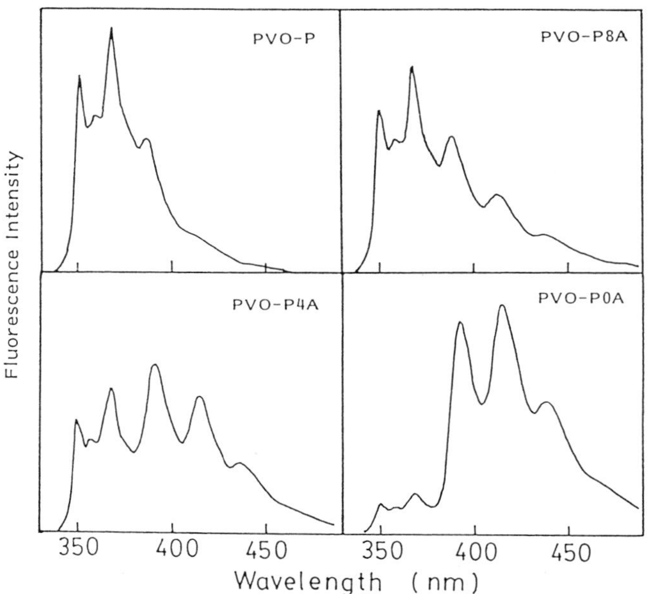

Fig. 4. Fluorescence spectra of PVO-P and PVO-P(n)A films. Excitation wavelength is 298 nm.

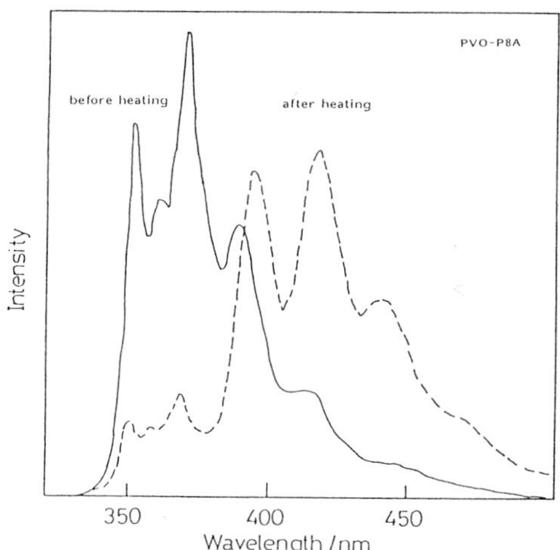

Fig. 5. Change of fluorescence spectrum for PVO-P8A before and after heating up to 100°C.

the deposition [15]. PVO-P4A, P6A, and P8A show an increase of energy transfer efficiency a few days after the deposition. After a few days, the increase slowed down. In this study, structural changes of polymer LB films by heating were examined by the interlayer energy transfer technique. We used samples that were slightly relaxed at room temperature for a week after the deposition. Fig. 5 shows the spectral change before heating and after heating up to 100 ^{o}C for PVO-P8A. Phenanthrene chromophore was selectively photoexcited at 298 nm. Before heating the main component of the fluorescence spectrum is phenanthrene fluorescence, but after heating anthracence fluorescence becomes the main component. This indicates that the layered structure is disordered, and the distance between the pair of P and A layers may become closer and this makes the energy transfer efficiency higher. Then, the thermal history of the layered structure was followed by measuring the energy transfer efficiency I_A/I_P. Fig. 6 and Fig. 7 show the change of energy transfer efficiencies I_A/I_P against temperature for PVO (Tg = 25 ^{o}C) and PVB (Tg = 63 ^{o}C) films, respectively. The energy transfer efficiency began to increase markedly above a certain critical temperature (Tc) with the first rise in temperature, where Tc is 38-40 ^{o}C for PVO films and ca. 60 ^{o}C for PVB

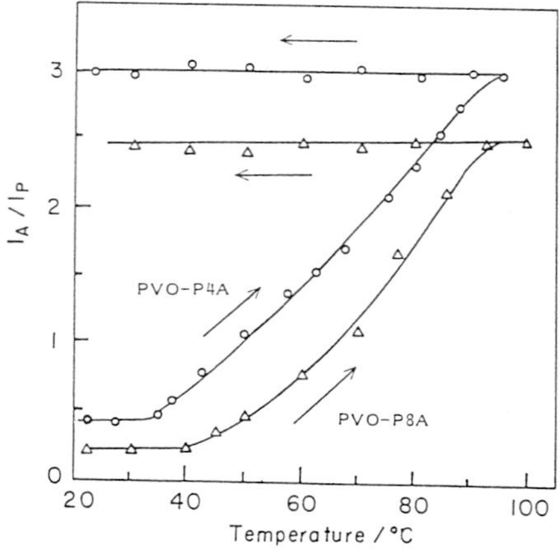

Fig. 6. I_A/I_P vs. heating temperature; (o) PVO-P4A and (Δ) PVO-P8A.

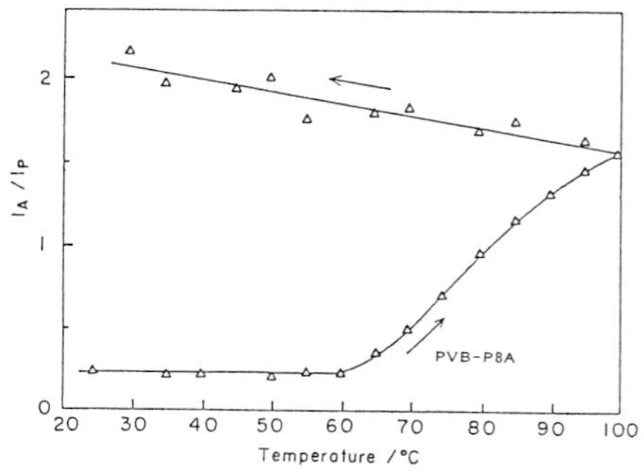

Fig. 7. I_A/I_P vs. heating temperature for PVB-P8A.

Table 2
Energy transfer efficiencies (I_A/I_P) before and after heating

Sample	I_A/I_P before heating	I_A/I_P after heating
PVO-P4A	0.5	3.0
PVO-P6A	0.4	2.5
PVO-P8A	0.1	2.5
Cast film	2.4	2.5

films. Once the sample was heated to 100°C, the increased efficiency was maintained during successive cooling and heating cycles. This means that the structural relaxation of LB films is irreversible. These figures show that the critical temperature (Tc) depends on the glass transition temperature (Tg) of the polymer: the higher the Tg of the polymer, the higher the Tc. The final reaching value after the layered structure was disordered is ca. 2.5 for PVO films and this value is close to the value of spin-coated films containing the P and A polymers with the same compositions (Table 2). These results show that the layered structure is disordered, and P polymer and A polymer are mixed with each other.

Fluorescence decay curves of the donor emission give more direct information on the energy transfer process. Fig. 8 shows the semilogarithmic plot of fluorescence intensities vs. time after a pulsed excitation at 298 nm. Fluorescence decay curves were measured by a single photon counting method using a picosecond laser system as the excitation light pulse. The instrumental response function was ca. 500 ps (fwhm). The details of measurements have been described elsewhere [17]. The PVO-P8A film before heating shows a nearly straight line, i.e., a single-exponential decay curve with a slight fraction of a fast component. The donor and acceptor probes are separated by the spacer of ca. 8 nm thickness. The uniform distance of separation gives a constant energy-transfer rate for each donor, resulting in single-exponential decay. On the other hand, if the system contains molecules at various distances of separation between the P-A pairs, the donor probes exhibit a multi-exponential decay from the sum of different transfer rates as an ensemble

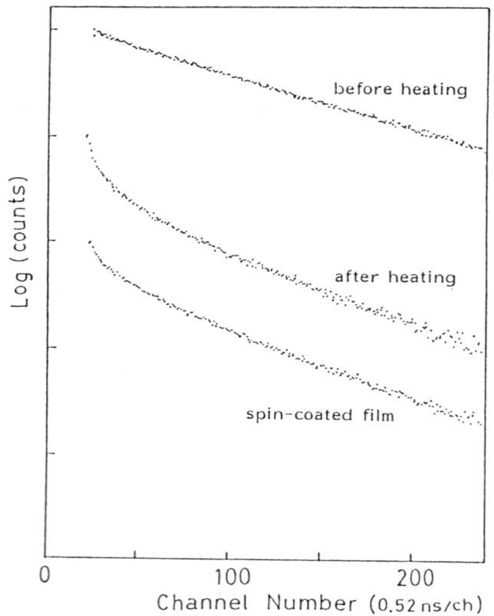

Fig. 8. Decay curves of P fluorescence (357 nm) for PVO-P8A film before and after heating, and for spin-coated film.

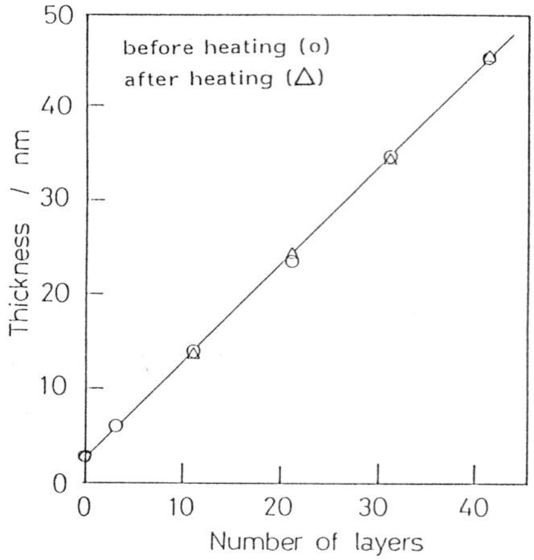

Fig. 9. Film thickness of PVO LB films before and after heating.

average, e.g., a stretched-exponential [18]. Such a case is the PVO-P8A film after heating and the spin-coated film. Both films show quite similar decay profiles that may be ascribed to a mixed polymer structure. The result shows that the layered structure is irreversibly disordered and mixed by the interlayer diffusion above a certain critical temperature Tc. Fig. 9 shows the relationship between the thickness of layered films and the number of layers for PVO films before heating and after heating, where the thickness was measured by ellipsometry. The result shows that the thickness hardly changes by heating although the layered structure is disordered. This means that the energy transfer method can detect sensitively structural changes, but the ellipsometry cannot.

This work was partially supported by a Grant-in-Aid for Scientific Research on Priority Areas, New Functionality Materials-Design, Preparation and Control (No.03205073) from the Ministry of Education, Science and Culture of Japan.

4. REFERENCES

1 K. Fukuda, M. Sugi (eds.), Langmuir-Blodgett Films, 4 (1989), Elsevier, London.

2 H. Kuhn, D. Möbius, and H. Bücher in Physical Methods of Chemistry; A. Weissberger and B.W. Rossiter (eds.), Wiley, New York, 1972, Vol 1, Part 3B, p 557.

3 R.H. Tredgold, Thin Solid Films, 152 (1987) 223.

4 T. Takenaka, K. Harada, M. Matsumoto, J. Colloid Interface Sci., 73 (1980) 569.

5 T. Kawaguchi, H. Nakahara, K. Fukuda, Thin Solid Films, 133 (1985) 29.

6 S.J. Mumby, J.D. Swalen, J.F.Rabolt, Macromolecules, 19 (1986) 1054.

7 H. Ringsdorf, G. Schmidt, J. Schneider, Thin Solid Films, 152 (1987) 207.

8 J. Schneider, C. Erdelen, J.F. Rabolt, Macromolecules, 22 (1989) 3475.

9 G. Duda, A.J. Schouten, T.Arndt, G.Leiser, G.F. Schmidt, C. Bubeck, G. Wegner, Thin Solid Films, 159 (1988) 221.

10 K. Naito, J.Colloid Interface Sci., 131 (1989) 218.

11 M. Matsumoto, T.Itoh, T.Miyamoto, in Cellulosics Utilization, H.Inagaki, G.O. Phillips (eds.), Elsevier, London, 1989, p 151.

12 K. Oguchi, T. Yoden, Y. Kosaka, M. Watanabe, K. Sanui, N. Ogata, Thin Solid Films, 161 (1988) 305.

13 S. Ito, H. Okubo, S. Ohmori, M. Yamamoto, Thin Solid Films, 179 (1989) 445.

14 S. Ohmori, S. Ito, and M. Yamamoto, Macromolecules, 23 (1990) 4047.

15 S. Ohmori, S. Ito, and M. Yamamoto, Macromolecules, 24 (1991) 2377.

16 I.B. Berlman, Energy Transfer Parameters of Aromatic Compounds, Academic, New York, 1973.

17 S. Ito, K. Takami, Y. Tsujii, and M. Yamamoto, Macromolecules, 23 (1990) 2666.

18 J. Klafter and M.F. Shlesinger, Proc. Natl. Acad. Sci. USA, 83 (1986) 848.

Dynamics and Mechanisms of
Photoinduced Transfer and Related Phenomena
N. Mataga, T. Okada and H. Masuhara (Editors)

Simulation of Primary Process in Photosynthesis by Monolayer Assemblies

Masamichi Fujihira

Department of Biomolecular Engineering, Tokyo Institute of Technology, 4259 Nagatsuta, Midori-ku, Yokohama 227, Japan

Abstract

In the photosynthesis, solar energies harvested by antenna pigments are funneled to special pairs in the reaction centers, where multistep electron transfer reactions proceed to separate electron-hole pairs far apart across the lipid bilayer thylakoid membrane. The well-organized and asymmetric molecular arrangement across the membrane play an important role in the charge separation of the photosynthetic reaction center. In a series of studies, the simulation of the light harvesting and the succeeding charge separation processes with Langmuir-Blodgett (LB) films were examined. These molecular devices can be used for the photoelectric conversion and are called molecular photodiodes. Our recent research on simulation of the photosynthetic primary process with LB films will be reviewed.

1. INTRODUCTION

In biosystems molecules organize themselves into complex functional entities with cooperating components of molecular dimensions. For example, well-organized molecular assemblies in thylakoid membranes play an important role in photosynthetic processes of plants and bacteria [1]. In 1984, the atomic structure of the reaction center, i.e. the basic machinery for the beginning events of photosynthesis, of a purple bacterium *R. viridis* was determined with x-ray diffraction [2]. All reaction centers are complexes containing protein subunits and donor-acceptor molecules. These centers span an inner membrane in the plant or bacterial cell, and their donor-acceptor complexes perform charge separation that creates a potential gradient across the membrane. Before the charge separation can take place at any reaction center, solar energy must be harvested by light-absorbing antenna pigments and transmitted to the center.

To design artificial photosynthetic molecular systems [3-5] for solar energy conversion, it is of great interest to mimic the elaborate molecular machinery for the light harvesting and the charge separation. The consideration of the structure and function of the asymmetric spatial arrangement of electron donors and acceptors in the charge separation unit across the thylakoid is most essential. The Langmuir-Blodgett (LB) film is one of the most appropriate artificial material by which the spatial arrangement of the various functional moieties across the film can be constructed readily at atomic dimensions. In a series of studies, we have attempted to simulate the elemental processes of the photosynthesis by taking advantage of LB monolayer assemblies. In this paper, recent developments from our group will be reviewed.

2. MOLECULAR PHOTODIODE WITH HETEROGENEOUS A/S/D LB FILMS

The first molecular photodiode was fabricated with a molecularly ordered film on a gold optically transparent electrode (AuOTE) prepared by the LB method [6] as shown in Figure 1(a), where hydrophilic parts and hydrophobic units are indicated by circles and squares, respectively. With their amphiphilic properties, three functional compounds tend to orient regularly in the heterogeneous LB films. Another interesting and fascinating application of LB films is their use as controlled–thickness spacers or "distance keepers". Therefore, the distances between the three functional moieties, i.e. A, S, and D, can be closely controlled at known values. The electron

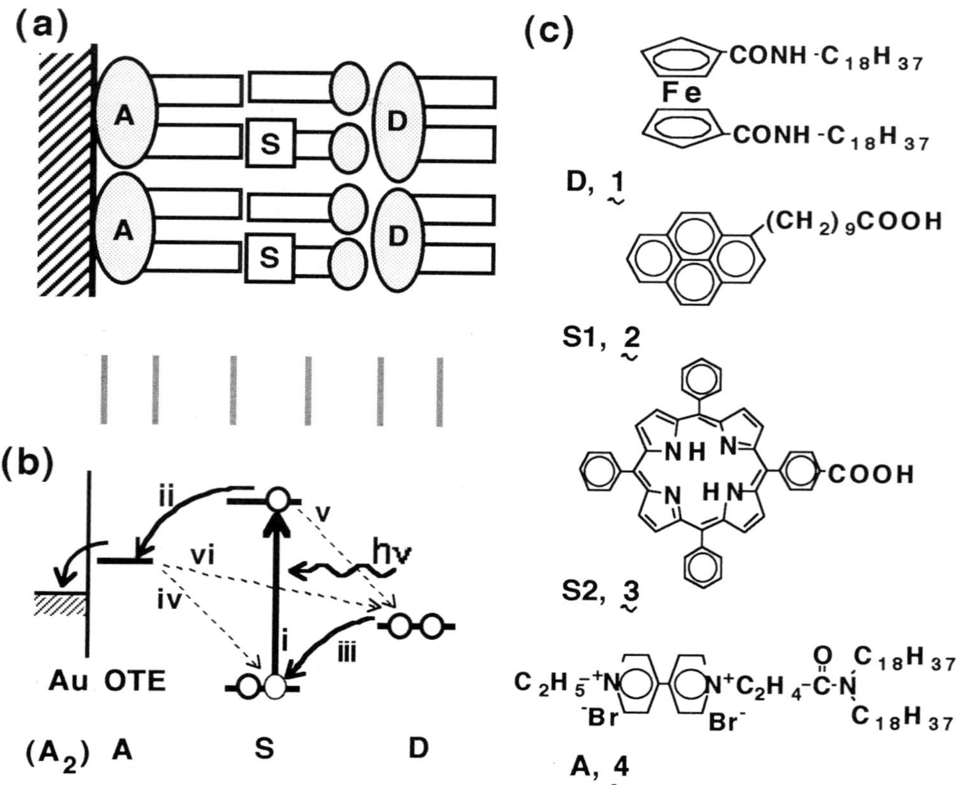

Figure 1. Molecular photodiode with heterogeneous A/S/D LB film on AuOTE: (a), structure; (b), energy diagram; (c), structural formulae of A, S and D.

transfer process in such molecular assemblies is free from any complication due to diffusion. Kuhn and Möbius [3,4] have previously studied the distance dependence of the rate of photoinduced electron transfer in LB films. Their proposed dependenence agrees with the experimental [7–9] and theoretical [10] results for non-adiabatic electron transfers, which are described by the following equations [10]:

$$k = k(r)\exp(-\Delta G^{*}/RT) \tag{1}$$

$$k(r) = k_{o}\exp\{-\beta(r - r_{o})\} \tag{2}$$

In addition to the distance dependence, the effect of the standard free energy difference $\Delta G°$ for the electron transfer is another important factor in determining the rate of electron transfer. Relationship (3) was first introduced by Marcus where the free energy barrier for the reaction, i.e. ΔG^{*} in eq.(1), is given in terms of the reorganizantion energy λ [10]:

$$\Delta G^{*} = \lambda/4(1 + \Delta G°/\lambda)^{2} \tag{3}$$

The changes in bond lengths of the reactants and the changes in solvent orientation coordinates in the electron transfer are related to λ[10]. In Marcus's original theory, the motion of the nuclei was treated classically. There have been several attempts to treat the nuclear coordinates quantum mechanically and to modify the equation for the energy gap $\Delta G°$ [10,12]. In connection with the design of the proper energy diagram for the molecular photodiode, the inverted region, where the rate decreases with an increase in a large excess of $-\Delta G°$, predicted by eq.(3) is most important. The presence of the inverted region has been confirmed experimentally by the use of internal electron transfer systems with rigid spacers [13].

Keeping the distance and the $\Delta G°$ dependence in mind, we considered how to design a better molecular photodiode. In Figure 1(b), the energy diagram of the A/S/D molecular photodiode is depicted as a function of distance across the LB film. If the forward processes indicated by arrows with solid lines are accelerated and the backward processes with dashed lines are retarded by setting the distances and the energy levels appropriately, the photoinduced vectorial flow of electrons can be achieved. Namely, the acceleration by setting $-\Delta G°$ equal to λ is assumed for the forward electron transfer processes ii and iii, while the retardation, as a consequence of the inverted region, is assumed for the back-electron transfer processes iv and v. Once an electron-hole pair is separated successfully, the recombination of the pair across the large separation by LB film (process vi) is hindered.

The three kinds of functional amphiphilic derivatives used for the first A/S/D type molecular photodiode [6] are shown in Figure 1(c) together with porphyrin sensitizer **3** used later. By depositing these three amphiphiles on AuOTE, as shown in Figure 1(a), and by use of the resulting electrode as a working electrode in a photoelectrochemical cell [7], the photoinitiated vectorial flow of electrons was achieved and detected as photocurrents. The AuOTE is a metal electrode and hence does not by itself possess a rectifying ability as does a semiconductor electrode [7]. In spite of the inability of the substrate electrode to rectify, the photocurrent had opposing directions depending on the spatial arrangement of A/S/D or D/S/A. The direction was in accordance with the energy profile across LB films in Figure 1(b).

Much higher photocurrents were observed for stacks of multilayers of each component, e.g. in the form of A,A,A/S,S,S,S/D,D,D and D,D,D/S,S,S,S/A,A,A [6,14]. The direction of the photocurrent also agreed with what we would expect for these multilayered systems. An amphiphilic porphyrin derivative **3** was also used as a sensitizer[14].

In such heterogeneous LB films, the long alkyl chains intervened between the A and S and between the S and D moieties. As a result, part of the excited sensitizers were deactivated by the emission of photons without qunching by electron transfer. To cope with this problem, polyimide LB films consisting of A, S, and D units were used for constructing more efficient moleclar photodiodes in collaboration with Kakimoto, Imai, and their co-workers [15]. They had reported the preparation and properties of polyimide LB films [16]. Since polyimide LB films have no long alkyl spacer between the layers (monolayer thickness 0.4 – 0.6 nm), electrons should be more readily transferred. For example, sub-μA order of photocurrents were observed for AuOTE's which were coated with six layers of A, two layers of S, and six layers of D. These magnitudes are ca. 10 times larger than those for photodiodes with conventional LB films.

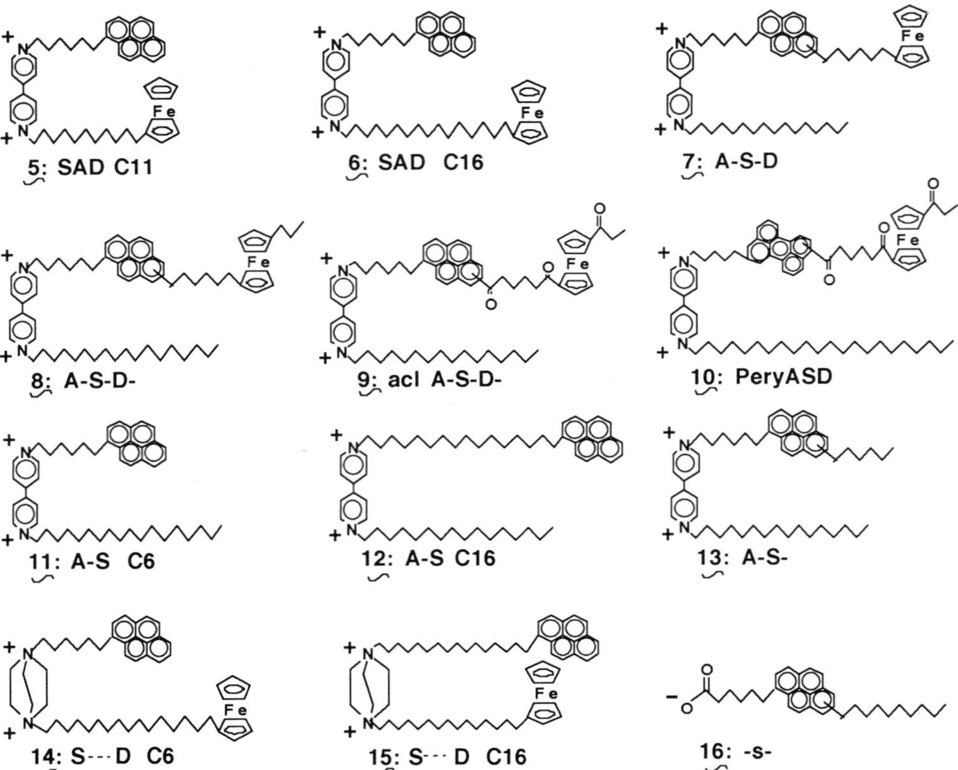

Figure 2. Structural formulae of folded type S–A–D and linear type A–S–D triads and their reference A–S, S–D and –S– compounds.

3. MOLECULAR PHOTODIODE CONSISTING OF FOLDED TYPE S–A–D AND LINEAR TYPE A–S–D TRIADS

Another approach to shortening the distances between the functional moieties is the use of unidirectionally oriented amphiphilic triad monolayers [6,17]. Each triad contained an A, S, and D moiety as its functional subunits. Other groups studied also the two–step photodriven charge separation and back electron transfer reactions of triad molecules of the S–A_1–A_2 [18] or A–S–D [19,20] types. They succeeded in retarding charge recombination in homogeneous solution, but did not attempt to orient these triad molecules in one direction and thus to conduct a direct photoelectric conversion. In our first amphiphilic triad **5** in Figure 2 [6], A, S, and D corresponded to viologen, pyrene, and ferrocene moieties, respectively. The viologen moiety is hydrophilic, whereas the pyrene and ferrocene moieties are hydrophobic. Subunits A and S were linked together with a C_6 alkyl chain, while subunit D was linked to subunit A with a longer C_{11} alkyl chain. Later, a modified triad **6** in Figure 2 was synthesized [17] in order to improve the balance of the two distances between A and S and between S and D. In Figure 2 are also shown the structures of "linear" type A–S–D triads and their reference compounds, A–S, S–D, and –S–.

Due to their amphiphilic properties, the A, S, and D moieties were considered to be arranged spatially in this order, owing to the difference in length of two alkyl chains,

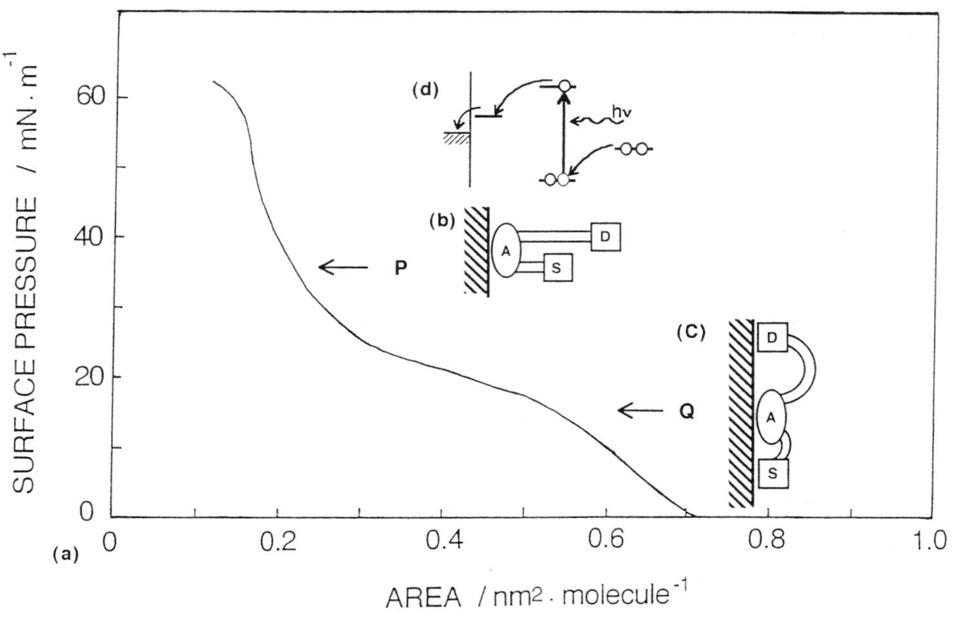

Figure 3. Surface pressure – area isotherm (a) of a mixed monolayer of triad **5** with AA; schematic representation of oriented (b) and non–oriented (c) triad **5**; corresponding energy diagram (d).

from the electrode to the electrolyte solution perpendicularly as vizualized in Figure 3(b). This arrangement was confirmed by observing anodic photocurrents, whose direction was in accordance with the photoinduced vectorial flow of electrons expected from the energy diagram of the oriented triad, as depicted in Figure 3(d). From the shape illustrated in Figure 3(b), S–A–D is called "folded" type triad. The photodiode function of these folded type S–A–D triads was studied in detail [17] in terms of i) the wavelength of the incident photons, ii) the surface pressure applied for monolayer deposition, iii) the applied electrode potentials, iv) the distances between A and S and between S and D, and v) the role of the D moiety.

The photocurrent spectra in terms of the formal quantum efficiency [7] were recorded for the mixed monolayers (triad **5** and arachidic acid (AA) (1:2)) which were deposited on AuOTE's at the two different surface pressures indicated by P and Q in Figure 3(a). The most important and interesting point was that the photocurrent increased by a factor of ca. 20 when the surface pressure for film deposition was changed only from 15 (at Q) to 35 mN m^{-1} (at P). The increase in the surface concentration by compression from Q to P was insufficient to explain such a dramatic increase in the photocurrent. Rather, the non-oriented structure of triad **5** and the unfavorable relative location of the A, S, and D moieties (Figure 3(c)) in the monolayer at the low surface pressure (Q) might be responsible for the small photocurrent. By contrast, the much larger photocurrents observed at the high surface pressure (P) support the more favorable orientation of the folded type S–A–D triad molecule postulated in Figure 3(b).

A higher efficiency of triad **6** than that of triad **5** was also observed, which might be attributed to a better matching in concurrent electron transfer reactions between A and S and between S and D owing to an improved balance between the A–S and S–D distances. Another amphiphilic compound **11** without the D moiety in Figure 2 confirmed a positive contribution of the D moieties in triads **5** and **6**.

In order to improve the orientation of triad molecules in the monolayer, the linear type A–S–D triads in Figure 2, were synthesized [21]. The mixed monolayer of one of the linear type A–S–D triads (**7**) with behenic acid (BA) exhibited a much higher photocurrent than did that of the folded type S–A–D triads. This indicates that a more ideal spatial arrangement of the A, S, and D moieties was attained for the linear triad molecules in the mixed monolayer.

4. KINETICS OF PHOTO-INDUCED ELECTRON TRANSFER AND THE EFFECT OF ELECTRICAL DOUBLE LAYER

As described in section 2, kinetics of photoinduced multi-step electron transfer plays a crucial role in efficiencies of the molecular photodiodes. To clarify the energy gap and the distance dependence of the photoinduced electron transfer in heterogeneous LB films or triad monolayers, nanosecond and picosecond laser photolyses were carried out.

As a reference in photoinduced multi-step electron transfer in the A/S/D and D/S/A LB films, the luminescence decay curves were recorded for sensitizers such as pyrene and $Ru(bpy)_3^{2+}$ derivatives confined as one monolayer in LB films. These films also contained other monolayers of acceptor or donor amphiphiles which were deposited apart from the sensitizer monolayer by a fixed distance as shown in Figure 4. The structures of the sensitizers and those of amphiphilic acceptors and donors

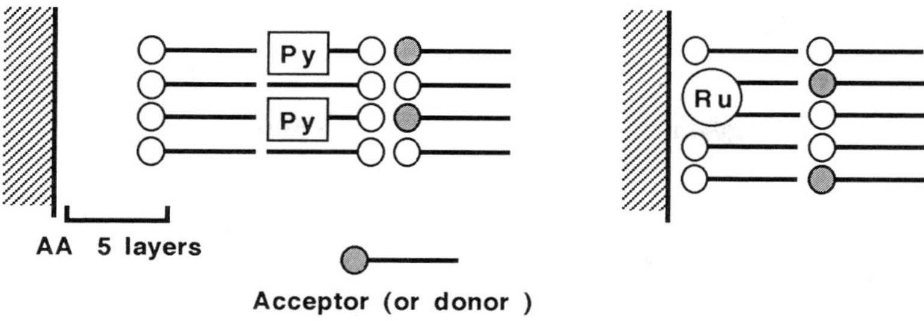

Figure 4. Heterogeneous LB films used for kinetic study of photo-induced electron transfer: a, pyrene decanoic acid (**2**) and A (or D); b, RuC19 (**21**) and A (or D).

are shown in Figures 1(c) and 5. In addition to the difference of 1.0 V in the oxidation potentials between the excited pyrene **2** and the $Ru(bpy)_3^{2+}$ **17** derivatives, the redox potentials of four types acceptors **4, 18 – 20** ranged widely up to 1.8 V. This enabled us to examine the energy gap ΔG° dependence of the photoinduced electron transfer rate of the A/S LB films [22]. It was concluded from the results that the possibility of a Marcus-type inverted region exists at highly negative ΔG°, although it is not definite yet because of the limited data.

To examine the energy gap law for S/D LB films, four kinds of amphiphilic ferrocene derivatives **21 – 24** as electron donors with different standard redox potentials E°'s and with the same alkyl chain spacer were synthesized [23]. The comparison of the energy gap ΔG° dependence of the reductive quenching of the $Ru(bpy)_3^{2+}$ derivative **17** with the ferrocene derivatives for three different systems, i.e. the LB films, the micellar, and the solution systems, implies that the electrical potential difference between the hydrophilic head-groups in LB films has to be taken into account for estimation of the effective energy gap ΔG°. Namely, an efficient photoinduced electron transfer quenching happens even in a S/D LB film in which the reaction was expected to be up-hill ($\Delta G° > 0$) on the basis of half-wave potentials and thus too slow to be detected. The result contradicted also with that of the corresponding solution system.

The effect of the inner potential difference in the electrical double layer was further studied by using three kinds of amphiphilic ferrocene derivatives, **21**, **25**, and **26**, with an anionic, a cationic, and a nonionic head-group, respectively [24] (Figure 5). All the redox potentials of three ferrocene derivatives were ca. 0.9 V vs. SCE and more positive than the reduction potential of excited $Ru(bpy)_3^{2+}$ of 0.6 V vs. SCE. The luminescence decay curves for **17** in three types of S/D systems were recorded together with that in a reference LB film in which the pure monolayer of AA was deposited in place of the ferrocene donor layer. Only the decay curve for **17/21** showed a fast decay component, while the other two curves are very similar to that for the reference LB film and showed almost a single exponential decay without appreciable quenching.

The change in sign of the head-group charge was expected to vary signs of the electrical double layer and thus the effective energy gap ΔG°. The potential difference will readily become a few hundreds millivolts as observed for monolayers at the

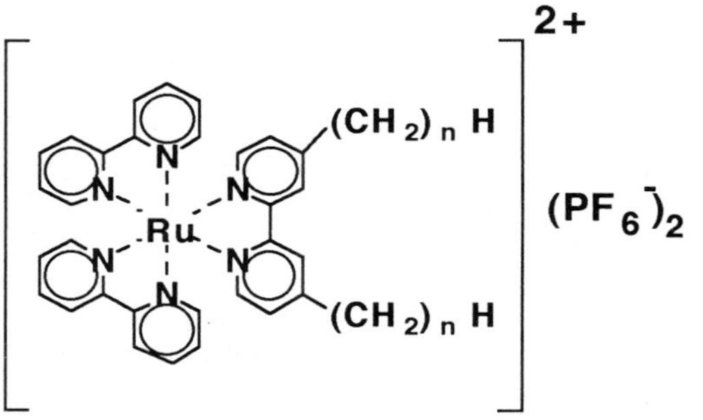

17 Ru C19 (n = 19)

18

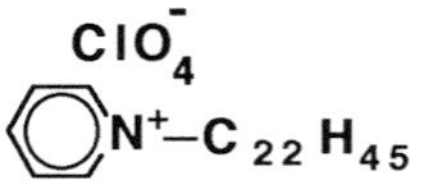

19

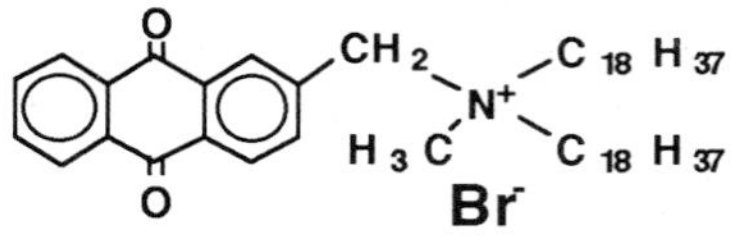

20

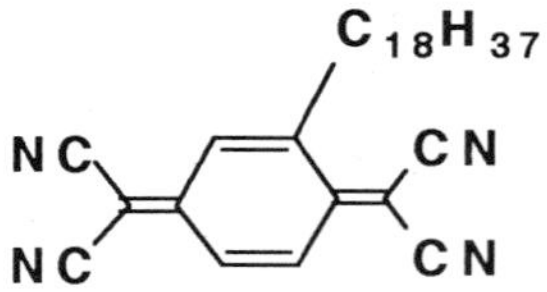

21

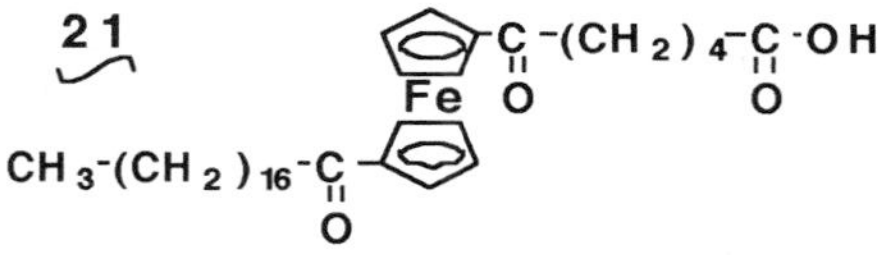

22

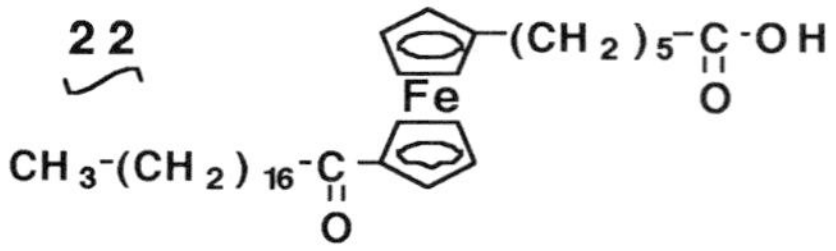

23

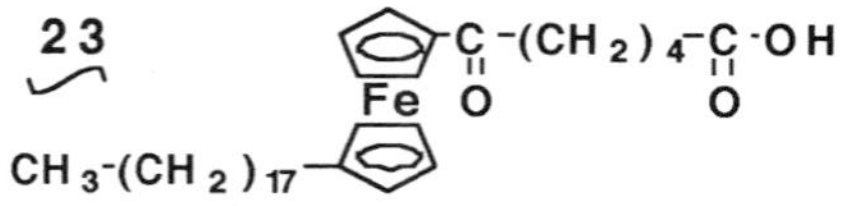

24

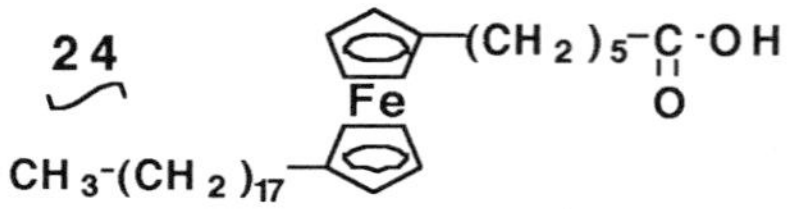

25

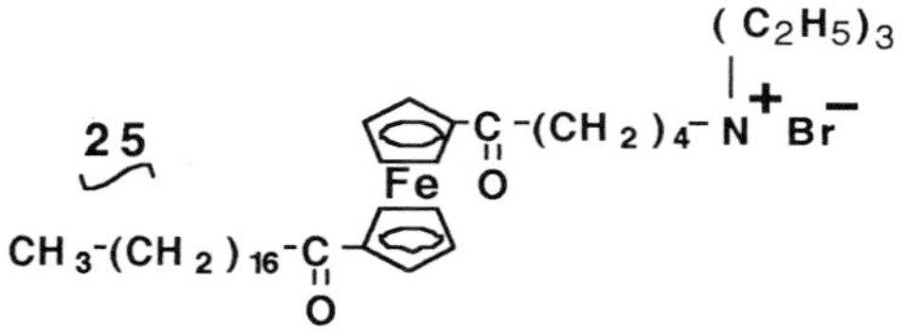

26

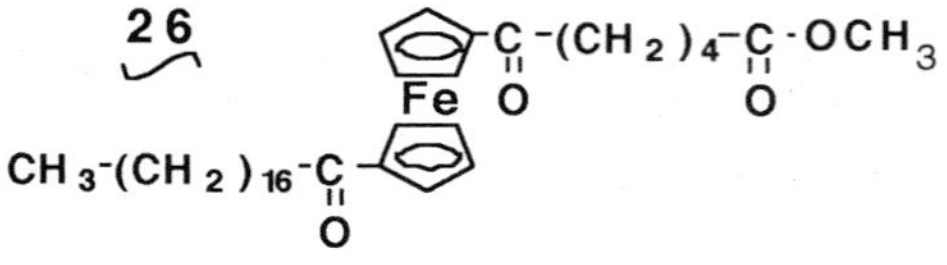

Figure 5. Structural formulae of amphiphilic $Ru(bpy)_3^{2+}$ sensitizer and electron acceptor and donor amphiphilic derivatives used in the kinetic study.

air–water interface [25] and for micellar surfaces [26]. If this potential difference is taken into account, appreciable electron transfer quenching, observed specifically in the $Ru(bpy)_3^{2+}$/the anionic ferrocene derivative system, is rationalized. It is noteworthy that the electrostatic potential effect was proposed to explain why the photochemical reaction follows the L pathway in the reaction center with almost C_2 symmetry [27].

In order to increase the response time and efficiency of the molecular photodiode consisting of folded type S–A–D triad molecules, it is important to know the distance and orientation dependence of the electron transfer quenching in A–S and S–D monolayers. For this purpose, A–S and folded type S–D amphiphiles with alkyl chain spacer of different carbon numbers shown in Figures 6(a) and (b), respectively, were prepared. The luminescence decay of these LB films were measured with the picosecond laser photolysis [28]. The linear decrease in log k_q with an increase in carbon number was observed for the A–S monolayers (Figure 6(c)), while the regular dependence of log k_q on the alkyl chain lengths could not be obtained for folded type S–D systems. This result indicates that the alkyl chains in the A–S diad molecules are extended in their monolayers and therefore the distance between A and S can be controlled by changing the carbon number, while this is not the case for the folded type S–D diads as visualized in Figure 6(d).

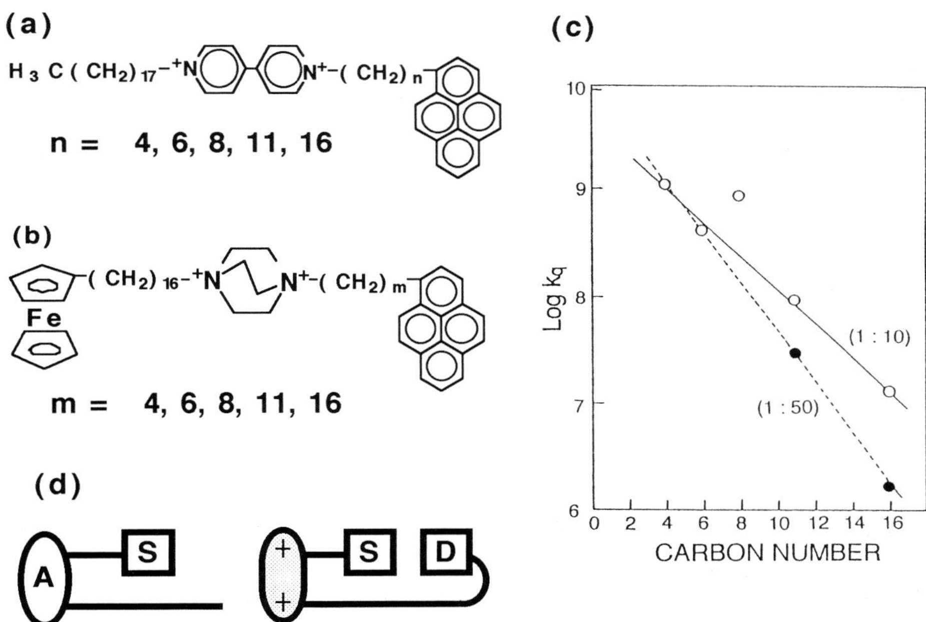

Figure 6. Structural formulae of A–S (a) and S–D (b) compounds used to study the effect of alkyl chain lengths on kinetics of electron transfer quenching in mixed monolayers of A–S (or S–D) and AA.

5. SIMULATION OF THE PRIMARY PROCESS IN THE PHOTOSYN THETIC REACTION CENTER BY MIXED MONOLAYER WITH TRIAD AND ANTENNA MOLECULES

In the next step, we simulated the light harvesting and succeeding charge separation processes by a monolayer assembly consisting of synthetic antenna pigments and triad molecules, as illustrated in Figure 7 [5,29]. For the light–harvesting (H) antenna pigments, an amphiphilic pyrene derivative was used [30]. For the amphiphilic linear triad molecule, a perylene moiety, as the S unit, and viologen and ferrocene moieties, as the A and the D units, respectively, were used. The structures of the antenna **16** and the triad **10** molecule are also shown in Figure 2. Because of the overlap of the emission spectrum of the antenna pyrene and the absorption spectrum of the sensitizer perylene moiety of the triad (Figure 8 (a)), light energies harvested by the antenna molecules were efficiently transferred to the sensitizer moiety of the triad. Thus, the excitation energy of the perylene moiety should be converted to electrical energies via multistep electron transfer across the monolayer as described above.

Figure 8(b) shows the photocurrent spectrum of the mixed monolayer of the triad and the antenna with a molar ration of 1:4. Maxima of anodic photocurrents at ca.

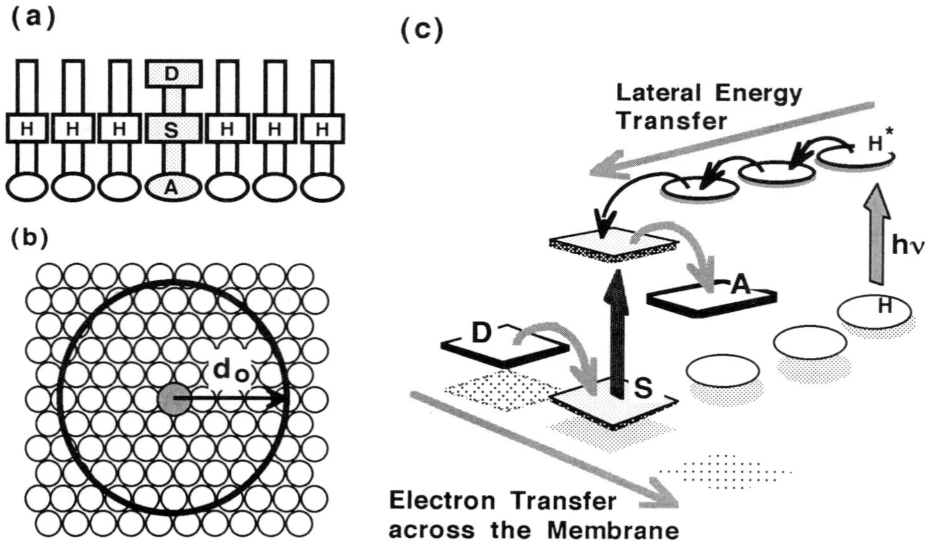

Figure 7. Schematic representation of the artificial photosythetic reaction center by a monolayer assembly of Pery A–S–D triad **10** and antenna –S– **16** molecules for light harvesting (H), energy migration and transfer, and charge separation via multistep electron transfer: (a), side view of monolayer assembly; (b), top view of a triad surrounded by antenna molecules; (c), energy diagram for photoelectric conversion in a monolayer assembly.

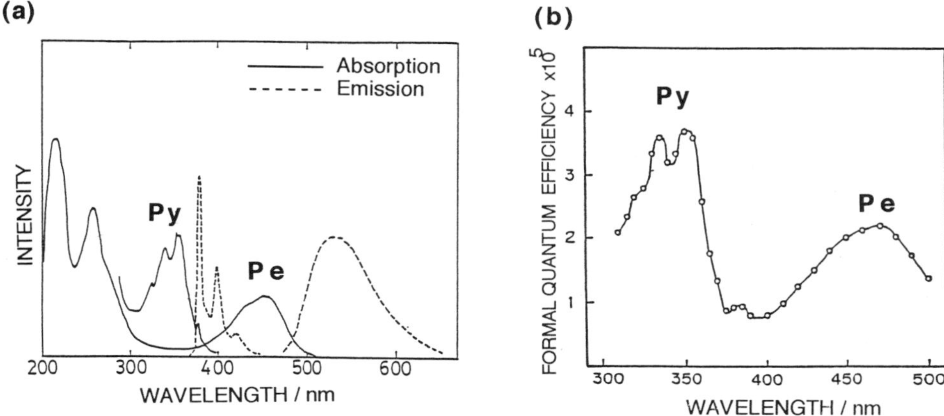

Figure 8. UV and visible absorption and emission spectra (a) of antenna pyrene **16** and perylene triad **10** derivatives in ethanol and photocurrent spectrum (b) of the mixed monolayer of **10** and **16** (1:4) deposited on AuOTE.

350 and 470 nm are found. These correspond to the adsorption maxima of pyrene and acylated perylene. The anodic direction of the photocurrent agrees with the energy diagram and the orientation of the triad shown in Figure 7. In contrast, negligible photoresponse was obtained with the pure antenna monolayer. The result indicates that charge separation in the triad molecules was initiated by light absorption both with perylene sensitizer itself and with the pyrene antennas followed by the energy transfer.

6. ACKNOWLEDGEMENTS

I would like to thank all my colleagues who collaborated in this research. This research was supported by a Grant-in-Aid for Scientific Research 61470076 and those on Priority Areas 63604534 and 63612506 from the Ministry of Education, Science, and Culture and by a Grant from the Nissan Science Foundation.

7. REFERENCES

1 L. Stryer, Biochemistry, 3rd ed. (Freeman, New York, 1988).
2 J. Deisenhofer, O. Epp, K. Miki, R. Huber and H. Michel, J. Mol. Biol. 180 (1984) 385; Nature 318 (1985) 618.
3 H. Kuhn, J. Photochem. 10 (1979) 111; Thin Solid Films 99 (1983) 1.
4 D. Möbius, Ber. Bunsenges. Phys. Chem. 82 (1978) 848; Acc. Chem. Res. 14 (1981) 63.
5 M. Fujihira, Mol. Cryst. Liq. Cryst. 183 (1990) 59.
6 M. Fujihira, K. Nishiyama and H. Yamada, Thin Solid Films 132 (1985) 77.

7 T. Osa and M. Fujihira, Nature 264 (1976) 349; M. Fujihira, N. Ohishi and T. Osa, Nature 268 (1977) 226; M. Fujihira, T. Kubota and T. Osa, J. Electroanal. Chem. 119 (1981) 379.
8 J. W. Verhoeven, Pure Appl. Chem. 58 (1986) 1285.
9 J. R. Miller, J. V. Beitz and R. K. Huddleston, J. Am. Chem. Soc. 106 (1984) 5057; J. R. Miller, Nouv. J. Chim. 11 (1987) 83.
10 R. A. Marcus and N. Sutin, Biochim. Biophys. Acta 811 (1985) 265.
11 R. A. Marcus, J. Chem. Phys. 43 (1965) 2654; P. Siders and R. A. Marcus, J. Am. Chem. Soc. 103 (1981) 748.
12 J. Ulstrup and J. Jortner, J. Chem. Phys. 63 (1975) 4358.
13 J. R. Miller, L. T. Calcaterra and G. L. Closs, J. Am. Chem. Soc. 106 (1984) 3047.
14 M. Fujihira, K. Nishiyama and H. Yoneyama, J. Chem. Soc. Jpn. (1987) 2119.
15 Y. Nishikata, A. Morikawa, M. Kakimoto, Y. Imai, Y. Hirata, K. Nishiyama and M. Fujihira, J. Chem. Soc., Chem. Commun. (1989) 1772.
16 M. Suzuki, M. Kakimoto, T. Konishi, Y. Imai, M. Iwamoto and T. Hino, Chem. Lett. (1986) 395; M. Kakimoto, M. Suzuki, T. Konishi, Y. Imai, M. Iwamoto and T. Hino, Chem. Lett. (1986) 823; Y. Nishikata, M. Kakimoto, A. Morikawa, I. Kobayashi, Y. Imai, Y. Hirata, K. Nishiyama and M. Fujihira, Chem. Lett. (1989) 861.
17 M. Fujihira and H. Yamada, Thin Solid Films 160 (1988) 125.
18 S. Nishitani, N. Kurata, Y. Sakata, S. Misumi, A. Karen, T. Okada and N. Mataga, J. Am. Chem. Soc. 105 (1983) 7771; N. Mataga, A. Karen, T. Okada, S. Nishitani, N. Kurata, Y. Sakata and S. Misumi, J. Phys. Chem. 88 (1984) 5138.
19 T. A. Moore, D. Gust, P. Mathis, J. C. Mialoeq, C. Chachaty, R. V. Bensasson, E. J. Land, J. C. Doizi, P. A. Liddel, W. R. Lehman, G. A. Nemeth and A. L. Moore, Nature, 307 (1984) 630.
20 M. R. Wasielewski, M. P. Niemczyk, W. A. Svec and E. B. Pewitt, J. Am. Chem. Soc. 107 (1985) 5562.
21 M. Fujihira and M. Sakomura, Thin Solid Films 179 (1989) 471.
22 M. Fujihira, K. Nishiyama and K. Aoki, Thin Solid Films 160 (1988) 317.
23 T. Kondo, H. Yamada, K. Nishiyama, K. Suga and M. Fujihira, Thin Solid Films 179 (1989) 463; T. Kondo and M. Fujihira, Kobunshi Ronbunshu 47 (1990) 921.
24 T. Kondo and M. Fujihira, Chem. Lett. (1991) 191; T. Kondo, M. Yanagisawa and M. Fujihira, Electrochim. Acta in press.
25 V. Vogel and D. Möbius, Thin Solid Films 132 (1985) 205; 159 (1988) 73; J. Colloid Interface Sci. 126 (1988) 408.
26 C. Wolff and M. Grätzel, Chem. Phys. Lett. 52 (1977) 542; M. Grätzel, Heterogeneous Photochemical Electron Transfer (CRC Press, Florida, 1989).
27 J. R. Norris and M. Schiffer, C&EN News July 30 (1990) 22.
28 M. Fujihira, H. Yamada, and H. Ohtani, submitted to J. Phys. Chem.
29 M. Fujihira, 2nd International Symposium on Bioelectronic and Molecular Electron ic Devices R&D Association for Future Electron Devices, Dec.12–14, 1988, Fujiyoshida, Japan, pp. 35–38; M. Fujihira, M. Sakomura and T. Kamei, Thin Solid Films 180 (1989) 43.
30 M. Fujihira, T. Kamei, M. Sakomura, Y. Tatsu and Y. Kato, Thin Solid Films 179 (1989) 485.

Part 4:
Electron and Energy Transfer in Photosynthesis and Related Phenomena

Dynamics and Mechanisms of
Photoinduced Transfer and Related Phenomena
N. Mataga, T. Okada and H. Masuhara (Editors)

MULTIPHOTON EVENTS IN LIGHT HARVESTING ANTENNAE

Anne M. Brun and Anthony Harriman

Center for Fast Kinetics Research, The University of Texas at Austin, Austin, Texas 78712, U.S.A.

Abstract

A series of covalently-linked porphyrin (linear) trimers and (random) pentamers has been studied by laser spectroscopy under conditions of widely differing light intensity. The trimeric porphyrins, which are closely-coupled and exhibit strong exciton-coupling effects, appear to function as a single chromophore. It has proved extremely difficult to attach more than one photon to the trimer. The pentameric porphyrins, which are linked through highly-flexible dialkoxy sidechains, readily accept multiple photons onto the cluster but exciton annihilation is surprisingly slow.

1. INTRODUCTION

Extensive research effort has been directed toward the construction of supramolecular systems capable of mimicking some of the essential features of the natural photosynthetic apparatus. Particular attention has been given to the design of closely-spaced, donor-acceptor moieties that facilitate rapid, vectorial electron transfer upon illumination with visible light and many elaborate model systems have been described. These studies have extended our understanding about the mumerous parameters that control the rate of forward and reverse electron transfer processes but none of the models formulated to-date genuinely mimic the natural system. Less attention has been paid to replicating the light harvesting antennae that form an integral part of photosynthesis, although there have been many reports describing electronic energy transfer between covalently-linked dimers and higher oligomers. Assembling a more complete supramolecular system, comprising both light harvesting and electron transferring subunits, has been achieved only in a solitary case [1,2]. This latter system utilized four zinc porphyrins (as photon collectors) covalently linked to a central free-base porphyrin which acted as energy acceptor and electron donor.

One inevitable problem associated with the construction of

light harvesting antennae concerns energy-dissipating exciton annihilation processes. These reactions, which plague the natural organisms at high illumination intensities [3-5], compete with photon migration to the reaction centre complex (where electron transfer reactions occur) and minimize the energy storage capabilities of the system. As part of an extended programme intended to evaluate the importance of exciton annihilation in model systems we have considered both triplet-triplet and singlet-singlet annihilation processes in covalently-linked porphyrin clusters. It is shown that the nature of the spacer function plays a domineering role in determining the efficiency of exciton annihilation.

2. EXPERIMENTAL

All porphyrins were synthesized and purified as described elsewhere: the pentameric porphyrins were available from a previous study [1] whereas the trimeric porphyrins were prepared specially for this project [6]. All compounds gave satisfactory analyses and were further purified by tlc prior to making the photophysical measurements. Benzene and N,N-dimethylformamide (DMF) (Aldrich spectroscopic grade) were fractionally-distilled from CaH_2. Solutions were prepared fresh and protected from undue illumination.

Absorption spectra were recorded with a Hewlett Packard 8450A diode array spectrophotometer. Luminescence spectra were recorded with a Perkin Elmer LS5 spectrofluorimeter and were corrected for wavelength responses of the detector. Fluorescence quantum yields, calculated from integrated spectra, were measured relative to zinc tetraphenylporphyrin (ZnTPP) as standard [1]. Singlet excited state lifetimes were measured by the time-correlated single-photon-counting technique using a mode-locked Nd-YAG laser (Antares 76S) synchronously pumping a cavity-dumped Rhodamine 6G dye laser (Spectra Physics 375B/244) as excitation source. Narrow bandpass filters were used to isolate fluorescence from scattered laser light. A Hamamatsu microchannel plate was used to detect emitted photons, for which the instrumental response function had an fwhm of ca. 50 ps. Data analyses were made according to O'Connor and Phillips [7] using computer deconvolution to minimize reduced chi-square parameters. Alternately, a Hamamatsu streak-camera (temporal resolution ca.2 ps) was used in conjunction with a frequency-doubled, mode-locked Nd-YAG laser excitation source. In this case, narrow bandpass filters were used to isolate fluorescence, which was attenuated with neutral density filters. All solutions for fluorescence studies were optically-dilute (absorbances <0.08 at the excitation wavelength) and were air-equilibrated.

Conventional flash photolysis absorption studies were made

with a frequency-doubled, Q-switched Quantel YG481 Nd-YAG laser (pulsewidth 10-ns, 380 mJ). Solutions were adjusted to possess absorbances at 532 nm of ca. 0.1. Deoxygenation was achieved by N_2-purging. Differential absorption spectra were recorded point-by-point with 5 individual laser shots being averaged for each determination. Kinetic measurements were made at selected wavelengths with 50 individual laser shots being averaged for each analysis. Data analyses were made by computer least-squares iterative procedures.

Improved time resolution was achieved by employing a frequency-doubled, mode-locked Quantel YG402 Nd-YAG laser (pulsewidth 30-ps) as excitation source. Averaging procedures were used in which 300 laser shots were accumulated for each determination. Solutions were adjusted to possess absorbances of ca. 0.6 at the excitation wavelength. Residual 1064 nm output from the laser was focussed into 1/1 H_2O/D_2O to produce a white light continuum for use as the analyzing beam. Variable delay times in the range 0-6 ns were selected in a random sequence and transient spectra recorded with an Instruments SA UFS200 spectrograph interfaced to a Tracor Northern 6200 MCA and a microcomputer. Kinetic analyses were made by overlaying up to 30 individual spectra and fitting data at selected wavelengths using computer least-squares iterative procedures.

Laser intensities were attenuated as follows: First, the laser was carefully aligned so as to give a linear distribution across the beam profile, as evidenced by the uniform discolouration of photographic printing paper. The laser beam was collimated and directed through a 3-stage Glan-Taylor high-intensity polarizer. The first polarizer served to ensure correct polarization of the laser and the second and third polarizers were crossed to different degres so as to provide variable attenuation. Output intensities were measured with a calibrated Laser Precision Rj-7600 powermeter.

3. RESULTS AND DISCUSSION

3.1. Bimolecular Exciton Annihilation with monomeric Porphyrins

It is well documented [8] that porphyrins undergo triplet-triplet annihilation in deoxygenated fluid solution. We have re-investigated this phenomenon with several simple (monomeric) porphyrins with a view to discriminating between bimolecular and unimolecular exciton annihilation processes that might occur within the supramolecular systems. Experiments were made with free-base meso-tetraphenylporphyrin

(H_2TPP), with the corresponding zinc(II) complex (ZnTPP), with free-base octaethylporphyrin (H_2OEP), and with its zinc(II) complex (ZnOEP) in deoxygenated benzene solution at room temperature. The experimental studies are illustrated by reference to ZnTPP and parameters for the other compounds are compiled in Table 1.

Table 1
Properties of the triplet excited state of the monomeric porphyrins in benzene solution

Compound	Φ_t	$k_1/10^3$ (s^{-1})	$k_2/10^9$ (M^{-1} s^{-1})	$\epsilon/10^3$ (M^{-1} cm^{-1})	λ_{max} (nm)
H_2TPP	0.67	1.9	3.2	67	440
ZnTPP	0.83	3.1	3.4	73	470
H_2OEP	0.54	2.7	3.5	44	440
ZnOEP	0.65	3.5	3.0	54	435

A deoxygenated solution of ZnTPP (15 μM) in benzene was irradiated with a single 10-ns laser pulse at 532 nm. Figure 1 shows a plot of the differential absorbance measured at 470 nm, which is the observed peak maximum (λ_{max}; ϵ = 73,000 M^{-1} cm^{-1}) [8], as a function of incident laser power. Saturation occurs at a point where there is essentially 100% conversion of ground state ZnTPP into the triplet species. The half-life of the triplet state shows a marked dependence on laser power and decay profiles change from first-order at very low power to second-order at high power. Throughout the entire power-dependence plot, the decay profile can be fit to mixed first- and second-order kinetics with a first-order rate constant (k_1) of 3.1 x 10^3 s^{-1} and a second-order rate constant (k_2) of 3.4 x 10^9 M^{-1} s^{-1}. Similar behaviour was observed for the other monomeric porphyrins and the derived k_1 and k_2 values, together with the differential molar extinction coefficients measured at the appropriate maxima (ϵ), are compiled in Table 1. Using ZnTPP as reference [8,9], quantum yields for formation of the triplet state (Φ_t) were measured at low laser power and are collected in Table 1.

Each of the porphyrins fluoresce in fluid solution and fluorescence quantum yields (Φ_f) and lifetimes (τ_f), measured by single-photon-counting, are given in Table 2. Differential transient absorption spectra were recorded by the pump-probe technique for the singlet excited state of each of the porphyrins in fluid benzene solution. Absorption maxima and differential molar extinction coefficients, derived from these spectra at low laser power, are collected in Table 1. Using a mode-locked Nd-YAG laser (FWHM = 30 ps) as excitation source and streak-camera detection, we were unable to detect a reduction in fluorescence lifetime or a change in decay profile at high (>20 mJ) laser intensities. However, integrating the fluorescence decay profiles indicated that the fluorescence quantum yield decreased at high power. This effect is attributed to two-photon absorption by the excited singlet state [10], which absorbs appreciably at 532 nm; for ZnTPP, the molar extinction coefficient for the first excited singlet state at 532 nm is 8,600 M^{-1} cm^{-1} compared to that of the ground state species of 6,500 M^{-1} s^{-1}. This effect, although noticable, is significant only at high laser powers. The two-photon absorption cross-sections σ_{02} are given also in Table 2. Taking the maximum instantaneous laser intensity I(t) to be $\approx 10^{27}$ photons cm^{-2} s^{-1} together with the calculated one-photon absorption cross-sections at 532 nm σ_{01}, the relative rate of two-photon absorption (R) can be expressed as follows:

$$\sigma_{02} I(t)^2 / \sigma_{01} I(t) = R \qquad (1)$$

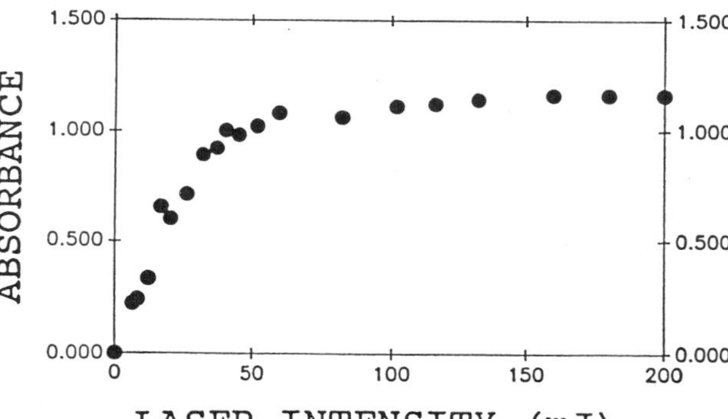

Figure 1. Effect of laser intensity on the triplet absorbance at 470 nm for ZnTPP in benzene solution.

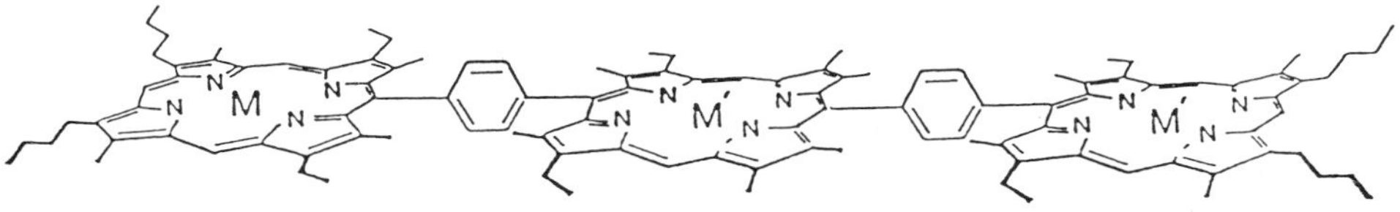

M = ZINC(II) IONS OR PROTONS

Figure 2. Structures of the linear trimeric porphyrins.

Table 2
Properties of the excited singlet state of the monomeric porphyrins in benzene solution

Compound	Φ_f	τ_f (ns)	λ_{max} (nm)	$\epsilon/10^3$ (M^{-1} cm^{-1})	R
H_2TPP	0.130	10.6	440	54	5×10^{-6}
ZnTPP	0.033	2.0	450	75	8×10^{-4}
H_2OEP	0.115	10.2	425	40	9×10^{-6}
ZnOEP	0.040	1.9	445	52	2×10^{-5}

3.2. Photophysical Properties of the Trimeric Porphyrins

Structures of the two trimeric porphyrins are displayed in Figure 2. Using X-ray crystallographic data obtained for the comparable bis-copper linear dimeric porphyrin [6], it is expected that the tetrapyrrolic rings adopt a flat arrangement with a centre-to-centre separation distance of 12.7 Å. There is some opportunity for rotation around the linking benzene ring but the pyrrole substituents prevent full rotation. The compounds can be regarded, therefore, as linear trimeric porphyrins of fixed distance but slightly variable interplanar orientation. As such, an important feature of these compounds concerns the degree of communication between the three porphyrin rings.

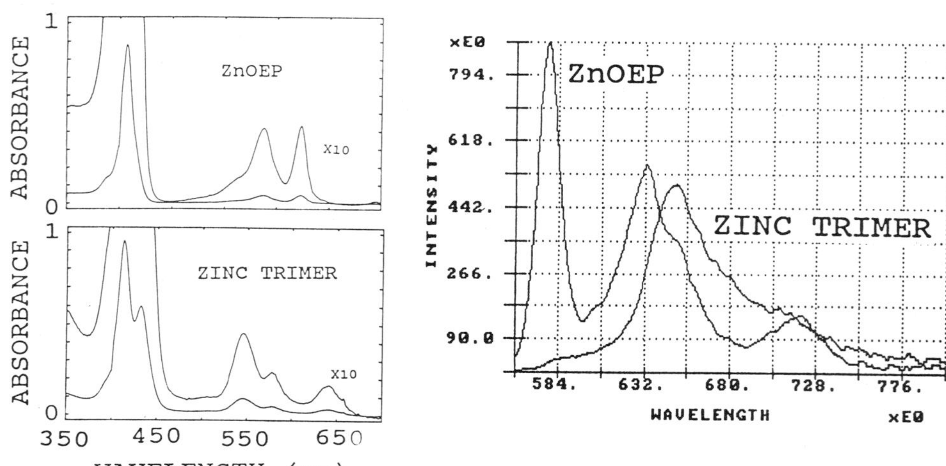

Figure 3. Comparision of (a) absorption and (b) fluorescence spectra recorded for ZnOEP and ZnTRI in benzene solution.

The free-base analogue (H_2TRI) is modestly soluble in benzene whereas the zinc complex (ZnTRI) is more soluble in DMF and photophysical studies were made in these solvents. Absorption spectra recorded for both compounds exhibit considerable exciton coupling effects, especially in the Sôret region around 400 nm. This phenomenon is particularly significant for ZnTRI where the Soret band is clearly split into two components and a new absorption band appears around 650 nm; Figure 3 compares (a) absorption and (b) fluorescence spectra recorded for ZnOEP and ZnTRI. Fluorescence spectra recorded for the trimers are similarly modified compared to those of the relevant monomers. For H_2TRI the bands are red-shifted (≈12 nm) and somewhat broadened whereas for ZnTRI the spectrum is changed extensively. The observed spectral shifts cannot be accounted-for in terms of simple exciton theory [11] due, possibly, to the co-existence of multiple orientations.

Photophysical properties were measured for the trimers under conditions of low laser intensity and the derived values are collected in Table 3. Transient absorption spectra recorded for the excited singlet and triplet states retain similar features to those characteristic of H_2OEP and ZnOEP but there are important differences. The most pronounced change is observed in the Sôret region of the triplet spectra. For the trimers the bleaching region does not match the ground state absorption spectrum and, for ZnTRI, it is clear that only one of the exciton transitions is bleached upon excitation. This effect, which may be seen from Figure 4, is being studied in more detail and will be reported elsewhere.

Fluorescence from the trimers is quenched slightly compared to the corresponding monomers and the triplet states, which are formed in reduced quantum yield, are significantly shorter-lived. These effects most probably arise from the exciton coupling effect providing a means for rapid nonradiative deactivation of the excited states [12].

Table 3
Photophysical properties recorded for the linear trimeric porphyrins in fluid solution

Compound	Φ_f	τ_f (ns)	Φ_t	$k_1/10^5$ (s^{-1})	τ_{an} (ps)
H_2TRI	0.100	9.2	0.45	2.0	230
ZnTRI	0.033	1.5	0.50	1.9	200

Experiments were performed at much higher laser intensity with a view to evaluating the possibility of depositing two (or three!) photons onto one trimer molecule: the feasibility

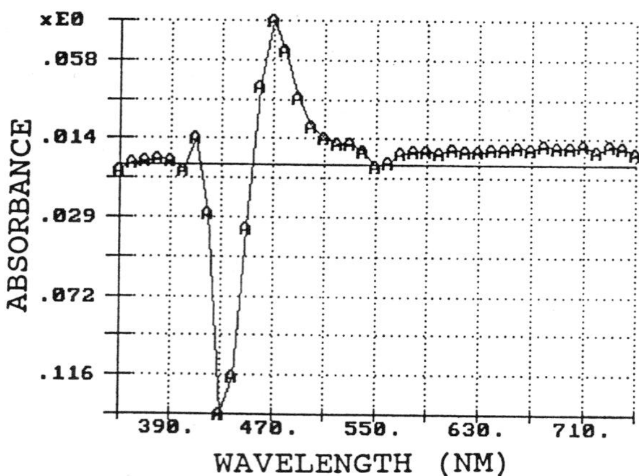

Figure 4. Triplet absorption spectrum recorded for ZnTRI in benzene solution.

of this occurrence depends on the degree of electronic coupling between the three tetrapyrrolic rings. Excitation with a 10-ns laser pulse at 532 nm allows the concentration of the triplet state to be recorded. As shown in Figure 5, at low laser intensity there is a linear dependence between transient differential absorbance at the appropriate maximum and the incident laser power. Saturation occurs at relatively low intensity and, for both trimeric porphyrins, corresponds to only 30-35% conversion of ground state into triplet. Increasing the laser intensity above the saturation value gives only slight increases in absorbance. Furthermore, the observed triplet lifetime in deoxygenated solution remains independent of laser intensity and there is no indication of a short-lived (τ>10 ns) component in the decay records. At the low concentrations used, and taking into account the rather short inherent triplet lifetimes, bimolecular processes are unimportant in these studies. This behaviour is consistent with the tetrapyrrolic rings in these trimeric porphyrin molecules acting in a cooperative fashion and virtually restricting excitation to a single photon per molecule. The photon may be considered as being delocalized over the whole molecule and not concentrated at an individual porphyrin subunit. The small, but discernible, increase in absorption at post-saturation intensities may result from some nonlinear optical effect due to the high laser powers used. No changes in absorption spectra or kinetics arise because of this effect.

Similar experiments were made to determine the fluorescence lifetime of the trimeric porphyrins at different laser intensity using streak-camera detection. At low intensity ($I(t) \approx 10^{22}$ photons cm^{-2} s^{-1}), single-exponential decay profiles were observed which corresponded to lifetimes comparable to those derived from single-photon-counting studies. At much higher laser intensities ($I(t) \approx 10^{25}$ photons cm^{-2} s^{-1}), non-exponential decay profiles were observed which contained a short-lived component in addition to the regular lifetime. These decay profiles could be analysed in terms of a two-exponential fit, although the analyses were less than satisfactory: analysis in terms of a $t^{1/2}$ dependence gave no real improvement in the quality of the fit [13]. As the laser intensity was increased the fractional amplitude of the shorter-lived component increased but the derived lifetime (τ_{an}) remained constant within experimental error (Table 3). [There is a corresponding decrease in fluorescence quantum yield with increasing laser intensity.] The shorter-lived component accounts for about 20% of the total fluorescence intensity at the maximum laser intensity ($I(t) = 10^{27}$ photons cm^{-2} s^{-1}). It is unreasonable to interpret this component in terms of singlet-singlet exciton annihilation arising from two photon absorption because of the slow timescale of the process. Indeed, from the average decay time and using the crystal structure data, the average exciton diffusion coefficient would be only $\approx 10^{-4}$ cm^2 s^{-1}. Instead, it is more probable that the effect arises from stimulated emission, although this proposal was not probed since it is only a minor part in the overall process, even at the highest intensity.

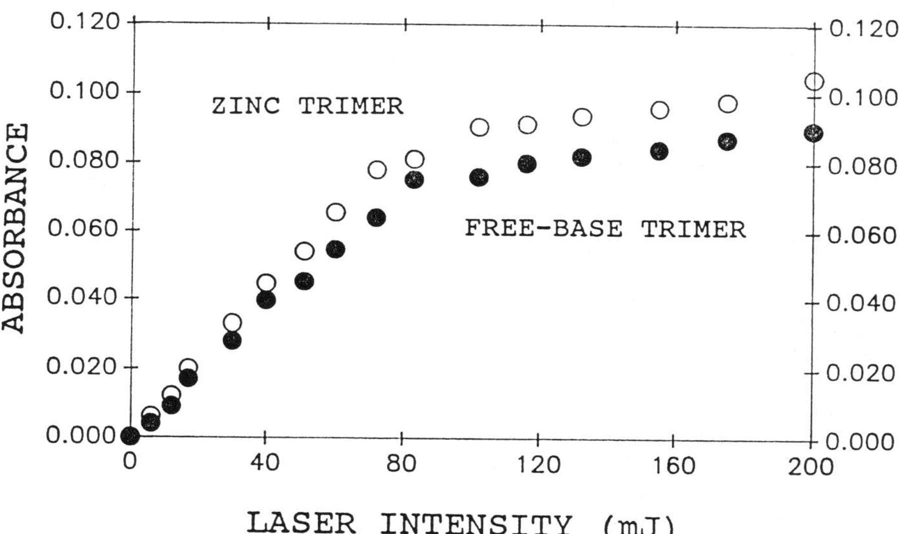

Figure 5. Effect of laser intensity on the triplet absorbance for ZnTRI in DMS solution.

3.3. Photophysical Properties of the Zinc Pentameric Porphyrin

R = $-C_6H_4-CH_3$

M = S = Zn Zn_5
M = Zn S = 2H Zn_4H_1
M = 2H S = Zn Zn_1H_4

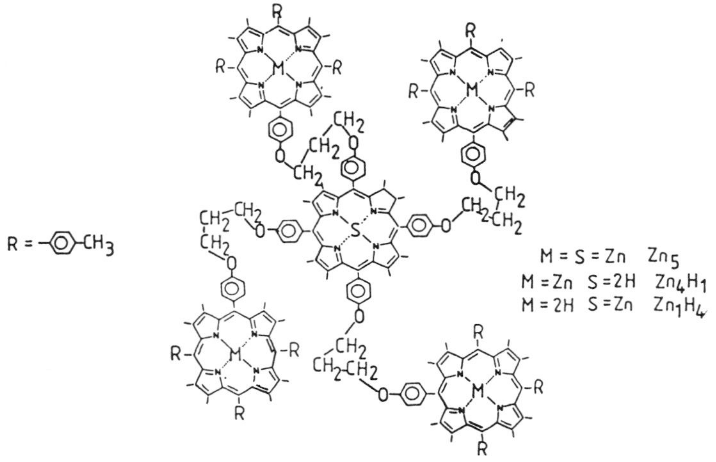

Figure 6. Structures of the porphyrin pentamers studied in this work.

The zinc pentameric porphyrin (Figure 6) was studied in benzene solution. The absorption spectrum indicates no exciton coupling between the rings and the five porphyrins appear to be independent of each other. Fluorescence is readily observed (Φ_f = 0.031 ; τ_s = 1.5 ns) which is slightly quenched relative to ZnTPP and a long-lived triplet excited state is formed in high yield (Φ_t = 0.65) which retains the characteristic differential absorption spectrum described for ZnTPP. At low levels of excitation with a 10-ns laser pulse at 532 nm, the triplet decays via first-order kinetics (τ_t = 110 ≤s) in deoxygenated benzene solution. Using a dilute solution (2.21 ≤M) in order to eliminate bimolecular effects, the triplet state properties were studied as a function of laser intensity. It was observed that with increasing laser intensity the triplet decay profile became dual-exponential (Figure 7). In addition to the regular lifetime, a faster decaying component (τ_{an}) could be resolved at higher intensity; the fractional contribution of this component increases linearly with increasing intensity. At all laser intensities, the shorter-lived component had a lifetime of 24 μs and retained the same differential absorption spectrum as that of the triplet. The decay profiles were well-described in terms of two competing first-order processes but could be fit also, but not so well, to competing mixed first- and second-order processes.

Monitoring at 470 nm, the absorbance increases linearly with increasing laser intensity without reaching saturation. At the highest laser intensity used, an average of 2.5 photons have been added to each pentamer such that, presumably, 2 or 3 porphyrins per cluster are present in the triplet excited state. At lower concentration (1 μM), complete conversion into the triplet state (i.e. addition of 5 photons per pentamer molecule) was achieved such that each of the 5 porphyrins is present as a triplet. The shorter-lived component in the decay records can be assigned, therefore, to triplet-triplet exciton annihilation, for which the first-order rate constant is 3.3 x 10^4 s^{-1}. This rate constant is derived on the basis of a static system but the individual porphyrins possess considerable degrees of freedom and can diffuse on the timescale of the experiment [14]. In fact, with a diffusion coefficient of 10^{-6} cm^2 s^{-1}, the diffusional length is 3,000 Å in 100 μs. Since the average edge-to-edge separation distance is expected to be of the order of about 40 Å, for which the diffusional time constant is only 40 ns, triplet-triplet annihilation is very inefficient.

Fluorescence from the zinc pentamer was studied as a function of laser intensity using streak-camera detection. Over a wide intensity range (10^{22}<I(t)<10^{27} photons cm^{-2} s^{-1}) the observed decay profiles could be analysed satisfactorily in terms of a single-exponential process with a lifetime of 1.5 ns and there was no change in the fluorescence quantum yield. Under these conditions, it was calculated that an average of 3 photons were added to each cluster. The absence of fluorescence quenching suggests that singlet-singlet exciton coupling in this system occurs too slowly to compete with the inherent deactivation processes. If both singlet and triplet excitons possess similar diffusion coefficients we would not expect to see singlet annihilation on the timescale of the experiment.

Similar results were obtained with the corresponding free-base porphyrin pentamer. Again, triplet exciton but not singlet exciton annihilation was observed. The apparent triplet exciton annihilation rate constant was found to be 2.5 x 10^4 s^{-1}.

3.4. Triplet Exciton Annihilation in Mixed Porphyrin Pentamers

We assume that energy migration may occur among the various porphyrin molecules within a cluster, although this may be a fairly inefficient process due to the poor overlap integral. As an extension of the above studies, similar experiments were made with pentamers comprising mixed zinc and free-base porphyrins (Figure 6). In view of the above results, only triplet state processes were studied. Because of the energetics of the system [1,2], photons absorbed by a zinc porphyrin will be transferred rapidly and irreversibly to an adjacent free-base porphyrin. From single-photon-counting studies conducted in benzene solution it was deduced that singlet energy transfer from zinc to free-base porphyrins occurred with rate constants of 11.1 and 9.4 x 10^9 s^{-1}, respectively, for Zn_1H_4 and Zn_4H_1. As a result of this process, addition of a single photon onto each pentamer will result in population of the triplet state of the free-base porphyrin, regardless of which porphyrin absorbs the incident photon [1,2]. Addition of a further photon can populate a second triplet which will be associated with a free-base porphyrin in Zn_1H_4 and a zinc porphyrin in Zn_4H_1.

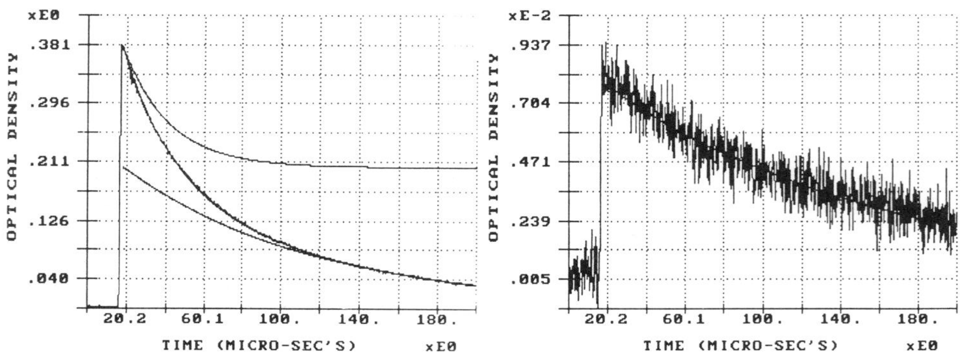

Figure 7. Decay profiles recorded for the triplet excited state of the zinc porphyrin pentamer at low (12 mJ) and high (120 mJ) laser intensities.

Excitation of Zn_1H_4 with a 10-ns laser pulse at 532 nm results in formation of a free-base porphyrin triplet state which decays with a lifetime of 190 μs at low laser intensity. As the intensity is increased a shorter-lived component appears in the decay profiles, as described for the zinc porphyrin pentamer. This shorter-lived component retains the same lifetime (τ_{an} = 38 μs) at all laser intensities but its fractional contribution increase with increasing power. The differential absorption spectrum does not change with laser intensity, at least up to the point where 4 photons are added to each pentameric molecule. As above, the short-lived component is attributed to triplet-triplet exciton annihilation, for which the rate constant is 2.1 x 10^4 s^{-1}. This value is similar to that obtained for the free-base pentamer such that annihilation appears to occur between the peripheral porphyrins, via diffusional migration, without involving the central porphyrin.

Similar experiments were conducted with Zn_4H_1 in dilute benzene solution. At low laser intensity, the free-base porphyrin triplet excited state is populated and decays via first-order kinetics with a lifetime of 420 ≤s. At higher laser intensity. a shorter-lived component appears in the decay profiles but its lifetime (τ = 120 μs) and differential spectrum suggests that it is an isolated zinc porphyrin triplet. There is no indication of mixed triplet-triplet annihilation between free-base and zinc porphyrin triplet states. Addition of more photons onto the cluster results in the appearance of a third component in the decay profiles (τ_{an} = 22 μs) which is assigned to triplet-triplet exciton annihilation between peripheral zinc porphyrins; the rate constant for this process being 3.7 x 10^4 s^{-1}. Thus, this system provides a good example of a non-linear optical effect.

4. CONCLUDING REMARKS

Two different types of porphyrin-based clusters have been studied; namely, closely-spaced linear trimers and flexibly-linked pentamers. The trimers are unsuitable for use in light harvesting antennae since they appear to function as a single chromophore. The pentamers are much more appropriate choices for light harvesters. Each of the five porphyrins within the operates independently and exciton migration is slow. Efforts are underway to extend these pentamers into large 3-dimensional arrays by adding a successive layer of porphyrins.

5. ACKNOWLEDGEMENT

This work was supported by the Texas Advanced Technology Program. We thank Vincent L. Capuano for a sample of the free-base trimeric porphyrin. The Center for Fast Kinetics Research is supported jointly by the Biomedical Research Technology Program of the Division of Research Resources of the N.I.H. (RR00886) and by the University of Texas at Austin.

6. REFERENCES

1 J. Davila, A. Harriman and L.R. Milgrom, Chem. Phys. Lett., 136 (1987) 427.

2 A. Harriman, Supramolecular Photochemistry, V. Balzani (ed.) NATO ASI Series, 214 (1987) 207.

3 D. Mauzerall, Biophys. J., 16 (1976) 87.
4 T.G. Monger and W.W. Parson, Biochim. Biophys. Acta, 460 (1977) 393.
5 J. Breton and N.E. Geacintov, Biochim. Biophys. Acta, 594 (1980) 1.
6 J.L. Sessler, M.R. Johnson and T.-Y. Lin, Tetrahedron, 45 (1989) 4767.
7 D.V. O'Connor and D. Phillips, Time Correlated Single Photon Counting, Academic Press, London, 1984.
8 L. Pekkarinen and H. Linschitz, J. Am. Chem. Soc., 82 (1960) 2407.
9 J.K. Hurley, N. Sinai and H. Linschitz, Photochem. Photobiol., 38 (1983) 9.
10 A.J. Campillo and S.L. Shapiro, Photochem. Photobiol., 28 (1978) 975.
11 D.E. LaLonde, J.D. Petke and G.M. Maggiora, J. Phys. Chem., 93 (1989) 608.
12 F.K. Fong, M.S. Showell and A.J. Alfano, J. Am. Chem. Soc., 107 (1985) 7231.
13 J.A. Altmann, G.S. Beddard and G. Porter, Chem. Phys. Lett., 58 (1978) 54.
14 N. Tamai, T. Yamazaki, I. Yamazaki and N. Mataga, Chem. Phys. Lett., 120 (1985) 24.

DISCUSSION

Fukumura

1. You showed us the saturation of triplet state yield and explained it with a complete excitation of the ground state to the triplet state. If you know the value of the T–T abs. coefficient, you could estimate the concentration of the triplet state from the value of the initial optical density. Could you check it?

2. Can you really neglect the effect of S_1–S_1 annihilation, simultaneous multiphotonic process, or the other non–linear processes under such high power irradiation?

Harriman

1. Yes, we know accurately the molar extinction coefficients for the various triplet excited states. By varying the total concentration of chromophore and laser intensity, we are able to confirm the achievement of complete conversion of ground state species into triplet. This could not be done for the trimers, if the trimer is considered to consist of three separate porphyrin molecules.

2. We have not been successful in detecting S_1–S_1 exciton annihilation in these molecules despite much investigation. A large amount of heat is deposited in the system during T_1–T_1 annihilation and could contribute towards the observed effects.

Wasielewski

1. What solvents were used for your trimer studies?

2. Can you explain why you do not see a bleach of both exciton components of the band in the trimer upon excitation? Is it possible that these are not exciton components, but independent bands due to solvation of one or two porphyrins that differs from the third.

Harriman

1. Benzene, toluene and N,N–dimethylformamide were used with comparable results. The free–base analogue dissolves only in benzene–type solvents.

2. I have no satisfactory explanation for the non–linear bleaching. It is observed for zinc, free–base and palladium trimers, some of which do not bind axial ligands. The exciton bands can be described reasonably well in terms of simple exciton theory and appear to arise from exciton–splitting.

Faure

Several years ago you have shown that $S_1 \rightarrow S_n$ and $T_1 \rightarrow T_n$ absorption of

ZnTPP occur in the same region (500 nm). Is it the same behavior for Au–Porphyrin and if yes how can you discriminate triplet and singlet decays?

Harriman

Gold(III) porphyrins were selected because their excited singlet states are extremely short (<0.5 ps) due to the internal heavy atom effect. Consequently, these compounds can be considered to act only through the triplet manifold.

Steiner

Usually triplet–triplet annihilation leads to the porphyrin of higher singlet states and it can be probed by delayed fluorescence. Did you investigate this phenomenon in connection with your porphyrin oligomers?

Harriman

Delayed fluorescence was confirmed only for palladium porphyrins. We did not study this phenomenon for zinc porphyrins and the various gold porphyrins do not luminesce in fluid solution.

Jortner

The "transition" from exciton dynamics in a supermolecular assembly (strong interactions) to isolated excitations (weak interactions) requires the elucidation of the magnitude of triplet exciton splitting relative to dynamic disorder (due to thermal nuclear motion) and to static disorder (inhomogeneous broadening). This issue pertains to exciton localization, which is of considerable interest. From the experimental point of view it will be instructive to obtain high–resolution spectra in low–temperature solids and possibly also of isolated molecules in supersonic jets. From the theoretical point of view, previous success in the calculation of triplet exciton band structure in molecular crystals in conjunction with information on electronic wavefunctions of porphyrins, calls for the calculations of triplet exciton splitting in the molecular assemblies of porphyrins.

Harriman

I agree completely that better quality (i.e. non–optical) spectroscopic methods should be applied to such systems. We are trying to do low temperature studies at present which may provide better data.

Mobius

You said that your case put 5 photons in the Zn–porphyrin pentamers, but only 1 photon in the trimer. Do you know a reason for this difference in behavior?

Harriman

The trimers appear to function as a single supramolecular assembly rather than as three isolated porphyrins. We cannot, at this time, explain this behavior in terms of the molecular electronics.

Dynamics and Mechanisms of
Photoinduced Transfer and Related Phenomena
N. Mataga, T. Okada and H. Masuhara (Editors)

DYNAMICS OF ENERGY TRANSFER AND TRAPPING IN PHOTOSYNTHETIC BACTERIA

Villy Sundström[a] and Rienk van Grondelle[b]

[a]Department of Physical Chemistry, University of Umeå, Sweden.

[b]Department of Biophysics, Physics Laboratory, Free University, Amsterdam, The Netherlands.

Abstract

With infrared transient absorption spectroscopy we have studied how energy migrates through the light-harvesting antenna of photosynthetic purple bacteria, and how the energy is trapped by the reaction center. In Bchl *a*-containing purple bacteria the light-harvesting (LH) antenna is highly heterogeneous, consisting of several spectroscopically distinct pigments. The energy transfer through the antenna is fast ≈10 ps in these bacteria, and trapping is relatively slow ≈ 35 ps at 77 K (and probably not much faster at room temperature). For the Bchl *b*-containing purple bacterium *Rps.viridis* the results show that the antenna is homogeneous, and trapping appears to be much faster than suggested by calculated Förster spectral overlap for energy transfer from antenna to the special pair.

1. PIGMENT ORGANIZATION

Purple photosynthetic bacteria all contain either bacteriochlorophyll *a* (Bchl *a*) or bacteriochlorophyll *b* (Bchl *b*) as the dominating pigment in both the light-harvesting antenna and the reaction center. Of the Bchl *a* -containing bacteria many different species have been characterized and studied, whereas for Bchl *b* -bacteria only one species *Rps. viridis* has been reasonably well studied. From the high resolution crystallographic structures of *Rps. viridis* (1) and *Rb. sphaeroides* (2) reaction centers it turns out that the reaction centers of the different purple bacteria are quite similar. The reaction center is surrounded by the LH1 core antenna, which consists of pairs of α and β transmembrane polypeptides to which the Bchl pigment molecules are associated (about 30 Bchl/RC), probably via a histidine residue. The existing models for polypeptide organization in the membrane places the centers of the Bchl molecules in the same plane with the Q_y transition dipoles in the plane of the membrane. EM micrographs of *Rps. viridis* (3) show that at least for this species LH1 occurs in a highly ordered structure with a six-fold symmetry. Whether this is also the case for the Bchl-*a*-containing species is not known, but similar arrangements have been proposed. From measurements of induced fluorescence (4) and singlet-singlet excitation annihilation (5) of *R. rubrum* and *Rps. viridis* it has been concluded that the LH1 antenna forms an extended network connecting many (approx. 20) reaction centers into a so called domain consisting of about 1000 Bchl molecules at room temperature. At low temperature (4 K) a much smaller domain size (approx. 100 Bchl) was deduced from the annihilation data (5), and concluded to be a result of decreased rate of energy transfer between individual Bchl molecules.

Various models have been suggested for the organization of the α and β polypeptides of the LH1 antenna. It is believed that pairs of α and β polypeptides form a minimum unit which aggregates further to form the fully functional in vivo antenna (6). However, there is no general agreement as to the composition of this minimum unit and its aggregation. The isolation of a subunit form, B820, of LH1 of *R. rubrum* (7) yielded additional knowledge about the aggregation state of LH1. From absorption and CD spectra it was concluded that two excitonically interacting Bchl *a* molecules are responsible for the spectral properties of B820 (8). Molecular weight determination with gel filtration showed that B820 had a considerably lower molecular weight than the associated B873 form, $(\alpha\beta)_2$ or $(\alpha\beta)_3$ were considered to be likely forms (7). Triplet minus singlet spectroscopy and measurements of singlet-triplet annihilation (9) also suggested that the B820 absorption is due to a dimer of Bchl *a*, and the B820 complex was concluded to consist of only one αβ-heterodimer. This dimer was proposed to be the building block of the LH1 antenna.

Several species of purple bacteria also contain a peripheral LH2 antenna, at higher excited-state energy, in addition to the LH1 core antenna. The size of the LH2 antenna depends on growth conditions and it is often the dominating pigment form. Light absorbed by the LH2 antenna is transported to the reaction center via the LH1 antenna. Models similar to those discussed above for LH1 have been suggested for the organization of LH2 (10). Hence, the antenna consists of α and β transmembrane polypeptides to which binds three Bchl molecules /αβ-dimer. No agreement exists on the aggregation state and the minimum functional unit of LH2. A minimum unit with the composition $(\alpha\beta)_2$ containing six Bchl molecules and three carotenoids explains much of the spectroscopic properties of LH2, but the in vivo aggregation is probably more extensive. Preliminary crystallographic data at low resolution (11) suggests an aggregation of $(\alpha\beta)_2$ into a larger $(\alpha\beta)_6$ unit with a three-fold symmetry, whereas electron microscopy data on two-dimensional crystals of detergent-isolated LH2-complexes suggests an aggregation of five $(\alpha\beta)_2$ to $(\alpha\beta)_{10}$ (12). Whether either of these structures represent the relevant aggregation state in vivo is presently not known.

2. ENERGY TRANSFER IN PURPLE BACTERIA CONTAINING ONLY THE LH1 ANTENNA

R. rubrum and *Rps. viridis* are two examples of photosynthetic purple bacteria containing only LH1 as their major light-harvesting pigment. These are the two species of this type of bacteria for which most of the reported time-resolved work has been performed.

2.1 Rhodospirillum (R). rubrum

LH1 of *R. rubrum* contains Bchl *a* and is characterized by an infrared absorption band peaking at 880 nm. With photochemically active ("open") reaction centers at room temperature the antenna exciton lifetime is 60-70 ps, as measured with both picosecond absorption (13) and fluorescence techniques (14). This represents the total trapping time by the active reaction centers. When the reaction center is in its closed state, with the primary donor oxidized (P^+), the exciton lifetime lengthens to about 200 ps, showing that the reaction center efficiently quenches the excitation energy also with the primary donor in its oxidized state. Figure 1 shows the results of picosecond transient absorption measurements of chromatophores of *R. rubrum* with "open" (A) and "closed" (B) reaction centers at room temperature.

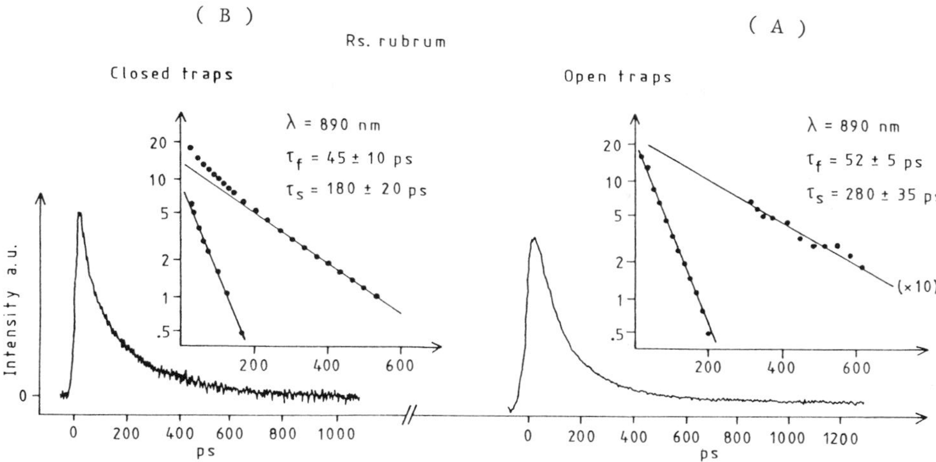

Figure 1. Absorption kinetics of *R. rubrum* at room temperature.

The kinetic trace of Fig. 1B also displays a faster 30-50 ps decay component in addition to the 200 ps quenching time due to the closed reaction center. This faster decay is a result of energy equilibration between the main LH1 antenna and a minor low-energy antenna component denoted B896. Energy transfer kinetics of purple bacteria measured at room temperature is often observed to be non-exponential. Due to the small energy separation (as compared to kT) between many of the antenna pigments the observed kinetics reflects energy equilibration of the excitation energy, rather than a unidirectional energy transfer from one antenna component to another. The kinetics observed for *R.rubrum* at room temperature is an example of this behaviour. By measuring the kinetics at 77 K effectively unidirectional rates can be obtained, and the spectral resolution is much improved as a result of narrower absorption bands.

Low temperature absorption (15,16) and fluorescence kinetics of *R.rubrum* show that LH1 is heterogeneous and that excitation energy is transferred from the main LH1 antenna to the minor B896 pigment with a time constant of approximately 20 ps, Fig. 2. That spectrally distinct pigments are involved in this process is concluded from the fact that the two kinetic components have unique isosbestic wavelengths and thus different absorption spectra.

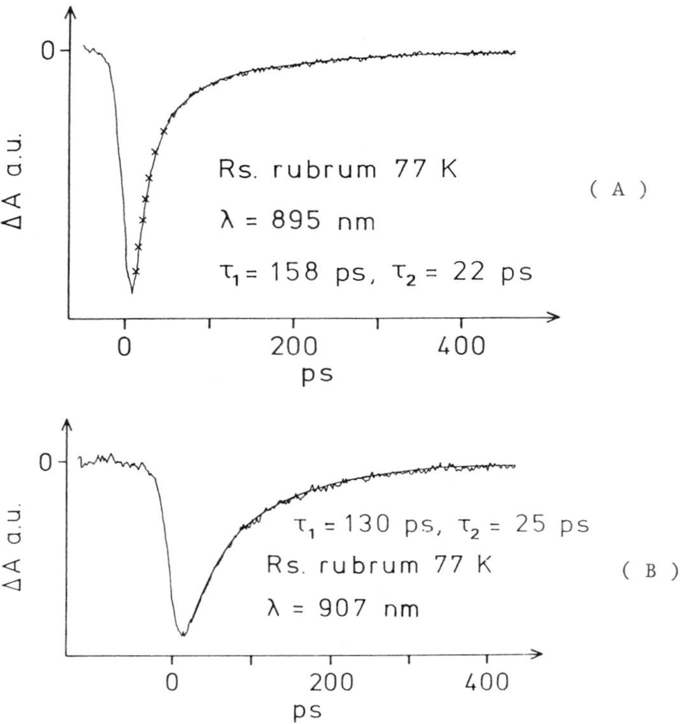

Figure 2. Absortion kinetics of *R. rubrum* at 77 K.

When absorption kinetics is measured at 895 nm (Fig. 2A) the decay is dominated by the 20 ps time constant due to B880 --> B896 energy transfer, since this wavelength is close to the isosbestic wavelength of B896. At longer wavelengths (Fig. 2B) the amplitude increases of the slower ≈ 150 ps component, and at $\lambda > 910$ nm this is the dominating contribution to the decay. As was mentioned earlier this slower component represents the quenching of B896 excitons by the closed reaction center.

By selective excitation and probing of B896 the trapping of the B896 excitons by the photochemically active reaction center was found to occur with a time constant of 35 ps in the temperature range 100-177 K (17,18). At lower temperature (77 K) this time increased to about 70 ps, probably due to decreased spectral overlap between B896 and the special pair of the reaction center (P870). The observed 35 ps trapping time was estimated to correspond to a 30 Å distance between the antenna B896 and P molecules, which is in good agreement with what can be expected on the basis of size and shape of the reaction center. The trapping time at room temperature was concluded to be not more than a factor of two shorter than the value measured in the temperature range 100-177 K (18). This result shows that trapping is a slow process in these bacteria, which probably explains the "need" for a minor B896 component, consisting of only a few Bchl molecules per reaction center, which concentrates the energy to the vicinity of the reaction center and provides a special entry to P.

2.2 Rhodopseudomonas viridis

Until very recently there was an almost complete lack of knowlwdge about the energy transfer dynamics in the LH1 antenna of *Rps. viridis*. The main reason for this was the the lack of suitable picosecond laser sources in the wavelength range of Bchl *b* absorption, 960 - 1050 nm. However, this has changed and presently mode-locked dye lasers and titanium-sapphire lasers are available, that operate at these wavelengths.

Using low-intensity infrared (960-1020 nm) picosecond pulses we have measured the antenna exciton lifetime of *Rps.viridis* at various RC redox states (19). It was shown that the exciton lifetime in the light-harvesting antenna with photochemically active reaction centers at room temperature is 60 ps, Fig. 3A. Prereducing the secondary electron acceptor of the reaction center (PIQ_A^-) prior to excitation results in a somewhat longer antenna lifetime 90 ps (Fig. 3B), and closing the reaction center by preoxidizing the primary donor ($P^{+}960$) lengthens this time to about 150 ps (Fig. 3C). Measurements of time-resolved absorption anisotropy and isotropic absorption kinetics at several wavelengths at 77 K strongly suggested that the LH1 antenna of *Rps. viridis* is homogeneous and that the Q_y transitions of the antenna Bchl *b* molecules have a circularly degenerate orientation in the plane of the membrane, in agreement with earlier steady-state measurements (20). The observation of a low (≤ 0.1) and constant on the picosecond time-scale absorption anisotropy is consistent with a rapid subpicosecond energy transfer between nearest-neighbor Bchl *b* antenna molecules.

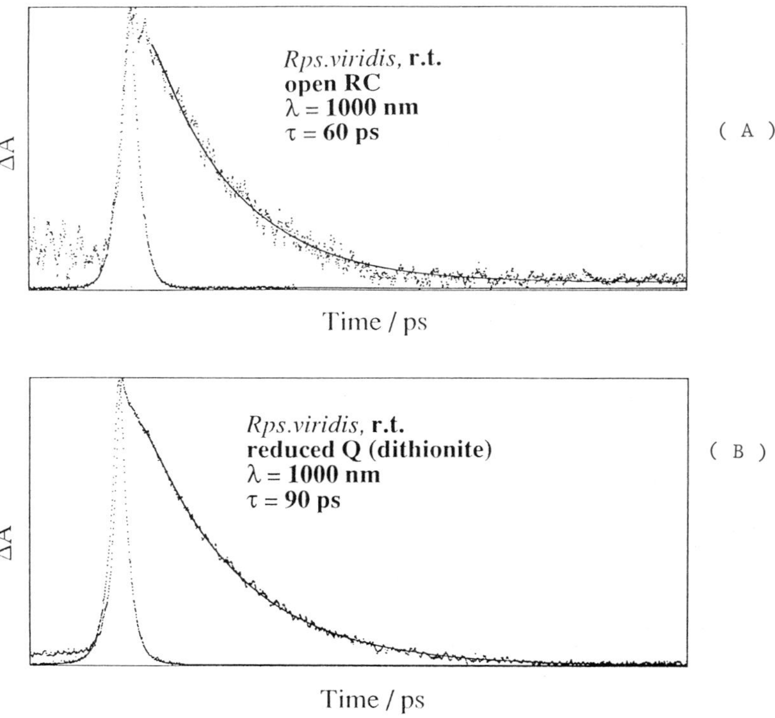

Figure 3 A,B. Absorption kinetics of *Rps. viridis* at room temperature and different RC redox states.

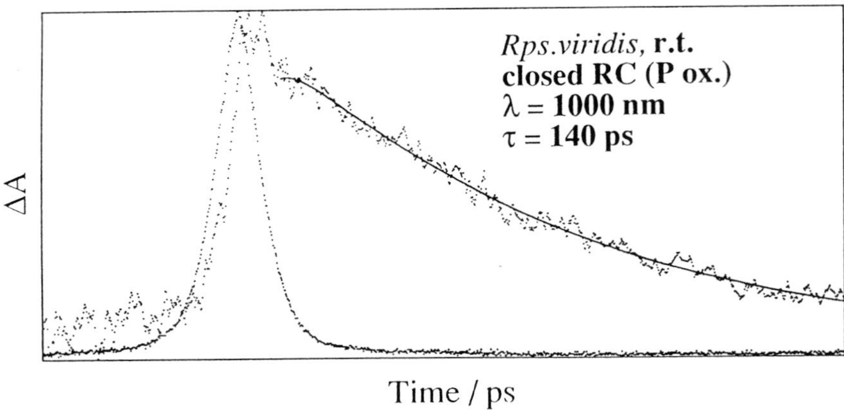

Figure 3 C. Absorption kinetics of *Rps. virids* at room temperature with closed RC.

Using these results as input to a random-walk model shows that: i. The longer antenna exciton lifetime upon Q_A prereduction is a result of a factor of two decreased rate of primary charge separation; and ii. The measured trapping time of 60 ps is consistent with the estimated spectral overlap between antenna fluorescence and special pair absorption, only if the antenna and reaction center are considerably more tightly coupled than in the Bchl *a* bacteria (15 Å for the molecules surrounding the reaction center, as compared to 30 Å for the Bchl *a* species (17)), and the initial deactivation of the special pair excited state occurs considerably faster than the P^*I --> P^+I^- charge separation observed in isolated reaction centers; the formation of an intradimer charge transfer state, as suggested on the basis of hole-burning experiments (21) and Stark spectra (22) of isolated reaction centers, may be such a process. Similar conclusions about the trapping was also reached by Trissl et al (23) on the basis of a comparison between the electrically measured trapping time and a model calculation of the trapping. Thus, it appears that despite the observed similarities of the overall trapping times in Bchl *a* and Bchl *b* containing bacteria there are substantial differences on the molecular level.

3. ENERGY TRANSFER IN LH2-CONTAINING PURPLE BACTERIA

Rhodobacter (Rb.) sphaeroides is the by far best studied LH2-containing purple bacterial species. We will therefore here concentrate on results obtained for this species. When the antenna Bchl excited-state decay of *Rb. sphaeroides* with open reaction centers is measured at room temperature with picosecond absorption (13) or fluorescence (24) techniques, a non-exponential decay is generally observed composed of two lifetime components of about 30 and 70-100 ps, Fig. 4A. This biexponential decay is a consequence of the more complex antenna structure of this bacterium, and reflects the equilibration of energy between LH1 and LH2. Using a simple two-step equilibrium model (Fig. 4B) involving LH1, LH2 and the reaction center it was shown that the measured apparent trapping time of about 100 ps corresponds to an actual trapping time of 60 ps, i.e. the same value as for trapping in *R. rubrum*.

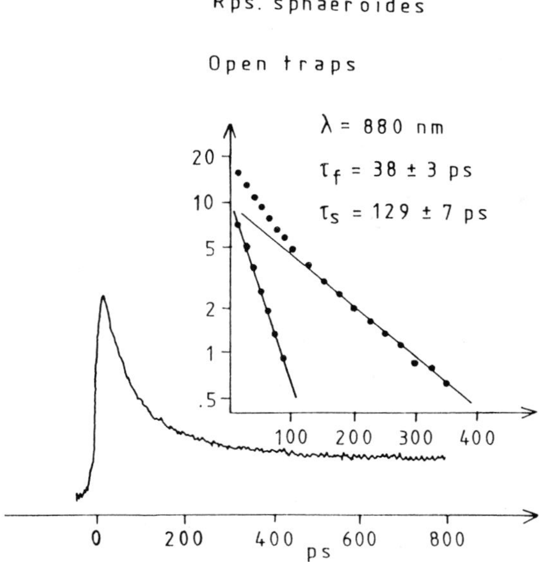

Figure 4 A. Absorption kinetics of *Rb. sphaeroides* with open RC.

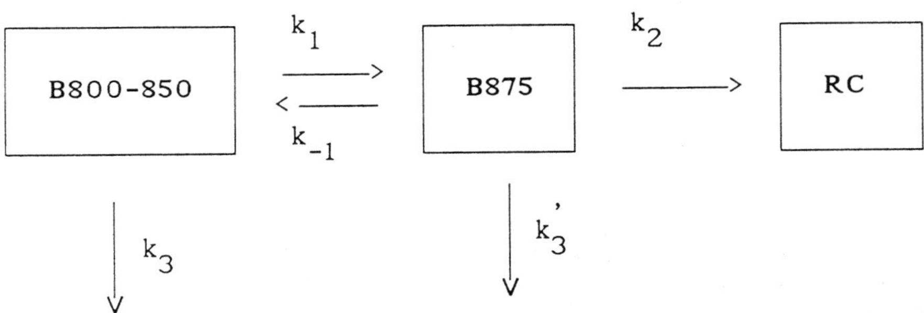

Figure 4 B.Model to account for observed kinetics in *Rb. sphaeorides.*

LH2 contains two types of Bchl *a* chromophores absorbing at 800 and 850 nm, respectively. The stoichiometric ratio is one Bchl800 to two Bchl850. Upon excitation of Bchl800 the energy is rapidly, within 1-2 ps, transferred to Bchl850. This is shown in Fig. 5 for a detergent isolated B800-850 complex.Very similar kinetics were observed for this transfer step in chromatophore membranes (13), and at room temperature the transfer is probably somewhat faster, 0.5 - 1 ps due to better spectral overlap. Absorption anisotropy revealed that there is a very fast subpicosecond energy transfer between identical Bchl850 molecules, causing efficient depolarization of the absorption signal (r≤0.1). Contrary to this, the Bchl 800 kinetics is highly polarized, showing that only limited energy transfer occurs among the Bchl 800´s.

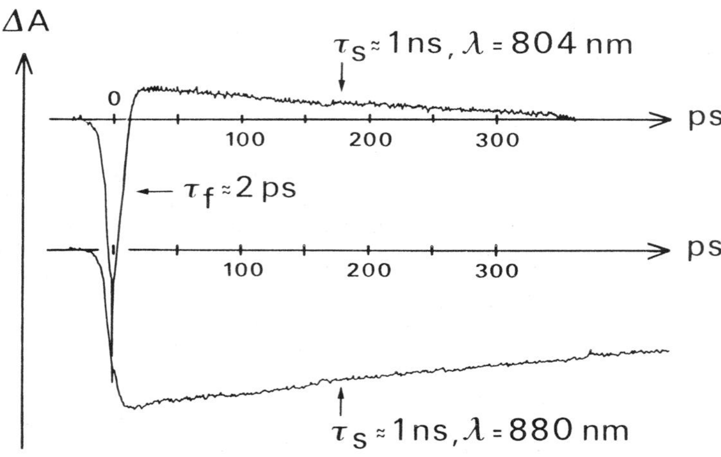

Figure 5. Absorption kinetics of detergent isolated B800-850. The measurement at 804 nm shows the ultrafast B800--->850 energy transfer.

In order to study the pathways of energy flow through the light-harvesting antenna of *Rb.sphaeroides* and the degree of pigment heterogeneity, transient absorption measurements were performed at 77 K with independently tunable excitation and probing wavelengths in the wavelength interval 790 - 910 nm (25). The most direct way to measure the total energy transfer time through the light-harvesting antenna is to excite the high energy pigments (B800) and probe the arrival of the excitons to the lowest energy antenna pigments (B896). When this was done by measuring the rise-time of the ground state bleaching of B896 the excitations were seen to arrive to B896 with an average time constant of about 12 ps, Fig.6.

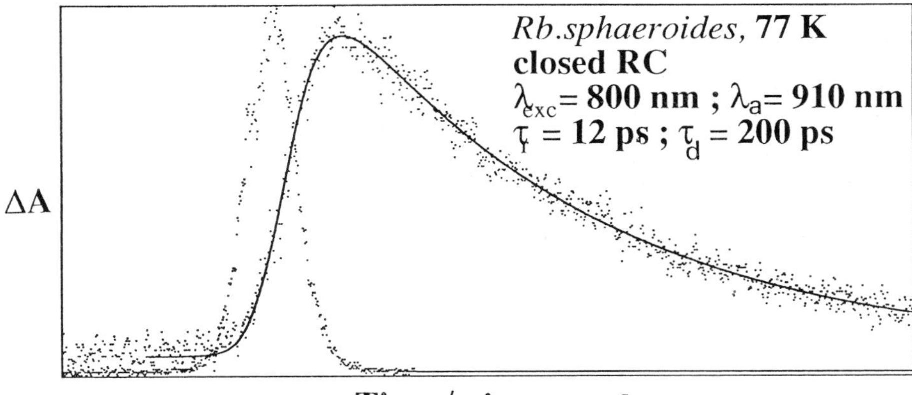

Figure 6. Dual-wavelength absorption kinetics of *Rb. sphaeroides* showing 12 ps rise-time for B800-->B896 energy transfer.

Selective excitation of B850 in the wavelength range 830-860 nm and probing the arrival of the excitations to B896, resulted in an excitation wavelength dependent rise-time for the B850 --> B896 process, Fig. 7 A. Excitations originating in the blue wing of B850 are transferred to B896 with an apparent time constant of 40 ps, while excitations originating in the central and red parts of the B850 absorption band are transferred with a 9 ps time constant. The most likely interpretation of this result is that B850 consists of two (or several) spectroscopically different components which transfer energy with different rates to LH1. A two component model, where the two components are characterised by the limiting time constants 9 and 40 ps observed for the appearance of the B896 excited-state, results in the decomposition of the B850 absorption band shown in Fig. 7B.

The analysis also showed that the 12 ps rise-time obtained for B800 --> B896 energy transfer is best interpreted as the weighted average of the above mentioned 9 ps (80%) and 40 ps (20%) transfer times. This results in a kinetic model (Fig. 8) with two channels for energy transfer from LH2 to LH1, one major fast channel with a total transfer time of about 9 ps and a minor slower one with a 40 ps transfer time. The description of the wavelength dependent energy transfer from B850 in terms of two kinetic components with different transfer times does not necessarily imply that on the molecular level there are two different pigments. A perhaps more likely interpretation of this result is that there is a peripheral fraction of LH2 which is differently connected or organised and therefore has a slightly different spectrum and slower transfer time, due to more and slower transfer steps. Time-resolved fluorescence suggest a similar picture (25,26), and Friberg et al (26) proposed a model with two parallel pools of B800-850, i.e. an inhomogeneity of both B800 and B850.

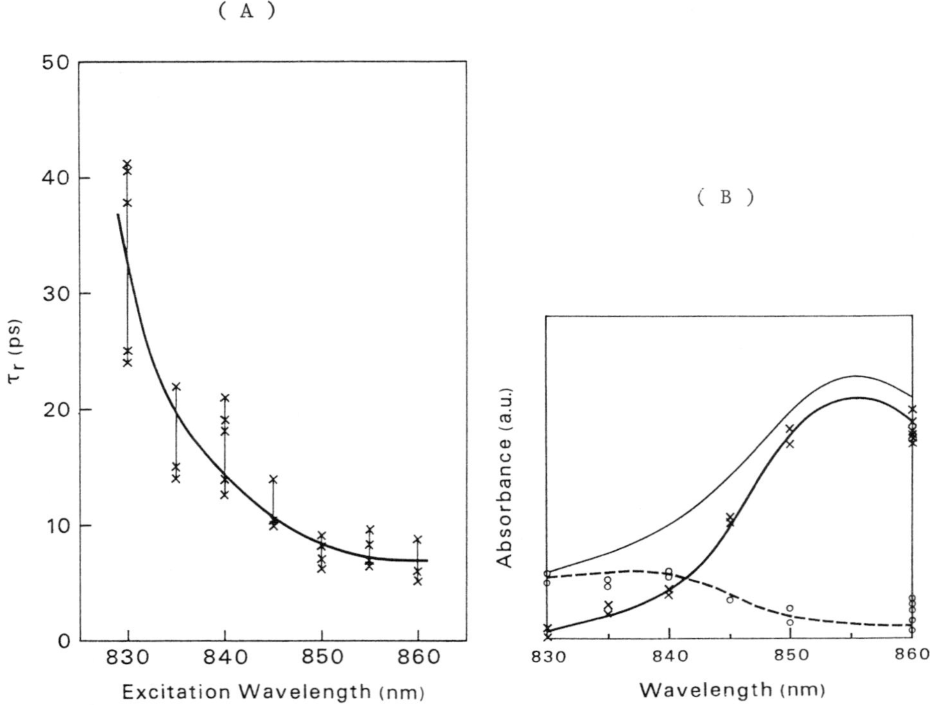

Figure 7. Excitation-wavelength dependence of B896 excited-state rise-time after excitation in the B850 band (A). Component spectra resulting from a two-component analysis of the wavelength dependence (B)

However, from the available experimental results there seem to be no indications of B800 being inhomogeneous. As judged from absorption and fluorescence spectra of a LH2-only containing mutant of *Rb.sphaeroides* (27) and absorption kinetics measured for *Rb.sphaeroides* WT chromatophores, there may be a B870 pigment coupling LH1 and LH2.

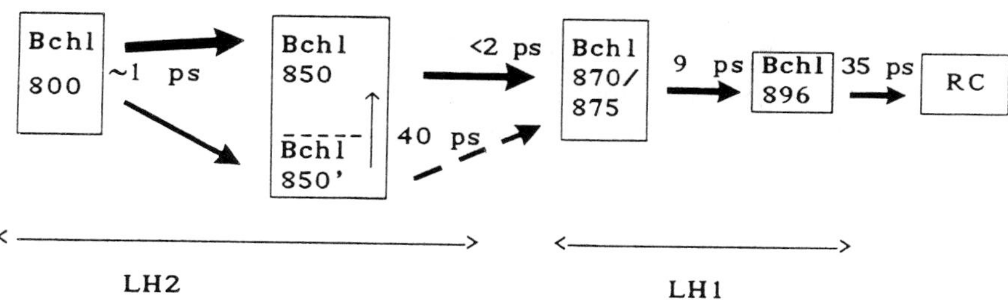

Figure 8. Kinetic model for the pathways of energy transfer in *Rb. sphaeroides.*

The 9 ps rise-time observed for B896 excitons upon B850 excitation, is mainly due to the transfer step B875 --> B896 within LH1. This was concluded from the facts that directly excited B875 decays with an approximately 10 ps time constant, and that B850 --> B875 energy transfer occurs within less than 2 ps. The latter result was obtained by exciting *Rb.sphaeroides* chromatophores at 850 nm and probing the rise and decay of the B875 excited state at 885 nm (Fig. 9), where there is no absorbance change due to B896 (isosbestic wavelength of B896).

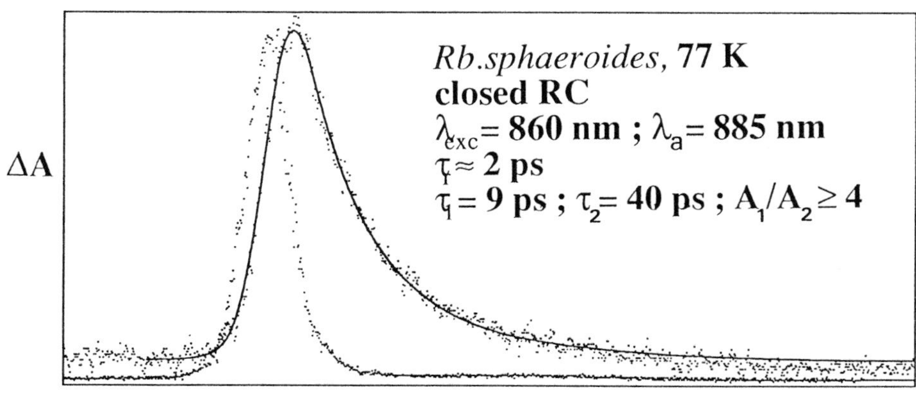

Figure 9. Kinetics of the B875 excited state following excitation of B850.

4. ENERGY TRANSFER IN ISOLATED LIGHT-HARVESTING COMPLEXES AND MUTANTS.

4.1 LH1

To elucidate details of the energy transfer, isolated light-harvesting complexes and mutants lacking one or several pigments are helpful. Picosecond absorption and absorption anisotropy kinetics on a detergent isolated LH1 complex of *Rb. sphaeroides* confirmed the inhomogeneous nature of LH1 (16), as had been previously suggested from both time-resolved and steady-state measurements on intact systems. A similar study on two mutants of *Rb. sphaeroides*, one containing only LH1 (i.e. no LH2 and RC) and another containing LH1 + RC further supported the view that LH1 is inhomogeneous and consists of at least two components, B875 and B896. Light absorption by the main pigment B875 results in fast energy transfer to B896 with a time constant of 10-15 ps(28). In the isolated complex and the LH1-only mutant the B896 excited state then decays slowly (650 and 450 ps, respectively), since there is no energy acceptor present in these cases. Time-resolved absorption anisotropy of detergent isolated LH1 (16) and a LH1-only mutant (28) showed that subpicosecond depolarization occurs, probably as a result of very fast ($\geq 10^{12}$ s^{-1}) energy transfer within a minimum unit of LH1.

4.2 LH2

Several different LH2 complexes have been studied, and they all demonstrate essentially similar energy transfer characteristics. The most conspicous feature is the energy transfer from B800 to B850. From experiments with picosecond time-resolution it was estimated that this transfer occurs with a 1-2 ps time constant at 77 K (29) and somewhat faster at room temperature in the B800-850 complex of *Rb. sphaeroides* and *Rps. acidophila* 7750. In a similar complex but with the 850 nm band shifted to 820 nm, the energy transfer from B800 was concluded to occur within less than 0.5 ps, consistent with the better spectral overlap for B800-->B820, as compared to B800-->B850 transfer. In a transient absorption experiment with 240 fs time resolution Trautmann el al (30) could resolve the energy transfer from Bchl 800 to Bchl850 and found a 2.5 ps transfer time, thus confirming the earlier estimates based on experiments with lower time resolution. Hole-burning experiments performed within the Bchl800 band at 4 K (31) similarly found a 2 ps relaxation time which was attributed to Bchl800-->Bchl850 energy transfer. Measurement of picosecond absorption anisotropy revealed very fast subpicosecond energy transfer among the Bchl850's (29) while, the corresponding measurement for B800 demonstrated only limited Bchl800<-->Bchl 800 energy transfer. These results are largely consistent with the model of a B800-850 minimum unit proposed by Kramer el al (10). The low Bchl 850 anisotropy is consistent with the proposed circularly degenerate orientation of the Bchl850 Q_y dipoles within the plane of the membrane, and the high Bchl800 anisotropy agrees with the near perpendicular orientation of the two Bchl800 molecules within the minimum unit, which should result in a relatively slow energy transfer between the Bchl 800's.

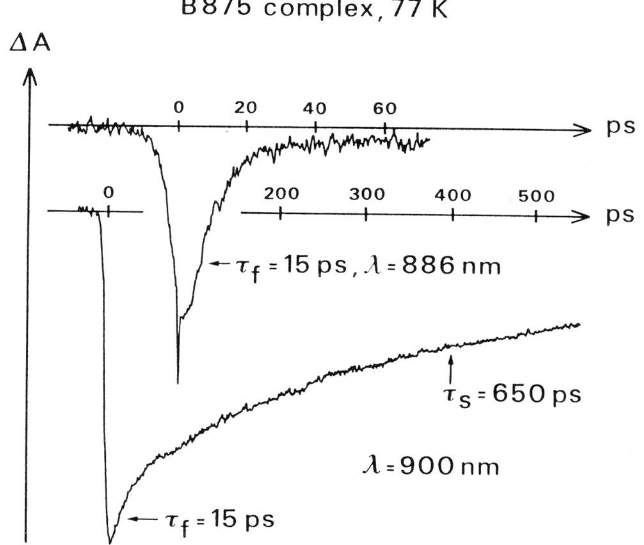

Figure 10. Decay kinetics of B875, reflecting B875-->B896 energy transfer.

REFERENCES

1 J. Deisenhofer, O. Epp, H.K. Miki, R. Huber and H. Michel, Nature 318 (1985) 612.

2 J.P. Allen, G. Feher, T.O.Yeates, H. Komiya, D.C. Rees, Natl. Acad. Sci USA 24 (1987) 5730.

3 W. Stark, F. Jay and K. Muehlethaler, Arch. Microbiol. 146 (1986) 130.

4 W.J. Vredenberg and L.N.M. Duysens, Nature, 197 (1963) 355.

5 M., Vos, R. van Grondelle, F.W. van der Kooy, D. van de Poll, J.Amesz, and L.N.M. Duysens, Biochim. Biophys. Acta 850 (1986) 501.

6 H. Zuber, Trends Biochem. Sci 11 (1986) 414.

7 J.F. Miller, S.B. Hinchigeri, P.S. Parkes-Loach, P.M. Callahan, J.R.Sprinkle, J.R. Riccobono and P.A. Loach, Biochemistry 26 (1987) 5055.

8 P.S. Parkes- Loach, J.R. Sprinkl and P.A. Loach, Biochemistry 27 (1988) 2718.

9 F. van Mourik, K. vand den Oord, K.J. Visscher, P.S. Parkes-Loach, P.A. Loach, R.W. Visschers, and R. van Grondelle. To be published 1991.

10 H.J.M. Kramer, R. van Grondelle, C.N. Hunter, W.M.J. Weterhuis and J. Amesz, Biochim. Biophys. Acta 765 (1984) 156.

11 R.J. Cogdell, unpublished results.

12 R. van Grondelle, unpublished results.

13 V. Sundström, R., van Grondelle, H. Bergström, E. Åkesson and T.Gillbro, Biochim. Biophys. Act*a*. 851 (1986) 431.

14 A.P. Razjivin, R.V. Danielius, R.A. Gadonas, A.Yu. Borisov and A.S. Piskanskas, FEBS Lett. 143 (1982) 40.
15 R. van Grondelle, H. Bergström, V. Sundström, and T. Gillbro, Biochim. Biophys. Acta 894 (1987) 313.
16 H. Bergström, W.H.J. Westerhuis, V. Sundström, R.A. Niederman and T. Gillbro, FEBS Lett 233 (1988) 12.
17 H. Bergström, R. van Grondelle, and V. Sundström, FEBS Lett 250 (1989) 503.
18 K.J. Visscher, V. Sundström, H. Bergström, and R. van Grondelle, Photosynth. Res. 22 (1989) 211.
19 Fu Geng Zhang, T. Gillbro, R. van Grondelle and V, Sundström, Biophys. Journal 1991, submitted.
20 J. Breton, D.L. Farkas, and W.W. Parson, Biochim. Biophys. Acta 808 (1985) 421.
21 D. Tang, R. Jankowiak, G.J. Small and D.M. Tiede, Chem. Phys. 131 (1989) 99.
22 D.J. Lockhart and S.G. Boxer, Proc. Natl. Acad. Sci USA 85 (1988) 107.
23 H.W. Trissl, J. Breton, J. Deprez, A. Dobek and W. Leibl, Biochim. Biophys. Acta 1015 (1990) 322.
24 K. Schimada, M. Mimuro, N. Tamai, I. Yamazaki, Biochim. Biophys. Acta. 975 (1989) 72.
25 Fu Geng Zhang, R. van Grondelle and V. Sundström, Biophys. Journal 1991 in press.
26 A. Freiberg, V.I. Godik, T. Pullerits K. and Timpman, Chemical Physics 128 (1988) 227.
27 C.N. Hunter and R. van Grondelle, In: Photosynthetic light-harvesting systems. Organization and function. Scheer, H. and Schneider, S., eds.Walter de Gruyter, Berlin, New York, 1988, 247.
28 C.N. Hunter, R. van Grondelle, H. Bergström, and V. Sundström,. Abstracts Fifth European Bioenergetics Conference Aberystwyth, UK (1988) 20.
29 J. Amesz and H. Vasmel, Fluorescence properties of Photosynthtetic Bacteria. In: Light Emission By Plants and Bacteria, Ch. 15, Govindjee, Amesz, J., Fork, D.C. eds., Academic Press, New York (1986) 423.
30 J.K. Trautman, A.P. Shreve, C.A. Violette, H.A. Frank, T.G. Owens and A.C. Albrecht, Proc. Natl. Acad. Sci. USA 87 (1990) 215.
31 H. van den Laan, Th. Schmidt, R.W. Visschers, K.J. Visscher, R. van Grondelle and S. Völker, Chem. Phys. Lett. 170 (1990) 231.

DISCUSSION

Fleming

1. Have you observed an excitation wavelength dependence in Rps. viridis?
2. Does the fluorescence spectrum of viridis imply that vibrational relaxation is complete on the time scale of the trapping?

Sundstrom

1. We look for it in the wavelength range 980–1010 nm but could not observe any. If a minor antenna component exists it should be at the short-wavelength region of the antenna absorption band. Thus we feel we should have observed it if present.

2. We have not investigated that matter in detail, but on the time scale of trapping i.e. 60 ps it appears likely to be so.

Itoh

Is your discussion Rps. viridis case also applicable to the case in Rb. sphaeroides or to the long-wavelength minor antenna chlorophyll components in photosystems I and II of green plant?

Sundstrom

The situation in Rb. sphaeroides and also green plants is different from that in Rps. viridis these organisms have heterogeneous antenna consisting of several pigments including low energy pigment B896 in Rb. sphaeroides. This pigment couples effectively the energy to P and the observed trapping a rate correlates well with the expected antenna - P distance about 30 Å.

Yoshihara

Why decay kinetics of the open and closed form of reaction centers are relatively similar like only several times?

Sundstrom

It is generally observed in many different photosynthetic organism to be such that the "closed" inactive reaction center quenches the antenna excitations quite efficiently, at a rate which is not more than a few times slower than the actual trapping.
There is no clear understanding as to the mechanism of this quenching, but energy transfer is possible.

Dynamics and Mechanisms of
Photoinduced Transfer and Related Phenomena
N. Mataga, T. Okada and H. Masuhara (Editors)

NON-EXPONENTIAL DECAY OF FLUORESCENCE OF TRYPTOPHAN AND ITS MOTION IN PROTEINS

Fumio Tanaka[a] and Noboru Mataga[b]

[a]Mie Nursing College, 100 Torii-cho, Tsu 514, Japan

[b]Department of Chemistry, Faculty of Engineering Science, Osaka University, Toyonaka 560, Japan

Abstract

To elucidate the non-exponential nature of the intensity decay and to analyze the motional mode of tryptophan in proteins, theoretical expressions for the decays of fluorescence intensity and anisotropy were derived based on a model where indole ring as an asymmetric rotor rotates about covalent bonds in a spherical macromolecule, and the fluorescence quenching constant depends on the rotational angles. Numerical calculations on some model ststems by means of the present theory predict that both decays correlate with each other and are non-exponential depending on the rotational diffusion coefficients of the internal motion and the potential energy for the internal rotation. On the basis of such analysis, dynamics of tryptophan in the proteins containing single tryptophan was examined in the case of some protein systems. Time-resolved intensity and anisotropy of fluorescence of single tryptophan of erabutoxin b were quantitatively analyzed with this theory.

1. INTRODUCTION

It has been demonstrated by means of a picosecond time-resolved fluorescence spectroscopy that the fluorescence of tryptophan of many proteins containing a single tryptophan decays with a non-exponential decay function, and at the same time the time-resolved anisotropy also does with a non-exponential function. It is important to elucidate the origin of this non-exponentiality in the intensity decay of those proteins, not only to see the microenvironment of tryptophan residue in the proteins, but also to understand a dynamic nature of the protein structure. The followings are, so far, considered to be the origin of the non-exponentiality in the intensity decay ;

i) a heterogeneous distribution of indole ring in a protein with a few discrete fluorescence lifetimes [1] or continuous lifetimes [2],
ii) a solvent relaxation of the surroundings of tryptophan,
iii) a rotational motion of tryptophan in proteins [3-5].

In the previous works [4] we have derived theoretical expressions for both intensity and anisotropy decays based on a model where the indole ring internally rotates in a protein, and applied them to erabutoxin b [5]. In those works, however, some terms are neglected in the theoretical expressions for both decays. In the present work we have obtained the complete theoretical expressions for the time-resolved intensity and anisotropy, and have examined the observed data on both decays, qualitatively and also quantitatively, on the basis of them.

2. THEORETICAL MODEL

We model the indole ring as a completely asymmetric rotor covalently bound to a spherical macromolecule. The fluorophore possesses a motional freedom of internal rotation around covalent bonds in the macromolecules. The geometrical arrangement of tryptophan and macromolecule are illustrated in Figure 1. The internal motion is described by a modified Smoluchowski equation as before [4], which is obtained by introducing a rotation-dependent quenching constant of the fluorophore into the Smoluchowski equation as given by Favro [6].

$$\frac{\partial}{\partial t} G(\Omega'\omega\Omega\omega' t) = - \{k_1 + k_q(\omega)\} G(\Omega'\omega\Omega\omega' t)$$

$$- (D_p \mathbf{J}_\Omega^2 + \mathbf{J}_\omega \cdot \mathbf{D}_i \cdot \mathbf{J}_\omega) G(\Omega'\omega\Omega\omega' t)$$

$$- \frac{1}{2kT} [\mathbf{J}_\omega \cdot \mathbf{D}_i \cdot \mathbf{J}_\omega V(\omega)] G(\Omega'\omega\Omega\omega' t)$$

$$- \frac{1}{(2kT)^2} [\mathbf{J}_\omega V(\omega)] \cdot \mathbf{D}_i \cdot [\mathbf{J}_\omega V(\omega)] G(\Omega'\omega\Omega\omega' t) \qquad (1)$$

where $G(\Omega'\omega\Omega\omega' t)$ denotes a Green function which represents the rotational motion of protein from an experimental laboratory system to the protein system with Euler angles shown by Ω, and the rotational motion of indole ring from the protein system to indole system with Euler angles $\omega = (\alpha\beta\gamma)$. $\mathbf{J}_\Omega$ and $\mathbf{J}_\omega$ are angular momentum operators for spherical and symmetric rotors. D_p and $\mathbf{D}_i$ are diffusion coefficient of spherical rotor and diffusion coefficient tensor of asymmetric rotor. k_1 and $k(\omega)$

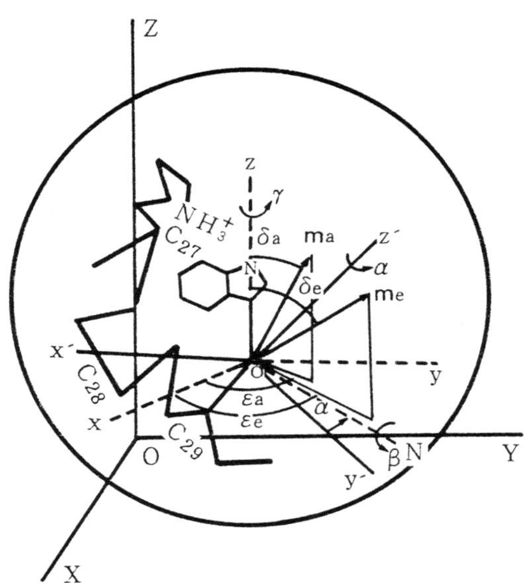

Figure 1. Geometry of tryptophan in protein. Coordinate system of spherical protein and tryptophan are shown by (x'y'z') and (xyz). Origin of (x'y'z') is chosen at CH_2 group connecting peptide bond and indole ring. Internal rotation of tryptophan from the (x'y'z') to (xyz) system is denoted by $\omega=(\alpha\beta\gamma)$. The transition moments of absorption and emission are shown by $\mathbf{m}_a$ and $\mathbf{m}_e$.

are fluorescence quenching constant independent of the internal motion and dependent on it. $V(\omega)$ is potential energy for the internal rotation.

The Green function may be expressed with spherical harmonics for macromolecule and eigenfunctions of asymmetric rotor. The eigenfunction of the asymmetric rotor can be expanded with those of symmetric rotors [4]. A product of the eigenfunctions for asymmetric rotor appeared in the Green function is represented as Eq (2).

$$\Psi_\kappa^*(\omega)\,\Psi_\kappa(\omega') = \sum_{r=0}^{\infty}\sum_{ss'=-r}^{r}\sum_{p=0}^{\infty}\sum_{qq'=-p}^{p} C_{ss'}^{\kappa r,*}\,C_{qq'}^{\kappa p}\cdot\Phi_{ss'}^{r*}(\omega)\,\Phi_{qq'}^{p}(\omega') \tag{2}$$

where $C_{qq'}^{\kappa p}$ is an expansion coefficient of asymmetric rotor with symmetric rotors, which is to be determined as an eigenvector corresponding to an eigenvalue by solving a set of linear simultaneous equations, as described in the previous work [4]. Cross terms, $C_{ss'}^{\kappa r,*}\,C_{qq'}^{\kappa p}$, $(r \neq p,\ s \neq q,\ s' \neq q')$, in Eq (2), which are neglected in the previous work [4], are fully taken into account.

The expressions for the decays of fluorescence intensity and anisotropy are derived according to the method described in the previous work [4].

3. NUMERICAL RESULTS OF THE DECAYS OF INTENSITY AND ANISOTROPY IN MODEL SYSTEMS

General features of the decays of intensity and anisotropy are examined on the basis of the numerical results calculated with the theoretical expressions for some model systems. We have assumed that i) the quenching constant depends on a solid angle between movable z-axis and a vector formed by axial origin and quencher, and accordingly, it is maximum when z-axis directs toward quencher and minimum when z-axis directs opposite to the quencher, and ii) the potential energy depends on a solid angle between z-axis of protein system in the X-ray structure and the movable z-axis, and accordingly, it can be neglected when the movable z-axis is coinsident with the X-ray structure, and it is maximum when the solid angle is at 180^0. The dependences of the quenching constant and potential energy on the angles are approximated by Eqs (3) and (4).

$$k(\omega) = k_q^0 \left[1 + \frac{4\pi}{3} \sum_{m=-1}^{1} Y_{1m}^*(\alpha_q \beta_q)\, Y_{1m}(\alpha\beta)\right] \quad (3)$$

$$V(\omega) = p \left[1 - \frac{4\pi}{3} \sum_{m=-1}^{1} Y_{1m}^*(\alpha_p \beta_p)\, Y_{1m}(\alpha\beta)\right] \quad (4)$$

where k_q^0 and p are averaged quenching constant over the rotational angles and height of potential barrier. $(\alpha_q\beta_q)$ indicates the location of quencher measured in (x'y'z') system and $(\alpha_p\beta_p)$ does that of z-axis in the X-ray structure. The second terms of right hand sides of Eqs (3) and (4) represent the solid angles presented above.

Figure 2 shows the dependences of decays of the intensity and anisotropy on the diffusion coefficient, D_{zz}. In these calculations the potential energy was neglected. The dependences of the decays of fluorescence intensity and anisotropy on the height of potential energy are also examined.

From these numerical results, the following feature of both decays may be pointed out ;

1) both decays correlate with each other and strongly depend on the diffusion coefficients of the internal rotation,
2) as internal motion becomes slower, deviation of intensity decay from a exponential function is more enhanced,
3) when the internal motion is slow enough, the decay of anisotropy could occur mainly by the rotational motion of the macromolecule,
4) the intensity always decays non-exponentially, whenever the internal rotation is observed by the anisotropy decay,
5) the intensity decay is quite sensitive to the presence of a potential barrier for the rotation, but the anisotropy is rather insensitive to it.

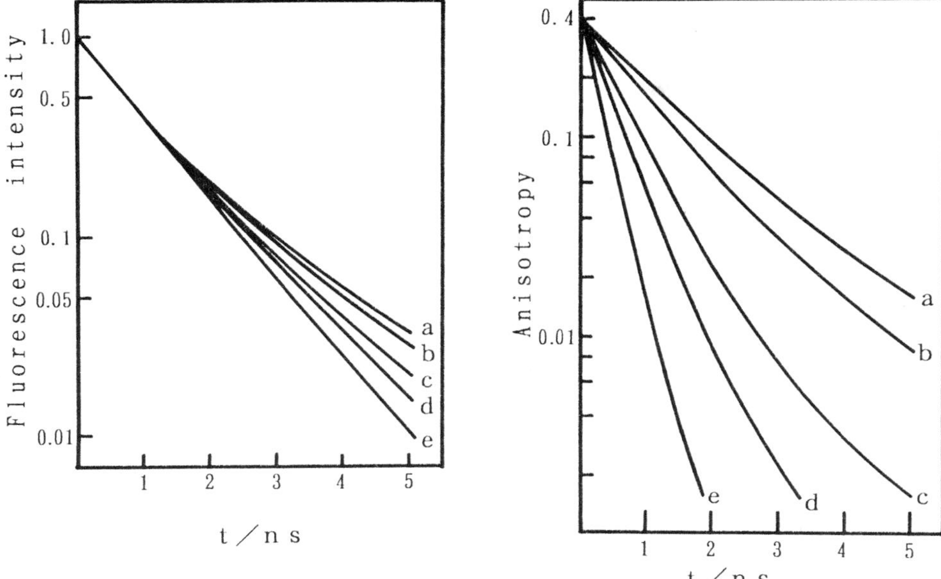

Figure 2. Calculated decay curves of intensity and anisotropy. Parameters used are ; k_q^0 = 1.0 (ns^{-1}), D_{xx} = D_{yy} = 0.05 (ns^{-1}), D_{zz} is 0.05 in curve a, 0.1 in b, 0.3 in c, 0.5 in d and 1.0 in e, in unit of ns^{-1}. The potential energy was neglected. The other parameters are listed in Table 2.

4. INTERPRETATION OF THE OBSERVED DECAYS OF FLUORESCENCE INTENSITY AND ANISOTROPY OF TRYPTOPHAN IN PROTEINS

The observed decay data of both the intensity and anisotropy of tryptophan in the proteins containing single tryptophan, which are found in the literatures, are listed in Table 1. These may be classified into the following cases ;

case 1 both intensity and anisotropy decay with single-exponential functions,

case 2 the anisotropy decays with a single-exponential function, but the intensity does with a non-exponential function,

case 3 the both decay with non-exponential functions,

case 4 the intensity decays with a single-exponential function, but the anisotropy does with a non-exponential function.

As we can see from Table 1, most proteins are in the case 3. A few proteins are in the case 1 or the case 2, but none is in the case 4. The dynamic nature of tryptophan in proteins of the case 1 to the case 4 may be interpretated as follows :

The case 1 ; the indole ring may be fixed in the proteins, and

Table 1
Decay parameters of tryptophan in proteins containing single tryptophan

Protein	Case	Intensity			Anisotropy		Conditions	Ref.
		τ_1	τ_2	τ_3	ϕ_1	ϕ_2	(T/°C)	
Nuclease B	1	5.1			9.9		20	8
Nuclease S	3	6.2	2.03		15.6	0.65	10	9
Myelin	3	4.7	1.97		1.26	0.09	20	10
Adrenocorticotropin	3	5.1	2.0		4.5	0.92	4	10
Glucagon	3	3.6	1.1		1.67	0.42	19	11
Albumin(human)	3	7.8	3.3		23	6.0	31	12
	2	7.5	0.98		44		10	9
Apolipoprotein C-I	3	6.8	2.9	0.4	3.6	0.2	15	13
Apoazurin(Pae)	1	5.2			4.94		20	14
Apoazurin(Åfe)	3	2.8	1.06		6.88	0.16	20	14
Holoazurin	3	4.2	0.10		11.8	0.51	20	8
Ribonuclease	1	4.0			6.0		20;pH5.5	8
	2	4.0	1.6		5.3		20;pH7.4	15
Parvalbumin	3	4.5	0.99		5.28	0.23	25;no Ca	16
	3	4.0	0.65		7.9	0.54	25;+Ca	16
Apoferritin (horse)	3	5.7	2.50	0.31	-	0.15	10;in SDS	9
Phospholipase A_2	3	6.8	2.81	0.52	8.4	0.43	10	9
Superoxide dismutase	3	6.1	2.08		13.3	0.70	10	9
Apocytochrome c (horse)*a*	3	5.4	2.99	1.12	5.33	1.09	20;1.8μM Em. 358 nm	17
Bacteriophage M13 coat	3	5.7	2.5	0.3	9.8	0.5	10;in 700 mM SDS	18
protein	3	5.5	2.2	0.8	8.0	0.5	10;in DMPC/DMPA	18
Melittin	3	4.7	2.6	0.53	1.54	0.33	20;monomer	19
	3	4.5	2.03	0.53	5.56	1.15	20;tetramer*b*	
	3	6.1	2.19	0.46	3.3	0.75	19;in DMPC	
	3	5.1	1.95	0.48	5.6	1.0	36;in DMPC	

τ_i and ϕ_i indicate fluorescence lifetime and rotational correlation time, respectively. These values are shown in ns unit.
a Additional lifetime component, 0.22 ns, and rotational correlation time, 0.14 ns, are present.
b Additional rotational time, 0.16 ns, is present.

hence the local structure surrounding it should be uniform over the all protein molecules leading to the intensity decay with a single exponential function. In this case, the internal motion cannot be observed by the anisotropy.
The case 2 ; the indole ring rotates slowly in the proteins compared to the rotational diffusion coefficients of internal motion, and hence, the intensity decays with a non-exponential function, despite of the single-exponential decay of the anisotropy due to the rotational motion of the spherical protein.
The case 3 ; the diffusion coefficients of the internal rotation may be comparable with the averaged quenching constant over the rotational angles, and hence the both decays become non-exponential, according to the numerical results.
The case 4 ; the intensity decay should be always non-exponential when the internal rotation is observed by the anisotropy decay, so that this case could not be found in the literatures, so far. If the internal motion is very fast compared to the averaged quenching constant over the rotational angles, as in a fluorophor in bulk solutions, it should be too fast to observe in the time region of the intensity decay by the anisotropy.

5. TIME-RESOLVED FLUORESCENCE OF SINGLE TRYPTOPHAN IN ERABUTOXIN B

Erabutoxin b is a neurotoxic protein from the venom of the sea snake *Laticauda semifasciata* and consists of 62 amino acid residues. The three-dimensional structure of erabutoxin b is shown in Figure 3. It contains a single-tryptophan (Trp-29). We have assumed the cation of Lys-27 to be a quencher for the fluorescence of Trp-29 as in the previous work [5]. Time-resolved fluorescence of Trp-29 of erabutoxin b is measured in the picosecond region. The observed decays of fluorescence anisotropy and intensity at 20 ^{0} C are shown in Figure 4. These are both non-exponential. The anisotropy was reproducible with a double-exponential function with rotational correlation times of 49 ps and 3 ns. These decays are simulated with the theoretical expressions derived. The quenching constant and the potential energy are approximated by Eqs (3) and (4). We have six diffusion coefficients, D_{xx}, D_{yy}, D_{zz}, D_{xy}, D_{yz}, D_{zx}, for the internal rotation of asymmetric indole ring. The independent coefficients of an asymmetric rotor are three, D_{xx}, D_{yy}, D_{zz} in a free rotation as in bulk solution. In the present system, however, we do not know which are independent coefficients among these six, since the internal motion should be restricted due to covalent bonds between indole ring and peptide chain of the protein. Accordingly, we have chosen some of them as variable parameters. In addition, a normalization factor, Fmax, the averaged quenching constant, k_q^0 in Eq (3), and height of potential energy, p in Eq (4), are also variable parameters. The directions of the transition moments of tryptophan are represented with polar coordinates, $(\varepsilon\delta)$, in

Table 2
Constants used for the analysis

		Location of quencher[c]		Location of energy minimum[d]		Direction of transition moment[e] absorption		emission	
k_1[a]	D_p[b]	α_q	β_q	α_p	β_p	ε_a	δ_a	ε_e	δ_e
0.1163	0.0556	-24.8	79.6	41.8	66.7	90	variable	90	50

a Rate constant independent of the internal rotation. The value was obtained from the longest lifetime of free tryptophan in alkaline solution and shown in unit of ns^{-1}.
b Rotational diffusion coefficient, represented in ns^{-1}.
c NH_3^+ group of Lys-27 is assumed to be the effective quencher. Coordinates of N atom of the ammonium group were obtained from X-ray data. Angles are represented in unit of degree.
d Location of the potential energy minimum is assumed to be one of Trp-29 obtained by X-ray data. Angles are represented in degree.
e It is assumed that the emission of Trp-29 is from L_a state, and absorption (excited at 295 nm) to both the L_a and L_b states. ($\varepsilon\delta$) indicate polar coordinates in (xyz) system and are represented in unit of degree. The location of the transition moment of L_a in indole ring is reported by Yamamoto and Tanaka [20].

Table 3
Best-fit parameters of the observed decays of fluorescence intensity and anisotropy of erabutoxin b with the rotational model

		Diffusion coefficient[c]							
k_q^0[a]	p[b]	D_{yy}	D_{zz}	D_{yz}	D_{zx}	δ_a[d]	χ_t^2[e]	χ_f^2[f]	χ_a^2[g]
1.02			0.66	0.50	0.50			1.632	
1.01		0.18	0.28	0.56				1.583	
0.99	0.187		0.50	0.39	0.39			1.617	
1.13	1.611		0.79	0.86				1.395	
1.05		0.38		0.44		78	3.230	1.781	5.147
1.02			0.46	0.51		16	2.008	1.592	2.558
1.02		0.31	0.13	0.53		35	1.902	1.587	2.412

a Averaged quenching constant over the rotational angles. Represented in unit of ns^{-1}.
b Hight of potential energy. Represented in unit of kT.
c Diffusion coefficients of internal rotation. Represented in unit of ns^{-1}.
d Angle between z-axis and the transition moment of absorption (excitation, at 295 nm) in the indole ring.
e The value of chi-square for total data of decays of the intensity and anisotropy.
f The value of chi-square for the intensity decay.
g The value of chi-square for the anisotropy decay.

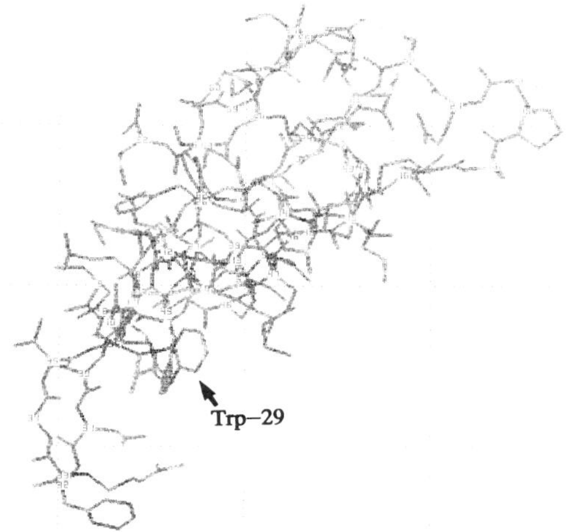

Figure 3. Three-dimensional structure of erabutoxin b [7]. The single tryptophan, Trp-29 locates at the interface, and may be exposed to water. The nearest amino acid residue is Thr-35. Ammonium group of Lys-27 is quite close to Trp-29.

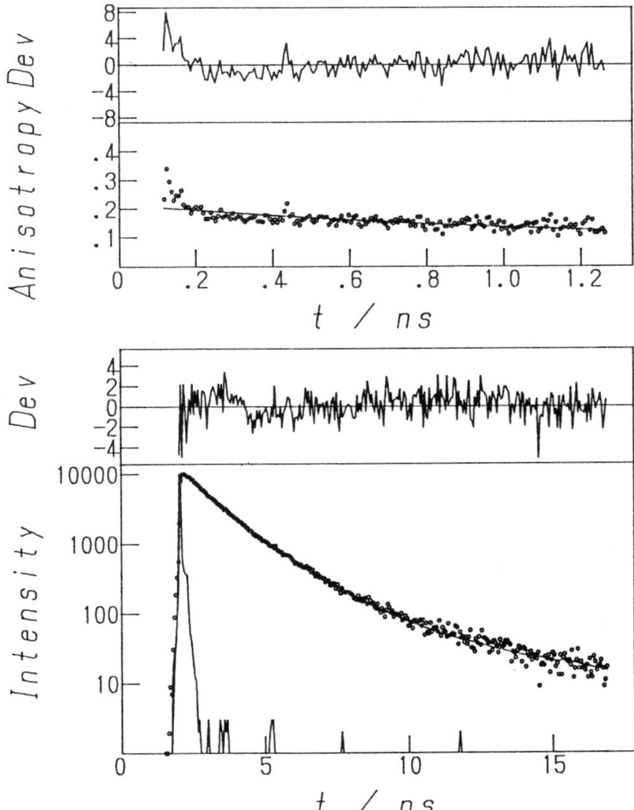

Figure 4. The observed and calculated decays of anisotropy and intensity. The parameters used and obtained are listed in Table 2 and Table 3, respectively.

the (xyz) system (see Figure 1). We have assumed that the emission of Trp-29 is from L_a state [21]. At the wavelength of the pulsed excitation (295 nm), transitions to both of L_a and L_b states [22] could take place. Accordingly, an averaged direction of the two transition moments may be shown as $(\delta_a, 90^0)$. δ_a may be another variable parameter for the anisotropy decay.

The theoretical decays of both the fluorescence intensity and anisotropy are calculated changing some of these parameters so as to fit with the corresponding observed decays, according to the method of non-linear least square based on Marquardt argorism. Constants used for the calculations are listed in Table 2. Best-fit parameters obtained are listed in Table 3. Difference between the calculated and observed decays markedly increased, whenever D_{xx} or D_{xy} among the diffusion coefficients were included as variable parameters. D_{zz} and D_{yz} were always greater than the others. The calculated decay of fluorescence intensity fit very well the observed one. The fitting was much improved as the potential energy for the internal rotation was introduced into the theoretical expressions. The both decays were also simulated simultaneously with common parameters. Rotational diffusion coefficient of the entire protein was estimated from the longer correlation time (3 ns) to be 0.056 ns^{-1}. In the simultaneous simulation the potential energy could not be introduced because it took much computational time. Best-fit was obtained when D_{yy}, D_{zz} and D_{yz} among the diffusion coefficients were chosen as variable parameters. The values of chi-square were 1.902, 1.587 and 2.412 for total data points of the intensity and anisotropy, those of intensity only and of anisotropy only, respectively. The obtained diffusion coefficients were D_{yy} = 0.31, D_{zz} = 0.13 and D_{yz} = 0.53, in unit of ns^{-1}. In this case k_q^0 = 1.02 (ns^{-1}) and δ_a = 35 ^{0}. Calculated decays of the anisotropy and intensity at the best-fit are shown with solid curves in Figure 4. Deviations from the observed data are also shown in the Figure. Although the fitting for the intensity decay was satisfactory, the fast decay in the anisotropy (49 ps in the correlation time) could not be reproducible at these parameters. It may be improved if potential energy is taken into account.

6. CONCLUDING REMARKS

The non-exponentiality in the intensity decay of tryptophan in proteins may be brought about by various factors, depending on the dynamic nature of the respective protein. It may be elucidated in terms of the motional-dependent quenching rate of the fluorescence of tryptophan, in some proteins. The decays of intensity and anisotropy become correlative upon the introduction of it into the motional equation of tryptophan. The both decays of intensity and anisotropy should be rationalized with a single model for the full understandings of the protein dynamics.

ACKNOWLEDGEMENT

We are grateful to Dr. N. Tamai (ERATO, Kyoto) and Prof. I. Yamazaki (Hokkaido University) for the measurement of time-resolved fluorescence of erabutoxin b. We also thank to The Computer Center, Institute for Molecular Science, for the use of HITAC M-680 computer.

7. REFERENCES

1 C. M. Hutnik and A. G. Szabo, Biochemistry 28 (1989) 3923.
2 J. R. Alcala, E. Gratton, F. G. Prendergast, Biophys. J. 51 (1987) 925.
3 F. Tanaka and N. Mataga, Biophys. J. 39 (1982) 129.
4 F. Tanaka and N. Mataga, Biophys. J. 51 (1987) 487.
5 F. Tanaka, N. Kaneda, N. Mataga, N. Tamai, I. Yamazaki, and K. Hayashi, J. Phys. Chem. 91 (1987) 6344.
6 L. D. Favro (R. E. Burgess, ed.) Fluctuation Phenomena in Solids, p.79, Academic Press, New York, 1958.
7 B. W. Low, H. S. Preston, A. Sato, L. S. Rosen, J. E. Searl, A. D. Rudko, and J. A. Richardson, Proc. Natl. Acad. Sci. USA 73 (1976) 2991.
8 I. Munro, I. Pecht, and L. Stryer, Proc. Natl. Acad. Sci. USA 76 (1979) 56.
9 T. Kouyama, K. Kinoshita, A. Ikegami, Eur. J. Biochem. 182 (1989) 517.
10 J. B. Ross, K. W. Rousslang, L. Brand, Biochemistry 20 (1981) 4361.
11 A. Grinvald and I. Z. Steinberg, Biochim. Biophys. Acta 427 (1976) 663.
12 A. van Hoek, J. Vervoort, A. J. W. G. Visser, J. Biochem. Biophys. Methods 7 (1983) 243.
13 A. Jonas, J. Privat, P. Wahl, J. C. Osborne, Jr., Biochemistry 21 (1982) 6205.
14 J. W. Petrich, J. W. Longworth and G. R. Fleming, Biochemistry 26 (1987) 2711.
15 L.X.-Q. Chen, J. W. Longworth and G. R. Fleming, Biophys. J. 51 (1987) 865.
16 S. T. Ferreira, Biochemistry 28 (1989) 10066.
17 M. Vincent, J-C. Brochon, F. Melora, W. Jordi and J. Gallay, Biochemistry 27 (1988) 8752.
18 K. P. Datema, A. J. W. G. Visser, A. van Hoek, C. J. A. M. Wolfs, R.B. Spruijt, and M.A. Hemminga, Biochemistry 26 (1987) 6145.
19 T. Kulinski and A. J. W. G. Visser, Biochemistry 26 (1987) 540.
20 Y. Yamamoto and J. Tanaka, Bull. Chem. Soc. Jpn. 45 (1972) 1362.
21 N. Mataga, Y. Torihashi, and K. Ezumi, Theor. Chim. Acta 2 (1964) 158.
22 B. Valeur and G. Weber, Photochem. Photobiol. 25 (1977) 441.

DISCUSSION

Kakitani

Besides the factors you considered, the fluctuation of protein folding can affect the time dependencies of fluorescence intensity decay and/or anisotropy. That is if its fluctuation takes place in the time scale of ns, it causes a distribution of the potential for the tryptophane motion, giving rise to a non-exponential decay in picoseconds time region. Such effect may be related to the one talked by Sumi in this conference.

Tanaka

We don't know about the presence of slow fluctuation. If this is the case, the fluctuation surely affects the both decays. The quenching rate should depend on the mutual configurations of the fluorophore and the quencher. If fluctuation of the quencher is very fast, it may be a function of the configuration only of the fluorophore. The effect of a time-dependent potential energy due to the slow fluctuation for the internal rotation on the both decays is also important problem. It may be very difficult to treat this problem properly.

Yamamoto

Experimentally, we have one variable temperature. Temperature change may tell the validity of the model.

Tanaka

That is good point. We shall try this kind of experiment, in some proteins.

Dynamics and Mechanisms of
Photoinduced Transfer and Related Phenomena
N. Mataga, T. Okada and H. Masuhara (Editors)

Primary electron transfer events in bacterial photosynthesis

Joshua Jortner and M. Bixon

School of Chemistry, Tel-Aviv University, 69978 Tel-Aviv, Israel

Abstract

Recent experimental work on fsec electron transfer kinetics in the reaction center (RC) contribute towards the elucidation of the nature of central energy conversion processes in biology. We explore the mechanism of the primary process, focussing on the special role of the bacteriochlorophyll monomer (B) located between the primary donor ($^1P^*$), a bacteriochlorophyll dimer (P), and a bacteriopheophytin (H). We consider a kinetic scheme, which combines two parallel pathways of electron transfer: a unistep superexchange channel mediated via electronic interactions with P^+B^-H, and a two-step sequential channel involving a P^+B^-H chemical intermediate. In this kinetic scheme we used microscopic nonadiabatic electron transfer rates, which were extended to incorporate the effects of medium-controlled dynamics. The results of the kinetic modelling are presented as a function of the free energy gap ΔG_1 between the equilibrium nuclear configurations of the donor $^1P^*BH$ and the (physically and/or chemically) mediating state P^+B^-H. The parallel sequential-superexchange mechanism reduces to the limit of a nearly pure sequential pathway for large negative ΔG_1 at all temperatures and to the limit of an almost pure superexchange pathway for large positive ΔG_1 at all temperatures and for moderate ΔG_1 at low temperatures. The fsec kinetic data at room temperature and at 10K raise the possibility that the mechanism involves the superposition of superexchange and sequential mechanisms at 300K and the dominance of superexchange at low temperatures. Auxiliary experimental information regarding magnetic data, i.e, the singlet-triplet splitting of the radical pair P^+BH^-, the kinetics of the charge separation in mutagenetically altered RCs, with tyrosine M208 being replaced by phenylalanine, and the unidirectionality of charge separation across the A branch of the RC are analysed in terms of the proposed mechanism. The prevalence of the parallel sequential and superexchange electron transfer routes for the primary charge separation would introduce an element of redundancy, which insures the occurrence of an efficient process which is stable with respect to the variation energetic parameters in different photosynthetic RCs.

1. INTRODUCTION

The understanding of the primary charge separation processes in photosynthetic bacterial reaction centres [1] is of considerable intrinsic interest because of several compelling reasons:

(1) The elucidation of the nature of a central energy conversion process, which transforms light energy into electrochemical energy, in biology.

(2) The establishment of structure-function relationships for an important series of biophysical electron transfer (ET) processes.

(3) Opening avenues for the structural and functional control of electron transfer by mutagenesis and by "chemical engineering" of the reaction center (RC).

(4) Allowing for the control of electron transfer in the RC by external fields.

(5) The elucidation of optimization principles for ensuring the efficiency of primary electron transfer in a variety of photosynthetic reaction centers.

(6) Providing the conceptual framework for the unified description of electron transfer processes

in biophysical systems, i.e., a protein medium and in chemical systems, e.g., solutions, glasses and solids.

The primary charge separation in photosynthetic bacterial RC is characterized by three unique features: (i) Ultrafast rate in the psec domain which precludes energy waste due to back transfer to the antenna; (ii) non-Arrhenius temperature dependence of the rate, which manifests optimal coupling to the medium nuclear motion, and (iii) unidirectionality across the A branch of the quasisymmetric RC.

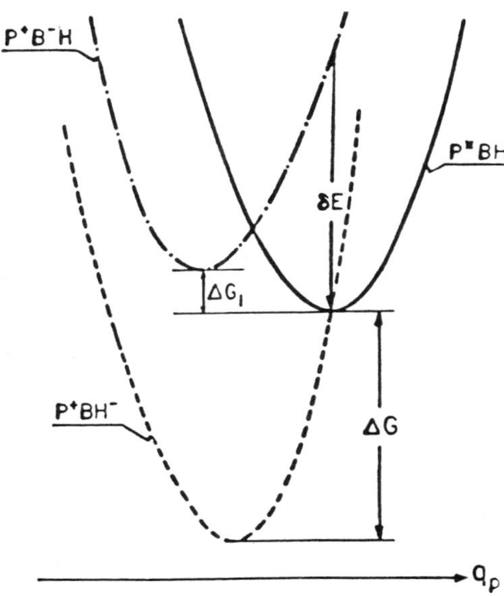

Figure 1. Nuclear potential energy curves for the parallel sequential-superexchange mechanism. The superexchange unistep occurs by the activationless crossing from the P^*BH curve (solid line) to the P^+BH^- curve (dashed line), while the first step (k_1) in the two-step sequential process occurs (in the classical limit) by thermal activation on the P^*BH curve to its crossing with the P^+B^-H curve (dash-dotted line) followed by curve crossing. ΔG and ΔG_1 are the free energy gaps between P^*BH and P^+BH^- and between P^*BH and P^+B^-H, respectively, while δE is the vertical energy difference.

All the mechanisms proposed for the primary electron transfer (ET) process from the singlet excited state ($^1P^*$) of the bacteriochlorophyll dimer (P) along the A branch attribute a special role to the accessory monomer bacteriochlorophyll (B), which is located [2] between P and a bacteriopheophytin (H). Two classes of mechanisms were advanced

(1) The one-step superexchange mechanism [3-9]

$$^1P^*BH \rightarrow P^+BH^- \quad . \tag{1}$$

The direct ET from $^1P^*$ to H is mediated by superexchange electronic interactions via the virtual vibronic states of P^+B^-H, which are located in energy above that of $^1P^*BH$ (Fig. 1).

(2) The two-step sequential ET [10-12], which involves a genuine chemical intermediate. Several alternatives were proposed for the chemical intermediate, e.g., P^+B^-H , with the sequential mechanism being

$$^1P^*BH \xrightarrow{k_1} P^+B^-H \xrightarrow{k_2} P^+BH^- \quad . \tag{2}$$

The analysis of the primary charge separation in terms of one distinct mechanism, involving either the superexchange (1) or the sequential (2) ET, may be questionable, as both mechanisms can prevail in parallel [7,8,10]. In a previous analysis of the superexchange mechanism we have emphasized [8] that an inevitable consequence of the unistep superexchange process is the occurrence of the parallel sequential ET process, which takes place at the intersection of the potential surfaces of $^1P^*BH$ and P^+B^-H. We now analyze the parallel sequential-superexchange kinetic scheme for the primary ET in the RC. This kinetic scheme is consistent with the available experimental information over a broad range of acceptable combinations of energetic parameters ranging from the purely sequential limit to a superposition of a thermally activated sequential and a superexchange mechanism.

The results of the kinetic analysis will be confronted with recent fsec spectroscopic studies. The experimental fsec data of Holzapfel et al. [13,14,15] for RCs of *Rb.sphaeroides* and *R.viridis* at 300K were analysed in terms of sequential ET, process (2). Concurrently, the fsec data of Kirmaier and Holten [16,17] were analysed in terms of a unistep primary ET in a heterogeneous system, with the rate being dependent on the interrogation wavelength. It is somewhat difficult to reconcile the effects of heterogenuity with an interrogation wavelength depedent rate, as one would rather expect that inhomogeneous kinetics will be manifested by nonexponential time evolution and by "kinetic hole burning" effects, i.e., the dependence of the kinetics on the excitation wavelength. Thus the interesting effects of heterogenuity on primary ET require further elucidation. The fsec spectroscopic studies of Holzapfel et al for RCs of *Rb.sphaeroides* [13-15] at 300K demonstrate that the primary ET kinetics involves two time constants τ_1 = 3.5±0.4psec and τ_2 = 0.9±0.3psec, which are consistent with the prevalence of the sequential process (2). On the other hand, the low temperature data at 10K for both *Rb.sphaeroides* and *R.viridis* [18] are consistent with the unistep process (1), this conclusion being further supported by electric field effects on fluorscence polarization at 80K [19,20]. The available direct and auxiliary experimental information raises the possibility that the mechanism of primary ET in the RC involves superposition of superexchange and sequential channels at 300K and the dominance of superexchange at low temperatures.

2. MODELLING OF PRIMARY ET KINETICS

The basic kinetic scheme for the primary process includes two ET pathways: (i) a superexchange channel with a rate constant k, (ii) a sequential channel with rate constants k_1 and k_{-1} for the direct and reverse first step, respectively, and k_2 for the second step. The reverse and direct steps are related by $k_{-1} = k_1 \exp(\Delta G_1/k_BT)$, where k_B is the Boltzmann constant. The back reaction k_{-1} can be significant at room temperature, and has to be incorporated, while back reactions which are characterized by large values of $\Delta G/k_BT$ can be neglected.

The kinetic analysis of the time dependent relative concentrations $[P^*BH]$, $[P^+B^-H]$ and $[P^+BH^-]$ is straightforward. The system is characterized by two time constants $\tau_1 = (-s_-)^{-1}$ and $\tau_2 = (-s_+)^{-1}$, where s_+ and s_-,

$$s_+ = -[k+k_1+k_{-1}+k_2 + \delta] / 2$$
$$s_- = -[k+k_1+k_{-1}+k_2 - \delta] / 2 \qquad (2.1)$$

with

$$\delta = \left[(k+k_1+k_{-1}+k_2)^2 - 4(kk_{-1}+kk_2+k_1k_2)\right]^{1/2} \qquad (2.2)$$

The maximum concentration of the intermediate P^+B^-H is obtained from the equation $d/dt[P^+B^-H](t=t_{max}) = 0$. The time t_{max} when the maximum concentration is achieved is

$$t_{max} = 1/(s_+-s_-)\ell n\left(\frac{s_-}{s_+}\right) , \qquad (2.3)$$

and the maximum concentration of P^+B^-H, i.e., $B_{max} = [P^+B^-H](t=t_{max})$, is

$$B_{max} = \frac{k_1}{s_+-s_-}\left[\left(\frac{s_-}{s_+}\right)^{\frac{s_+}{s_+-s_-}} - \left(\frac{s_-}{s_+}\right)^{\frac{s_-}{s_+-s_-}}\right] . \tag{2.4}$$

The fraction of reaction that proceeds through the sequential channel is given by:

$$F_{seq} = \int_0^\infty k_2[P^+B^-H](t)\, dt = \frac{k_1k_2}{s_+s_-} \tag{2.5}$$

which provides the branching ratio between sequential and superexchange channels.

3. MICROSCOPIC RATES

The conventional nonadiabatic ET rate constant, k^{NA}, is given by

$$k^{NA} = \frac{2\pi}{\hbar} V^2F \tag{3.1}$$

where V is the electronic coupling and F the thermally averaged nuclear Franck Condon Factor.
The electronic coupling V for the direct ET rates is given by $V_{PB} = \langle {}^1P^*BH|H|P^+B^-H\rangle$ for k_1, and by $V_{BH} = \langle P^+B^-H|H|P^+BH^-\rangle$ for k_2, where H is the Hamiltonian for the system. The superexchange interaction for the unistep rate k is given by

$$V_{super} = V_{PB}V_{BH}/\delta E \tag{3.2}$$

where δE is the vertical energy gap (Fig. 1).
The thermally averaged Franck-Condon factor, F, has to incorporate both low-frequency medium modes ($\hbar\omega_m << k_BT$), which are characterized by the medium reorganization energy E_m, and high-frequency intramolecular modes of (mean) frequency ω ($\hbar\omega >> k_BT$), which are characterized by an electron-vibron coupling S_c. The F factor is [21]

$$F(E_m,\Delta G,S_c,\hbar\omega,T) = (2\pi E_m k_BT)^{-1/2} \exp(-S_c) \sum_{n=0}^{\infty} \frac{(S_c)^n}{n!} \exp[-(\Delta G+n\hbar\omega+E_m)^2 / 4E_mk_BT] \tag{3.3}$$

where ΔG is the free-energy gap between the equilibrium nuclear configurations of the reactants and the products. Until recently, we have asserted that quantum nuclear effects on primary ET are minor, setting $S_c = 0$) in Eq. (3.3), which reduces to the classical limit. Recent analysis [21] has demonstrated that moderate couplings $S_c \simeq 0.5$-1.0, which are sensible in view of spectroscopic data for porphyrins [22], can go a far way in modifcation of temperature dependence and free-energy relations for strongly exoergic ET. For the present case of primary ET processes, where both $|\Delta G|$ and E_m are small, the quantum effects will somewhat modify the details of the branching ratios and the temperature dependence of the primary ET rates.
The conventional ET theory was modified to incorporate medium relaxation effects on ET dynamics with the ET rate, k_{ET}, being given by [23]

$$k_{ET} = k^{NA}/(1+\kappa) \quad , \tag{3.4}$$

where the adiabaticity parameter is

$$\kappa = \frac{4\pi V^2 \tau_s}{\hbar\lambda} \tag{3.5}$$

where τ_s is the longitudinal medium relaxation time induced by a constant charge. The value of $\tau_S \simeq 200$ fsec for the RC was inferred from molecular dynamics simulations for the RC at room temperature [24].

In the nonadiabatic limit ($\kappa << 1$), Eq. (2.10) reduces to the conventional result $k_{ET} = k^{NA}$, Eq. (2.6). In the medium controlled limit ($\kappa >> 1$) the maximal value for an ET rate corresponds to an activationless process ($E_a = 0$), being

$$k_{ET}^{MC} = \tau_s^{-1}(\lambda/16\pi\ k_B T)^{1/2} \tag{3.6}$$

allowing for the estimate of an upper limit for an ET rate. For $\lambda = 1000\ cm^{-1}$ and $T = 300K$ one obtains $k_{ET}^{MC} = 0.31/\tau_s$, which with $\tau_s = 200$ fsec results in the medium-controlled activationless rate of $k_{ET}^{MC} = (700\ \text{fsec})^{-1}$. Medium controlled dynamic effects may be significant for the fastest processes in the RC. The experimental observation of the rate $k_2 = (900\ \text{fsec})^{-1}$ [13-15] may reflect medium-controlled ET in the RC.

The four ET rate constants are given by

$$k = \left(\frac{2\pi}{\hbar}\right)\left(\frac{V_{PB}V_{BH}}{\delta E}\right)^2 F(\lambda, \Delta G, S_c, \hbar\omega, T) \tag{3.7}$$

$$k_1 = \left(\frac{2\pi}{\hbar}\right)\frac{V_{PB}^2}{(1+4\pi V_{PB}^2\tau_s/\hbar\lambda_1)} F(\lambda_1, \Delta G_1, S_{c1}, \hbar\omega, T) \tag{3.8}$$

$$k_{-1} = k_1 \exp(\Delta G_1/k_B T) \tag{3.9}$$

$$k_2 = \left(\frac{2\pi}{\hbar}\right)\frac{V_{BP}^2}{(1+4\pi V_{BH}^2\tau_s/\hbar\lambda_2)} F(\lambda_2, \Delta G_2, S_{c2}, \hbar\omega, T) \ . \tag{3.10}$$

These rate constants correspond to the classical limit for medium modes, which is realized when $k_B T \geq \hbar\omega_m$. Taking $\hbar\omega_m = 80\text{-}100\ cm^{-1}$ the classical limit prevails for $k_B T \geq 100K$. Thus numerical calculations were performed in the range $T = 300\text{-}100K$.

4. MODEL CALCULATIONS

The kinetic scheme involves a proliferation of microscopic parameters. The analysis of the kinetic scheme was performed with the following "reasonable" parameters. The nuclear, electronic and dynamic parameters are assumed to be temperature independent. We shall start the analysis by setting all $S_c = 0$. The input parameters used in the calculation of the elementary rates were

ΔG = -2000 cm^{-1}
λ = 2000 cm^{-1} (assuming that k is activationless)
$\lambda_1 = \lambda_2$ = 1000 cm^{-1} (λ_1, $\lambda_2 < \lambda$)
$E_a = E_{a2} = 0$
τ_s = 200 fs.
V_{BH}/V_{PB} = 4.
The central energetic parameter ΔG_1, which essentially determines the mechanism of the primary ET, was varied over the broad range ΔG_1 = -800cm^{-1} to +600cm^{-1} .

The only missing parameter is now the electronic coupling V_{PB}. This was determined by incorporating experimental kinetic information by demanding that the long effective relaxation time $\tau_1 = (-s_-)^{-1}$ at room temperature should be equal to the experimental (rate determining) primary rate, i.e., $(-s_-)^{-1} = 3.3 \times 10^{-12}$ sec at 300K.

Eqs. (3.7)-(3.10) were used to calculate the elementary rates k, k_1, k_{-1} and k_2, which were then incorporated into Eqs. (2.1)-(2.5) for the kinetic scheme. Once V_{PB} was fixed (for a given value of ΔG_1) at 300K the same electronic couplings were used for the calculation of kinetic information at other temperatures.

5. RESULTS

We shall focus on the relevant observables, which can be confronted with experimental data. All the results of the kinetic modelling will be displayed vs the free energy gap ΔG_1 for several temperatures. The output data involve

(i) The elementary rate constants;
(ii) The effective relaxation times τ_1 and τ_2, whose ratio (with $\tau_1 = 3.3 \times 10^{-12}$ sec at 300K) is given in Fig. 2;
(iii) The maximum concentration B_{max}, of [P^+B^-H], which appears at time t_{max} (Fig. 2);

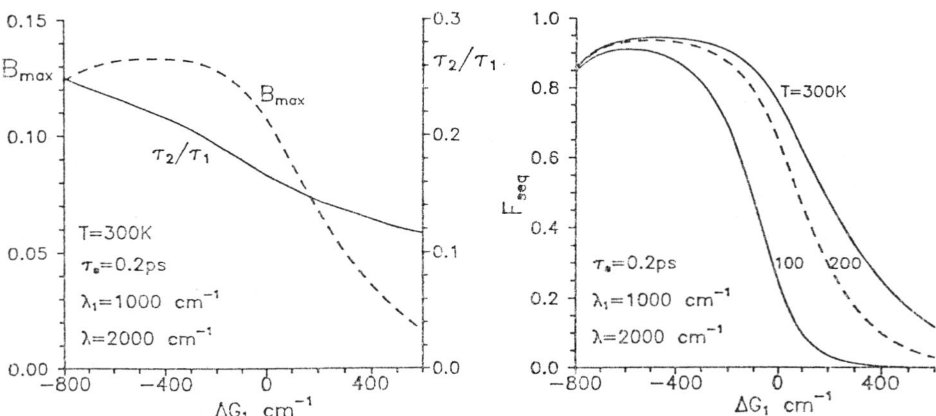

Figure 2. The free energy gap (ΔG_1) dependence of the ratio of the relaxation times τ_2/τ_1 and of B_{max} for several values of τ at 300K. For all these data τ_1 = 3.3psec.

Figure 3. The ΔG_1 dependence of the branching ratio F_{seq} between the sequential and superexchange channels.

(iv) The temperature dependence of the ET rates;
(v) The branching ratio F_{seq} for the sequential channel which is displayed in Fig. 3 for several temperatures;
(vi) The ratio of the slow to fast amplitudes in the decay of $[P^*BH]$.

The dependence of $|V_{PB}|$ on ΔG_1 is presented in Fig. 4. Both $|V_{PB}|$ and $|V_{BH}| = 4|V_{PB}|$ increase with increasing ΔG_1. All the attributes were calculated in the temperature range T = 300K-100K (with V_{PB} fixed at 300K). The lowest temperature T = 100K provides a reasonable description of the low temperature kinetic data in the RC.

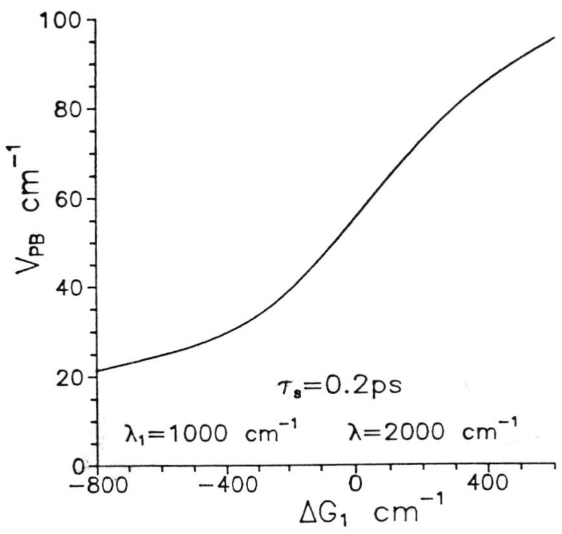

Figure 4. The dependence of the electronic coupling $|V_{PB}|$ on ΔG_1, which is subject to the experimental constraint τ_1 = 3.3 psec at 300K.

From the foregoing results of the kinetic analysis extensive information emerged regarding the competition of the two-step sequential ET and the unistep superexchange mechanisms over a broad range of ΔG_1. Furthermore, the kinetic scheme implies that the relative contributions of the two parallel mechanisms are temperature dependent. Accordingly, we can utilize the temperature dependence of the branching ratio F_{seq} (Fig. 3) for the classification of the mechanisms which are dominant in various energy domains. The following energy domains can be distinguished

Range (I). Sequential mechanism at all temperatures. $\Delta G_1 \leq -400\text{cm}^{-1}$.

Range (II). Superposition of sequential and superexchange mechanism at all temperatures. $-400\text{cm}^{-1} \leq \Delta G_1 \leq 0$.

Range (III). Superposition of sequential and superexchange mechanisms at room temperature and superexchange mechanism at low temperature. $0 \leq \Delta G_1 \leq 400\text{cm}^{-1}$.

Range (IV). Superexchange mechanism at all temperatures. $\Delta G_1 \geq 400\text{cm}^{-1}$.

The available direct kinetic data allow for a preliminary determination of the permissable free energy ΔG_1 domain.

(1) Two decay lifetimes. The experimental results [13] τ_2/τ_1 = 0.26±0.10 at 300K are consistent with our model calculations over the range $\Delta G_1 = -800\text{cm}^{-1}$ to 200cm^{-1}.

(2) Temperature dependence of τ_1. The temperature dependence of the relaxation time τ_1 was found to be weak, exhibiting the features of activationless ET. The ratio r_1 = $\tau_1(100K)/\tau_1(300K)$ calculated for S_c = 0 was found to be r_1 = 1.7 at ΔG_1 = 0. Incorporation of

quantum effects reduces this ratio to $r_1 = 1.2$ for $S_c = 1.5$ and $\hbar\omega = 1500cm^{-1}$. A quantitative account for the experimental result $r_1 = 0.5$ requires the further incorporation of thermal contraction effects of the protein medium.

(3) The magnitude of B_{max}. Holzapfel et al [13], have reported $B_{max} \simeq 0.15$ at 300K. There seems to be a substantial uncertainty in the determination of B_{max}, which rests on limited experimental information regarding absorption coefficients of the prosthetic groups and their negative ions as well as of $^1P^*$. This rather uncertain room temperature result, $B_{max} \simeq 0.15$, is consistent with our model calculations (Fig. 2) over the range $\Delta G_1 = -800cm^{-1}$ to $100cm^{-1}$.

(4) The overall temperature dependence and general characteristics of the kinetics. The room temperature fsec spectroscopic data seem to reveal two lifetimes with $\tau_1/\tau_2 = 4.0\pm1.5$ for *Rb.sphaeroides* [13-15]. If these observations and analyses will be borne out by subsequent work one can conclude that the primary process in *Rb.sphaeroides* corresponds to ranges (I) or (II) or (III), i.e., $-800cm^{-1} \leq \Delta G_1 \leq 200cm^{-1}$. A similar conclusion emerges from the analysis of the experimental absolute values of B_{max} within the present kinetic model. The lower limit of ΔG_1 can be somewhat reduced in view of the analysis of the energetics [19] which gives $\Delta G_1 \geq -600cm^{-1}$. Thus the permissable free energy domain which is consistent with the room temperature kinetic data is $-600cm^{-1} \leq \Delta G_1 \leq 200cm^{-1}$. The low temperature (10K) fsec data for both *Rb.sphaeroides* and *R.viridis* [18] were analysed in terms of a single exponential decay of $^1P^*$ and buildup of P^+BH^-, which implies that the kinetics is either biexponential with $\tau_1 \sim \tau_2$ within experimental uncertainty, or that the kinetics is characterized by a single lifetime. Provided that the occurrence of a single-lifetime kinetics will be confirmed by further experiments then the avialable low temperature data seem to indicate that the primary kinetics corresponds to ranges (III) or (IV). From the combination of the room temperature results and the low temperature data there is a distinct possibility that the primary mechanism falls within range (III), with $\Delta G_1 = -100$ to $200cm^{-1}$ for *Rb.sphaeroides*. Thus, the primary mechanism may involve parallel sequential and superexchange mechanisms at high temperature and essentially superexchange at low temperature. This conclusion is, of course, preliminary, requiring extensive further critical experimental scrutiny.

6. SOME IMPLICATIONS OF MUTAGENESIS

Functional and structural control of ET by mutagenesis provides a novel powerful tool for the modification of electronic coupling and nuclear Franck-Condon factors. Experimental fsec kinetic data on mutagenetically altered RCs of *Rb.spaeroides* are expected to infer on the functional and structural importance of specific amino acid residues. The exchange of the polar tyrosine M208 (Tyr) residue by the nonpolar phenylalanine (Phe) in *Rb.sphaeroides* results in the slowing of the primary ET rate to $1/\tau_1^M = (16psec)^{-1}$ [25], as compared with $1/\tau_1 = (3.5\pm0.4psec)^{-1}$ for the native RC [13]. This observation can be rationalized in terms of the modification of the energetics of the P^+B^- ion pair state. According to recent energetic calculations [26] the replacement of Tyr M208, by the Phe results in the increase of the energy of the P^+B^- state by $\sim100cm^{-1}$. Such a change in the value of ΔG_1 has the following consequences: (i) The reduction of the nuclear Franck-Condon factor in k_1 resulting in the decrease of the rate determining step in the sequential process. (ii) The increase of the vertical energy gap δE, which will decrease V_{super}, Eq. (II.7), resulting in the decrease of the superexchange rate k. Thus both parallel channels will be retarded. To investigate the consequences of the energetics of Tyr M208 → Phe mutagenesis on ET kinetics we have utilized the kinetic scheme of Section 2 and the nuclear parameters of Section 3 together with two modifications: (i) The free energy gap for k_1 was modified to be $\Delta G_1^M = \Delta G_1 + \delta G_1$, where $\delta G_1 = 1000cm^{-1}$ is the shift of the energy of P^+B^-. (ii) The electronic couplings V_{PB} and V_{BH} are taken from the results for the native RC, being given by the data of Fig.4.

In Fig. 5 we present the lifetimes τ_1^M for the rate determining ET step in this mutagenetically modified RC vs the free energy gap ΔG_1 of the native system. These results reveal a marked lengthening of τ_1^M relative to τ_1 for the native RC, the room temperature data ranging from $\tau_1^M/\tau_1 = 5$ for $\Delta G_1 = -600cm^{-1}$ to $\tau_1^M/\tau_1 = 8$ for $\Delta G_1 = 0$. These results for the permissable range of ΔG_1 are in accord with the experimental result $\tau_1^M/\tau_1 = 6$ at 300K. This analysis predicts a weak temperature dependence of τ_1^M in the range of the permissable gap $\Delta G_1 = -200cm^{-1}$ to $100cm^{-1}$. The present kinetic analysis of the ET kinetics in the mutagenetically modified RC invokes the dangerous assumption of structural invariance of the RC with respect to the substitution of M208. Of course, even small configurational changes of the relative location of P and B across the A branch may grossly modify the electronic couplings and the ET dynamics.

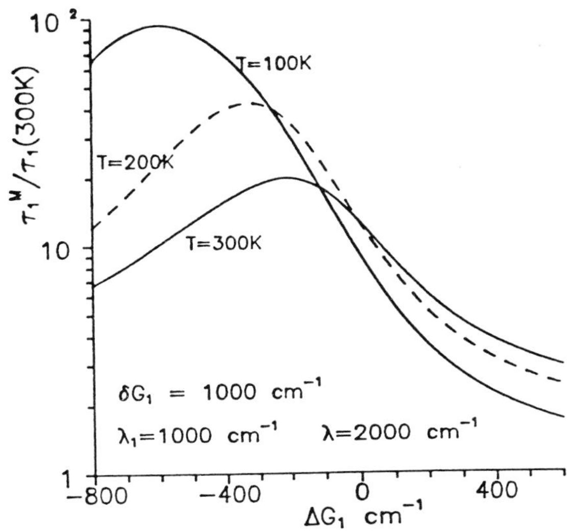

Figure 5. The dependence of the lifetime τ_1^M of the rate determining ET step in tyrosine M208→ phenylalanine mutagenetically modified RC, on the (free) energy gap ΔG_1 for the corresponding wild RC. Data are given for the temperature range T = 300K-100K.

7. UNIDIRECTIONALITY OF CHARGE SEPARATION

The unidirectionality of the primary charge separation in the RC across the A branch constitutes a central dynamic phenomenon in photosynthesis. This effect was attributed to symmetry breaking [6,7], which may originate from the distinct electronic and nuclear contributions to the elementary ET rates, which in turn modify the experimental lifetimes.
(i) Modification of the electronic coupling terms V_{PB} and V_{BH} along the A and B branches [7]. Utilizing the intermolecular overlap approximation, these electronic coupling terms were computed for the atomic coordinates of *R.viridis* at a resolution of 2.6Å. According to recent calculations based on the final refinement (2.2Å) of the atomic coordinates of *R.viridis* RCs (M.Plato, private communication).

$$V_{PB}(A)/V_{PB}(B) = 2.0 \pm 0.4$$
$$V_{BH}(A)/V_{BH}(B) = 6.5 \pm 1.5$$
$$V_{BH}(B)/V_{PB}(B) = 1.1 \pm 0.2 \quad . \qquad (7.1)$$

Thus the contributions of the electronic couplings to the sequential rate k_1 and the

superexchange rate k are as follows:
Electronic contribution to $k_1(A)/k_1(B) = |V_{PB}(A)/V_{PB}(B)|^2 = 4\pm1$;
Electronic contribution to $k(A)/k(B) = |V_{PB}(A)V_{BH}(A)/V_{PB}(B)V_{BH}(B)|^2 = 170\pm80$.
(ii) Modification of the nuclear Franck-Condon factor. The microscopic environment of the B groups on the A and B branches of the RC is different with B_A located near the polar tyrosine M208 and B_B located near a nonpolar Phe. Energetic calculations for the RC have established that the energies of the ion pairs $P^+B_A^-$ and $P^+B_B^-$ are different with the latter being higher in energy [26]. The protein environment of the B branch of the RC where B_B is located near the Phe residue bears a close analogy to the A branch of the mutagenetically modified (Tyr M208 → Phe) RC. To obtain an estimate of the effects of unidirectionality on the parallel sequential superexchange mechanism, we have again performed a kinetic simulation for ET across the B branch according to the scheme of section 2 and nuclear parameters (3) together with the following modifications: (i) The electronic couplings $V_{PB}(B)$ and $V_{BH}(B)$ were scaled according to Eq. (5.1). (ii) The free energy gap between $^1P^*$ and P^+B^- across the B branch was modified to be $\Delta G_1(B) = \Delta G_1(A) + 1000cm^{-1}$. The lifetimes $\tau_1(A)$ and $\tau_1(B)$ were calculated neglecting crossing between the A and B branches. As $\tau_1(B)/\tau_1(A) >> 1$, this assumption is a-posteriori justified.

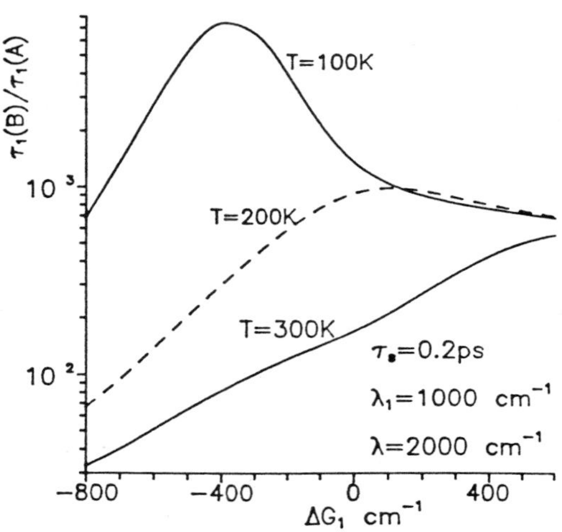

Figure 6. Kinetic modelling of the unidirectionality of charge separation. The ratio $\tau_1(B)/\tau_1(A)$ of the rate determining steps for ET across the A and B branches of the RC are in the temperature range 300K-100K.

In Figure 6 we present the results for the ratio $\tau_1(B)/\tau_1(A)$ over the relevant range of $\Delta G_1(A)$, spanning the temperature domain T = 300-100K. The total contribution to the unidirectionality can be separated to the nuclear term δ_N and electronic term δ_e, so that $\tau_1(B)/\tau_1(A) = \delta_e\delta_N$. Both δ_N and δ_e contribute in the same direction, towards the enhancement of the lifetime ratio. The nuclear contribution is given by the lifetime ratio for the mutant which was calculated in section (5.A). The electronic contribution δ_e is larger in the free energy and the temperature domain, where the contribution of superexchange becomes dominant and thus the ratio $\tau_1(B)/\tau_1(A)$ increases at lower temperatures. The calculated value of $\tau_1(B)/\tau_1(A)$ over the permissable domain of ΔG_1 is 3-500 at room temperature (300K) and 500-7000 at a low temperature (100K). The lower experimental limit $\tau_1(B)/\tau_1(A) \geq 25$ at 90K [27] is consistent with the results of these model calculations.

8. EPILOGUE

The present analysis rests on three implicit assumptions. First, the temperature dependence of the nuclear, electronic and dynamic parameters was disregarded. Second, we asserted that configurational relaxation of the medium occurs on the subpicosecond time scale. Third, we assumed that the protein medium is microscopically homogeneous.

The main conclusion emerging from our kinetic modelling is that the fsec kinetic data at high [13-15] and at low [18] temperatures are consistent with a primary charge separation model which involves the superposition of sequential and superexchange ET at room temperature and a superexchange mechanism at 10K. In general, the explicit distinction between a superexchange and a sequential mechanism may be impossible, as both mechanisms can prevail in parallel over a broad range of the energetic, nuclear and electronic parameters. This kinetic scheme for the primary ET reduces to the limit of a nearly pure sequential mechanism for large negative ΔG_1 at all temperatures and to the limit of an almost pure superexchange mechanism for large positive ΔG_1 at all temperatures and for moderate ΔG_1 at low temperatures.

The peaceful coexistence of the sequential and superexchange ET routes for the primary ET in the RC of some purple bacteria, allows for the occurrence of efficient charge separation over a rather broad range of the energy gap ΔG_1 (and possibly of other nuclear and electronic parameters). In all these systems kinetic optimization is essential, with the primary process being sufficiently fast to compete with energy waste due to backtransfer to the antenna. What is required is an efficient primary process, which is stable with respect to (moderate) variations of the energetic parameters. This energetic stability criterion is satisfied by the parallel sequential-superexchange mechanism, which constitutes an efficient and ultrafast charge separation over a rather broad energy range ΔG_1 of ~800cm^{-1} (2-3kcal mole^{-1}). The occurrence of the parallel mechanism for the primary ET introduces an element of redundancy, which might be essential to ensure the prevalence of efficient primary charge separation in different photosynthetic RCs.

REFERENCES

1 M.E. Michel-Beyerle (ed.), Reaction Centers of Photosynthetic Bacteria, Springer Verlag, Berlin, 1990.

2 J. Deisenhofer and H. Michel, EMBO J., 8 (1989) 2149-2170.

3 N.W. Woodbury, M. Becker, D. Middendorf and W.W. Parson, Biochemistry, 24 (1985) 7516-7521.

4 J. Jortner, and M.E. Michel-Beyerle, in: Antennas and Reaction Centers of Photosynthetic Bacteria, M.E. Michel-Beyerle (ed.), pp. 345-365, Springer, Berlin, 1985.

5 J.R. Norris, D.E. Budil, D.M. Tiede, J. Tang, S.V. Kolaczkowski, C.H. Chang and M. Schiffer, in: Progress in Photosynthetic Research, J. Biggens (ed.), I, pp.1.4.363-1.4.369, Martinus Nijhoff, Dordrecht, 1987.

6 M.E. Michel-Beyerle, M. Plato, J. Deisenhofer, H. Michel, M. Bixon and J. Jortner, Biochim. Biophys. Acta, 932 (1988) 52-70.

7 M. Plato, K. Möbius, M.E. Michel-Beyerle, M. Bixon and J. Jortner, J. Am. Chem. Soc., 110 (1988) 7279-9285.

8 M. Bixon, J. Jortner, M.E. Michel-Beyerle and A. Ogrodnik, Biochim. Biophys. Acta, 977 (1989) 273-286.

9 R.A. Friesner and Y. Won, Biochim. Biophys. Acta, 977 (1989) 99-122.

10 R.A. Marcus, Isr. J. Chem. 28 (1988) 205-213.

11 R. Haberkorn, M.E. Michel-Beyerle and R.A. Marcus, Proc. Natl. Acad. Sci. USA, 70 (1979) 4185-4188.

12 S. Creighton, J.-K. Hwang, A. Warshel, W.W. Parson and J. Norris, Biochem., 27 (1988) 774-781.

13 W. Holzapfel, U. Finkele, W. Kaiser, D. Oesterhelt, H. Scheer, H.U. Stilz and W. Zinth,

Chem. Phys. Lett., 160 (1989) 1-7.
14 W. Holzapfel, U. Finkele, W. Kaiser, D. Oesterhelt, H. Scheer, H.U. Stilz and W. Zinth, Proc. Natl. Acad. Sci. USA, 87 (1990) 5168-5173.
15 K. Dressler, U. Finkele, C. Lauterwasser, P. Hamm, W. Holzapfel, S. Buchanan, W. Kaiser, H. Michel, D. Oesterhelt, H. Scheer, H.U. Stilz and W. Zinth, in: Reaction Centers of Photosynthetic Bacteria, M.E. Michel-Beyerle (ed.), pp. 135-140, Springer Verlag, Berlin, 1990.
16 C. Kirmaier and D. Holten, Proc. Natl. Acad. Sci. USA, 87 (1990) 3552-3558.
17 C. Kirmaier and D. Holten, in: Reaction Centers of Photosynthetic Bacteria, ed. M.E. Michel-Beyerle (ed.), pp. 113-125, Springer Verlag, Berlin, 1990.
18 J. Breton, J.-L. Martin, G.R. Fleming and J.-C. Lambry, Biochem., 27 (1988) 8276-8284.
19 D.J. Lockhart, R.F. Goldstein and S.G. Boxer, J. Chem. Phys., 89 (1988) 1408-1415.
20 A. Ogrodnik, U. Eberl, R. Heckmann, M. Kappl, R. Feick and M.E. Michel-Beyerle, in: Reaction Centers of Photosynthetic Bacteria, M.E. Michel-Beyerle (ed.), pp. 157-168, Springer Verlag, Berlin, 1990.
21 M. Bixon and J. Jortner, J. Phys. Chem., 95 (1991) 1941-1944.
22 U. Even and J. Jortner, J. Chem. Phys., 77 (1982) 4391-4398.
23 I. Rips and J. Jortner, J. Chem. Phys., 87 (1987) 2090-2099.
24 H. Treutlein, K. Schulten, J. Deisenhofer, H. Michel, A. Brunger, and M. Karplus, The Photosynthetic Bacterial Reaction Center. Structure and Dynamics, J. Breton and A. Vermeglio (eds.), NATO ASI Series A 149 139-150. Plenum Press, New York, 1988.
25 K.A. Gray, J.W. Farchaus, J. Wachtveitl, J. Breton, U. Finkele, C. Lauterwasser, W. Zinth and D., Oesterhelt, in: Reaction Centers of Photosynthetic Bacteria, M.E. Michel-Beyerle (ed.), p. 252, Springer Verlag, Berlin, 1990.
26 W.W. Parson, Z.-T. Chu and A. Warshel, Biochim. Biophys. Acta, (1990) (in Press).
27 W. Aumeier, U. Eberl, A. Ogrodnik, M. Volk, G. Shiedel, R. Feick, M. Plato and M.E. Michel-Beyerle, in: Current Research in Photosynthesis, M. Baltscheffsky (ed.), Kluwer Academic Publishers, Dordrecht, 1990.

DISCUSSION

Itoh

Please explain the relation of super–exchange and sequential mechanism in different types of reaction centers. Does it change?

Jortner

The energy gap $\Delta G_1 = -100 \pm 100$ cm^{-1} determined for the combination of our theory and the experimental results of Fleming and Norris may be, of course, smeared out due to inhomogeneous broadening effects. ΔG_1 is not a universal constant. It is conceivable that this energy gap will be different in different photosynthetic reaction centers (bacteria, plants, etc.). The occurrence of the sequential and superexchange mechanisms in parallel insures the existence of an efficient primary charge separation in different reaction centers. Thus our mechanism insures the universality of an ultrafast efficient primary charge separation in photosynthesis.

Kakitani

What is the origin of the temperature dependence of the fraction of sequential pathway?

Jortner

The temperature dependence of the branching ratio between sequential and superexchange pathway originates for the simultaneous contributions of;
(i) the enhancement of the superexchange channel at low temperature, due to the pseudo activationless nature of this process,
(ii) the retardation of the thermally activated sequential channel at low temperature.

Hynes

Your successful application of your approximate combination rule for the electronic coupling according to the superexchange mechanism is very impressive. Is it obvious that the combination rule would be different for the *direct* mechanism?

Jortner

Thank you. One can formulate combination rule for the direct process. Consider the two direct process $D^-A \longrightarrow DA^-$ and $D^-B \longrightarrow DB^-$ which are assumed to be both induced by the direct exchange terms V_{DA} and V_{DB}, respectively. Then the process $B^-DA \longrightarrow BDA^-$ will involve a superexchange contribution determined by the coupling $(V_{DA}\,V_{DB}/\delta E)$. This problem deserves a further examination.

Kakitani

In order to reproduce the nearly temperature independent rate for the $P^+H^- \longrightarrow PH$ reaction, you introduced a high frequency vibrational mode in addition to a soft mode of protein environment. Is it always true for any energy gap? If so, the traditional argument that the optimum energy gap law condition should be satisfied if the ET rate is almost temperature–independent is not correct. Why do people assume that this optimum condition is satisfied for $PH^* \longrightarrow P^+H^-$? Is there no possibility that the energy gap changes with temperature for this reaction in a manner consistent with a small increase of the rate as temperature is lowered?

Jortner

We have recently shown [M. Bixon and J. Jortner, J. Phys. Chem. 95, 1941 (1991)] that a surprisingly weak temperature dependence of the electron transfer rate prevails over a broad range of the free energy gap in the inverted region, which is induced by moderate values (S_c>0.5 for a protein medium) of the electron–quantum mode coupling. The weak temperature dependence of the rate is realized in the broad range

$$1-(2\hbar\omega_m/E_m)^{1/2} \leq -\Delta G/E_m \leq 20{,}000\ (\mathrm{cm}^{-1})/E_m$$

where ΔG is the free energy gap, E_m is the protein reorganization energy, and $\hbar\omega_m$ its characteristic frequency. for the primary process in the photosynthetic reaction center we should imagine what are the values of ΔG and E_m. For R_b spheroids we know for energetic data that ΔG=−2000 cm^{-1} for the gap between P^*BH and P^+BH^-. E_m cannot be too small in a protein medium a reasonable guess is $E_m \approx 1200$ cm^{-1} which is the value for a nonpolar hydrocarbon solvent. Accordingly, $-\Delta G/E_m$ is close to unity and we are still in the pseudo activationless domain where the nuclear Franck Condon factor is close to its maximum value.

Dynamics and Mechanisms of
Photoinduced Transfer and Related Phenomena
N. Mataga, T. Okada and H. Masuhara (Editors)

Exchange of the Acceptor Phylloquinone by Artificial Quinones and Fluorenones in Green Plant Photosystem I Photosynthetic Reaction Center

Shigeru Itoh and Masayo Iwaki

Division of Bioenergetics, National Institute for Basic Biology,
38 Nishigonaka, Myodaijicho, Okazaki 444, Japan

ABSTRACT

In photosystem I photosynthetic reaction centers of green plant (spinach), the constituent phylloquinone (2-methyl-3-phytyl-1,4-naphthoquinone; vitamin K_1), which functions as the secondary electron acceptor A_1 (Q_ϕ), was replaced by various quinones and fluorenones. The phylloquinone-binding site shifts the redox midpoint potential (E_m) values of semiquinone$^{\bullet-}$ to the negative direction by 0.3 V from $E_{1/2}$ in dimethylformamide. Reaction rate between $P700^+$ and quinone$^{\bullet-}$ or fluorenone$^{\bullet-}$ was almost temperature independent and was slightly affected when the energy gap between quinone/fluorenone and P700 was changed. This suggests that the reaction is coupled to some distribution of high-frequency vibrations with a small reorganization energy.

1 INTRODUCTION

Electron transfer reaction in reaction center (RC) protein-pigment complexes in photosynthetic organisms has been the unique experimental system to test theories of electron transfer. Nature has created two different types of RCs which are represented by two RCs functioning in series on the inner membranes of green plant chloroplasts: photosystems (PS) I and II. The isolated PS II RC complex [1] seems to have a structure essentially common with that of purple photosynthetic bacterial RCs, whose tertiary structures have been determined by X-ray crystallography [2, 3]. PS II/purple bacterial RC complexes are composed of essentially two polypeptides of molecular mass 30 kDa on which four to six chlorophylls, two pheophytins and two quinones are attached [1–4]. The quinones function as the primary (Q_A) and secondary (Q_B) electron acceptors as shown in Fig. 1. On the other hand, PS I RC complex has different prosthetic groups with different energy levels as shown in Fig. 1, and seems to have somewhat different structure. It is composed of larger polypeptides of about 80 kDa, more than 50 chlorophyll a molecules (including reaction center chlorophyll P700 and primary acceptor

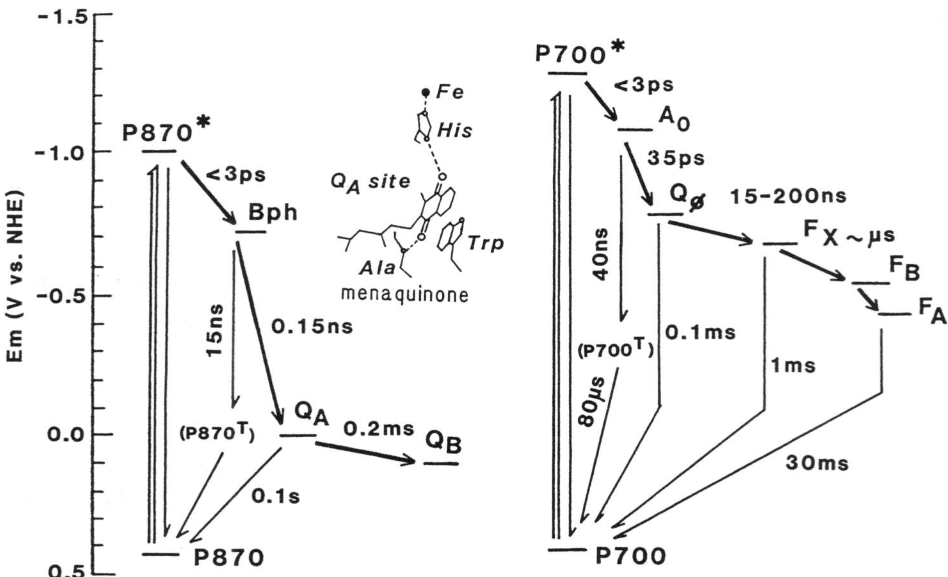

Fig. 1. Electron transfer pathway in purple bacterial (left) and PS I RC (right). P870, P870* and P870^T ; ground, lowest singlet excited and triplet state of the primary electron donor bacteriochlorophyll dimer, respectively. Bph; bacteriopheophytin. Insert figure shows structure of Q_A site of *Rps. viridis* RC reproduced after refs. 2 and 3. Reaction times are obtained from refs. 4, 5, 12 and 13.

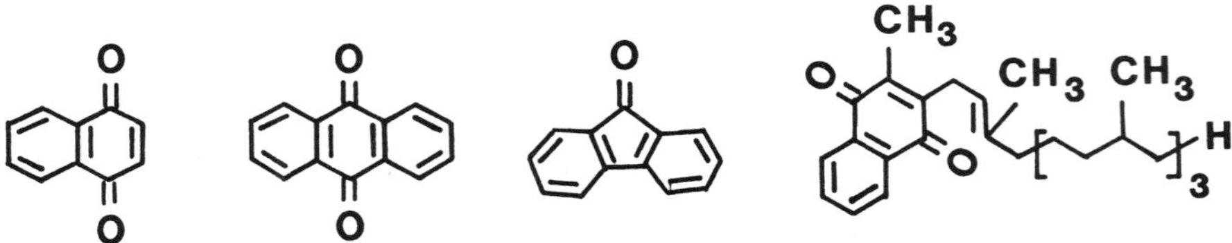

Fig. 2. Structures of molecules studied in this work as the electron acceptor in PS I RC. From left to right: 1,4-naphthoquinone, 9,10-anthraquinone, 9-fluorenone and phylloquinone.

chlorophyll A_0), 2 phylloquinones (2-methyl-3-phytyl-naphthoquinone) and a tertiary acceptor 4Fe-4S center F_X[4,5]. Electrons flow out from F_X to two 4Fe-4S centers F_A and F_B (F_{AB}) on another small polypeptide. The structure of this RC complex seems to be partially common with those of RCs of green sulfur bacteria, of which our knowledge is still limited. It is, thus, highly plausible that all the photosynthetic RCs use quinones as the secondary electron acceptor. Electron transfer mechanism in the reaction of quinone in these two different RCs is an interesting subject to be studied.

PS I RC contains two molecules of phylloquinones and one of them functions as the secondary electron acceptor Q_ϕ which has long been called as A_1, and mediates electrons to iron sulfur center F_X[4-7]. Function of the second phylloquinone is not yet clear. We first reported the extraction and reconstitution method of phylloquinone and gave a direct evidence for the chemical identity of A_1 [7] as later confirmed in cyanobacterial membranes [8]. The functional phylloquinone binding site (Q_ϕ site), binds various artificial quinones as well as herbicide [9-11] with affinities different from those reported in the Q_A site of purple bacterium *Rb. sphaeroides* [10].

We here describe the reaction of various quinones and fluorenones (see Fig. 2) introduced into the Q_ϕ site in spinach PS I RC. Most of quinones and fluorenones function as the secondary electron acceptor and restore the forward electron transfer to Fe-S centers. Relation between the electron transfer rate, and the molecular structure on energy levels of introduced compounds are studied. Various quinones have been shown to be incorporated into the purple bacterial RCs and the relation between the energy level and the reaction rate has been studied [12-15]. The study presented here shows that this type of analysis can also be done in green plant PS I RC.

2 MATERIALS AND METHODS

Lyophylized photosystem I particles, obtained from spinach chloroplasts, were twice extracted with a 1:1 mixture of dried and water-saturated diethylether; to completely remove the 2 phylloquinone molecules [9,10]. The phylloquinone-depleted PS I particles were also depleted of about 85 % of the antenna chlorophyll complement and all carotenoids but not other electron transfer components [7,16]. The extracted particles were solubilized as reported [9,10], and suspended in 50 mM Tris-Cl buffer, (pH 7.5) containing 30 % (v/v) glycerol and Triton X-100 (0.001 % (v/v))to give a final P700 concentration of 0.25 μM. All extraction procedure steps were done below 4 °C. To reconstitute quinones or fluorenones, the suspension of the extracted PS I particles was incubated for a day at 0 °C in the dark with quinones or fluorenones (see Fig. 2) dissolved in dimethylsulfoxide. The quinone-reconstituted RCs were stable for several days when stored below 0 °C. 10 mM ascorbate and 0.1 mM dichloroindophenol couple was added to the reaction medium to provide seconds time scale reduction of the small amount of $P700^+$ not rapidly reduced by intrinsic components.

The activity of the reconstituted PS I particles was assayed by measuring the flash-induced absorption change of P700 at 695 nm in a split-beam spectrophotometer with 1 μs time response [9] or by multichannel detector with 3 ns time response at 7 °C [17]. The intensity of actinic flash (532 nm, 10 ns FWHM, 0.7 Hz) from a frequency doubled Nd-YAG laser (Quanta-Ray, DCR-2-10), was attenuated to excite about a quarter of RCs to avoid sample damage (at 0.4 mJ). Signals were averaged between 32 and 128 scans as required in each case.

3 RESULTS

3.1 Nanosecond kinetics of primary donor P700 and primary acceptor A_0 in PS I RC containing an artificial quinone

In PS I RC in which intrinsic phylloquinone is extracted with diethyl ether, laser excitation induces an absorption change peaking at 693 nm showing a rather wide bandwidth (Fig. 3a). This absorption change is due to the biradical state ($P700^+$ A_0^-) which decays with a $t_{1/2}$ of 40 ns as reported elsewhere. The charge recombination

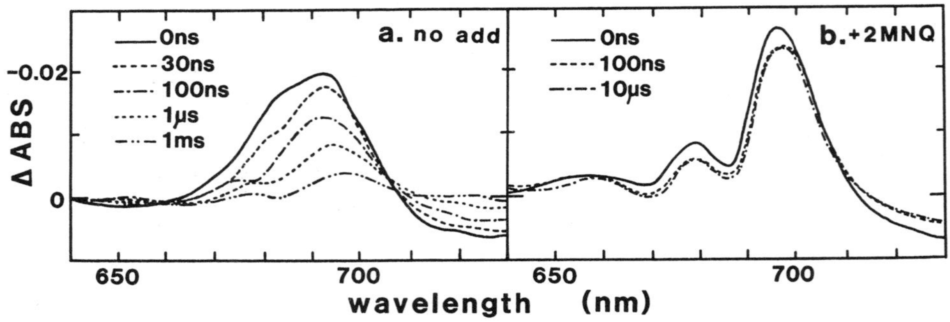

Fig. 3. Time resolved difference absorption spectra of phylloquinone-extracted PSI RC in the presence and absence of reconstituted quinone in ns time range at 280 K. a, no additions. b, in the presence of 10 μM 2-methyl-naphthoquinone (2MNQ). Difference absorption changes were measured at times indicated in Figure after the 10 ns laser-flash excitation.

between $P700^+$ and A_0^-, then, produces triplet state $P700^T$, as seen as the narrower red-shifted spectrum at 1 μs, or delayed fluorescence from P700* [17]. In PS I RC reconstituted with 2-methyl-naphthoquinone (2MNQ) in place of intrinsic phylloquinone, $P700^+$, which has a characteristic absorption peak at 696 nm, was detected at 0 ns (Fig. 3b). The spectrum does not change until 10 μs suggesting that A_0^- is rapidly oxidized by 2MNQ in a time range shorter than the response time of the measurement system (with a $t_{1/2}$ of 150 ps according to Kim et al. [18]). This indicates that the artificial quinone can oxidize A_0^- and stabilize $P700^+$ until μs–ms time range and that the effectiveness of quinones or other compounds to replace the function of phylloquinone in this RC can also be studied by the measurement of $P700^+$ at a μs time range.

3.2 Kinetics of flash-induced P700 oxidation and dark re-reduction in PS I RCs containing various quinones and fluorenones.

In the ether-extracted PS I RCs, only a small extent of $P700^+$ was detected in the μs–ms time range after the laser excitation due to the charge recombination (represented by a broken line). This produces a small signal of triplet state $P700^T$ at 695 nm, which decays with a $t_{1/2}$ of 80 μs (this $t_{1/2}$ is longer than that of a typical few μs decay seen in intact PS I RC, due to the co-extraction of carotenoids in the present preparation [7])(Fig. 4). A small amount of $P700^+$ is also induced. This decays slowly ($t_{1/2}$ = 30–100

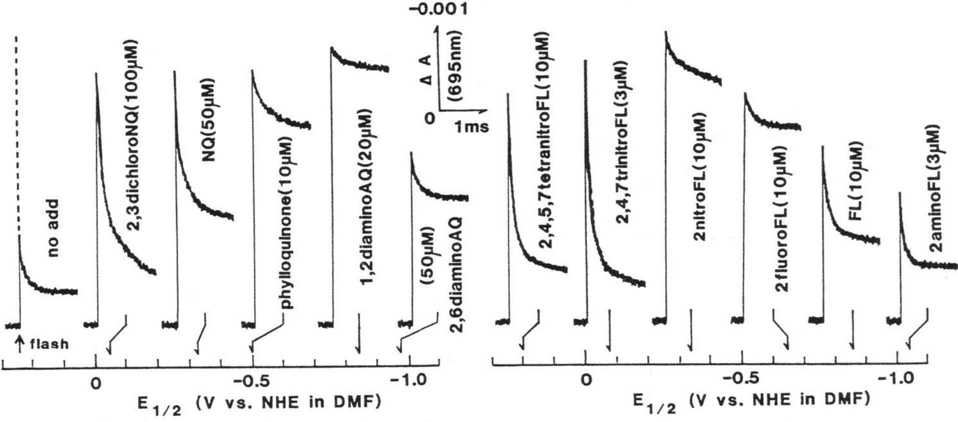

Fig. 4. Flash-induced absroption kinetics of P700 in the ether extracted PS I particles reconstituted with quinones (left) and fluorenone derivatives (right) in a μs–ms time rage at 280 K. $E_{1/2}$ values of quinones and fluorenones, indicated on the horizontal axis, are obtained from refs. 27–29 as mentioned in ref. 20.

ms), due to the reduction either by an added ascorbate-dichloroindophenol couple or by partially photo-reduced iron-sulfur centers.

The flash-induced $P700^+$ extent in the μs-ms range was increased when various naphthoquinones (NQ) and anthraquinones (AQ) were substituted for the intrinsic phylloquinone in the PS I RC (Fig. 4 left). This indicates that these quinones can oxidize A_0^- at a rate rapid enough to compete with the charge recombination between A_0^- and $P700^+$, presumably with a $t_{1/2}$ of subnanosec. Benzoquinones were rather poor in this activity except for duroquinone (not shown)[9]. The concentration of each quinone used in Fig. 2 was adjusted to provide the maximum level of reconstitution.

Re-reduction kinetics of the flash-induced $P700^+$, on the other hand, varied depending on the redox potential of the reconstituted quinone represented by the polarographically measured $E_{1/2}$ value of the semiquinone$^{\bullet-}$/quinone couple in organic solvent. This suggests that the midpoint potential (E_m *in situ*) value of quinone at the Q_ϕ site with respect to those of A_0 and F_X, is crucial in determining the reaction kinetics and that the E_m *in situ* is related to $E_{1/2}$ in dimethylformamide (DMF) as recently shown [19,20].

A group of quinones induced a slow decay ($t_{1/2}$ = 30–100 ms) following the high extent of flash-induced $P700^+$ as in the case of reconstitution of intrinsic phylloquinone [21]. The slow decay time indicates that $P700^+$ is not re-reduced by A_0^- or A_1^-, but by F_{AB}^- or by external reductant (see Fig. 1). Phylloquinone and most of anthraquinones belong to this group. The reduction of F_{AB} can actually be detected at 405 nm, which is an isosbestic wavelength of $P700^+$/P700; *i.e.*, these quinones fully reconstituted the function of the secondary acceptor. The E_m values of these quinones *in situ* in the Q_ϕ site are, thus, expected to be low enough to reduce F_X [9, 19, 20].

With an extremely low potential 2,6-diamino-AQ, both the initial extent of flash-induced $P700^+$ and the extent of the slow decay phase were small, so that this quinone is estimated to have an E_m comparable to or lower than that of A_0. Higher potential quinones, on the other hand, promoted a high initial extent of $P700^+$ followed by the fast decay ($t_{1/2}$ = 0.1–0.2 ms). The extent of the fast decay increased with the raising of the $E_{1/2}$ value. This suggests that these quinones have E_m values more positive than that of F_X and cannot fully reduce F_X and therefore, F_{AB}. The remaining semiquinone$^{\bullet-}$ is estimated to directly reduce $P700^+$ as represented by the fast decay.

P700 kinetics in PS I RC reconstituted with fluorenone derivatives are also studied [20]. Surprisingly most of fluorenone derivatives also restored the high $P700^+$ extent upon the flash excitation indicating that they can accept electrons from A_0^-(Fig. 4 right). The kinetics of $P700^+$ re-reduction depended on $E_{1/2}$ of fluorenones in a similar way as seen with quinone derivatives. Rapid decay of $P700^+$ with a $t_{1/2}$ of 60–100 μs, which is faster than the corresponding rate seen with quinones, was observed with high potential tetra- and tri-nitro-fluorenones. It is concluded that fluorenone derivatives, although they have only one carbonyl, show the same activity as that of quinone in PS I RC if their $E_{1/2}$ determined in organic solvent are in a similar range.

3.3 E_m values of quinone/fluorenone *in situ* at the Q_ϕ site.

Kinetics of P700 was further analyzed with 14 quinones and 7 fluorenones. Relative extent of the slow decay phase with respect to the initial extent of $P700^+$ was plotted against $E_{1/2}$ values (Fig. 5). The slow phase represents the relative amount of F_{AB} reduced via F_X mediated by these compounds. If the quinone and F_X equilibrirate rapidly before F_X^- reacts with F_{AB}, then, this curve, fitted with the theoretical one electron Nernst's curve, represents the redox titration of F_X by using quinones of different redox potential. The curve indicates that the quinone with an $E_{1/2}$ of –0.4 V in DMF gives

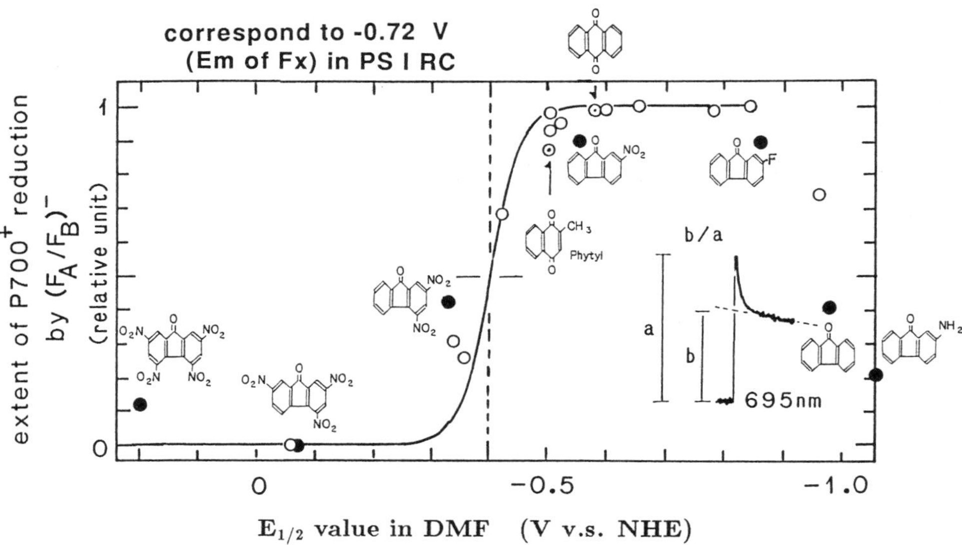

Fig. 5. Dependence of the relative extent of the slow-decay phase of $P700^+$ (corresponds to the extent of electron flow to F_X on the $E_{1/2}$ value of quinone/fluorenone. Solid line represents one electron Nernst's theoretical curve calculated with a –0.4 V E_m value. Quinones used were 1; 2,3-diCl-NQ, 2; 1-NO_2-AQ, 3; NQ, 4; 2-Me-NQ (menadione), 5; 2,3-diMe-NQ, 6: menaquinone-4, 7; phylloquinone, 8; 1-Cl-AQ, 9; AQ, 10; 2-Me-AQ, 11; 1-NH_2-AQ, 12; 2-NH_2-AQ, 13; 1,2-diNH_2-AQ, 14; 2,6-diNH_2-AQ. Fluorenones are shown by chemical formulas. NQ:1,4-naphthoquinone, AQ: 9,10-anthraquinone.

the half maximal increase of the slow phase. This suggests that a quinone with an $E_{1/2}$ of –0.4 V shows the same E_m as that of F_X (E_m = –0.72 V [22]) in the PS I RC protein, and that E_m values of quinones *in situ* in PS I RC are shifted about 0.3 V to the negative side from those in DMF. The E_m *in situ* value of phylloquinone, which shows an $E_{1/2}$ of –0.5 V, then, can be estimated to be 0.1 V more negative than that of F_X, *i.e.*, –0.82 V. On the other hand, the fall of initial extent of $P700^+$ observed at the negative end of $E_{1/2}$ (at –0.85 V) reflects the increase of $P700^T$ formation and suggests that E_m of semiquinone$^{\bullet-}$ becomes comparable to E_m of A_0. Then, E_m of A_0 is estimated to be –1.15 V. Similar plots of data with fluorenones also fits to this curve. When the amount of F_{AB}^- was plotted against $E_{1/2}$ of quinones, similar curve was also obtained [19].

The linear relationship between the E_m value at the Q_ϕ site of quinone and its $E_{1/2}$ in DMF, suggested above, contrasts with the results in the *Rb. sphaeroides* Q_A site at which $E_{1/2}$ in DMF of the reconstituted quinone was only roughly related to its *in situ* E_m value [13]. More work may be required since the present method of E_m estimation at the Q_ϕ site is not accurate for the quinone whose E_m significantly differs from that of F_X.

3.4 Free energy difference and the temperature dependence of the reaction rate between P700$^+$ and quinone$^{\bullet-}$/fluorenone$^{\bullet-}$

Quinone/fluorenone compounds introduced into PS I RC complex function as the secondary electron acceptor even at low temperature. Reaction rate at different temperature was calculated from $t_{1/2}$ of $P700^+$ reduction in the PS I RC containing reconstituted quinones and fluorenones (Fig. 6). With NQ ($E_{1/2}$ = –0.34 V) reaction rate of 3–5 $\times 10^3$ s^{-1} was obtained at temperatures between 77 and 290 K showing very weak temperature dependence, on the other hand, with trinitro-fluorenone ($E_{1/2}$ =–0.07 V), reaction rates were 2–3 times faster than those observed with naphthoquinone. Weak temperature dependence in the reaction rates was detected between 77 and 290 K also with trinitro-fluorenone. Similar measurements were done with various quinones, and rate constants at 273 and 77 K were plotted against the energy gap between P700 and quinone/fluorenone in Fig. 7. Most of quinones gave similar rate constants at both temperatures. Nitro-fluorenones gave higher rate constants at both 77 and 273 K.

4 DISCUSSION

4.1 Redox properties of quinones and their ability to function as the secondary electron acceptor

The thermodynamic relationship between the quinone and its reaction partners depends on the E_m values of the semiquinone$^{\bullet -}$/quinone couple *in situ* in the Q_ϕ site with respect to those of A_0 and F_X (Fig. 1).

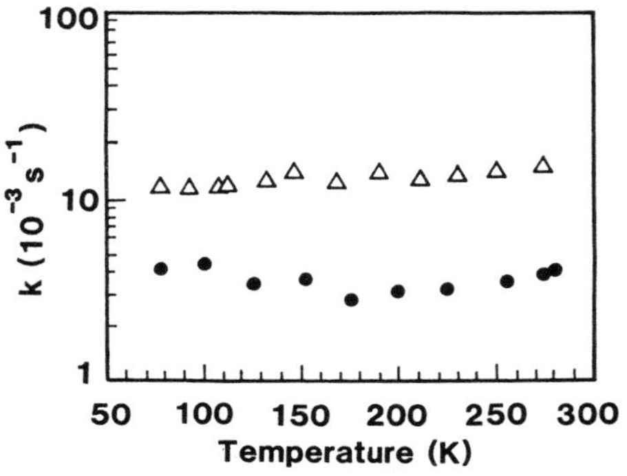

Fig. 6. Temperature dependence of the rate of reaction between P700$^+$ and naphthoquinone$^{\bullet -}$ (●) and trinitrofluorenone$^{\bullet -}$ (△).

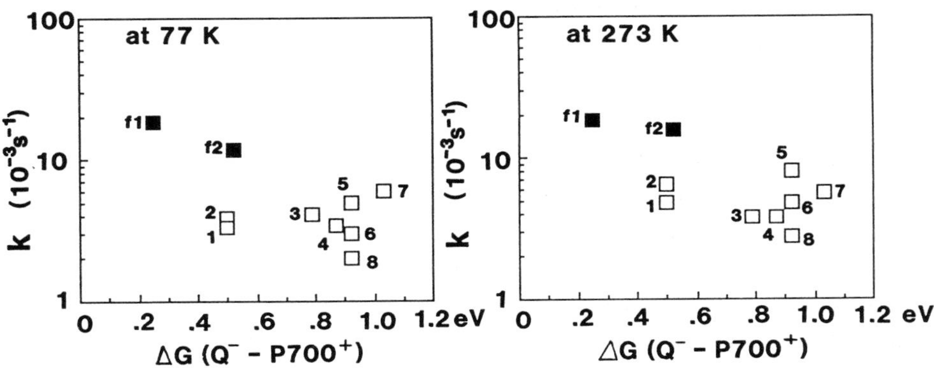

Fig. 7. Dependence of reaction rate between P700$^+$ and quinone$^{\bullet -}$s (fluorenone$^{\bullet -}$s) on $-\Delta G_0$. Quinones and fluorenones used were, 1; 2,3-diI-NQ, 2; 2,3-diBr-NQ, 3; NQ, 4; 2-Me-NQ, 5; 2,3-diMe-NQ, 6; Menaquinone-4, 7; AQ. 8; the rate with phylloquinone in untreated PS I preparation.

The E_m *in situ* values of A_0 and phylloquinone have never been determined directly due to their extremely negative E_m values. The former has been assumed to be similar to that of chlorophyll a$^-$ /chlorophyll a couple which is at about –1.0 V [23], and the latter to be around –0.9 V (*i.e.* intermediate between A_0 and F_X)[4, 5]. The E_m value of phylloquinone can be estimated to be –0.82 V from the results in Fig. 4. This E_m is about 0.3 V more negative than the $E_{1/2}$ measured in DMF. Moreover the E_m value for phylloquinone is 0.7 V more negative than that of menaquinone (which differs from phylloquinone only in the structure of the hydrocarbon tail, and gives similar $E_{1/2}$ in DMF), in the bacterial Q_A site [13] (see Fig. 1). Menaquinone almost fully replaced the function of phylloquinone in the PS I Q_ϕ site and is estimated to show similar E_m as phylloquinone. These observations indicate that the stability of semiquinone$^{\bullet -}$ is significantly lower at the Q_ϕ site than at the Q_A site due to the difference in proteinaceous environments. The E_m of A_0 is estimated to be around –1.1 V.

The estimated E_m value of phylloquinone *in situ* allows us to calculate the energy gaps between A_0 and phylloquinone, and between phylloquinone and F_X to be 0.3 and 0.1 eV, respectively. The energy gap between A_0 and phylloquinone is somewhat lower than the corresponding energy gap between bacteriopheophytin and Q_A in *Rb. sphaeroides* RC, which is calculated to be more than 0.5 eV (see Fig. 1).

4.2 Quinone/fluorenone structure and binding affinity for the Q_ϕ site

Various quinones and fluorenones bind to the PS I Q_ϕ site in place of intrinsic phylloquinone as reported elsewhere [10]. The phytyl tail or naphthoquinone ring of phylloquinone, are not essential for their function, although these properties contribute to the tight binding of the phylloquinone to the Q_ϕ site with a significant contribution from hydrophobic interactions [10] as reported for the bacterial Q_A site [24]. We have indicated that the addition of one methyl group increases the binding energy by 4 kJ/mole, and that one aromatic ring, by 10 kJ/mole due to the increase of the hydrophobicity [10]. This is similar to the results in the Q_A site in *Rb. sphaeroides* RC (P. L. Dutton, personal communication). On the other hand, fluorenones show one order lower affinities than those of anthraquinones indicating that the deletion of a carbonyl group from anthraquinone decreases the binding energy by about 6–8 kJ/mole at the Q_ϕ site site. This suggests the hydrogen bonding of quinone carbonyls to the protein amino acid residues [20].

Hydrogen bonding of quinone carbonyl, however, is expected to stabilize semiquinone and to shift E_m to the positive direction. The shift of E_m of fluorenone derivatives from their $E_{1/2}$ in organic solvent, however, can be estimated to be similar to those of quinones. The loss of carbonyl, then, does not seem to significantly affect the E_m. The negative shift of E_m values of quinone/fluorenone at the Q_ϕ site, therefore, seems to be due to other features of the quinone-binding niche that destabilize the semiquinone radical, such as the existence of a negative charge nearby quinone or the electronic interaction between π orbitals of quinone and the nearby aromatic amino acid residues as suggested recently [10].

4.3 Energy gap and the temperature dependence of the quinone reaction in PS I RC complex

The reaction rate between P700$^+$ and semiquinone$^{\bullet -}$ only weakly depended on temperature or on the free energy gap between them. Current theories of electron transfer propose that vibrations in the redox states and their surroundings are the source of the dependence of rate on $-\Delta G_0$ and on temperature. The classical limits of the theory connects the rate to the expression [25,26],

$$k = \frac{2\pi}{\hbar} \frac{|V(r)|^2}{\sqrt{4\pi\lambda k_b T}} exp - \left[\frac{(\lambda - \Delta G_0)^2}{4\lambda k_b T} \right]$$

where λ is the reorganization energy of the reaction coordinates and V(r), the electronic exchange matrix element between the reaction partners, $\hbar$ Planck's constant, k_b Boltzmann's constant and T absolute temperature. V(r) is expected to fall off rapidly as the exponential function of distance r. If $-\Delta G_0$ matches λ, then we can expect no apparent Arrhenius type activation energy according to this expression. Data in Figs. 6 and 7 indicate that the reaction rates between P700$^+$ and Q_ϕ^- are virtually temperature independent between 77 and 290 K with the variation of $-\Delta G_0$ between -0.22 and -1.12 eV. The results are inconsistent with the prediction of above equation and suggest the coupling to the quantum mode, i.e., coupling with some distribution of high-frequency vibrations [25,26]. The λ value is estimated to be lower than the lower limit of $-\Delta G_0$, 0.2 eV. The situation resembles those observed in the reactions between Q_A and pheophytin$^-$ and between Q_A^- and bacteriochlorophyll dimer cation in *Rb. sphaeroides* RC [12,15]. The temperature independence and the weak dependence on $-\Delta G_0$,thus, seem to be the general feature of the long distance intraprotein electron transfer in RC proteins. It does not seem to significantly depend on the difference in the extent of energy gaps between the prosthetic groups in these two different types of RCs as far as quinone function is concerned. Different prosthetic groups and protein structure do not seem to significantly change the general feature of the mechanism. The higher rate of quinone reactions in PS I RC compared to those in purple bacterial RC, and the higher rate of nitro-fluorenones than quinones shown in this study, on the other hand, may reflect the contribution of geometrical factors. Accumulation of more data on the lower $-\Delta G_0$ side and at the lower temperatures will give quantitative expression of dependence of the reaction rate on the variation of $-\Delta G_0$ at different temperatures and of the coupling of the different frequency modes in PS I RC. This will enable more precise discussion of the electron transfer mechanism in the two types of RCs.

Our results indicate that the reaction of quinones or fluorenones at the Q_ϕ site depends on the E_m *in situ* of their semiquinone$^{\bullet -}$/quinone couple but not significantly on their molecular structure. This confirms the conclusion of the quinone-reconstitution studies at the Q_A site in the *Rb. sphaeroides* RC [12-15]. The quinone reconstitution and replacement studies have given information about the dynamic relationship between the structure and function of the purple bacterial RC [12-15] and opened a new field for the study of intraprotein electron transfer mechanisms. This kind of approach can now be adopted for the green plant PS I RC complex, whose tertiary structure still remains to be characterized.

◇ LIST OF SYMBOLS ◇

A_0, photosystem I primary electron acceptor chlorophyll a.

A_1 (Q_ϕ), photosystem I secondary electron acceptor (phylloquinone).

DMF, dimethylformamide.

$E_{1/2}$, half wave reduction potential polarographically measured.

E_m, midpoint potential.

F_X, F_A and F_B, photosystem I electron acceptor iron sulfur centers, respectively.

P700, photosystem I primary electron donor chlorophyll a.

PS, photosystem.

Q_A and Q_B (sites), primary and secondary electron acceptor quinones (binding sites) in PS II or purple bacterial reaction center.

RC, reaction center.

◇ **ACKNOWLEDGMENT** ◇

The authors thank Dr. P. L. Dutton of Univ. Pennsylvania and Dr. S. Kakitani of Nagoya Univ. for their stimulating discussions, Drs. A. Osuka of Kyoto Univ., and W. Oettmeier of Rhur Univ. Bochum, for their kind gifts of quinones. The work is supported by Grants-in-Aid for Scientific Research (B) and for Co-operative Research (A) to S. I. from the Japanese Ministry of Education, Science and Culture.

REFERENCES

[1] Nanba, O. & Satoh, K. (1987) *Proc. Natl. Acad. Sci. USA*, 84, 109.

[2] Deisenhofer, J. & Michel, H. (1989)*EMBO J.*, 8, 2149.

[3] Michel, H., Epp, O. & Deisenhofer, J. (1986)*EMBO J.*, 8, 2140.

[4] Andréasson, L.-E., & Vängard, T. (1988)*Ann. Rev. Plant Physiol. Plant Mol. Biol.*, 39, 379.

[5] Golbeck, J. H. (1987) *Biochim. Biophys. Acta*, 895, 167.

[6] Takahashi, Y., Hirota, K. & Katoh, S. (1985) *Photosynth. Res.*, 6, 183.

[7] Itoh, S., Iwaki, M. & Ikegami, I. (1987) *Biochim. Biophys. Acta* 893, 508.

[8] Biggins, J. & Mathis, P. (1988) *Biochem.*, 27, 1494.

[9] Iwaki, M. & Itoh, S. (1989) *FEBS Lett.*, 256,11.

[10] Iwaki, M. & Itoh, S. (1991) *Biochem., 30, in press.*

[11] Itoh, S. & Iwaki, M. (1989) *FEBS Lett.*, 250, 441.

[12] Gunner, M. R., Robertson, E. E. & Dutton, P. L. (1986) *J. Phys. Chem.*, 90, 3783.

[13] Woodbury, N. W., Parson, W. W., Gunner, M. R., Prince, R. C. & Dutton, P. L. (1986) *Biochim. Biophys. Acta*, 851, 6.

[14] Warncke, K. & Dutton, P. L. In *Current Research in Photosynthesis* (Baltscheffsky, M. ed.) Kluwer Academic Publ., Dordrecht, (1990) I, p. 157.

[15] Gunner, M. R. and & Dutton, P. L. (1989) *J. Am. Chem. Soc.*, 111, 3400.

[16] Ikegami, I. & Itoh, S. (1987) *Biochim. Biophys. Acta*, 893, 517.

[17] Itoh, S. & Iwaki, M. (1988) *Biochim. Biophys. Acta*, 934, 32.

[18] Kim, D., Yoshihara, K., & Ikegami, I. (1989) *Plant Cell Physiol.* 30, 679.

[19] Iwaki, M. & Itoh, S. (1991) In *Advances in Chemistry Series No. 228, Electron transfer in Inorganic, Organic and Biological Systems (Bolton, J. Ed.)*. American Chemical Society, *Washington, DC (in press)*

[20] Itoh, S. & Iwaki, M. (1991) *Biochem., 30, in press.*

[21] Itoh, S. & Iwaki, M. (1989) *FEBS Lett.*, 243, 47.

[22] Chamorovsky, S. K. & Cammack, R. (1982) *Photobiochem. Photobiophys.*, 1982, 4, 195.

[23] Fujita, I., Davis, M. S. & Fajer, J. (1978) *J. Am. Chem. Soc.*, 100, 6280.

[24] Warncke, K., Gunner, M. R., Braun, B. S., Yu, C.-A., & Dutton, P. L. In *Progress in Photosynthesis Research*, (Biggins, J. ed.), (1987) I, p. 225.

[25] DeVault, D. (1980) *Q. Rev. Biophys.* 13, 387

[26] Marcus, R. A. & Sutin, N. (1985) *Biochim. Biophys. Acta* 811, 265.

[27] Kuder, J. E., Pochan, J. M., Turner, S. R., & Hinman, D.-L. F. (1978) *J. Electrochem. Soc.* 125, 1750.

[28] Gunner, M. R., Tiede, D. M., Prince, R. C. & Dutton, P. L. in *Function of Quinone in Energy Conserving Systems, (Trumpower, B. L. ed.), Academic Press, New York,* (1982) p. 265.

[29] Dutton, P. L., Gunner, M. R. & Prince, R. C. In *Trends in Photobiology, (Helene, C., Chalier, M., Montenay-G., T. & Laustriat, G. eds.), Plenum Press, New York* (1982) p. 561.

DISCUSSION

Fujihira

What is the mechanism or origin of the negative shift in redox potentials E in PSI? Hydrogen bonding can be origin of the positive shift of $E_{R/OX}$. But we don't know the factor causing negative shift except electrostatic effect?

Itoh

One possibility to shift E_m to the negative is an existence of negative amino acid residue nearby quinone. It will shift E_m to the negative through electrostatic interaction. This also explains the low binding affinity of halogenated quinones. The second possibility is the π–π interaction between quinone aromatic ring and nearby aromatic residues. This will also shift E_m. There two mechanisms will be only possible inside protein but will be very difficult to work in organic solvents.

Dynamics and Mechanisms of
Photoinduced Transfer and Related Phenomena
N. Mataga, T. Okada and H. Masuhara (Editors)

Intramolecular Photoinduced Electron Transfer in Pyromellitimide-linked Porphyrins

Atsuhiro Osuka[a], Satoshi Nakajima[a], and Kazuhiro Maruyama[a], Noboru Mataga[b] and Tsuyoshi Asahi[b]
Iwao Yamazaki[c] and Yoshinobu Nishimura[c]

[a]Department of Chemistry, Faculty of Science, Kyoto University, Kyoto 606, Japan

[b]Department of Chemistry, Faculty of Engineering Science, Osaka University, Toyonaka 560, Japan

[c]Department of Chemical Process Engineering, Faculty of Engineering, Hokkaido University, Sapporo 060, Japan

Abstract

Characteristic absorption of the anion radical of pyromellitimide around 715 nm was demonstrated to be particularly useful for studies on photoinduced electron transfer reactions. Charge recombination in pyromellitimide-linked porphyrin was greatly retarded in benzene due to the large energy gap and small reorganization energy. A long-lived charge separated state (τ = 2.5 μs) was formed in the case of a diporphyrin-porphyrin-pyromellitimide triad system.

1. INTRODUCTION

The appearance of the X-ray structure of bacterial photosynthetic reaction center (RC) [1] has exerted a great impact on mechanistic as well as synthetic approaches toward this natural charge separation (CS) apparatus [2,3]. In the RC, six tetrapyrrolic pigments and two quinones are positioned at precise distances and orientations to achieve a highly efficient CS. In order to better understand the detailed mechanisms in the RC, ultrafast laser photolysis investigations on the electron transfer (ET) reactions of elaborated models consisting of acceptor-linked oligomeric porphyrins are highly desirable. However, it is not necessarily easy to identify a charge-separated ion pair (IP) state when the acceptor is quinone, since the absorption spectra of the porphyrin cation radical are quite similar to those of S_1-state or T_1-state of porphyrin and the spectral characteristics of the quinone anion radical are rather obscure. One remarkable exception is the model system of carotenoid-linked porphyrins, in which the carotenoid cation radical exhibits strong absorption around 980 nm in CH_2Cl_2 which has been used to detect IP states [4].

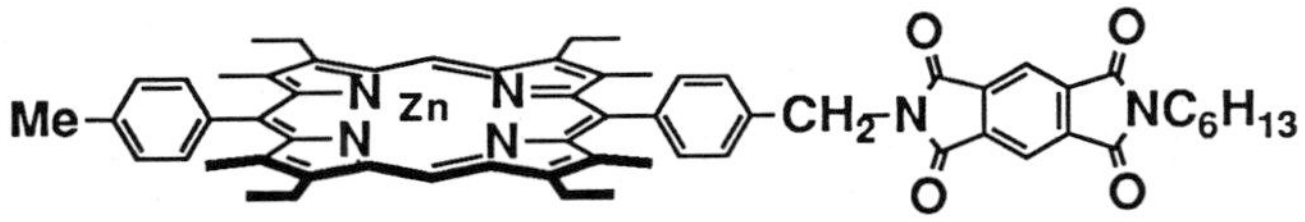

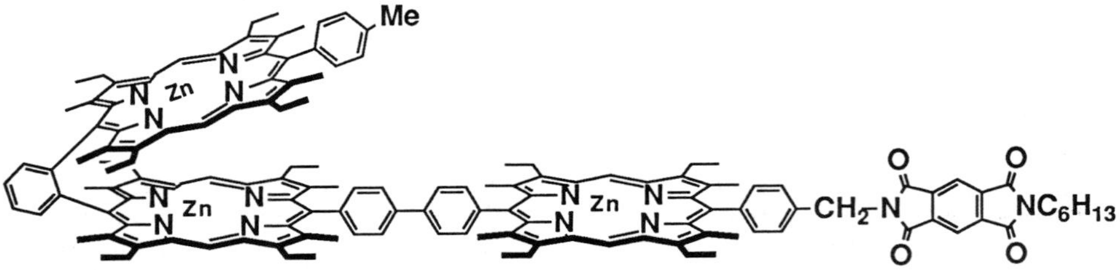

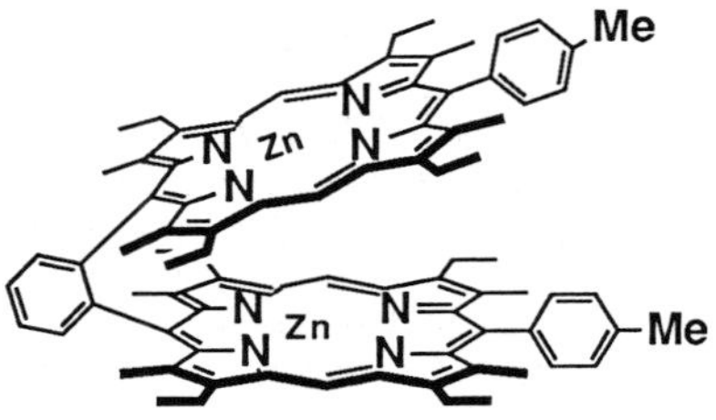

We report here the picosecond dynamics of pyromellitimide-linked porphyrins **1** and **2**, in which the sharp absorption around 715 nm due to the anion radical of pyromellitimide (Im) greatly facilitates the analysis of ET kinetics [5]. Although characteristic sharp absorption of $(Im)^-$ has long been recognized, it has not been utilized in the study of photoinduced ET kinetics. Sanders et al. also conducted picosecond laser photolysis and time-resolved absorption spectral measurements on the pyromellitimide-capped porphyrins only in a narrow range of 420-540 nm, and thus missed observing the characteristic absorption of $(Im)^-$ [6]. In the model **1**, the center-to-center distance between the zinc porphyrin and pyromellitimide is restricted to be ca. 10 Å (estimated from Corey-Pauling-Koltun model), while a 1,2-phenylene-bridged bis-zinc diporphyrin (D) is connected via a 4,4'-biphenyl linkage with the pyromellitimide-linked zinc porphyrin (M) in the model **2**. The X-ray analysis of the reference D compound **3** has revealed that the two porphyrin rings take a nearly parallel conformation with average interplanar distance of 3.45 Å and dihedral angle of 6.6° and instead of an exact "stacked" conformation the dimer **3** has a slipped offset conformation with rotation angle of 22.7° [7]. On the other hand, the ^{1}H-NMR spectrum of **3** in $CDCl_3$ contains only half the number of signals predicted by the crystal structure, indicating a very rapid interconversion between the two enantiomeric conformers in solution. The strong interactions between the two porphyrins in the D moiety results in lowering of the S_1-excitation energy by 0.19 eV and also in the decrease of one-electron oxidation potential by 0.17 V. These properties make the D moiety a promising excitation energy acceptor from the M moiety and a favorable electron donor toward the cation radical of the M in the model **2**.

2. EXPERIMENTAL

The model compounds **1**, **2**, and **4** were prepared according to our improved porphyrin cyclization method [8-10]. Synthetic details will be described elsewhere. The dimer **3** was prepared according to the reported procedure [7]. These model compounds were purified by repetitive TLC and recrystallization, and their purities were checked by 400 MHz ^{1}H-NMR prior to use.

Absorption spectra were recorded with a Shimadzu UV-3000 spectrometer and the steady-state fluorescence studies were carried out with a Shimadzu RF-502 spectrofluorimeter at room temperature. Fluorescence lifetimes were measured on 10^{-7} M air-saturated solution by picosecond time-correlated single photon counting system [11]. Picosecond transient absorption spectra were measured on ca. 10^{-4} M nitrogen-bubbled solution by means of a micro-computer controlled double-beam ps spectrometer with a repetitive mode-locked Nd^{3+}/YAG laser photolysis system [12]. The second harmonic of a single pulse (24 ps duration) was used as the excitation light.

3. RESULTS AND DISCUSSION

The absorption spectra of **1** and **2** are almost identical to the corresponding acceptor-free compounds, indicating that the attachment of Im to the porphyrin chromophores does not significantly perturb the electronic structure of the porphyrin. In figure 1, the absorption spectra of **1**, **2**, and **3** are presented. It is immediately apparent that the absorption spectrum of the trimeric model **2** is a superposition of the spectra of **1** and **3**.

In contrast to the absorption spectra, the steady-state fluorescence spectrum of the acceptor-free trimeric model **4** is quite different from the superposition of the spectra of the individual chromophores. The fluorescence intensity of the M (λ_{em} = 586 nm and 638 nm) decreases significantly whereas that of the D (λ_{em} = 680 nm) is enhanced,

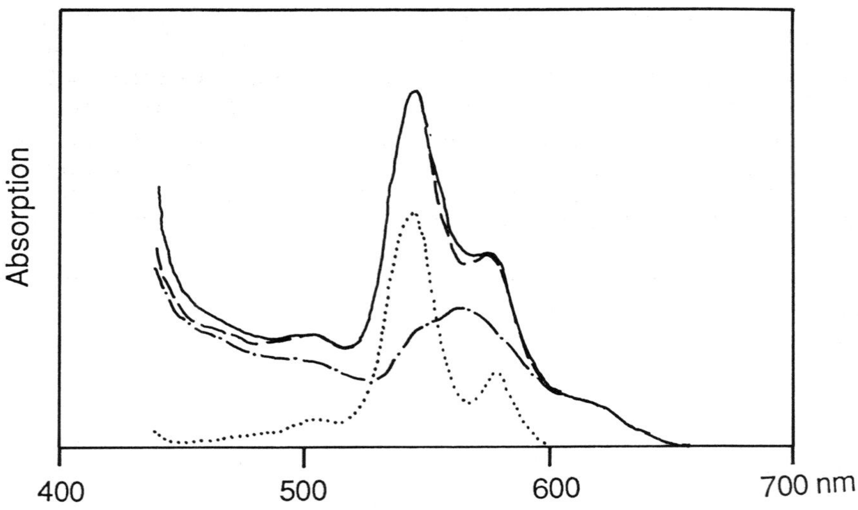

Figure 1. Absorption spectra of the models **1**, **2**, and **3** in THF. **1**, (·····); **2**, (——); **3**, (–·–); and **1** + **3**, (– – –), respectively.

4

indicating the occurrence of the efficient intramolecular excitation energy transfer (EN) from the M to the D in **4**. The rate of this energy transfer (k_E) was calculated from the following Eq. 1,

$$k_E = (\tau(M))^{-1} - (\tau_0(M))^{-1} \quad (1)$$

where τ(M)(68 ps) is the fluorescence lifetime of the ^{1}(M)* in **4** and τ_0(M) (1.5 ns) is the fluorescence lifetime of the reference acceptor-free-zinc porphyrin monomer. The k_E value thus calculated is 1.4×10^{10} s^{-1}.

In Fig. 2(a), the picosecond time-resolved transient absorption spectra of **1** observed in THF by exciting with 532 nm laser pulse were indicated. As the $S_n \leftarrow S_1$ absorption band of ZnP at 460 nm decays with τ = 60 ps, a sharp absorption band at 715 nm appears with rise time of 60 ps and decays with τ = 100 ps. A broad absorption band around 670 nm exhibits the same time-profile as that of the 715 nm band. By referring to the spectra obtained by electrochemical oxidation or reduction of the relevant chromophores, the absorption bands at 670 nm and 715 nm

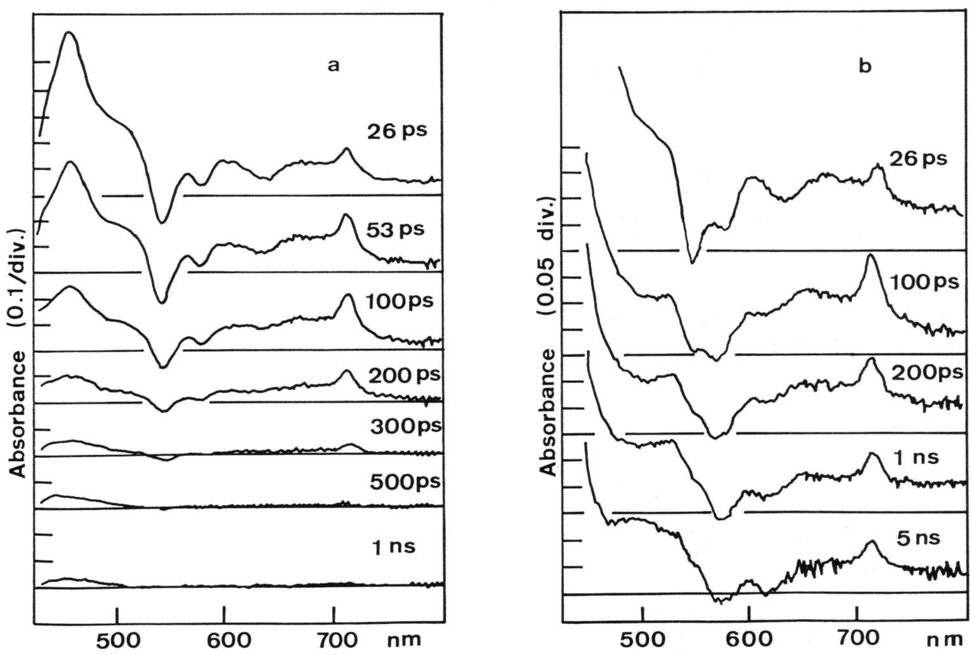

Fig. 2. Picosecond transient absorption spectra of **1**(a) and **2** (b) in THF at 25°C excited at 532 nm.

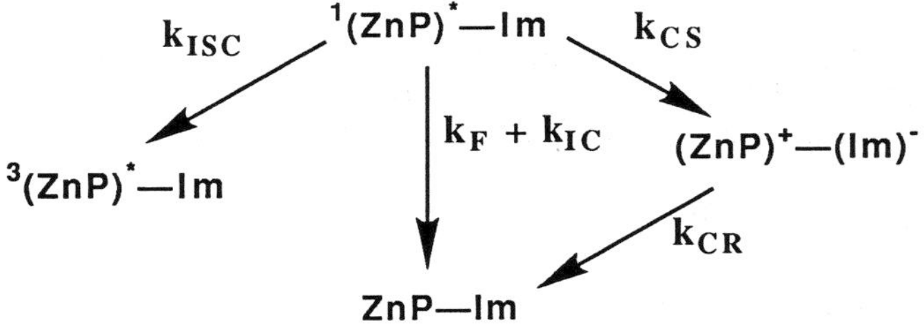

Scheme 1. Reaction scheme for the model **1**

have been assigned to $(ZnP)^+$ and $(Im)^-$, respectively. The long-lived absorption band around 460 nm can be assigned from its band shape to the $^3(ZnP)^*$, which may be produced by the intersystem crossing competing with the CS between the $^1(ZnP)^*$ and Im. We have obtained similar time-resolved transient absorption spectra also in benzene and dimethylformamide (DMF). IP state formed from **1** is long-lived in benzene but is rather short-lived in DMF.

$$A_\lambda(t) = \alpha \exp(-t/\tau_S) + \beta \exp(-t/\tau_{IP}) + \gamma \tag{2}$$

From the above results, the reaction scheme for **1** may be depicted as summarized in scheme 1. This reaction scheme gives Eq. 2 for the time dependence of the absorbance $A_\lambda(t)$ at a wavelength of λ, where $1/\tau_S = k_{CS} + 1/\tau_0$, $1/\tau_0 = k_F + k_{IC} + k_{ISC}$, $1/\tau_{IP} = k_{CR}$, and α, β, and γ are constant independent of time (t) and include initial concentration of $^1(ZnP)^*$, intersystem crossing yield of $^1(ZnP)^*$, yield of IP state formation, etc. By simulation of the observed time profiles at 715 nm and 460 nm with Eq. 2 as indicated in Fig. 3, k_{CS} and k_{CR} values have been obtained as indicated in Table 1, where the free energy gap $-\Delta G^0_{IP}$ for CR reaction estimated by using the oxidation and reduction potentials measured in DMF and corrected term for the ion solvation energies by Born formula [13] for the other solvents are also given. It is evident from Table 1 that k_{CS} does not show large solvent polarity dependence but k_{CR} shows a large decrease with decrease of the solvent polarity. In general, with decrease of the solvent polarity, the free energy gap for the CR reaction of IP state increases and that for the photoinduced CS reaction decreases, while the solvent reorganization energy decreases. Therefore, k_{CS} is not much affected by solvent polarity on the one hand, while k_{CR} shows a rather drastic decrease with the decrease of the solvent polarity. Consequently, the IP state from 2 has a long lifetime in nonpolar benzene solution. It should be noted here that such effect is very important in regulating the photoinduced CS processes even in the case of uncombined and combined (with flexible chains) porphyrin-quinone systems [14]. In the present work, this effect has been most clearly demonstrated by using a fixed distance porphyrin-pyromellitimide molecule **1**.

Table 1
Electron transfer rate constants and free energy gaps between the IP state and ground state

Solvent	k_{CS} /s^{-1}	k_{CR} /s^{-1}	$-\Delta G^0_{IP}$/eV
Benzene	1.9×10^{10}	3.6×10^8	2.23
THF	1.7×10^{10}	9.3×10^9	1.60
DMF	8.0×10^9	$\sim 5 \times 10^{10}$	1.37

By extending the system to **2**, we have observed long-lived absorption due to $(Im)^-$ in THF as indicated in fig 2(b). Deconvolution of the time dependence of the absorbances at 715 nm reveals the decay curve to be composed of two exponentials with lifetime of 70 ps and 2.5 μs, which can be ascribed to CR process from IP states, D—$(M)^+$—$(Im)^-$ and $(D)^+$—M—$(Im)^-$, respectively. The latter IP state is produced from the former by hole transfer. Based on the above results, the rate of the hole transfer from the $(M)^+$ to the (D) in **2** in THF is estimated to be $4.3 \times 10^9\ s^{-1}$. In benzene, however, we did not observe long-lived component in the transient absorption at 715 nm due to $(Im)^-$. In benzene, the transient absorbance of $(Im)^-$ at 715 nm in **2** decays with lifetime of ca 2.5 ns, which is almost identical with the lifetime (2.8 ns) observed for **1** in

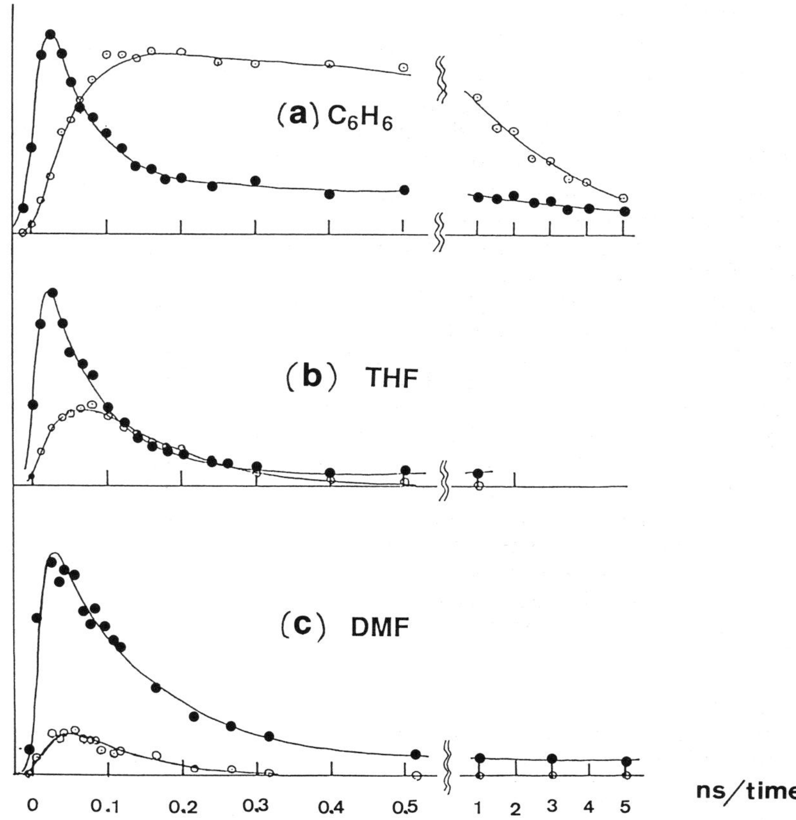

Figure 3. Time-profiles of the transient absorbances at 460 nm (●) and at 715 nm (○) in THF and DMF and at 720 nm (○) in benzene. (a), in benzene; (b), in THF; and (c), in DMF, respectively. The solid lines represents the results of simulation according to Eq. 2.

benzene. This result suggests that the hole transfer from the $(M)^+$ to the (D) does not occur in benzene solution. The failure of the hole transfer in nonpolar benzene solution would be ascribed to the much reduced electrostatic stabilization energy acquired by the IP state $(D)^+$—M—$(Im)^-$ compared with that in THF. Probably, the IP state $(D)^+$—M—$(Im)^-$ is higher in energy than the initially formed IP state D—$(M)^+$—$(Im)^-$ in benzene solution. Another important result is that selective excitation at the D moiety (λ_{ex} = 620 nm) in **2** does not lead to the formation of $(Im)^-$. This result implies that long-distance ET from the $^1(D)^*$ to the (Im) does not take place in the trimeric model **2**. Detailed features of the energetics and picosecond to microsecond photochemical dynamics of the triad **2** and its related models will be reported elsewhere.

Acknowledgment

The authors thank professor T. Ohno and Dr. K. Nozaki of Osaka University for measurements of nanosecond transient absorption spectra. This work was supported by a Grant-in-Aid for Specially Promoted Research (NO. 0201005) from the Ministry of Education, Science and Culture of Japan.

4. REFERENCES

1 J. Deisenhofer, O. Epp, K. Miki, R. Huber, and H. Michel, J. Mol. Biol., 180 (1984) 385.
2 M. R. Wasielewski, "Distance Dependencies of Electron Transfer Reactions," in "Photoinduced Electron Transfer," ed by M. A. Fox and M. Chanon, Elsevier, Amsterdam (1988), Part A, pp.161-206.
3 K. Maruyama and A. Osuka, Pure Appli.Chem., 62 (1990)1511.
4 D. Gust, T. A. Moore, A. L. Moore, L. R. Makings, G. R. Seely, X. Ma, T. T. Trier, and F. Gao, J. Am. Chem. Soc., 110 (1988) 7567.
5 For a preliminary report of this work, see: A. Osuka, S. Nakajima, K. Maruyama, N. Mataga, and T. Asahi, Chem. Lett., (1991) in press.
6 R. J. Harrison, B. Pearce, G. S. Beddard, J. A. Cowan, and J. K. M. Sanders, Chem. Phys., 116 (1987) 429.
7 A. Osuka, S. Nakajima, T. Nagata, K. Maruyama, and K. Toriumi, Angew. Chem., Int. Ed. Engl., 30 (1991) in press.
8 A. Osuka, K. Ida, T. Nagata, K. Maruyama, I. Yamazaki, N. Tamai, and Y. Nishimura, Chem. Lett., (1989) 2133.
9 T. Nagata, A. Osuka, and K. Maruyama, J. Am. Chem. Soc., 112 (1990) 3054.
10 A. Osuka, T. Nagata, F. Kobayashi, and K. Maruyama, J. Heterocycl. Chem., 27 (1990) 1657.
11 I. Yamazaki, N. Tamai, H. Kume, H. Tsuchiya, and K. Oba, Rev. Sci. Instr., 56 (1985) 1187.
12 H. Miyasaka, H. Masuhara, and N. Mataga, Laser Chem., 1 (1983) 357.
13 A. Weller, Z. Phys. Chem., NF, 133 (1982) 93.
14 N. Mataga, A. Karen, T. Okada, S. Nishitani, N. Kurata, Y. Sakata, and S. Misumi, J. Phys. Chem., 88 (1984) 4650 and 5138.

Dynamics and Mechanisms of
Photoinduced Transfer and Related Phenomena
N. Mataga, T. Okada and H. Masuhara (Editors)

Excitation Energy Transfer Processes in a Green Photosynthetic Bacterium *Chloroflexus aurantiacus*: Studies by Time-Resolved Fluorescence Spectroscopy in the Pico-second Time Range

Mamoru Mimuro

National Institute for Basic Biology, Myodaiji, Okazaki, Aichi 444, Japan

Abstract

Excitation energy transfer processes in a green photosynthetic bacterium *Chloroflexus aurantiacus* were kinetically studied by means of time-resolved fluorescence spectra in the pico-second time range and associated convolution calculations. Sequential energy flow in multi-decay and multi-spectral component systems was confirmed by matching kinetic parameters of the donor and acceptor. These results were not obtained by decay analysis on particular wavelengths but after separation of spectral components. The transfer mechanism is considered based on transfer time, together with the temperature effect on those processes.

1. INTRODUCTION

Excitation energy transfer is one of the possible relaxation processes of excited state molecules. This process is important for the collection of light energy for photosynthesis in plants and photosynthetic bacteria. They develop highly ordered pigment systems to gather and transfer light energy efficiently to photochemical reaction centers (RC) to drive photochemical energy conversion. Pigments in apo-proteins take up a definite molecular structure by ligation with or interaction with side chains of amino acids and spatial arrangement of these pigments is determined by a higher order structure of pigment-protein complexes in the membranes.

The analysis of energy transfer mechanisms was theoretically developed about 40 years ago; Förster's coulombic interaction theory [1] and Dexter's exchange interaction theory [2]. The basic concept between a donor and an acceptor is expanded to a molecular assembly, such as a photosynthetic pigment system. Experimental analysis, accompanied by the development of laser optics has meant that biological systems have become good targets for the analysis of energy transfer. Some pigment-protein complexes are crystallized [3, 4] and several different kinds of molecular interaction between pigments are known to be present in photosynthetic pigment systems. In this sense, biological systems have the advantage for analysis of energy transfer mechanisms in general.

The photosynthetic pigment system of a thermophilic green bacterium *Chloroflexus aurantiacus* (Fig. 1A) consists of bacteriochlorophyll (Bchl) *c*, found in membrane-associated antenna structures called chlorosomes and Bchl *a*, found in the chlorosome baseplate and in B808-866 complex in the cytoplasmic membranes (figures correspond to locations of the absorption maxima at physiological temperature, Fig. 1B). Each chlorosome contains about 10,000 molecules of Bchl *c*. Approximately 20 molecules of Bchl *c* take a self-aggregated form without any proteinous component [5]. The baseplate Bchl *a* is located at the interface between the chlorosome and cytoplasmic membranes in an amount of about 400 per chlorosomes. The B808-866 complex is coupled with the reaction center (RC) in the membranes. Steady state spectroscopy indicates energy flow in the order Bchl *c*, baseplate Bchl *a*, B808-866 Bchl *a* and finally to RC (Fig. 1B). These processes were kinetically investigated to obtain the exact transfer pathways and transfer mechanism, including detection of a short-lived and/or kinetically inhomogeneous component(s), which can be found only by time-resolved spectroscopy. The temperature effect on these processes was also investigated.

2. MATERIALS AND METHOD

C. aurantiacus was grown under the photo-heterotrophic condition [6] at 55°C for 2 days and used without any treatment. The time-resolved fluorescence spectra of intact

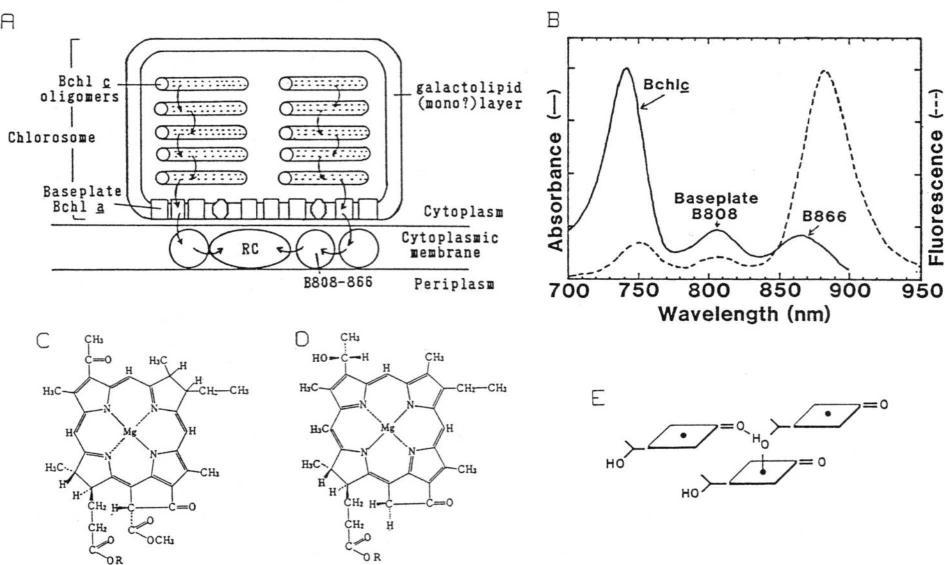

Fig. 1. Schematic cross section of photosynthetic pigment systems of *C. aurantiacus* (A) and absorption and fluorescence spectra at 50°C (B), molecular structure of Bchl *a* (C) and Bchl *c* (D), and a model for Bchl *c* oligomer (E).

In (A), rods in chlorosome represent Bchl *c* oligomer. Wavy arrows indicate the energy flow to RC. In (C) and (D), R refers phytol ($C_{20}H_{39}$) and fernesol ($C_{15}H_{25}$), respectively.

cells at 50°C were measured by the use of a time-correlated single-photon counting method [7]. The excitation pulse was at the wavelength of 715 nm with a duration of 6 ps and pulse intensity of about 10^9 photons/cm^2. Fluorescence was detected by a cooled micro channel-plate photomultiplier (Hamamatsu Photonics, R1564U-05). The time-resolution was 3 ps with associated convolution calculation [8] and the spectral resolution was better than 2 nm. Spectral sensitivity of the apparatus in the wavelength region for current measurements was almost constant [9]. For the measurements at −196°C, 15 % poly(ethylene)glycol 4000 was added to obtain a homogeneous ice.

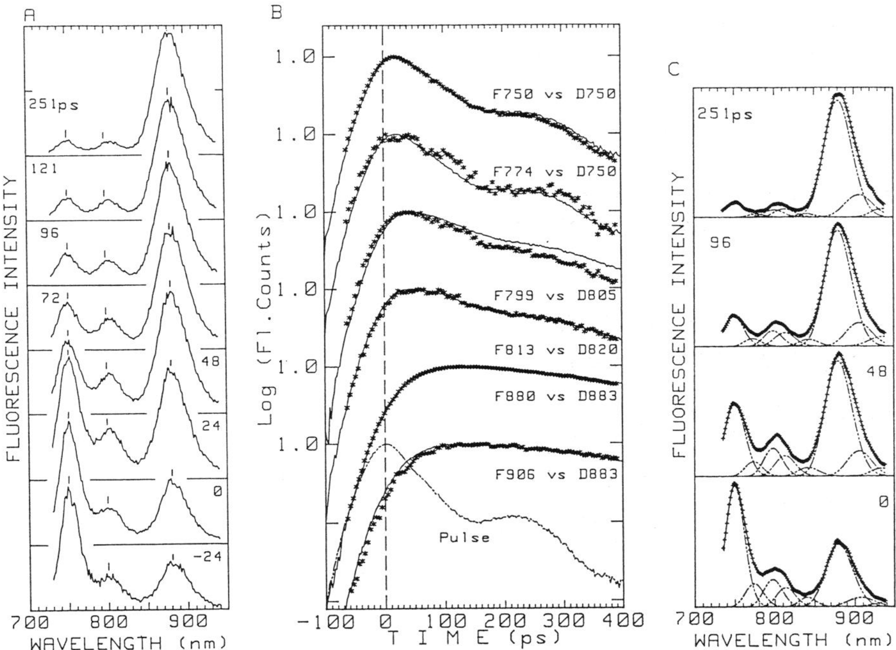

Fig. 2. Time-resolved fluorescence spectra of *C. aurantuacus* at 50°C and kinetic analysis.

Excitation pulse was at the wavelength of 715 nm with a duration of 6 ps. (A), Normalized time-resolved fluorescence spectra. Small bars on the spectra indicate locations of the fluorescence maxima at 0 ps. In (B), full lines show the rise and decay curves measured at particular wavelengths (i.e., D750) and asterisks, the rise and decay curves of resolved components (i.e., F750). (C) Deconvolution of the time-resolved fluorescence spectra into components assuming a Gaussian band shape. Plus signs are observed data, alternate lines, component spectra and broken line, a sum of component spectra.

3. RESULTS AND DISCUSSION

3.1. Time-resolved fluorescence spectra at 50°C.

Upon selective excitation of Bchl *c* in chlorosomes, energy flow from Bchl *c* to B808-866 complex was clearly observed by progressive shift of the fluorescence maxima in the order of Bchl *c*, baseplate Bchl *a* and B866 in the B808-866 complex (Fig. 2A) [7]. Within 100 ps after the pulse, the emission from the B866 was preferentially detected. These changes in the spectra clearly indicate a very fast energy transfer from Bchl *c* to B866 in the membranes. One noticeable feature is the shift of the fluorescence maximum of the baseplate from 800 to 810 nm, indicating plural components of the baseplate. Locations of the fluorescence maxima of other components were constant.

3.2. Estimation of lifetimes by convolution calculations.

Energy transfer pathways were kinetically investigated by the estimation of lifetimes of individual components. In the time-resolved spectra, three components were discernible; Bchl *c* at 750 nm, baseplate Bchl *a* around 805 nm and B866 at 883 nm. Convolution calculations were performed on the decay curves measured at those wavelengths assuming an exponential decay (Fig. 2B and Table 1A). For the criteria of the best fit, χ^2 and the Durbin-Watson parameter were adopted [8]. Channel shift treatment of the excitation pulse was also applied. The main decay component of Bchl *c* was 17 ps with a minor component of 190 ps. On the other hand, at 805 nm (baseplate Bchl *a*), a rise term corresponding to the decay of Bchl *c* was not found and plural decay terms were resolved; 40 ps (75 %) and 190 ps (25 %). The 40-ps decay corresponds to the rise of the B866, indicating direct energy flow from the baseplate Bchl *a* through non-rate limiting transfer ($\leq$1 ps) from B808 to B866 in the same complex. The 190-ps decay of the baseplate Bchl *a* may correspond to the presence of an uncoupled component as shown by the shift of the fluorescence maximum (see later). The decay of the B866 was 246 ps (99.8 %), indicating RC being closed under our excitation conditions.

Table 1. Lifetimes of fluorescence components in *C. aurantiacus* at 50°C.
A. Decay parameters estimated on the decay at particular wavelengths.

Components	Rise term	Decay term			
	τ (ps)	τ_1 (ps)	A_1	τ_2 (ps)	A_2
Bchl *c* 740	$\leq$3	17	0.97	190	0.03
Bchl *a* 795	$\leq$3	40	0.75	190	0.25
Bchl *a* 808-866	40	246	0.998	145000	0.002

B. Decay parameters after deconvolution to components.

Components	Rise term	Decay term					
	τ (ps)	τ_1 (ps)	A_1	τ_2 (ps)	A_2	τ_3 (ps)	A_3
F750		7	0.45	34	0.54	190	0.01
F774		9	0.65	41	0.33	195	0.02
F799	7	41	0.91	182	0.09		
F813	7	48	0.714	205	0.286		
F880	40	246	0.998	1030	0.002		
F906	61	245	0.99	1250	0.01		

τ; lifetimes and A; amplitudes.

3.3. Decay kinetics after resolution of component bands.

Two ambiguous points on the baseplate Bchl *a* remain in the above estimation. One is absence of the rise term and the other, presence of plural components which were not clearly resolved by decay analysis. The shift of the fluorescence maximum from 800 nm to 810 nm is indicative of plural components. These components can be spectrally resolved, thus deconvolution of spectra were carried out with the assumption of Gaussian band shape and lifetimes were estimated by rise and decay curves of individual components.

Eight components were necessary to fit the spectra in a whole time range of measurements (Fig. 2C); those are located at 750, 774, 799, 813, 840, 880, 906 and 937 nm (hereafter these are shown with a prefix of F, i.e. F750). The F750 and F880 correspond to Bchl *c* and B866, respectively. The F799 and F813 are resolved baseplate Bchl *a*. The presence of F774 (Bchl *c*) was suggested by linear dichroism spectra of isolated chlorosomes (Matsuura, personal communication). The F906 might correspond to the longer wavelength antenna which are known to have a lower energy level than that of RC and commonly found in photosynthetic purple bacteria [9–11]. The F840 and F937 might be vibrational bands of some components. By using these components, the spectra were consistently fit in the whole time range of measurements with changes only in intensities of individual components.

Rise and decay curves of individual components were shown by asterisks in Fig. 2B. The decay curves of the F880 was identical with that measured at 883 nm. However, consistency is not always observed on another components. The rise of the F750 was faster than that of the decay curve measured at 750 nm; this is probably due to separation of an overlapping component with a slow lifetime. On the resolved rise and decay curves, the convolution calculations were carried out; in this case, channel shift treatment was not applied.

Table 1B shows resolved lifetimes of individual components. Note that a fast decaying component was resolved in the Bchl *c* (7 ps, F750) and the second Bchl *c* (F774) also has a short-lived component with a lifetime of 9 ps. These decay terms correspond to the rise terms of the baseplate Bchl *a* (7 ps for F799 and F813), indicating energy flow from Bchl *c* to the baseplate Bchl *a*. The lifetime of the main decay component of the F799 was 41 ps which is almost the same as the rise term of the B866 (40 ps, F880). Thus, it is reasonable to assign the energy flow in the order of F750, F799 and F880 (B866). This is consistent with the results obtained by the kinetic analysis at particular wavelengths (Table 1).

The decay time of the F813 (48 ps) was in the same range as that of the F799 (40 ps) and the 48 ps transfer time was also close to the rise term of the B866. Thus the F813 most probably participates in the energy flow to B808-866. These results show that both of the baseplate Bchl *a* (F799 and F813) are similar to each other in their time behaviour. The presence of a long-lived component is an intrinsic characteristic of the baseplate Bchl *a* and a higher fraction of a long-lived component in F813 is responsible for the shift of the fluorescence maxima.

The other feature is the difference in the kinetics of the F906 from that of the F880; the rise of the F906 was slower by 20 ps than that of the F880, even though the decay components were identical. These phenomena can be explained by a fast equilibrium among F880, F906 and RC under the condition that all the RC is closed. A slow rise of the F906 indicates that this component is a direct energy donor to the RC, together with its estimated absorption maximum at 888 nm at 50°C. *C. aurantiacus* is known to have a RC similar to that of photosynthetic purple bacteria [12], thus the presence of the F906 confirms the ubiquitous presence of this type of antenna pigment in photosynthetic bacteria.

In summary, kinetic analysis of individual components indicate plural transfer pathways from Bchl *c* to B866. One is in the order of F750, F799 and F880, and the other, F774, F813 (or F799) and F880. The F906 mediates the energy transfer between F880 and RC. Since the absorption component corresponding to F774 is minor (cf. Fig. 1B), the former is assined to be the major transfer pathways in chlorosome. Supplemental pathway might arise from structural difference in chlorosomes (cf. Fig. 4).

Our previous estimation of lifetimes by convolution calculations on particular wavelengths cannot resolve the presence of the rise terms of the baseplate Bchl *a* (Table 1A). However, after separation into component spectra, it became clear (Table 1B), indicating that a difference in the methodology can be very critical in this type of analysis. The presence of the rise term in the baseplate Bchl *a* was consistent with recent global analysis of the decay curves; it gave a rise term both in intact cells (16 ps) and isolated chlorosomes (8 ps) [13]. Our estimation on the intact cells is faster (7 ps), however discrepancy of the results due to methodological difference can be resolved.

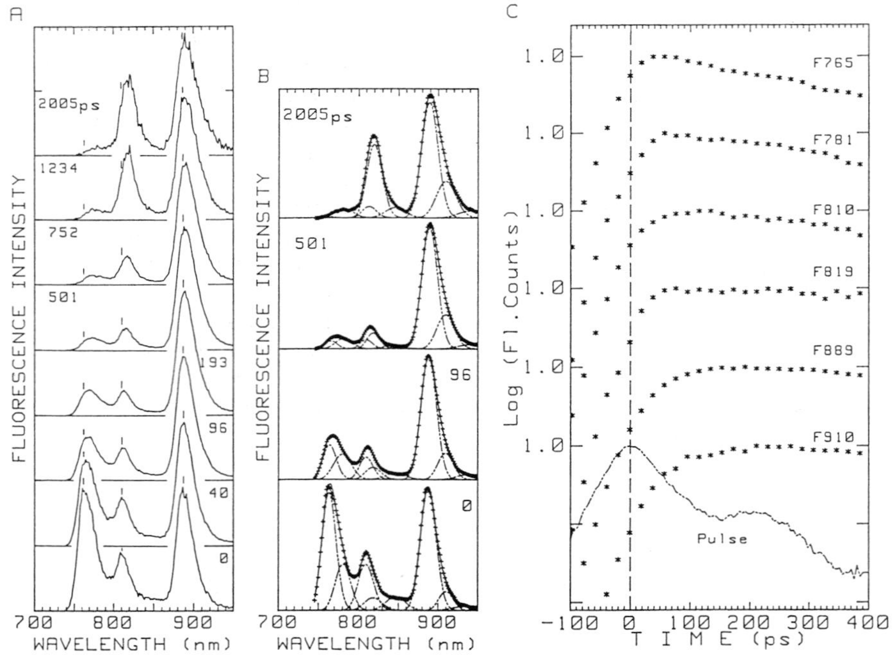

Fig. 3. Time-resolved fluorescence spectra measured at −196°C and kinetic analysis.

Measuring conditions were the same as in Fig. 2. (A), Normalized time-resolved fluorescenec spectra, (B) deconvolution of spectra into components, (C) rise and decay curves of resolved components (i.e., F765). Signs are the same as shown in Fig. 2.

3.4. Time-resolved fluorescence spectra at –196°C.

By lowering the temperature, individual antenna components will have a narrower band width, which consequently affects the spectral overlap between component spectra and transfer time. Thus, we measured the time-resolved fluorescence spectra at –196°C (Fig. 3A) conditions under which a higher spectral resolution is also expected.

Upon selective excitation of Bchl *c* in chlorosomes (715 nm), a progressive shift of the fluorescence maxima was clearly observed, giving rise to the same energy flow from Bchl *c* to B866 in the membranes as that observed at 50°C. Notable features were appearance of the B886 emission in an early time after the pulse and spectral shift with time in three wavelength regions; that is, Bchl *c* (760–780 nm), baseplate Bchl *a* (810–820 nm) and B866 Bchl *a* (888–893 nm). These indicate the presence of plural components, as found in the spectra at 50°C (Fig. 2C). A higher intensity around 820 nm in a later time range (longer than 1 ns) was remarkable and is consistent with the results by Brune et al. [14].

Deconvolution of the time-resolved fluorescence spectra into components were carried out with the same assumption as in the case of 50°C. Resolved components are F765, F781, F810, F819, F845, F889, F910 and F939; these correspond to the components found at 50°C except for a red shift of the fluorescence maxima. Their time behaviour are shown in Fig. 3C. A fast rise and decay was found in F765 and a very long-lived component was evident in F819. Again, the rise of the F910 was slower than that of the F889, indicating that the F910 is the longer wavelength antenna component.

Lifetimes of individual components were estimated by convolution calculations; several remarkable difference from the decay kinetics at 50°C became clear. Even for the F765 and F781, a rise term was resolved; –20 ps and –30 ps, respectively. This indicates the presence of a short-lived component which is a donor to both or either of F765 and F781. This donor might correspond to the short-lived component reported by Griebenow et al. [5]. The main decay of the F765 was 40 ps, which in turn corresponds to the rise of one of the baseplate Bchl *a* (F810), and the decay of the F810 is the same as the rise of the F889. These consistency clearly indicates sequential energy transfer in the order of F765, F810 and F889. Since the absorption fraction of the F765 is largest, this transfer pathway is assigned to the main flow in this bacterium.

The decay of the F781 was 50 ps and the corresponding rise was not found in any components. Also the decay of the F819 does not correspond to rise of the expected acceptor (F889). Therefore, even if these components participate in the energy flow, its fraction must be very small.

Table 2. Decay parameters for individual fluorescence components at –196°C.

Components	Rise term	Decay term					
	τ (ps)	τ_1 (ps)	A_1	τ_2 (ps)	A_2	τ_3 (ps)	A_3
F765	20	45	0.82	124	0.09	235	0.09
F781	30	69	0.79	273	0.19	475	0.02
F810	44	65	0.78	140	0.11	342	0.11
F819	42	95	0.48	540	0.43	4380	0.07
F889	66	155	0.63	337	0.37		
F910	85	281	0.99	890	0.01		

The rise term of F889 was 60 ps, and this is faster by 20 ps than that of the F910. This is again the same situation as that at the physiological temperature. The estimated location of absorption maxima of F889 and F910 are 867 and 889 nm at –196°C; this energy difference was almost the same as that at 50°C (866 vs 888 nm). Thus, the equilibrium at –196°C is to be shifted to the RC side. Even under this condition, the transfer rates are less affected by temperature, indicating that the structural factor is highly effective in the interaction between the longer wavelength atnenna and RC.

In summary, the energy transfer pathways at –196°C were essentially identical to those at 50°C. A temperature effect was observed only on the transfer times.

3.5. Transfer mechanism and temperature effect on it.

Consistency of kinetic parameters between Bchl *c* in chlorosomes and the baseplate Bchl *a* is a clear indicative of energy flow between them. Considering its transfer time (7 ps at 50°C and 45 ps at –196°C), the resonance transfer is the most probable mechanism. This is also the case between the baseplate Bchl *c* and B808-866 complexes in the membranes (40 ps at 50°C and 65 ps at –196°C). In this sense, the transfer mechanism itself is the same at the two temperature conditions. The temperature effect is evident as shown by a slower transfer time, probably due to a smaller spectral overlap between components.

A critical point is energy migration among oligomeric Bchl *c* molecules in chlorosomes. It is reported that each oligomeric Bchl *c* assembly consists of about 20 molecules, thus energy migration among them is expected to be shorter than 1 ps for an individual transfer step at 50°C, suggesting a strong (exciton) interaction between molecules. This is consistent with the observation based on the circular dichroism spectrum [15] and absorption spectrum [16]. On the other hand, the main Bchl *c* species (F765, A744) shows a rise term at –196°C, suggesting the presence of donor Bchl *c*. The difference in the transfer time between the rise and decay of the F765 at –196°C was 25 ps (Table 2). If this time corresponds to the energy migration time among Bchl *c* oligomeric species, it becomes slower by 3 fold, compared with that at 50°C. This might originate from difference in molecular structure or molecular interaction at –196°C.

As shown by kinetic analysis, there are two transfer pathways in whole cells of *C. aurantiacus* under both temperature conditions (Fig. 4). One involves the main Bchl *c* component (A740), thus it is assigned to be the main energy flow. The other is a minor pathway, however its transfer time is comparable to that of the main flow. These multiplicities may arise from differences in the molecular arrangement of pigment systems.

3.6. Difference in methodology for kinetic analysis.

For kinetic analysis, we adopted two methods; one is the convolution calculation on the decay curves measured at particular wavelengths where respective components share their maximum fraction in the spectra. The other method is the combination of the resolution of spectral components and convolution calculation. As shown in this study, the former cannot give an accurate number of components nor their time behaviour. On the other hand, the latter method reveals the presence of some components (F774, F813 and F906 at 50°C) which were hardly detected even by the global analysis [13, 17]. Furthermore, the latter method gives short-lived components. In this sense, the validity

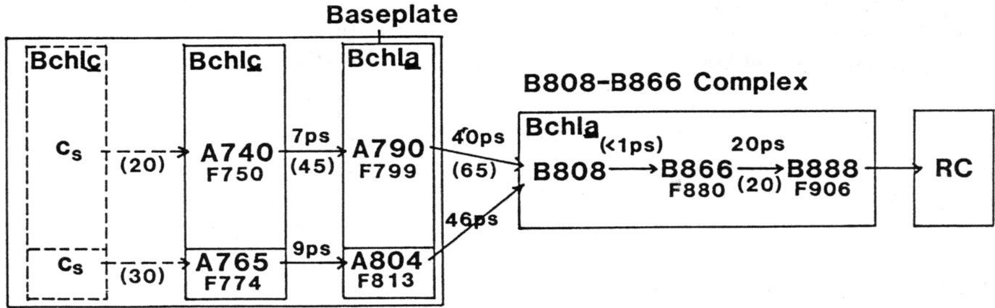

Fig. 4. Schematic energy flow in *C.aurantiacus* at two temperature conditions.

A and B represent components shown by locations of absorption maxima and F, by fluorescence maxima at 50°C. Dotted square in chlorosome is an energy donor Bchl *c* suggested by rise terms of decay analysis. The C_s refers a short-lived Bchl *c* components found at –196°C. Figures above arrows are transfer times at 50°C and those in parentheses below arrows, those at –196°C.

of our second method is higher in the analysis of the decay kinetics.

An important point is that the convolution calculation at the particular wavelength(s) is only valid when the experimental system consists of a single component or number of components with their decay components resolved. For a multi-decay and multi-spectral component system, like a photosynthetic pigment system, global analysis or combination of resolution of spectral component and kinetical analysis on those is necessary for an accurate analysis.

4. ACKNOWLEDGEMENTS

This work was carried out by a collaborative study with Prof. I. Yamazaki, Hokkaido University. The author thanks Dr. N. Tamai for his operation of the pico-second apparatus and also thanks Dr. T. Nozawa, Tohoku University, Drs. K. Matsuura and K. Shimada, Tokyo Metropolitan University, Prof. R.E. Blankenship, Arizona State University and Prof. R.S. Knox, University of Rochester for their discussions. Thanks are also due to Dr. A. J. Bell for his critical readings of the manuscripts.

5. REFERENCES

1. T. Förster (1948) Ann. Phys. Leipzig, 2, 55.
2. D. L. Dexter (1953) J. Chem. Phys., 21, 836.
3. J. Deisenhoffer, O. Epp, K. Miki, R. Huber and H. Michel (1985) Nature, 318, 618.
4. T, Schirmer, W. Bode, R. Huber, W. Sidler and H. Zuber (1985) J. Mol. Biol., 184, 257.

5. K. Greibenow and A.R. Holzwarth (1990) In *Molecular biology of membrane-bound complexes in phototrophic bacteria*, (G. Drews and E.A. Dawas, eds.), pp. 383, Plenum Press, New York.
6. B. K. Pierson and R. W.Castenholz (1974) Arch. Microbiol., 100, 283.
7. M. Mimuro, T. Nozawa, N. Tamai, K. Shimada, I. Yamazaki, S. Lin, R.S. Knox, B.P. Wittmershaus, D.C. Brune and R.E. Blankenship (1989) J. Phys. Chem., 93, 7503.
8. N. Boens, N. Tamai, I. Yamazaki and T. Yamazaki (1990) Photochem. Photobiol., 52, 911.
9. K. Shimada, M. Mimuro, N. Tamai and I. Yamazaki (1989) Biochim. Biophys. Acta, 975, 72.
10. K. Shimada, N. Tamai, I. Yamazaki and M. Mimuro (1990) Biochim. Biophys. Acta, 1016, 266.
11. R. van Grondelle, H. Bergstrom, V. Sundstrom and T. Gillbro (1987) Biochim. Biophys. Acta, 894, 313.
12. R. E. Blankenship, D. C. Brune, J. M. Freeman, G. H. King, J. D., McManus, T. Nozawa, J. T. Trost and B. P. Wittemrshaus (1988) In *Green Photosynthetic Bacteria* (J. M. Olson, J. G. Oremerod, J. Amesz, E. Stockebrandt, and H. G. Truper eds.), pp. 57, Plenum Press, New York.
13. T. Causgrove, D.C. Brune, J. Wang, B.P. Wittmershaus and R.E. Blankenship (1990) Photosynthesis Res., 26, 39.
14. D. C. Brune, G. H. King, A. Infosino, T. Steiner, M. L. W. Thewalt and R. E. Blankenship (1987) Biochemistry, 26, 8652.
15. R. J. van Dorssen, H. Vasmel and J. Amesz (1986) Photosynthesis Res., 9, 33.
16. D. C. Brune, T. Nozawa and R. E. Blankenship (1987) Biochemistry, 26, 8644.
17. A. R. Holzwarth, M. G. Muller and K. Griebenow (1990) J. Photochem. Photobiol., B5, 457.

AUTHOR INDEX

SUBJECT INDEX